Critical acclaim for *Splintering Urbanism* already includes:

Splintering Urbanism is the first analytical geography of the network society. It skilfully blends up-to-date information on metropolitan development, theoretical insights, and a good knowledge of debates in the field. It demonstrates that electronics-based networks segregate as much as they connect, and that they do so selectively. It is required reading for students of spatial transformation and is on the cutting edge of research in urban studies.

Manuel Castells, University of California at Berkeley,
author of *The Information Age*

Splintering Urbanism is an enormously important book. Graham and Marvin have a very specific angle into questions of infrastructure and cities that makes them stand apart from many other authors on the subject. The book's contribution is its mixing of matters deeply urban and material with the digital, its analysis of the particular type of fragmenting of urban space this can engender, and its connecting of these new aspects with conventional infrastructural conditions and challenges.

Saskia Sassen, University of Chicago and LSE, author of
Globalization and its Discontents and *The Global City*

Splintering Urbanism is a crucial text for architects and urban designers who are interested in the roles of network infrastructure – particularly new digital telecommunications infrastructure – in shaping the future of our cities. It synthesizes a vast amount of relevant material, develops a range of critical perspectives on it, and provides some clear starting points for exploring possible design interventions.

William Mitchell, School of Architecture and Planning, MIT,
author of *City of Bits* and *E-topia*

Splintering Urbanism is a truly path-breaking book and a tremendous achievement. In many ways it is even more impressive than its predecessor, *Telecommunications and the City*. Graham and Marvin have really brought the subject of urban infrastructure to life. *Splintering Urbanism*'s great strengths are its global perspective, its wide range of material and case studies, and the ways these are integrated to form a new way of looking at cities. This book could do for networks what Doreen Massey's *Spatial Division of Labour* did for industrial geography back in the 1980s.

Steven Pinch, University of Southampton,
author of *Worlds of Welfare*

Splintering Urbanism will be one of the most widely read and cited books in urban studies for some time. The book delivers an original, state-of-the-art and comprehensive analysis of changing infrastructure networks – especially telecommunications – in contemporary urban areas. It offers a fresh way of viewing and understanding city metamorphosis on a rapidly urbanising planet. The book clearly shows how networked infrastructures are set in place and how they help explain the economic, social and political power of urban areas. The book is so innovative, interdisciplinary and contemporary that it is basically without competition.

James Wheeler, University of Georgia,
co-editor of *Cities in the Telecommunications Age*

Inspiring! *Splintering Urbanism* is the most comprehensive book to date on the socio-cultural history of urban infrastructure. It includes impressive global coverage, historical foundations and insightful analysis of the most recent urban-technology dynamics. A 'must' read for scholars and practitioners in city planning, history of technologies and urban geography.

Yuko Aoyama, Clark University,
co-editor of *Cities in the Telecommunications Age*

Graham and Marvin, whose *Telecommunications and the City* became an instant classic, repeat their earlier success by focusing on urban infrastructure in the digital age. In so doing they bring to the fore a long neglected but critical foundation of cities that makes the 'space of flows' possible, revealing lucidly its connections to urban planning, transportation and telecommunications, and cyberspace. Rescuing infrastructure from simplistic metaphors driven by technological determinism is one of the book's finest contributions. In an age of unchallenged neoliberalism the shape of cities is being powerfully reworked by private capital with little regard for the social externalities urban development inevitably generates. *Splintering Urbanism* shows powerfully how our notions of time and space reflect the ways in which the geography of cities is periodically torn apart and reconstituted. At scales ranging from the local to the global, including the frequently overlooked developing world, they reveal the urban infrastructure as a contested field of class and gender relations, ideologies, environmental movements, and community groups.

Barney Warf, Florida State University,
co-editor of *Cities in the Telecommunications Age*

Splintering Urbanism's comprehensive analysis of the impact on cities of the privatization and unbundling of infrastructure networks, especially telecommunications, is highly original, timely, and deeply provocative. Urban designers, policy makers and architects will find compelling evidence here of a new challenge to the role of cities. The authors document how networked cities, far from equalizing opportunities, are increasingly fragmenting into cellular clusters of globally connected high-service enclaves and network ghettoes. They locate this discussion within a wide variety of contemporary theoretical discourse on cities, technology, economic and social development, reminding us of just how fundamental infrastructure is to the design, organization, and life of cities.

Ellen Dunham-Jones, Director of the Architecture Program,
Georgia Institute of Technology

SPLINTERING URBANISM

Splintering Urbanism presents a path-breaking analysis of the nature of the urban condition at the start of the new millennium. Adopting a global and interdisciplinary perspective, it reveals how new technologies and increasingly privatised systems of infrastructure provision – telecommunications, highways, urban streets, energy and water – are supporting the splintering of metropolitan areas across the world. The result is a new 'sociotechnical' way of understanding contemporary urban change, which brings together discussions about:

- Globalisation and the city
- The urban and social effects of new technology
- Urban, architectural and social theory
- Social polarisation, marginalisation and democratisation
- Infrastructure, architecture and the built environment
- Developed, developing and post-communist cities

Splintering Urbanism brings together a broad range of international case studies, boxed examples, over 100 illustrations and a comprehensive glossary. These take the reader on global journeys encompassing finance districts in Tokyo and New York; e-commerce spaces in Jamaica and northern England; new media enclaves in San Francisco and London; logistics and airport cities in Asia and the United States; malls in Atlanta and Singapore; gated communities in Istanbul, São Paulo, Mumbai and Johannesburg; new highway spaces in Melbourne, Manila and Los Angeles; and network ghettoes in the United States, the United Kingdom and the developing world. *Splintering Urbanism* will be essential reading for urbanists, geographers, planners, architects, sociologists, researchers in science and technology and communications studies, and all those seeking a definitive statement of the contemporary urban condition.

Stephen Graham is Reader in the Centre for Urban Technology at Newcastle University's School of Architecture, Planning and Landscape. **Simon Marvin** is Professor of Sustainable Urban and Regional Development and Co-Director of the Centre for Sustainable Urban and Regional Futures, University of Salford.

SPLINTERING URBANISM

*networked infrastructures,
technological mobilities and the urban condition*

STEPHEN GRAHAM and SIMON MARVIN

LONDON AND NEW YORK

First published 2001
by Routledge
11 New Fetter Lane, London EC4P 4EE

Simultaneously published in the USA and Canada
by Routledge
29 West 35th Street, New York, NY 10001

Routledge is an imprint of the Taylor & Francis Group

Typeset in Galliard by Keystroke, Jacaranda Lodge, Wolverhampton
Printed and bound in Great Britain by TJ International Ltd, Padstow, Cornwall

British Library Cataloguing in Publication Data
A catalogue record for this book is available from the British Library

Library of Congress Cataloging in Publication Data
A catalogue record for this book has been requested

ISBN 0–415–18964–0 (hbk)
ISBN 0–415–18965–9 (pbk)

FOR ANNETTE AND NICOLA

DEDICATED TO THE MEMORY OF
RICHARD H. WILLIAMS (1945–2001)

CONTENTS

Plate 1 New highway system in central Shanghai, China.
Photograph: Stephen Graham

PLATES

FIGURES

TABLES

BOXES

ACKNOWLEDGEMENTS

Between us, this book is the culmination of nearly thirty years' fascination with the complex intersections of cities and networked technologies. *Splintering Urbanism* has had an exceptionally long, tortuous and sometimes extremely painful gestation over a period of nearly a decade. It would not have been completed without the unstinting help, support and kindness of a wide range of people and institutions.

First, and most important, we must offer our loving thanks to Annette Kearney and Nicola Turner for their remarkable encouragement and support throughout the book's lengthy, and often intrusive, emergence. Without them we wouldn't have moved past the first page.

Second, we have received on-going support and encouragement from many friends and colleagues to develop and hone our thinking about cities, technology and infrastructure networks over the period of the last decade or so. Thanks to everyone who has inspired, helped and provided us with ideas and support.

At Newcastle University we owe a particular debt to Simon Guy, who provided key inputs into early stages of Chapter 2, as well as broader intellectual inputs into the book. Both of us would also particularly like to thank James Cornford, Alan Gillard, Andy Gillespie, John Goddard, Patsy Healey, Mark Hepworth, Ranald Richardson, Elizabeth Storey, Geoff Vigar, and John Wiltshire and all our colleagues for their support and encouragement over the past decade.

Further afield, we could not have got started on this project without the long-term encouragement of Manuel Castells, Olivier Coutard, Ken Ducatel, Peter Hall, Richard Hanley, Patrick Le Galès, Doreen Massey, Margit Mayer, Jean-Marc Offner, Jane Summerton and Erik Swyngedouw. More recently the book has benefited immensely from the feedback, ideas and support of Yuko Aoyama, Ash Amin, Neil Brenner, Manuel Castells, Ellen Dunham-Jones, Jenn Light, Bill Mitchell, Steven Pinch, Saskia Sassen, Barney Warf and James Wheeler. Thanks to you all for your criticism, comments, endorsements, enthusiasm and suggestions!

At a more individual level, Stephen Graham would like to express his gratitude to the following members of the School of Architecture and Planning at MIT, for their support and kindness during his visiting professorship there in 1999–2000: Sue Delaney, Joe Ferreira, Bernie Frieden, Duncan Kincaid, Malo Hutson, Ceasar McDowell, Turi McKinley, Bill Mitchell, Bish Sanyal, Qing Shen, Mike Shiffer and Anthony Townsend. The support of the School, and especially the Department of Urban Studies and Planning (DUSP), contributed much to the final completion and writing up of *Splintering Urbanism*. Thanks also to colleagues at Newcastle for covering for my absence, and at Clarke University, King's College London, Hull University, Leeds Metropolitan University, New York University, the New York Academy of Sciences, Brooklyn Technical College (CUNY) and the University of California at Berkeley for constructive and helpful comments on presentations of various parts of this book's contents.

Simon Marvin would like to thank all his former colleagues at the Centre for Urban Technology, the School of Architecture, Planning and Landscape and other departments and research centres at Newcastle University for ten happy and stimulating years of collaboration on cities and technology. Many thanks also to Peter Brandon, Ian Cooper, Michael Goldsmith, Alan Harding, Michael Harloe and Tim May for creating an exciting new context for interdisciplinary work on cities and technology at the research centre for Sustainable Urban and Regional Futures (SURF) at Salford University. Simon would also like to thank Steve Connor, Walter Menzies, Joe Ravetz and Simon Shackley for many interesting discussions about the environment, cities and regions. Finally, he offers his gratitude to United Utilities, and particularly the former research and development director, Roger Ford, for supporting the endowment that allows him to spend time thinking about cities and technologies.

This book, more than most, has been made possible only by drawing on and synthesising a huge body of work by a wide range of scholars. The debts we have built up here are too numerous to list in full. But our special thanks are due to the following for finding time for friendship, illuminating discussion or for producing work of the very highest quality that has helped us so much in the production of various parts of the book: Asu Aksoy and Kevin Robins on restructuring Istanbul, Ash Amin on democratisation, Alessandro Aurigi on IT and public space, Karen Baker on water networks, Hugh Barton on environmental neighbourhoods, Eran Ben-Joseph on the modern ideal, Neil Brenner on scales and state theory, Stann Brunn on defensive landscapes, Tim Bunnell on the Multimedia Super Corridor, Manuel Castells on all aspects of the 'network society', Ken Corey on IT in Singapore, Olivier Coutard on network regulation and airport links, Mike Crang on urban cyberspaces, Susan Drucker and Gary Gumpert on communication and public space, Paul Drewe on networks and planning, Ellen Dunham-Jones on post-industrial landscapes, Gabriel Dupuy for his pioneering work on 'network urbanism', Robert Evans on the sociology of knowledge, Joe Ferreira on Internet discrimination, Steven Flusty on postmodern urban theory, Matthew Gandy on histories of the networked metropolis, Andy Gillespie on geography and networks, Barbara Graham and David Holmes on Melbourne CityLink, Simon Guy on the environmental sociologies of cities, Patsy Healey on planning, power and the 'multiplex city', Graham Haughton on urban environmental issues, Thomas Hughes on urban sociotechnical change, Malo Hutson on labour market exclusion and IT, Maria Kaika on water and urban history, John Langdale on telecommunications and globalisation, Kent Larson on the 'smart' home, Jennifer Light on the history of home security, Kristen Little on 'wiring the barrio', everyone involved in the CNRS Groupement de Recherche Réseaux at LATTS in Paris, for paving the way, David Lyon on surveillance, Ceasar McDowell on IT and cultural politics, Neill Marshall on cities and teleservices, Andy Merrifield on public space and resistance, Bill Mitchell on infrastructure, fragmentation and rebundling, Pat Mokhtarian on telecoms and transport, Mitchell Moss on the Internet and the city, Tim Moss, Morten Elle and Suzanne Baslev on the restructuring of infrastructure, Clive Norris on cities and CCTV, Jean Marc Offner on networks and territoriality, Susan Owens on energy and cities, Jorge Otero-Pailos on architectural 'bigness', Joe Painter on urban democratisation, Marcus Power on the City of London, Joe Ravetz on sustainable cities, Ranald Richardson on back offices, Peter Rimmer on transport and telecom flows in the Pacific and restructuring in Jakarta, Jon Rutherford on planning and telecommunications, Saskia Sassen on global cities and global city networks, Sueli Ramos-Schiffer and Ricardo Toledo Silva on utility restructuring in Brazil, Andres Rodriguez-Pose on inward investment in Brazil, Suzanne Speak on marginalised neighbourhoods, Jane Summerton on sociologies of technology, Gerry Sussman on global information labour, Erik Swyngedouw on spatiality and water and social power in the 'cyborg' city, Joel Tarr on urban history, Anthony Townsend on cities and IT and the invaluable telecom-cities listserve, Emy Tseng on broadband and Internet routing, Jo Twist on IT and community, Mark Wilson on global back offices and Matt Zook on urban Internet geographies. All faults in our reading of this body of work do, of course, rest with us!

Third, we owe a considerable debt to everyone who helped in the production side of the book. We owe a special thank-you to Gwyneth Ashworth, Graphic Designer in the Reprographic Unit at Salford University. Gwyneth did a fantastic job skilfully transforming the often crumbled, damaged and blurred plates, tables and figures into the superb illustrations presented here. We would also like to thank Pam Allen at SURF for help in tracing the numerous permissions for the illustrations. At Routledge, Tristan Palmer was extremely supportive during the early stages; Sarah Lloyd and Sarah Carty helped keep the project alive during long dormant phases; Andrew Mould and Ann Michael have provided superb support during what inevitably was a very complex production process.

Fourth, some paragraphs or passages from the following chapters of *Splintering Urbanism* have appeared in slightly modified form in published journal articles. Chapter 1 – S. Graham (2000), 'Introduction: cities and infrastructure networks', *International Journal of Urban and Regional Research*, 24 (1), 114–20; S. Marvin, S. Graham and S. Guy (1999), 'Privatised networks, cities and regions in the UK', *Progress in Planning*, 51 (91–165). Chapter 3 – S. Marvin and N. Laurie (1999), 'An emerging logic of water management, Cochabamba, Bolivia', *Urban Studies* 36 (2), 341–57. Chapter 4 – S. Guy, S. Graham and S. Marvin (1997), 'Splintering networks: cities and technical networks in 1990s Britain', *Urban Studies*, 34 (2), 191–216. Chapter 5 – S. Graham (1998), 'The end of geography or the explosion of place? Conceptualising space, time and information technology', *Progress in Human Geography*, 22 (2), 165–85. Chapter 6 – S. Graham (2000), 'Constructing premium networked spaces: reflections on infrastructure networks and contemporary urban development', *International Journal of Urban and Regional Research*, 24 (1), 183–200; S. Marvin and S. Guy (1997), 'Smart meters and privatised utilities', *Local Economy*, 12 (2), 119–32; S. Graham and S. Speak (2000), 'Service not included: marginalised neighbourhoods, private service disinvestment, and compound social exclusion', *Environment and Planning A*, 1985–2001. Chapter 7 – S. Graham (1999), 'Global grids of glass: on global cities, telecommunications and planetary urban networks', *Urban Studies*, 36 (5–6), 929–49. Chapter 8 – S. Marvin and S. Guy (1997), 'Infrastructure provision, development processes and the coproduction of environmental value', *Urban Studies*, 34 (12), 2023–36; S. Marvin, H. Chappells and S. Guy (1999), 'Pathways of smart metering development: shaping environmental innovation', *Computers, Environment and Urban Systems*, 23, 109–26. Chapters 9–10 – S. Graham (2000), 'Constructing premium networked spaces: reflections on infrastructure networks and contemporary urban development', *International Journal of Urban and Regional Research*, 24 (1), 183–200; S. Graham and P. Healey (1999), 'Relational concepts of space and place: issues for planning theory and practice', *European Planning Studies*, 7 (5), 623–46.

Finally, we would like to thank the following for permission to include plates, figures and tables. Cover picture, National City SW, CA, Maptech, www.maptech.com. Plate 2, Jerde Partnership International, Venice CA; Plate 5, the Johns Hopkins University Press, Baltimore

MD; Plate 3, Photo Oikoumene, World Council of Churches; Plate 6, P. Andreu and MIT Press, Cambridge MA; Plate 7, *Woman's Journal*, IPC Media, London; Plate 13, Vedel Thierry and CNRS–CEVIPOF, Paris; Plate 18, Hilton Judin and Ivan Vladislavic (eds), *Blank – : Architecture, Apartheid and After*, NAI Publishers, Rotterdam; Plate 20, Blackwell Publishers, Oxford. Figure 1.1, Inmarsat, London; Figures 1.2 and 8.1, *Boston Globe*, Boston MA; Figure 1.3, Sprint Corporation, Kansas City; Figure 1.4, Xdrive Inc, Santa Monica CA; Figure 2.1, the Johns Hopkins University Press, Baltimore MD; Figures 2.2 and 2.4, MIT Press, Cambridge MA; Figure 2.3, Smithsonian Institution, Washington DC; Figures 2.5, 6.17 and 7.9, Longman Group, Harlow; Figure 2.6, George Braziller Inc, New York; Figure 2.7, University of California Press, Berkeley and Los Angeles; Figures 2.8 and 3.2, *Flux*, CNRS Central IV, Paris; Figure 2.9, Queens Museum of Art, New York; Figures 2.10, 6.2 and 6.10–11, School of Architecture, Rice University, Houston TX; Figure 2.11, the Art Institute of Chicago; Figure 2.12, Perseus Book Group, New York; Figure 2.13, Editions Armand Colin, Paris; Figure 2.14, AT&T Archives and University of California Press; Figure 2.15, Bell Telephone Canada and University of California Press; Figure 2.16, Northern Electric & Gas, Newcastle upon Tyne; Figure 2.17, E. F. & N. Spon, London; Figure 2.18, Sage Publications, London; Figure 2.19, *Geographical Review*, American Geographical Society, New York; Figure 3.1, *Intermedia*, International Institute of Communications, London; Figures 4.1, 4.3, 6.1 and 6.13, *Urban Studies*, Carfax Publishing Co., Abingdon; Figures 4.2 and 6.7, *Toll Roads Newsletter*, Frederick MD; Figures 4.4–5, International Bank for Reconstruction and Development, Washington DC; Figures 5.1 and 7.10, Routledge/ Taylor & Francis, London; Figure 6.3, Visteon Automotive Systems, Dearborn MI, and *New York Times*; Figure 6.4, *Utility Week*, Reed Information Services, East Grinstead; Figure 6.5, Prentice Hall International, Hemel Hempstead; Figure 6.8, *City*, Next City, Canada; Figure 6.9, Ninety-one Express Lanes, Orange County CA; Figure 6.12, *Landscape and Urban Planning*, Elsevier Science, Barking; Figure 6.14, *Environment and Urban Planning A*, Pion, London; Figure 6.15, *Telecommunications Policy*, Pergamon Press, Oxford; Figure 6.16, IBM Home Director, Morrisville NC; Figure 7.1, HarperCollins Publishers, London; Figure 7.2, Media Partners International, *Amsterdam*; Figure 7.3, COLT Telecom Group, London; Figure 7.4, Pergamon Press, Oxford; Figure 7.5, *Building*, Building (Publishers), London; Figure 7.6, Dr Utis Kaothien and the National Economic and Social Development Board of Thailand, Bangkok; Figure 7.7, *San Francisco Bay Guardian*, San Francisco; Figure 7.8, New York City Economic Development Corporation and New York Telecom Exchange; Figure 7.11, the *Star*, Kuala Lumpur; Figures 7.12 and 7.16, Blackwell Publishers, Oxford; Figure 7.13, the Chinese University Press, Hong Kong; Figure 7.14, *Built Environment*, Alexandrine Press, Oxford; Figure 7.15, *National Productivity Review*, John Wiley & Sons, New York; Figure 7.17, US Immigration and Naturalization Service; Figure 7.18, Heathrow Express, London; Figure 7.19, GTE Interworking, Irving TX; Figure 7.20, Matthew Zook, University of California, Berkeley. Table 2.1, Belhaven Press, London; Table 3.1, John Wiley & Sons, New York; Table 3.2, *Urban Studies*, Carfax Publishing Co., Abingdon; Tables 4.1–7, International Bank for Reconstruction and Development, Washington DC; Table 6.1, *Environment and Urban Planning A*, Pion, London; Table 6.2, *Utility Week*, Reed Information Services, East Grinstead; Table 7.1, *Communications Week International*, EMAP Media, Peterborough.

Every effort has been made to contact copyright holders for their permission to reprint material in this book. The publishers would be grateful to hear from any copyright holder who is not here acknowledged and will undertake to rectify any errors or omissions in future editions of this book.

Stephen Graham (s.d.n.graham@ncl.ac.uk)
Simon Marvin (s.marvin@salford.ac.uk)
Newcastle upon Tyne, September 2000

How is one to conceive of both the organization of a city and the construction of a collective infrastructure?

(Michel Foucault, 1984, 239)

The town is the correlate of the road. The town exists only as a function of circulation and of circuits; it is a singular point on the circuits which create it and which it creates. It is defined by entries and exits: something must enter it and exit from it.

(Deleuze and Guattari, 1997, 186)

I should tell you of the hidden [city of] Berenice, the city of the just . . . linking a network of wires and pipes and pulleys and pistons and counter-weights that infiltrates like a climbing plant.

(Italo Calvino, 1974, 148)

Cities are like electrical transformers: they increase tension, accelerate exchanges, and are endlessly churning human lives.

(Fernand Braudel, 1967, cited in Paquout, 2000, 83)

Cities are the summation and densest expressions of infrastructure, or more accurately a set of infrastructures, working sometimes in harmony, sometimes with frustrating discord, to provide us with shelter, contact, energy, water and means to meet other human needs. The infrastructure is a reflection of our social and historical evolution. It is a symbol of what we are collectively, and its forms and functions sharpen our understanding of the similarities and differences among regions, groups and cultures. The physical infrastructure consists of various structures, buildings, pipes, roads, rail, bridges, tunnels and wires. Equally important and subject to change is the 'software' for the physical infrastructure, all the formal and informal rules for the operation of the systems.

(Herman and Ausubel, 1988, 1)

Cities accumulate and retain wealth, control and power because of what flows through them, rather than what they statically contain.

(Beaverstock et al., 2000, 126)

If the city is to survive, process must have the final word. In the end the urban truth is in the flow.

(Spiro Kostof, 1992, 305)

PROLOGUE

Tales of the networked metropolis

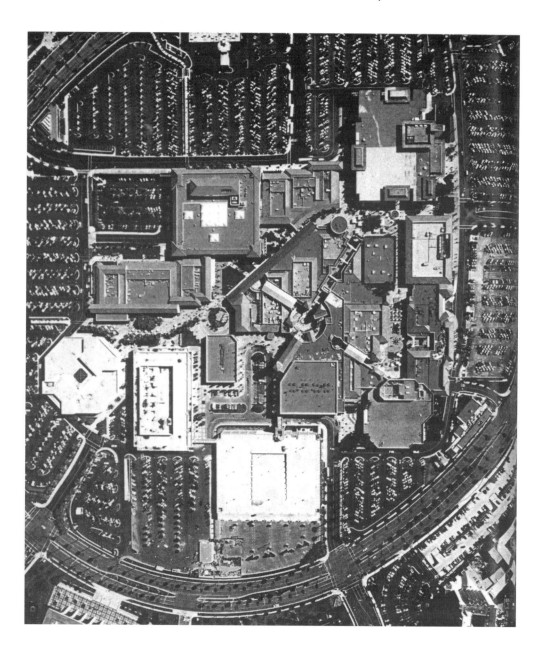

Plate 2 The Fashion Island mall, developed by the Jerde Partnership, in Los Angeles. *Source*: Jerde Partnership (1999), 82

DOWNTOWN CORES ON GLOBAL GRIDS OF GLASS

The US telecommunications firm WorldCom/MCI has built an optic fibre network covering only the core of central London. Only 125 km long, it carries fully 20 per cent of the whole of the United Kingdom's international telecommunications traffic. This is only one of a rapidly emerging global archipelago of urban optic fibre grids concentrated in the urban cores of the world's fifty financial capitals in Asia, Europe, Australasia, and North and South America. Such networks serve no other places. A widening global web of transoceanic and transcontinental fibre networks interconnects these high capacity urban grids, which are carefully located to serve the most communications-intensive international firms. However, whilst the cores of global financial centre spaces reach out to the globe with unprecedented power, increasing efforts are being made to 'filter' their connections with their host cities. In London, for example, the so-called 'ring of steel' supports electronic surveillance systems and armed guards on every entry point into the financial district. Cars entering have their number plates read automatically. Stolen cars are detected within three seconds. And the potential for the facial recognition of drivers, by linking automatically with digitised photographs on national licence records, exists in the system and has already been tested.

WALKING ON WATER

In many developing cities the ideal of distributing drinkable water and sewerage services to all has long been abandoned. Instead, highly dualistic systems are often in place. In the Indian megacity of Mumbai (Bombay), for example, residents of informal settlements actually use the water pipes which distribute drinkable water to affluent gated condominium complexes as perilous footways for transportation. But they have no access whatever to the water supplies within the pipe. Instead, such settlements are often forced to pay extremely high prices for bottled water that is brought in by tanker and sold by private entrepreneurs at huge profit margins.

CUSTOMISING INFRASTRUCTURES FOR INVESTMENT ENCLAVES

In the newly constructed tourist and manufacturing enclaves on Bintan island, Indonesia – a few miles to the south of Singapore – a telephone call or data communication link across the international boundary to Singapore is now counted as 'local'. One across the enclave walls to the surrounding Indonesian territory, however, is charged as 'international'. In Rio Grande do Sul, in southern Brazil, the state government has promised to build a new port, a dedicated canal link and utilities, rail and road links, in order to lure in a US$600 million General Motors car plant. All these expensive new infrastructures will be provided free of charge and will be used exclusively by the company. Because of the expense involved to the state government in this 'bidding war', basic water, energy and road infrastructures for people living in poverty across the state are at risk of being undermined or even withdrawn.

Collapsing technological systems

The doomsday scenarios about the collapse of infrastructure and technological systems due to the 'Y2K' bug were not, on the whole, matched by experience. But, almost without comment, the late 1990s saw the very real and widespread collapse of electricity, power and communications systems in Russia. One of the central modernisation efforts of the communist state had centred on the development of extensive and accessible electricity, telephone, water and heating systems within and between Russian towns and cities, initially to support industrialisation. Since the collapse of communism, however, many of these systems have decayed and collapsed. Sometimes this has been due to simple neglect and the lack of resources, spares and skilled technicians. In the northern cities of Russia, for example, the free municipal heating systems that made the climate bearable have sometimes ceased to function, a process that has significantly accelerated out-migration. But the more worrying trend is the large-scale theft of infrastructure networks, especially trunk electricity systems and communications grids. Over 15,000 km of electrical trunk cable have been stolen in recent years by criminal gangs and people in desperate poverty, to produce metals that can be sold on the black market for export. In a striking process of 'demodernisation', large parts of Russia now face power and electric outages for long periods of time as the tendrils that connected them with modernity are literally carted off and melted down for a quick buck. Not surprisingly, there have been devastating consequences for quality of life, economic development and essential services. This process, needless to say, has forced the wealthy and powerful to secure private and uninterruptible power and communications resources for the enclaved spaces where they live. However, those outside such increasingly defensive enclaves are not so fortunate.

Bypassing the airport crowds

New technologies are widely being adopted to allow favoured, rich and highly mobile travellers to pass seamlessly and quickly through ports, airports and rail terminals, whilst other passengers face traditional, and in many cases intensifying, scrutiny. Through the US Immigration and Naturalization Service Passenger Accelerated Service System (INSPASS), for example, frequent business travellers and diplomats travelling between the United States, Canada, Germany, the Netherlands and other advanced industrial nations can now obtain a smart card that is programmed with the unique biometric signature of the geometry of their hand. By swiping the card at Immigration and placing their hand on a scanner they are allowed 'fast track' routes through airports. Over 70,000 people enrolled in the trial; the Immigration and Naturalization Service are keen to make the system global. 'Not so long ago only strategic places under high surveillance, such as military intelligence agencies, were guarded by such a mechanism' (Mattelart, 1996, 305). Over 50,000 people were using the system in 1999.

My packets are more valuable than yours!

Whilst enormous investment is going into new optic fibre 'pipes' for the Internet, exponential increases in demand continually fill up any new space, creating Internet congestion. In response, companies like Cisco, which make the 'smart' routers and switches that organise flows on the Internet, are now devising ways of 'sifting' the most valued and important 'packets' of information from those that are deemed less important. The idea is that, in times of Internet congestion, the most valued 'packets' from the most profitable customers will be allowed to pass unhindered whilst the rest are blocked. Thus, beneath the rhetoric that the Internet is some egalitarian and democratic space, profound inequalities are being subtly and invisibly integrated into the very protocols that make it function.

From open grids to closed urban streets

Many urban streets in North America, Asia, Africa and Latin America are now privatised and self-contained rather than open and interconnected. Such streets act as entry points to 'gated' or 'master-planned' communities. These are carefully segregated and 'fortressed' from the rest of the city through walls, gates and high-technology surveillance systems, yet sustained through guarded, dedicated highway gates, customised water and energy connections, and telecommunications grids, that selectively connect them to the wider urban constellation and the universe beyond. The private governments of such spaces are actively exploring the delivery of their own water, transport and energy services to match their telecom networks, further removing them from involvement in the broader metropolitan fabric and enhancing their emergence as quasi-medieval city states. One gated community near Phoenix, Arizona, now even operates its own fleet of electric vehicles which cannot be used outside its boundaries on public highways (Kirby, 1998).

Skywalk cities and global citadels

At the same time, in the downtown cores of North American cities like Houston, Toronto and Minneapolis, the extending logic of 'skywalk' systems is bypassing the traditional street system. Skywalks link extending webs of office and shopping complexes downtown with carefully monitored, air-conditioned, and hermetically sealed pedestrian networks, what Boddy (1992) terms the 'analogous city'.

Airports, freight zones, retail malls, sports stadia and university, research, hospital, media and technology campuses are similarly emerging as zones of intense regional and global interchange whilst at the same time walls, ramparts and CCTV systems are constructed which actively filter their relationships with the local urban fabric. In Baltimore, for example, David Harvey notes the paradox that, whilst African American women cross these boundaries daily to clean some of the world's most famous hospitals (for example, Johns Hopkins), they are unable to access health services when they are ill because of lack of health insurance. Meanwhile 'life expectancy in the immediate environs of these internationally renowned hospital facilities is among the lowest in the nation and comparable to many of the poorer countries of the world' (2000, 136). Carlo Ezecieli (1998) calls these places 'global citadels'. Through such trends the physical fabric of many cities across the world is starting to fragment into giant cellular clusters – packaged landscapes made up of customised and carefully protected corporate, consumption, research, transit, exchange, domestic and even health-care spaces. Each tends to orient towards highway grids, global telecommunications connections, premium energy and water connections, whilst CCTV and security guard-protected 'public private spaces' mediate their relationships with their immediate environments. Thus they tend to turn their backs on traditional street fronts and the wider urban fabric, carefully filtering those 'undesirable' users deemed not to warrant access for work, play, leisure, residence or travel. The new American football stadium at Foxboro, Massachusetts, for example, is being built with an access road that is solely dedicated to owners and users of corporate boxes; all other fans must use the old public highway.

A tragic example of the starkness of such carefully designed local disconnections came on 14 December 1995 at the huge Walden Galleria Mall on the edge of Buffalo NY. An employee of the mall, Cynthia Wiggens, was trying to cut across a seven-lane highway from the public city bus stop when she was run down and killed by a ten-ton truck. City buses were not allowed to enter the mall, every aspect of which had been designed to attract high-spending middle- and upper-income consumers travelling exclusively by car (Gottdeiner, 1997, 132).

PRIVATE 'SMART' HIGHWAY CORRIDORS

In some cities, urban highways, too, are increasingly privatised, profit-oriented and customised to the needs of affluent commuters on particular urban corridors. In cities like Toronto, San Diego, Melbourne and Los Angeles new privately funded highways use completely automatic electronic tolling technologies to create entirely new transport and development corridors that are superimposed on old public highway grids. Land parcels are sold off by the highway corporation to create integrated corridors designed to serve affluent motorists. In some 'electronic highways' tariffs are electronically altered in 'real time' according to demand so that they can guarantee free flow and tempt in frustrated commuters from gridlocked public highway grids, allowing paying commuters to 'wormhole' through some of the most congested public highways in North America. Drivers' bank accounts are precisely debited according to the times and distances of travel. In Los Angeles commuters enjoy a saving of forty minutes compared with normal driving times along the ten-mile public highway.

THE ULTIMATE COMMUTE

Driven by fear of car-jacking and the inexorable gridlocking of the city's streets – a city with 8,500 murders a year, a rate ten times that of New York – the most privileged residents of the Brazilian megacity of São Paulo have discovered the ultimate means to escape the constraints of the highway, the street and even the terrestrial surface in their journeys around the city: a personal helicopter. At over 400 and increasing rapidly, the *New York Times* reports,

São Paulo's personal helicopter fleet is the fastest growing in the world, a powerful symbol of the almost surreal extremes of wealth and poverty in the city (15 February 2000, p. 1). 'Why settle for an armoured BMW when you can afford a helicopter?' asks Eric Wassen, a local dealer. At the same time the 3.7 million daily users of the city's 10,400 buses face heightening delays, pollution and violence amidst a chaotic, collapsing public transport system and heightening risks of violence.

MULTIPLYING UTILITY GRIDS

In the privatised utility markets of the United Kingdom people can now choose from dozens of gas suppliers, electricity companies and telecoms providers, and sometimes even water firms – firms whose headquarters are scattered all over the developed word. Singaporean cable. Dutch telecommunications. American energy. French water. In some cities 'multi-utilities' are emerging offering energy, water and telecoms on a 'one-stop shop' basis. Citizens can now back up their search for environmentally friendly food, transport and housing by paying extra for 'green' electricity inputted to the network by specialised companies from renewable sources. Housing tenants can similarly access 'red' electrons generated by socially conscious companies. For privileged consumers, new information technologies open up a virtual market place of different providers and value-added services. But for lower-income users the same technologies tend to be configured differently, to help distance suppliers from low-income people through the use of 'top up' smart and pre-payment cards. These involve no direct contact between supplier and consumer. They require users – who tend to be among the most immobile in society – to travel physically to 'top them up'. And they often trap people on higher tariffs and away from the benefits of competition.

1 INTRODUCTION

Networked infrastructures, technological mobilities and the urban condition

Plate 3 Electric power lines near Durban, South Africa, which run through Umlazi township without serving any of the people who live there. *Photograph*: Photo Oikoumene, World Council of Churches

A critical focus on networked infrastructure – transport, telecommunications, energy, water and streets – offers up a powerful and dynamic way of seeing contemporary cities and urban regions (see Dupuy, 1991). When our analytical focus centres on how the wires, ducts, tunnels, conduits, streets, highways and technical networks that interlace and infuse cities are constructed and used, modern urbanism emerges as an extraordinarily complex and dynamic sociotechnical *process*. Contemporary urban life is revealed as a ceaseless and mobile interplay between many different scales, from the body to the globe. Such mobile interactions across distances and between scales, mediated by telecommunications, transport, energy and water networks, are the driving connective forces of much-debated processes of 'globalisation'.

In this perspective, cities and urban regions become, in a sense, staging posts in the perpetual flux of infrastructurally mediated flow, movement and exchange. They emerge as processes in the distant sourcing, movement and disposal of water reserves and the remote dumping of sewage and waste. They are the hotbeds of demand and exchange within international flows of power and energy resources. They are the dominant sites of global circulation and production within a burgeoning universe of electronic signals and digital signs. They remain the primary centres of transnational exchange and the distribution of products and commodities. And they are overwhelmingly important in articulating the corporeal movements of people and their bodies (workers, migrants, refugees, tourists) via complex and multiple systems of physical transportation.

The constant flux of this urban process is constituted through many superimposed, contested and interconnecting infrastructural 'landscapes'. These provide the mediators between nature, culture and the production of the 'city'. There is the 'electropolis' of energy and power. There is the 'hydropolis' of water and waste. There is the 'informational' or 'cybercity' of electronic communication. There is the 'autocity' of motorised roadscapes and associated technologies. And so on. Importantly, however, these infrastructural 'scapes' are not separated and autonomous; they rely on each other and co-evolve closely in their interrelationships with urban development and with urban space.

How, then, can we imagine the massive technical systems that interlace, infuse and underpin cities and urban life? In the Western world especially, a powerful ideology, built up particularly since World War II, dominates the way we consider such urban infrastructure networks. Here, street, power, water, waste or communications networks are usually imagined to deliver broadly similar, essential, services to (virtually) everyone at similar cost across cities and regions, most often on a monopolistic basis. Fundamentally, infrastructure networks are thus widely assumed to be integrators of urban spaces. They are believed to bind cities, regions and nations into functioning geographical or political wholes. Traditionally, they have been seen to be systems that require public regulation so that they somehow *add cohesion* to territory, often in the name of some 'public interest'.

Infrastructure operators are assumed in this ideology to cover the territories of cities, regions and nations contiguously, like so many jigsaw pieces. They help to define the identity and development of their locality, region or nation in the process. The assumption, as Steven Pinch argued in his classic book *Cities and Services*, is that utility supplies (and sometimes public transport and telecommunications networks, too) are 'public local goods which are, generally speaking, freely available to all individuals at equal cost within particular local government or administrative areas' (1985, 10). The implication is that, compared with other 'point-specific' urban services like shops, banks, education and housing, they are of relatively little interest to urban researchers because, to all intents and purposes, they don't really *have* an urban geography in the conventional sense.

What, then, are we to make of the range of examples in the Prologue, which seem so contradictory to such assumptions? What is happening to the previously sleepy and often taken-for-granted world of networked urban infrastructure? How can we explain the emergence of myriads of specialised, privatised and customised networks and spaces evident in the above examples, even in nations where the ideal of integrated, singular infrastructures – streets, transport networks, water grids, power networks, telephone infrastructures – was so recently central to policy thinking and ideology? And what might these emerging forms of infrastructure development mean for cities and urban life across an urbanising planet?

To us, these stories, and the many other tales of networked infrastructure in this book, raise a series of important questions. Are there common threads linking such a wide range of cases? Are broadly common processes of change under way across so many places and such a wide range of different infrastructure networks? How can we understand the emerging infrastructure networks and urban landscapes of internationalising capitalism, especially when the study of urban infrastructure has been so neglected and so dominated by technical, technocratic or historical perspectives? How is the emergence of privatised, customised infrastructure networks across transport, telecommunications, energy and water – like the ones discussed above – interwoven with the changing material and socioeconomic and ecological development of cities and urban regions? And, finally, what do these trends mean for urban policy, governance and planning and for discussions about what a truly democratic city may actually mean?

The rest of this book will address these questions through an international and trans-disciplinary analysis of the changing relationship between infrastructure networks, the technological mobilities they support, and cities and urban societies. In this first chapter we set the scene for this discussion. We do so in six parts. First, we introduce the complex interdependences of urban societies and infrastructure networks. Second, we explore how contemporary urban change seems to involve trends towards uneven global connection combined with an apparently paradoxical trend towards the reinforcement of local boundaries. In the third and fourth parts we move on to analyse why Urban Studies and related disciplines have largely failed to treat infrastructure networks as a systematic field of study. We point out that, instead, it has widely been assumed that technologies and infrastructures simply and deterministically shape both the forms and the worlds of the city, and wider constructions of society and history. Fifth, we explore those moments and periods which starkly reveal the ways in which contemporary urban life is fundamentally mediated by such networks: collapses and failures. We close the chapter by drawing up some departure points for the task of the remainder of the book: imagining what we call a critical urbanism of the contemporary networked metropolis.

TRANSPORT, TELECOMMUNICATIONS, ENERGY AND WATER: THE MEDIATING NETWORKS OF CONTEMPORARY URBANISM

Our starting point in this book is the assertion that infrastructure networks are the key physical and technological assets of modern cities. As a 'bundle' of materially networked, mediating infrastructures, transport, street, communications, energy and water systems constitute the largest and most sophisticated technological artefacts ever devised by humans. In fact, the fundamentally *networked* character of modern urbanism, as Gabriel Dupuy (1991) reminds us, is perhaps its single dominant characteristic. Much of the history of modern urbanism can be understood, at least in part, as a series of attempts to 'roll out' extending and multiplying road, rail, airline, water, energy and telecommunications grids, both within and between cities and metropolitan regions. These vast lattices of technological and material connections have been necessary to sustain the ever-expanding demands of contemporary societies for increasing levels of exchange, movement and transaction across distance. Such a perspective leads us to highlight four critical connections between infrastructure networks and contemporary urbanism that together form the starting points of this book.

CITIES AS SOCIOTECHNICAL PROCESS

First, economic, social, geographical, environmental and cultural change in cities is closely bound up with changing practices and potentials for mediating exchange over distance through the construction and use of networked infrastructures. 'Technological networks (water, gas, electricity, information, etc.) are constitutive parts of the urban. They are mediators through which the perpetual process of transformation of Nature into City takes place' (Kaika and Swyngedouw, 2000, 1). As Hall and Preston put it, in modern society 'much innovation proves to depend for its exploitation on the creation of an infrastructural network (railways; telegraph and telephone; electricity grids; highways; airports and air traffic control; telecommunications systems)' (1988, 273).

In a sense, then, the life and flux of cities and urban life can be considered to be what we might call a series of closely related 'sociotechnical processes'. These are the very essence of modernity: people and institutions enrol enormously complex technological systems (of which they often know very little) to extend unevenly their actions in time and space (Giddens, 1990). Water and energy are drawn from distant sources over complex systems. Waste is processed and invisibly shifted elsewhere. Communications media are enrolled into the production of meaning and the flitting world of electronic signs. And people move their bodies through and between the physical and social worlds of cities and systems of cities, either voluntarily or for pleasure or, it must be remembered, through the trauma and displacement of war, famine, disaster or repression.

'ONE PERSON'S INFRASTRUCTURE IS ANOTHER'S DIFFICULTY': URBAN INFRASTRUCTURE NETWORKS AS 'CONGEALED SOCIAL INTERESTS'

Second, and following on from this, infrastructure networks, with their complex network architectures, work to bring heterogeneous places, people, buildings and urban elements into dynamic relationships and exchanges which would not otherwise be possible. Infrastructure networks provide the distribution grids and topological connections that link systems and practices of production with systems and practices of consumption. They unevenly bind spaces together across cities, regions, nations and international boundaries whilst helping also to define the material and social dynamics, and divisions, within and between urban spaces. Infrastructure networks interconnect (parts of) cities across global time zones and also mediate the multiple connections and disconnections within and between contemporary cities (Amin and Graham, 1998b). They dramatically, but highly unevenly, 'warp' and refashion the spaces and times of all aspects of interaction – social, economic, cultural, physical, ecological.

Infrastructure networks are thus involved in sustaining what we might call 'sociotechnical geometries of power' in very real – but often very complex – ways (see Massey, 1993). They tend to embody 'congealed social interests' (Bijker, 1993). Through them people, organisations, institutions and firms are able to extend their influence in time and space beyond the 'here' and 'now'; they can, in effect, 'always be in a wide range of places' (Curry, 1998, 103). This applies whether users are 'visiting' web sites across the planet, telephoning a far-off friend or call centre, using distantly sourced energy or water resources, shifting their waste through pipes to far-off places, or physically moving their bodies across space on highways, streets or transport systems.

The construction of spaces of mobility and flow for some, however, always involves the construction of barriers for others. Experiences of infrastructure are therefore highly contingent. 'For the person in the wheelchair, the stairs and door jamb in front of a building are not seamless subtenders of use, but barriers. One person's infrastructure is another's difficulty' (Star, 1999, 380). Social biases have always been designed into urban infrastructure systems, whether intentionally or unintentionally. In ancient Rome, for example, the city's sophisticated water network was organised to deliver first to public fountains, then to public baths, and finally to individual dwellings, in the event of insufficient flow (Offner, 1999, 219).

We must therefore recognise how the configurations of infrastructure networks are inevitably imbued with biased struggles for social, economic, ecological and political power to benefit from connecting with (more or less) distant times and places. At the same time, though, we need to be extremely wary of the dangers of assigning some simple causal or deterministic power to technology or infrastructure networks *per se* (Woolgar, 1991). Infrastructures and technologies do not have simple, definitive and universal urban 'impacts' in isolation. Rather, such large technological systems (Summerton, 1994a) or technical networks (Offner, 1993) are closely bound up within wider sociotechnical, political and cultural complexes which have contingent effects in different places and different times (see Tarr and Dupuy, 1988; Joerges, 1999a).

INFRASTRUCTURE NETWORKS AS EMBEDDED GEOPOLITICS

Third, infrastructure networks make up considerable portions of the material, economic and geopolitical fabric of contemporary cities and systems of cities. As capital that is literally 'sunk' and embedded within and between the fabric of cities, they represent long-term accumulations of finance, technology, know-how, and organisational and geopolitical power. New infrastructure networks 'have to be immobilised in space, in order to facilitate greater movement for the remainder' (Harvey, 1985, 149). This means that they can 'only liberate activities from their embeddedness in space by producing new territorial configurations, by harnessing the social process in a new geography of places and connecting flows' (Swyngedouw, 1993, 306).

The 'messy' practices of embedding, building and maintaining infrastructure networks beneath, through and above the fabric of cities thus infuses the politics of metropolitan areas. They require complex regulatory articulations between markets, national and local states and, increasingly, transnational bodies. Whilst there are global trends towards various types of privatisation and liberalisation in the development of networked urban infrastructures, the way in which the contested politics of network development are played out in each city, region or nation is still closely related to the broader constructions of governance, the state, and the market in each case (Lorrain and Stoker, 1997).

INFRASTRUCTURE NETWORKS AND CULTURES OF URBAN MODERNITY AND MOBILITY

Finally, infrastructure networks, and the sociotechnical processes that surround them, are strongly involved in structuring and delineating the experiences of urban culture and what Raymond Williams (1973) termed the 'structures of feeling' of modern urban life. Networked technologies of heat, power, water, light, speed and communications have thus been intrinsic to all urban cultures of modernity and mobility (Thrift, 1995). They are invariably invoked in images, representations and ideologies of urban 'progress' and the modern city by all sorts of actors – developers, planners, state officials, politicians, regulators, operators, engineers, real estate developers and appliance manufacturers, as well as artists, journalists, social scientists, futurists and philosophers (see Kaika and Swyngedouw, 2000).

Infrastructure networks have traditionally also tended to be central to the normative aspirations of planners, reformers, modernisers and social activists to define their notions of a desirable urban order: the good city (see Friedmann, 2000). Consider, for example, Le Corbusier's and Frank Lloyd Wright's utopias based on highways; the 1920s futurists' obsession with air, rail, cruise liners and motor travel; Ebenezer Howard's concern for municipal rail connections; or the centrality of boulevards and sewers within Haussmann's nineteenth century 'modernisation' of Paris (see Dupuy, 1991, 105). Think, too, of the more recent speculations about how the good city might finally be realised as a 'cybercity', a 'city of bits' or an 'e-topia', laced with the latest digital media technologies and networks (Mitchell, 1995, 1999; see also Wheeler *et al.*, 2000).

NETWORKED PARADOXES: GLOBAL CONNECTIONS AND LOCAL (DIS)CONNECTIONS

Of course, cities, metropolitan life and infrastructural connections with (more or less) distant elsewheres have been inextricably interwoven throughout the last 7,000 years of urban history (Soja, 2000). What has changed in the past century, however, are:

- the intensity, power, speed and reach of those connections;
- the pervasiveness of reliance on urban life based on material and technological networks and the mobilities they support;
- the scale of technologically mediated urban life;
- the duplicating, extending variety and density of networked infrastructures;
- the speed of sophistication of the more powerful and advanced infrastructures (see Urry, 2000b).

Today the majority of the population in the Western world, and an increasing proportion of the developing, newly industrialising and post-communist worlds, live in cities that represent the largest and most concentrated source of demand for water, energy and transport and communications services. Much of the material and technological fabric of cities, then, *is* networked infrastructure. At the same time, most of the infrastructural fabric *is* urban 'landscape' of various sorts. Almost every aspect of the functioning of infrastructure, the retrofitting of new networks and the renewal of older networks is focused on the needs of serving expanding urban areas, and the demands for communication of people, goods, raw materials, services, information, energy and waste within and between cities.

Vast networks connect users in almost every building with more or less distant power stations, sewage works, reservoirs, gasfields, transport grids and global communication systems. Enormous regional, national and international networks and powerful institutions have been constructed to suck resources into, and extract waste from, cities, and to exchange communication between predominantly urban centres over the globe. Networked infrastructure, in short, provides the technological links that make the very notion of a modern city possible (Tarr and Dupuy, 1988).

CONTEMPORARY INFRASTRUCTURAL MOBILITIES: GLOBALISATION AND LIBERALISATION

However, something quite profound is happening in the world of urban infrastructure. Internationally, all the major urban infrastructure networks – water and waste, energy, telecommunications and much of the transport infrastructure – are gradually being 'opened up' to private sector participation in the management and provision of services. In many cases public and private monopolies are being replaced by contested, profit-driven markets.

As a result, the infrastructure sector is now one of the most important sectors in international flows of finance, capital, technology and expertise as international infrastructure firms roam the world in search of healthy profits and high rates of return from lucrative niche markets or franchises. Across the planet, the era of the monopolistic provision of standardised services is being undermined as the World Trade Organisation, the Group of Eight, and regional economic blocs like the European Union in Europe, NAFTA in North America, ASEAN in South East Asia, and Mercosur in South America variously work, albeit at very different rates and in very different contexts, to support shifts towards the liberalisation of national and local infrastructure monopolies (McGowan, 1999).

As a result of such processes, acquisitions, mergers and strategic alliances between utility and infrastructure corporations present some of the fastest-moving scenes on international financial markets. Such events can dramatically change the infrastructural logics of cities and regions almost overnight (Curwen, 1999; McGowan, 1999). This is creating new competitive markets that complement or replace predictable and monolithic monopolies with highly fragmented and differentiated styles of service provision with highly complex, and often hidden, geometries and geographies.

URBAN FRAGMENTATION AND RECOMBINATION: INFORMATION TECHNOLOGIES AS 'HEARTLAND' TECHNOLOGIES

Information technologies are clearly developing as the crucial information infrastructures mediating our increasingly information-intensive urban economies, societies and cultures (Castells, 1996, 1998). But, as the 'heartland' technologies of contemporary economic, technological and cultural change, they are also being applied to help reconstitute how more 'traditional' transport, energy and water networks operate. Consider, for example, the growth of 'smart' highways, 'virtual' energy markets, access-controlled streets, CCTV-surveilled downtown skywalks, and personalised multimedia and communications services (see Freeman, 1990).

The powers of new information technologies support the complex restructuring of urban forms, lifestyles and landscapes. This is based on the parallel processes of fragmentation and recombination of urban uses and functions, within and between cities and systems of cities (Mitchell, 1999). Whilst some activities are scattering across geographical space to be integrated electronically – ATMs, back offices, e-commerce vendors, corporate sites – information technologies may also support the 'renucleation' of work, home and neighbourhood services for certain people and places – activities that were often separated into single-use zones during the development of the industrial, functional, city (ibid.).

PARADIGM CHALLENGES IN THE 'NETWORK SOCIETY'

Above all, the increasingly 'hybrid' nature of contemporary cities, where powerful digital connections elsewhere articulate every aspect of urban life, requires us continually to rethink the paradigms that we use when analysing cities. Such processes 'challenge the long-held privileged status of Cartesian geometry, the map, and the matrix or grid. Infrastructural links and connectors, as well as information exchanges and thresholds, become the dominant metaphors to examine the boundless extension of the regional city' (Boyer, 2000, 75).

Increasingly, as Manuel Castells (1996, 1997a, 1998) suggests, these processes are directly supporting the emergence of an internationally integrated and increasingly urbanised, and yet highly fragmented, *network society* that straddles the planet. New, highly polarised urban landscapes are emerging where 'premium' infrastructure networks – high-speed telecommunications, 'smart' highways, global airline networks – selectively connect together the most favoured users and places, both within and between cities. Valued spaces are thus increasingly defined by their fast-track connections elsewhere, as any examination of the intensifying transport, telecommunications and energy links between the dominant parts of 'global' cities reveals. At the same time, however, premium and high-capability networked infrastructures often effectively bypass less favoured and intervening places and what Castells calls 'redundant' users. Often such bypassing and disconnection are directly embedded into the design of networks, both in terms of the geographies of the points they do and do not connect, and in terms of the control placed on who or what can flow over the networks. Through such processes, Castells predicts that:

The global economy will expand in the twenty-first century, using substantial increases in the power of telecommunications and information processing. It will penetrate all countries, all territories, all cultures, all communication flows, and all financial networks, relentlessly scanning the planet for new opportunities of profit-making. But it will do so selectively, linking valuable segments and discarding used up, or irrelevant, locales and people. The territorial unevenness of production will result in an extraordinary geography of differential value making that will sharply contrast countries, regions, and metropolitan areas. Valuable locales and people will be found everywhere, even in Sub-Saharan Africa. But switched-off territories and people will also be found everywhere, albeit in different proportions. The planet is being segmented into clearly distinct spaces, defined by different time regimes.

(1997b, 21)

BEYOND THE TERRITORIALLY COHESIVE CITY: PROXIMITY ≠ MEANINGFUL RELATIONS!

Virtually all cities across the world are starting to display spaces and zones that are powerfully connected to other 'valued' spaces across the urban landscape as well as across national, international and even global distances. At the same time, though, there is often a palpable and increasing sense of local disconnection in such places from physically close, but socially and economically distant, places and people. Some have even interpreted this widespread pattern of development as signifying some form of convergence between developed, newly industrialised, post-communist and developing cities (Cohen, 1996).

Because of these dynamics, and the intensifying uneven development of infrastructures, physically close spaces can, in effect, be relationally severed (Graham and Healey, 1999). At the same time, globally distant places can be relationally connected very intimately. This undermines the notion of infrastructure networks as binding and connecting territorially cohesive urban spaces. It erodes the notion that cities, regions and nations necessarily have any degree of internal coherence at all. And it forces us to think about how space and scale are being refashioned in new ways that we can literally see crystallising before us in the changing configurations of infrastructure networks and the landscapes of urban spaces all around us.

In short, emerging urban landscapes, and the relationships between infrastructure networks and urban spaces, seem to embody powerfully the changing dynamics of global political economies and societies. As Carlo Ezecieli argues, from the point of view of US cities:

while markets are establishing systems of planetary interdependence and metropolitan regions become more and more directly related to a global dimension, there appears to be a paradoxical tendency toward the reinforcement of local boundaries. In crime-ridden American neighborhoods buildings tend to be fortified like military bases. In gated communities the protection of privileged circles through the erection of physical boundaries is marketed as an attractive amenity. Primary urban facilities like large hospitals, universities, and shopping malls, establish simulations of 'public' venues within physically bounded and access-controlled environments.

(1998, 4)

THE NEGLECT OF NETWORKED URBAN INFRASTRUCTURES AND TECHNOLOGICAL MOBILITIES IN TREATMENTS OF THE CITY

Study a city and neglect its sewers and power supplies (as many have), and you miss essential aspects of distributional justice and planning power.

(Star, 1999, 379)

Unfortunately for us, a major investigation of the complex relations between infrastructure, technology and contemporary cities such as this book is not well served by previous literature. Outside a few specialised debates on urban transport (see Hanson, 1993), urban history (see Tarr and Dupuy, 1988) and emerging information technologies (see Castells, 1989, Graham and Marvin, 1996), urban infrastructure networks and the mobilities they support have traditionally hardly been considered the most exciting focus of debate in urban studies and policy making. 'Because these systems include complex technological artifacts, they are often viewed as "engineers' stuff", not worth the interest of the social sciences' (Coutard, 1999, 1).

Why is this so? Why do disciplines which purport to understand the nature of the contemporary metropolis systematically neglect the networked infrastructures and technological mobilities that are so important in defining its nature, form and process? Five reasons can be identified.

PARALLEL DISCIPLINARY FAILINGS

First, the inertia of disciplinary and subdisciplinary boundaries has severely hindered understanding of a subject which intrinsically demands an interdisciplinary or transdisciplinary starting point. When literatures on networked urban infrastructure have emerged in Planning, Geography, Urban Studies, Engineering, Sociology and Architecture, they have often been inward-looking, technical, and overly specialist.

By way of illustration, we can identify parallel failings across geography, sociology and architecture which have contributed in different ways to these disciplinary failings to develop critical, cross-cutting perspectives on urban infrastructures and technological networks as a whole.

GEOGRAPHY: ASSUMPTIONS OF TECHNOLOGICAL 'NEUTRALITY'

Taking Geography first, Michael Curry (1998, 2) has suggested that, with a few notable exceptions, geographers (especially English-speaking ones) have not embraced the study of what he calls 'geographic technologies' like utilities and IT systems. This is for the simple reason that 'they have adopted the view, so widespread, that all technologies are natural and neutral' (ibid.). Nor has the obvious invisibility of most contemporary utilities and communications systems fitted with Geography's traditional emphasis on land use and the visualities of urban life. Curry also wonders whether, deep down, 'many geographers may not harbor fears that in the end some critics are right, and that these new technologies will lead to the death of space and place, and hence of their own discipline' (ibid.).

One subdiscipline within geography, that specialising in transport, has managed to emerge, but it has remained a fairly closed subdiscipline. Transport geography has only very limited connections with broader constructions of contemporary urban geography. Hamilton and Hoyle have lamented that broader debates about the city 'rarely give transport the coverage it deserves' (1999, 1).

SOCIOLOGY: THE LIMITATIONS OF CLASSICAL FORMULATIONS

In Sociology the early efforts of writers like Lewis Mumford (1934) to create overarching and historically informed treatments of the interplay of cities, mobilities and technologies have not been built on. According to Armand Mattelart, the sociology of communication, in particular, has largely failed in the necessary task, which is to 'do away with the separations between different areas and crossing the angles of vision of the disciplines in order to bring out the manifold logics by which the multiple forms of technology have molded, and in turn been molded by, the history of humankind, its mentalities, and its civilizations' (1996, ix). Recent work on the analysis of 'large technical systems' has, however, led to some progress here (see Mayntz and Hughes, 1988; LaPorte, 1991; Summerton, 1994a; Coutard, 1999).

As in Geography, the caprices of intellectual trends have continually rendered networked infrastructures, and the technological mobilities they support, unfashionable in Sociology. For example, despite the extraordinary motorisation of cities in the past thirty years or so, John Urry noted recently that 'sociology has barely noticed . . . automobility, or even the car more generally' in its preoccupation with the strolling and *flânerie* of the walking urbanite. Sociology has been even more neglectful of other networked infrastructures (1999, 5).

In response, Urry urges sociologists to look beyond the classical and often rather static twentieth century formulations centring on how class, gender, ethnicity and social mobility were constructed within individual 'societies' bounded by nation states. Instead he suggests the need to reformulate the discipline as a 'sociology of mobilities' (2000b). This would deal centrally with the 'postsocietal' nature of the contemporary world, with its 'diverse mobilities of peoples, objects, images, information, and wastes' and its 'complex interdependencies between, and social consequences of, these diverse mobilities' (2000a, 185). Urry's is one of the clarion calls to which we address this book.

ARCHITECTURE AND URBANISM: BEYOND THE BUILDING AS ISOLATED UNIVERSE

In the last thirty years or so, urbanists and architects, too, have tended to neglect networked infrastructures and the flows and mobilities that they support. They have tended to focus overwhelmingly on the designed spaces within building envelopes, rather than the networked infrastructures that knit buildings together, binding and configuring the broader spaces of metropolitan life.

As Jon Jerde, a well known architect of theme parks and entertainment complexes, suggests, 'architects rarely focus attention on the process that creates – and the conditions that surround – the object or building' that they are designing or deconstructing (Jerde Partnership, 1999, 203). Because architects 'rarely define sites in multiples', they tend not to see them 'in a way that will permit exploration of the organizational or network architecture' of buildings that combine closely with infrastructure or organisational networks across diverse spaces (Easterling, 1999a, 2). In the 1970s, in particular, many architects largely turned their backs on the problems of the wider metropolis (Wall, 1996, 158).

'THE FORGOTTEN, THE BACKGROUND, THE FROZEN IN PLACE': INFRASTRUCTURE NETWORKS AS THE 'CINDERELLA' OF URBAN STUDIES

Second, and following from this discussion, it is clear to us that urban infrastructure networks and the mobilities they support have very much been left as the 'Cinderella' of contemporary urban studies and urbanism. Most social analyses of cities still address urban sociologies, economic development, governance and politics, urban cultures and identities, and urban ecologies and environments, without seriously exploring the roles of networked infrastructures in mediating all. As Susan Star argues, such networks tend still to be 'the forgotten, the background, the frozen in place' (1999, 379).

Even discussions of the cultures, sociologies and geographies of urban 'modernity' often fail to assert the essential contribution of networked infrastructures of all types to the processes and experience of modern urbanism (see, for example, Savage and Warde, 1993). Urban studies, moreover, often tend towards static formulations of the nature of urban society and urban life. Only rarely do discourses of the city 'script the city as a process of flows' – an approach which tends to emphasise the roles of massive technological networks and infrastructural mobilities in mediating urban life (Kaika and Swyngedouw, 2000, 2).

Consider, for example, the ways in which urban and regional studies have begun to address consumption issues with considerable energy. Debates have sprung up surrounding the restructuring of public services (Pinch, 1989, 1997), the links between private and public consumption and quality of life (Rogerson *et al.*, 1996; Miller, 1995), and the transformation of many post-industrial city spaces into entertainment, leisure and consumption zones (see Hannigan, 1998a). However, infrastructure networks again tend to remain largely ignored in such debates, closed off within their inward-looking and technical subdisciplines. Very little urban research has addressed the important shifts now under way in the consumption and development of what we may call *distributive network services* that use technological networks to distribute power, communication, water and mobility services across space and time.

DIALECTICS OF INVISIBILITY AND MONUMENTALISM

Third, the hidden nature of much of the contemporary physical fabric of infrastructure in many cities has also contributed to their 'Cinderella' status (see Latour and Hermand, 1998). Many urban networks in the contemporary city remain 'largely opaque, invisible, disappearing underground, locked into pipes, cables, conduits, tubes, passages and electronic waves' (Kaika and Swyngedouw, 2000, 2). They seem 'by definition [to be] invisible, part of the background for other kinds of work' (Star, 1999, 380).

This invisibility has allowed the subterranean guide to emerge as a sub-genre of urban guide and photographic books, allowing those who want to look beyond the urban myths and legends that tend to surround the underground of cities to explore the full depth, complexity and history of a city's 'root system' (see, for example, Granick, 1947; Trench and Hillman, 1984; Greenberg, 1998). Such books help us to visualise the hidden background of urban networked infrastructures. Consider, for example, Robert Sullivan's introduction to Harry Granick's classic book *Underneath New York* (1947):

Imagine grabbing Manhattan by the Empire State Building and pulling the entire island up by its roots. Imagine shaking it. Imagine millions of wires and hundreds of thousands of cables freeing themselves from the great hunks of rock and tons of musty and polluted dirt. Imagine a sewer system and a set of water lines three times as long as the Hudson River. Picture mysterious little vaults just beneath the crust of the sidewalk, a sweaty grid of steam pipes 103 miles long, a turn-of-the-eighteeenth-century merchant ship bureau under Front Street, rusty old gas lines that could be wrapped twenty-three times around Manhattan, and huge, bomb-proof concrete tubes that descend almost eighty storeys into the ground.

(iv)

The tendency to obscure the management and development of infrastructures within highly technical and technocratic institutions, driven by the supposedly depoliticised, instrumental rationalities of engineering cultures, has served further to obfuscate the worlds of networked urban infrastructure. Transport, for example, is 'usually confined to a separate, substantive treatment which tends to leave to the transport experts the physical definition of its function and its location in specialized zones' (Solà-Morales, 1996, 14). Very often, infrastructure networks remain politically contained by the widespread and powerful assumption that state or private monopolies will simply provide services when, and where, they are needed, as public or quasi-public services to sustain urban life. Reflecting this, the whole of infrastructure is sometimes captured within catch-all terms like 'public works'.

However, it is important to note that a reverse tendency to infrastructural invisibility and political obfuscation does periodically emerge. Rather than being hidden, here infrastructure networks are revealed, celebrated and constructed as iconic urban landmarks, as embodiments of the 'phantasmagoria' of particular urban times and places (Kaika and Swyngedouw, 2000). Such is the case, for example, with contemporary satellite ground stations (Rio, Cologne, Tokyo, London Docklands, Roubaix, Bangalore), international airports (Hong Kong, Osaka, Denver and many others), high-tech bridges (Boston, Newcastle, Istanbul), private highways studded with 'public art' (Melbourne), fast train networks and stations (Europe's TGVs), and telecommunications towers (Barcelona). Such constructions are part of what Castells calls 'a new monumentality [which is] able to provide symbolic meaning to spatial forms' in times of unprecedented metropolitan fluidity, sprawl and the spread of relatively similar and indistinguishable 'generic' urban landscapes (1999c). Many such projects continue to embody national and local 'symbols of modernity and arrival' (Vale, 1999, 391).

In the last two centuries the construction of infrastructure as symbolic marker characterised the modernist highway networks of the post-World War II period, and the water towers, dams, power stations, reservoirs and water treatment stations of nineteenth century West European cities (see, for example, Trench and Hillman, 1984). In a curious process of recycling, many of the latter are now being reconstructed as art galleries and leisure centres, celebrating postmodern urban consumption whilst inadvertently also symbolising the metaphorical and physical shift of much of the industrial and productive fabric of the networked city beneath the urban scene (see Kaika and Swyngedouw, 2000). London's Tate Modern – an old electricity generating station – is a classic example.

THE BANALISATION OF TECHNOLOGICAL MOBILITIES: TENDENCIES TO 'BLACK-BOX' URBAN INFRASTRUCTURE NETWORKS

Fourth, and as a result of their general neglect, infrastructure networks have often remained taken for granted. To use the parlance of social studies of technology, they have been 'black-boxed'. For many Western urbanites, certainly, using a phone, driving a car, taking an airline or rail trip, turning a tap, flushing a toilet, or plugging in a power plug, is so woven into the fabric of daily life, and so 'normalised' and banal that (whilst they function adequately) it scarcely seems important.

Infrastructure services, and the huge technological networks that underpin them, seem immanent, universal, unproblematic – 'obvious' even. People tend not to worry where the electrons that power their electricity come from, what happens when they turn their car ignitions, how their telephone conversations or fax and Internet messages are flitted across the planet, where their wastes go to when they flush their toilets, or what distant gas and water reserves they may be utilising in their homes.

TECHNOLOGICAL DETERMINISM AND THE DOMINANCE OF EVOLUTIONIST TREATMENTS OF INFRASTRUCTURAL HISTORY

A final problem is that, with the exception of some of the work of a group of conceptually sophisticated French researchers (see Dupuy, 1991; Offner, 1993, 2000, and the French journal *Flux*, 1991–99), and a growing corpus of writers in the Anglo-American *Journal of Urban Technology*, critical research into urban infrastructure has recently tended to focus on historical rather than contemporary contexts (see Chant and Goodman, 1998; Goodman and Chant, 1999; Roberts and Steadman, 1999).

Like the mainstream of social research on technology, these historical analyses have often adopted narrow versions of technological determinism. Here, new infrastructural and technological innovations are seen to 'impact' linearly on cities and urban life (see, for example, Garrison, 1990). 'Infrastructural technology is often regarded as largely unproblematic and even autonomous in shaping the life and form of urban areas' (Aibar and Bijker, 1997).

This view reflects the classic, deterministic view of the role of networks like transport and telecommunications in which 'changes in [infrastructure] technology lead inexorably to changes in urban form' (Hodge, 1990, 87). In this view, new networked infrastructures like the Internet become little more than 'progenitor[s] of new urban geometries' (ibid., 87). A simple, linear, cause-and-effect chain is assumed where the technology itself is seen as the direct causal agent of urban change.

The subdiscipline of Urban History has made much more effort than most to explore the relations between cities and urban infrastructure networks and technologies (see Johnson-McGrath, 1997). But even here, Konvitz *et al.* argue that:

historians asserting the importance of their area's specialisation [in technology and infrastructure] have often failed to win the recognition of their co-practitioners. Just as urban historians often focused on one city, historians of urban technology often focused on one technology. Within this framework, a growing corpus of work provided sophisticated accounts of streetcar systems, railroad networks, and automobiles as distinctive subjects and as part of their individual relationships to urban change.

(1990, 288)

Such approaches, in turn, relate to the wider dominance of technological determinism, especially in Western culture. Even on the rare occasions when attention looks beyond one network, the reliance on such determinism, with its 'simple yet highly plausible before-and-after narrative structure' tends to prevail (Smith and Marx, 1995). Such a view often combines with a one-dimensional perspective where attention focuses on one city, or one set of supposedly homogeneous technological 'impacts' which are then posited for all cities everywhere.

Commonly, this intellectual device is quickly translated into the broader use of technological and infrastructural depictions of historical urban 'ages': from the 'hydraulic civilisations' of the first urban centres in Mesopotamia (Soja, 2000, 51) to the 'steam', 'electric', 'auto', 'nuclear' and 'information age' metropolises of the past three centuries (see, for example, Garrison, 1990). The problem with such approaches is that they tend to reify technologies as having overwhelming power in ushering in simple and discrete societal shifts which seem to amount to some natural process of urban evolution. The parallels between historical periods tend to be underplayed; the tendency of newer networks to overlie and combine with, rather than replace, earlier networks is often forgotten; and, once again, the forms and processes of city life tend to be simply read off as the deterministic result of the intrinsic nature of the new generation of technology. As Mattelart suggests:

only an evolutionist concept of history as cut up into successive, watertight stages might deceive us into believing that the memory of centuries does not continue to condition the contemporary mode of communication. As proof, one need only point out the kinship between the messianic discourses on the networks of steam and electricity in the nineteenth century, and those that in the twentieth century accompany the policies of economic and social recovery through information and high-tech.

(1996, xvi; see Marvin, 1988; Offner, 2000)

FLEETING GLIMPSES OF NETWORKED FRAGILITY: EXPERIENCES AND FEARS OF INFRASTRUCTURAL COLLAPSE

The normally invisible quality of working infrastructure becomes visible when it breaks: the server is down, the bridge washes out, there is a power blackout.

(Star, 1999, 382)

Clearly, then, when infrastructure networks 'work best, they are noticed least of all' (Perry, 1995, 2). Catastrophic failures, on the other hand, serve to reveal fleetingly the utter reliance of contemporary urban life on networked infrastructures. This is especially so where the entire economic system has been reconstructed around highly fragile networks of computers and information technology devices (see Rochlin, 1997).

More than ever, the collapse of functioning infrastructure grids now brings panic and fears of the breakdown of the functioning urban social order. 'Fear of the dislocation of urban services on a massive scale,' writes Martin Pawley, is now 'endemic in the populations of all great cities,' simply because contemporary urban life is so utterly dependent on a huge range of subtly interdependent and extremely fragile computerised infrastructure networks (1997, 162).

In fact, in all parts of the world the fragility of infrastructure networks is becoming ever more obvious, just as infrastructurally mediated connections across distance become more and more intrinsic to contemporary urban life (see Suarez-Villa and Walrod, 1999; Barakat, 1998; Rochlin, 1997). Natural disasters and famines, especially in developing nations, often underline the particular fragility of infrastructural connections in such places (see Figure 1.1). But in developed nations, too, 'the earthquakes in Kobe or Los Angeles remind us how fragile the ideology of progress can be' (Allen, 1994b, 13).

It is worth exploring some examples of the fears and failures that surround contemporary infrastructural collapse.

FEARS OF INFRASTRUCTURAL COLLAPSE AND THE Y2K PHENOMENON

The remarkable global debate about the feared impacts of the 'Y2K' computer bug at the dawn of the year 2000 was a particularly potent example of the fears of the comprehensive collapse of systems of technological mobility and flow (see Figure 1.2). Stoked up by an entire 'doomsday industry' of self-interested IT consultants, John Gantz, from the International Data Corporation, reckons that over US$70 billion of public and private money was actually wasted, largely in developed nations, altering systems that would not have collapsed any way. To some, it was little more than a complex and giant hoax based on exploiting the deep-seated cultural fears of technical collapse and social panic that lie deep within our infrastructurally mediated civilisation (James, 2000).

THE COMPLEX REALITIES OF TECHNICAL COLLAPSE

But the effects of infrastructure collapses, when they happen, are very real. Often, they are catastrophic. Such effects have been all too apparent in the past thirty years. Most familiarly, they have occurred through wars (Sarajevo 1984, Beirut 1978, Belgrade 1999), earthquakes

Figure 1.1 Satellite communication as saviour in the Developing World: an advertisement for Inmarsat. *Source*: Inmarsat

Figure 1.2 Our deep cultural fear of the collapse of networked infrastructures: the 'Y2K' phenomenon, after the (non-)event. *Source*: *Boston Globe*, 6 January 2000, A13

(Los Angeles 1996, Kobe 1995, Turkey and Taiwan 1999), ice storms (Montreal 1997–98), floods (Central America 1998), supply crises (oil in Western cities 1973) or societal revolutions (Russia and Eastern Europe 1989–).

Instances of technical malfunctioning also need to be considered. In developing cities these are often common and periodic, even with new and 'high-tech' infrastructure networks. In June 2000, for example, it was reported that the national optic fibre grid threaded within and between India's main 'hi-tech' cities was regularly collapsing owing to a bizarre culprit. Rats, living inside the network ducts, had developed a taste for the PVC casing of the fibres. They were even eating that hallowed symbol of the 'information age' – the glass optic fibres themselves – regularly breaking the network in the process.

Technical failures occur in developed cities, too, but with less frequency and more attention. For example, on 5 April 2000 the entire London stock exchange was forced to stop for eight hours owing to a 'software glitch', seriously undermining its reputation. In early 1998 the electricity supply to the city of Auckland in New Zealand collapsed for nearly a month, with devastating consequences, because the newly liberalised power market led to a lack of back-up connections. And in February 1975 a fire left a 300 block stretch of Manhattan's Lower East Side without a phone system for twenty-three days. This collapse led to everything from massive economic disruption to reports of increased isolation, alienation and psychological stress (Wurtzel and Turner, 1977).

SOCIETAL REVOLUTIONS AND INFRASTRUCTURAL COLLAPSE: THE 'DEMODERNISATION' OF POST-COMMUNIST SOCIETIES

In the cities of the post-communist world, the massive recent societal shifts show how previously taken-for-granted infrastructure systems can quickly decay or be withdrawn on a more or less permanent basis. In the cities of Siberia and the far north in Russia, for example, the collapse of many heavily subsidised municipal heating systems, along with the wider economic and social deterioration, has encouraged those who can to flee. Since 1989 over 100,000 people have left Murmansk alone.

In addition, major elements of Russia's power transmission and telecommunications systems are effectively being stolen by criminal gangs to be melted down and sold overseas on the black market for metals. More than 15,000 miles of power lines were pulled down between 1998 and 2000 alone, yielding 2,000 tons of high-quality aluminium, worth more than US$40 million on the international black market. Not surprisingly, this widespread collapse of Russia's infrastructure systems has plunged large parts of Russia into power outages for weeks or months at a time in what the mayor of the town of Kiselevsk called the 'crashing down of the whole technological system' (quoted in Tyler, 2000, A10). In such circumstances it is not surprising that the social and economic enclaves of the new capitalist and criminal elites are starting to adopt strategies of securing their own private infrastructure services that are more reliable.

WHEN TURNING OFF BECOMES SUICIDE: NETWORK COLLAPSES IN THE ALWAYS-ON DIGITAL ECONOMY

In Western and advanced societies, and increasingly in fast-computerising developing ones too, the pervasive importance of twenty-four-hour systems of electrically powered computer networks, in supporting all other infrastructures, makes electrical power cuts and outages particularly fearful. The explosive growth of electronic commerce, consumption, and distribution and production systems – infrastructures that are mediated at every level by electrically powered computers and telecommunications – means that these days we are all, in a sense, 'hostages to electricity' (Leslie, 1999, 175).

With the growing electronic mediation of society, the economy and culture, information and communications systems, along with the electricity systems that support every aspect of their operation, need to be as reliable and secure as possible twenty-four hours a day. The consequences of collapse and outages can be extremely expensive and economically catastrophic. 'The always-on economy, by definition, depends upon continuous energy. For a large business online, the cost of a power interruption can exceed $1 million per minute' (Platt, 2000, 116–28). For stock markets and electronic financial services firms the costs can be much greater still.

This point is not lost on the infrastructure firms themselves in their advertising, or in their increasing investment in duplicate and back-up power systems to protect on-line service

Figure 1.3 An advertisement of the Sprint telecommunications firm, stressing the reliability of its networks as a basis for e-commerce. *Source*: Sprint Corporation

providers, Internet backbones, cable television and phone companies (see Figure 1.3). Nor is it missed by leading IT and software entrepreneurs. Taking an unusually reflective and critical stance for a software engineer, Bill Joy, co-founder of Sun Microsystems, caused a furore among readers of the bible of the high-tech elite, *Wired*. He suggested that the mediation of human societies by astonishingly complex computerised infrastructure systems will soon reach the stage when 'people won't be able to just turn the machines off, because they will be so dependent on them that turning them off would amount to suicide' (2000, 239).

UNLEASHING NETWORKED COLLAPSES: INFRASTRUCTURAL WARFARE

Nor is the fragility of electronically and electrically mediated economies lost on those who, for the last 10,000 years of urban history, have always driven the leading edge of infrastructural and technological innovation: military strategists. In the burgeoning debates on 'cyberwar' or 'infowar', stress falls on the ways in which the orchestrated and systematic sabotage of an enemy's societal infrastructure networks may now be a useful complement to, or even replacement for, physical weapons of mass destruction (see Robins and Webster, 1999, chapter 7).

Of course the Kosovo war was very much about the physical reality of blowing people into small pieces. But the United States also deployed a new type of bomb which rains down graphite crystals to disable comprehensively electrical power and distribution stations. It was, the US military argued, a new method of disabling an enemy without the public relations embarrassments of 'collateral damage' that often follow carpet bombing and the use of so-called precision guided munitions (which still have a habit of killing civilians even when they hit their target). In an adaptation of the tactics of medieval siege warfare to the networked metropolis, freezing the elderly in their homes, disabling critical heath care systems and destroying running water are the new weapons of choice in media-driven 'cyber warfare' (Ignatieff, 2000).

CATCHING THE LOVE BUG: SABOTAGE, HACKING AND COMPUTER VIRUSES

One of the advantages of the new computerised economy was thought to be that it reduced capitalism's vulnerability to terrorism and theft. The use of computer viruses has removed this illusion.

(Lawson, 2000, 11)

But perhaps the most culturally potent image of the fragility of our technically networked civilisation comes from the phenomena of hacking, computer viruses and deliberate attempts at sabotage. Here the simple pressing of an 'enter' key thousands of miles away can launch a self-replicating virus across the Internet that can bring substantial parts of the international technological economy to an extraordinarily expensive standstill, all within a matter of hours.

A classic example was the 'I love you' or 'Love bug' virus, launched by a college student in the Philippines on 3 May 2000. This virus moved to infect 45 million computers in at least twenty nations across the world within three days, clogging and destroying corporate e-mail systems in its wake. Overall damage was estimated at well over US$1 billion and many *Fortune* 500 companies were substantially affected (see Figure 1.4). The virus also exposed some of the transnational tensions and inequalities that surround corporate IT. Some newspapers in the Philippines, for example, expressed national pride that the country could spawn a hacker who could bring the fragile computer communications systems of Northern corporations to (albeit temporary) collapse.

Many other examples of viruses have emerged recently. Earlier in the year 2000, coordinated attacks by computer hackers on major commercial Internet sites brought many to expensive collapse. In a separate case, a fifteen-year-old boy from suburban Montreal was arrested in April 2000 for bringing the Yahoo! and Microsoft Network e-mail and web systems to near collapse at the end of 1999. The response to this particular case says a great deal about how the punishment of individuals is unlikely to reduce such events when the corporate communications and e-commerce systems exhibit such glaring fragility. 'It is the ultimate absurdity,' writes Willson, 'that, having brought the entire corporate world into

On May 4, 2000, a virus spread through
the world, cleverly disguised as a love note.

In a matter of hours, it began to wipe out
attachments, documents and files on e-mail
servers everywhere.

It destroyed vital information.

It shut down multinational corporations.
It ruined futures. Perhaps careers.

In the midst of all this destruction, there were
a significant number of people who found
their files totally unaffected.

And their businesses and lives intact.

Zero casualties.

Figure 1.4 Exploiting the destructive wake of the 'Love bug' virus: Xdrive's advertisement for Internet storage and backup services. *Source: New York Times*, 8 May 2000, YN2.

a system so unstable and vulnerable that a child can throw mighty commercial enterprises into chaos, society believes the solution is to incarcerate the child' (2000, 15).

WAYS FORWARD: TOWARDS A CRITICAL NETWORKED URBANISM

Together, all these factors – disciplinary failings and the neglect of networked infrastructures, their hidden and taken-for-granted nature, assumptions of technological determinism, and the panic effects of networked collapses – mean that attention to infrastructure networks tends to be *reactive* to crises or collapse, rather than sustained and systematic (Perry, 1995, 2). In such a context the failure to analyse systematically the complex linkages between contemporary urban life and networked urban infrastructures as a whole is understandable.

The aim of this book is to reveal the subtle and powerful ways in which networked infrastructures are helping to define, shape and structure the very nature of cities, and, indeed, of civilisation. To begin the process, we would point to four crucial starting points for our task of constructing a critical urbanism of contemporary networked societies.

ADDRESSING THE COMPLEX INTERDEPENDENCES BETWEEN NETWORKED INFRASTRUCTURES

With the notable exceptions of Dupuy (1991) and Tarr and Dupuy (1988), and the French Journal *Flux*, the central question of how interlinked *complexes* of infrastructures are involved in the social production and reconfiguration of urban space and experiences of urban life tend to be ignored. But, as Thrift (1990) argues, transport, communications and other networked grids, cannot be easily split apart; as 'sociotechnical hybrids' they rely on each other and co-evolve in their interrelationships with urban development, urban life and with urban space (Urry, 1999). Chains of related innovations bind infrastructure networks closely to broader technological systems; these, in turn, are seamlessly woven into the fabric of social, economic and cultural life.

Only very rarely do single infrastructure networks develop in isolation from changes in others. By far the most common situation is where urban landscapes and processes become remodelled and reconstituted on the basis of their complex articulations with a variety of superimposed transport, communications, energy and water infrastructures (Gökalp, 1992). As Easterling suggests, 'many of the most interesting innovations and design inventions appear on the cusp of change from one network to another, when one system is being subsumed by another presumed to be more fit' (1999b, 114). This is the case with today's massive investment in computer communications systems, characterised by 'smart and flexible patterns of switching between heterogeneous components and multiple scales of activity' (ibid.), which are being overlaid upon older, electromechanical transport, street, energy, communication and water networks.

INFRASTRUCTURE NETWORKS AS SOCIOTECHNICAL ASSEMBLIES OR 'MACHINIC COMPLEXES'

Second, technologies and infrastructure networks must therefore be considered as sociotechnical *assemblies* or 'machinic complexes' rather than as individual causal agents with identifiable 'impacts' on cities and urban life (Thrift, 1995). For example, networked personal computers are useless without modems, Internet servers, functioning software, phone and cable networks or wired or wireless telephones or Internet channels. As we have just seen, all these, in turn, rely on extensive, reliable electricity infrastructures which provide essential support for a growing universe of electronic interactions and transaction systems. In the United States the Internet consumed 8 per cent of all electricity in 1999; by 2020, some estimates suggest, this will rise to a staggering 30 per cent!

Electronic generation and communications systems, in turn, interrelate closely with physical movements of people, freight and raw materials over roads, railways, airline networks and water and sewer systems. Automobiles and roads, similarly, now relate extremely closely to the use of mobile phones, as well as to proliferating electronic and digital infrastructures developed for managing, regulating and controlling highway use or enhancing drivers' safety, social power or entertainment (Urry, 1999). In similar ways, water, energy and communications networks are closely intertwined in supporting domestic and industrial life. These interrelationships of infrastructures, moreover, are multidimensional and bidirectional, making an open-minded, interdisciplinary position necessary before analytical progress can be made.

PHYSICAL SYNERGIES BETWEEN INFRASTRUCTURE SYSTEMS

Third, even the optic fibres within and between cities, which carry the bulk of the exploding range of electronic communications, are being laid along rights of way and conduits that tend closely to parallel infrastructural systems for physical movement (Graham and Marvin, 1994). This is not surprising when one considers that, typically, 80 per cent of the costs of starting a telecom business come with the traditional, messy process of getting cables in the ground to link up dispersed customers.

In central London, as in other so-called 'global' cities, dense webs of optic fibres are now threaded along the beds of 'industrial age' canals and long-disused hydraulic power systems, as well as through the underground subway system and water and energy conduits. In New York the energy company Consolidated Edison offers direct fibre connections to 2,000 buildings in Manhattan through its power conduits. And all across the world, highway, power, water and rail companies are both offering their ducts and conduits and rights of way to telecom companies and, in these times of liberalisation, starting to offer telecom services themselves. 'What makes a great railway franchise is what makes a great telecom franchise,' the chairman of one such company in Florida stated (quoted in Tanner, 2000, B3).

INFRASTRUCTURE = LANDSCAPE = ARCHITECTURE! TOWARDS ARCHITECTURES AND URBANISMS OF THE NETWORKED CITY

Architecture has been pitilessly absorbed into the metropolis. . . . The metropolis has replaced the city, and as a consequence architecture as a static enterprise has been displaced by architecture as a form of software.

(Lerup, 2000, 22–3)

As a final departure point we can begin to draw on some work which has resulted from a greater appreciation of urban networked infrastructures among architects and urbanists. As Rem Koolhaas, one of the world's most influential architectural critics, has suggested, for architects infrastructure is 'a relatively new subject . . . it allows architecture to be much less isolated in its own territory and to find a connection with subjects dangerous and glamorous, like demographics' (1998a, 94).

Mobility, infrastructure networks and flows are thus emerging as major emphases of contemporary architectural and urbanist theory and practice. This is being especially encouraged by the mass diffusion of information technologies and automobiles, along with the simultaneous production and organisation, through franchises, mass production techniques, modern logistics systems and corporate networks, of multiple and generic built spaces that are intimately coordinated across vast distances. Such strategies are about the architectural shaping of time as well as spaces. Through them 'generic specifications for assembling offices, airports, highways, and many different kinds of franchises are explicitly calibrated according to protocols of timing and interactivity', based on their seamless interlinkage through infrastructure networks (Easterling, 1999a, 3).

Notable urbanists like Koolhaas (1998a, b), Easterling (1999a, b) and Martin Pawley (1997) insist, in short, that in the contemporary city, more than ever, 'infrastructure, architecture, and landscape amalgamate to become one complex' (Angélil and Klingmann, 1999, 18–20). The city must now be understood as a 'continuous, topologically formed field structure, its modulated surface covering vast extensions of urban regions' (ibid.). Moreover, 'despite its inherent discontinuities, breaks and fragmented orders, a specific form of cohesion is attributed to the contemporary city, the urban landscape perceived as a connected tissue' (ibid.).

The implication of such views is that the conventional divisions of contemporary urban professions must be overcome if we are to understand an urban world where 'architecture is declared as landscape, infrastructure as architecture, and landscape as infrastructure' (ibid., 20). Architecture and urbanism thus now widely recognise, and even celebrate, the fact that:

the experience of the city is increasingly subject to the flows and interchange generated by the increased circulation of people, vehicles, and information. The rhythm of these flows, which changes the character and function of space over time, has come to have no less significance to the experience of the city than the height of its buildings, the width of its streets, and the disposition of its monuments. The traffic of people, vehicles, and information are also the environment and material of the city.

(Wall, 1996, 159)

THE AIM OF THIS BOOK: CONSTRUCTING A PARALLEL AND CROSS-CUTTING PERSPECTIVE ON URBAN AND INFRASTRUCTURAL CHANGE

In this book we therefore seek to respond to what we feel is an urgent need: to develop a more robust, cross-cutting, international, critical, dynamic and transdisciplinary approach to understanding the changing relations between contemporary cities, infrastructure networks and technological mobilities. The book constructs a new and broad framework for exploring the relations between contemporary cities, new technologies and networked infrastructures. It argues that a parallel set of processes are under way within which infrastructure networks are being 'unbundled' in ways that help sustain the fragmentation of the social and material fabric of cities. Such a shift, which we label with the umbrella term *splintering urbanism*, requires a reconceptualisation of the relations between infrastructure services and the contemporary development of cities. This book attempts to develop such a reconceptualisation.

Our perspective is deliberately very broad, extremely international and highly interdisciplinary. It is only through such a perspective, we believe, that an understanding of the parallel processes of infrastructural splintering and urban change may be achieved. We have constructed this perspective to help start breaking down three sets of barriers which, we believe, have tended strongly to inhibit sophisticated analyses of cities, technologies and infrastructures over the past thirty years or so.

BREAKING DOWN INTERDISCIPLINARY BARRIERS

First, we want to start to break down barriers between a range of largely separated debates about cities, technologies and infrastructure networks. We believe that such disciplinary barriers have long inhibited sophisticated treatment of the interplay between cities and the sociotechnical constructions of infrastructure networks and the diverse mobilities they underpin. In this book we therefore try to draw together relevant discussions and debates in Urban Studies, Geography, Planning, Sociology, Architecture, Urbanism, Urban History, Science, Technology and Society (STS), Engineering, Social Theory and Communications Studies into a single, integrating narrative.

One inspiration for this approach comes from the French pioneering communication theorist Armand Mattelart. His integrated analyses of space, technology, infrastructure networks and social power draw equally on many disciplines. He writes in the preface to his book *The Invention of Communication* that:

just as it was hardly obvious in the 1930s [for Lewis Mumford in his book *Technics and Civilization*] to make a link between the cannon and the telegraph as instruments of vanquishing space, it is still difficult today to legitimate a transdisciplinary approach that, for example, does not hesitate to trace the possible kinship between the first attempts by topographers of routes of waterways to control territories in the seventeenth and eighteenth centuries, the normalisation and classification of individuals and regions by the pioneers of 'moral statistics' according to indices of social pathology during the nineteenth century, and the targeting of 'consumption communities' by modern marketing in the twentieth century.

(1996, ix–x)

APPROACHING NETWORKED INFRASTRUCTURES AS A WHOLE

Second, we seek, through such a transdisciplinary perspective, to help shift the study of networked urban infrastructures *as a whole* to the centre of contemporary debates and analyses about cities and contemporary urban life. We want to help banish the partitioning off of networked urban worlds into the dry, technocratic and closed professional discourses of the 'technical' bodies which tend to run and manage them. We want to help 'open' the worlds of urban networked technologies to the gaze of critical urban research. And we want to assert that, far from being 'boring', 'dull' or 'banal', analysing the ways in which social, economic, cultural or environmental power becomes extended over the times and spaces of urban life – through the construction and use of infrastructure networks – offers us an opportunity to construct dynamic, sophisticated and synthesising appreciations of the nature of contemporary urban development.

Indeed, we believe that such perspectives are desperately needed. Because much of contemporary urban life is precisely *about* the widening and intensifying use of networked infrastructures to extend social power, the study of the configuration, management and use of such networks needs to be at the centre, not the periphery, of our theories and analyses of the city and the metropolis.

We strive throughout the book to overcome the network specialism in virtually all writing about urban infrastructure. Wherever possible, following writers like Dupuy (1991), Hall and Preston (1988), Thrift (1996a), Mattelart (1996), Offner (1996, 1999), and Tarr and Dupuy (1988), we try to treat telecommunications, transport, street, energy and water networks together and in parallel.

This is not to imply that all these networks are by any means identical to each other. Rather, it is to stress that broadly similar trends can be identified in each and to assert that insights into how contemporary urbanism and infrastructure networks are intertwined can best be achieved by exploring the bundle of modern urban infrastructures together.

TRANSCENDING DIVISIONS BETWEEN THE ANALYSIS OF 'DEVELOPED' AND 'DEVELOPING' CITIES AND BETWEEN 'LOCAL' AND 'GLOBAL' SCALES

Finally, we want to help transcend the still common divide between the study of so-called 'developed' cities and 'developing' cities. We believe, following Cohen (1996), Robinson (1999), King (1996) and others, that it is no longer tenable (if it ever was) to divorce the study of Western and developed cities from those in the rest of the world. Just as in the era of colonial urban systems, contemporary geographical divisions of power and labour on our rapidly urbanising planet wrap cities and parts of cities into intensely interconnected, but extremely uneven, systems. These demand an international, and multiscalar, perspective.

As Michael Peter Smith has argued, all urban places are now, in a sense, 'translocalities' with multifaceted and multiscaled links and connections elsewhere. This means that 'there is a need to expand the study of transnational urbanism to encompass the scope of transnational processes, as well as to focus future urban research on the local and translocal specificities of various transnational sociospatial practices' (1999, 133). To him 'future urban research ought to focus considerable attention on comparatively analyzing diverse cases of *transnational network formation* and *translocality construction*' (ibid., 134, original emphasis).

In a similar vein, but from an infrastructural perspective, Olivier Coutard argues that 'studies of supranational [infrastructure systems] (such as telecommunications, energy or air transport systems) and of urban technical networks (water supply and sewerage systems, for example) must be related, if only because of the unifying dynamics generated by regulatory reform in these industries' (1999, 13).

The final goal of the book is to address these two demands. We do this by arguing that practices of splintering urbanism are starting to emerge in virtually all cities across the globe, whether in the developed, developing, newly industrialising or post-communist worlds, as local histories, cultures and modernities are enrolled into internationalising capitalist political economies in various ways. Such practices, moreover, are closely related to the development and reconfiguration of infrastructure networks between cities.

Of course, this is not to argue that cities and infrastructure networks do not retain powerful differences and specificities. Far from it. But it is an assertion that cross-cutting analyses of changes in infrastructure, technology and urban development can be profitably made across these diverse ranges of urban contexts. As Michael Ogborn suggests, understanding the 'spaces of modernity' of contemporary and historical cities 'is always about traversing the ground that lies between totalisation and difference' (1999, 238). It is this line that we continually negotiate throughout the book.

THE STRUCTURE OF THE BOOK

In what follows we separate our discussion into three parts. The first brings together four chapters which piece together an historical, practical and theoretical understanding of processes of splintering urbanism. We begin our account in the next chapter, where we explore the construction of the previous dominant paradigm of infrastructural development, which we call the modern infrastructural ideal. Chapter 3 then explores the range of forces that are comprehensively unravelling this ideal. In Chapter 4 we look in more detail at the parallel practices through which infrastructure networks can be unbundled and urban landscapes fragmented. We complete the first part of the book in the fifth chapter, which develops a theoretical perspective to help explain the interlinked fragmentation of cities and splintering of infrastructures.

Part Two includes two thematic chapters which go on to explore processes of splintering urbanism across the world's cities in considerable empirical detail. Urban social landscapes are addressed in Chapter 6; the relationships between urban economies and 'glocal' infrastructure are explored in Chapter 7.

In Part Three, which incorporates the final, concluding chapter and a postscript, we take stock of the preceding discussions. We explore the ways in which the complex politics and spatialities of contemporary cities inevitably limit the degree to which the network spaces of cities can be totally segregated from each other. We analyse the limits and resistance which practices of splintering urbanism face. Finally, we draw out the book's implications for urban research and practice.

PART ONE

UNDERSTANDING SPLINTERING URBANISM

Plate 4 A skybridge walkway linking elements of the Prudential and Copley Square 'bundled city' complexes in central Boston, Massachusetts: a private, premium pedestrian network which bypasses the traditional street system. *Photograph*: Stephen Graham

By considering the city as an enormous artifact, the size and distribution of its streets, sidewalks, buildings, squares, parks, sewers and so on can be interpreted as remarkable physical records of the sociotechnical world in which the city was developed and conceived.

(Aibar and Bijker, 1997, 23)

Twentieth-century urbanism is a complex story of coming to terms with extensive mechanical relations, in which investments for land and infrastructure predominate. [It is] a story of regions opened up at the periphery under the auspices of state subsidies and far-flung highways, sewers, and water-supply systems. The ensuing geography became one of single-use enclaves wherever institutionalized planning took over, partitioning ways of life, productive activities, and architecture into exclusion zones.

(Waterhouse, 1993, 300)

The end of our millennium is witnessing the consummation of the crisis of the positivist idea of a 'necessary and continuous' progress without deviations, detours or retreats.

(Mattelart, 1996, 303)

The city is a gearbox full of speeds.

(Wark, 1998, 3)

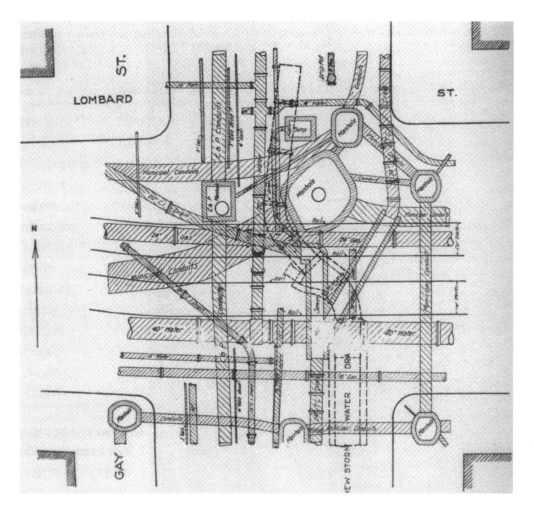

Plate 5 What Lewis Mumford called the 'invisible city' – underground pipes and conduits at the junction of Gay and Lombard Streets in Baltimore, Maryland, 1908. *Source*: Melosi (2000), 134

> A whole history remains to be written of spaces, which would at the same time be the history of powers – from the great strategies of geopolitics to the little tactics of the habitat, . . . from the classroom to the design of hospitals, passing via economic and political installations.
>
> (Foucault, 1980, 149, cited in
> Joerges, 1999b, 283)

THE LONG PATH TOWARDS (ATTEMPTED) STANDARDISATION

During the period between about 1850 and 1960 there was a general movement, particularly in Western cities, from the piecemeal and fragmented provision of networked infrastructures to an emphasis on centralised and standardised systems. This occurred through a major expansion in both public and private sector investment in infrastructure networks (Mattelart, 1996; Hughes, 1983). The growth of regulated network monopolies supported the rapid diffusion of water and sewerage systems throughout cities and systems of cities, and the improvement of transport networks to improve circulation. Initially, private sector accomplishments focused on the expansion of the telegraph network, the development of early gas distribution systems (from the early 1820s onwards), the rapid expansion of the telephone and electricity networks (from the 1880s onwards) and the deployment of electric streetcars (trams), subways and elevated lines. At the same, Western powers imposed adapted versions of the ideal of the standardised infrastructure network across Africa, Latin America, Asia and Australasia, but only for the urban spaces of the colonisers and their associated socioeconomic elites (a story we shall return to in the final part of this chapter).

During the period 1850–1960 much of urban politics was dominated by questions of infrastructural and technological investment. Western cities, in particular, were in transition from the older compact commercial city to the new industrial metropolis with a strong core and a ring of residential suburbs. Across the urban world, small, fragmented islands of infrastructure were joined up, integrated and consolidated towards standardised, regulated networks designed to deliver predictable, dependable services across (and, increasingly, beyond) the metropolis (Tarr and Dupuy, 1988; see Cox and Jonas, 1993, for a US example).

At the same time, the modern metropolis became a hotbed of innovation and a maelstrom of social, cultural and economic change as new notions of speed, light, power and communications were constructed (Thrift, 1996a). Standardised, compartmentalised notions of space and time were, in a sense, constructed through the rolling out of networks across wider and wider spaces (first cities, then urban systems and then international territories). Many commentators argued that the growth of regional and national railway and telegraph networks, in particular, seemed to require single systems for the management of time and space. 'If we all manage, throughout the entire network, to be punctual to the second,' wrote the French commentator Audibert in the 1850s, 'we will have endowed humanity with the most effective instrument for building a new world' (quoted in Gökalp, 1992; see also Kern, 1986).

Production, distribution and consumption thus became gradually reconstituted on mass, industrial scales, on the basis of networked exchange, through widening nets of superimposed and interconnecting pipes, tracks, roads, wires and conduits. As Pascal Preston suggests in the case of Britain:

by the end of the nineteenth century the material achievements of the past century ensured that the dreams and fantasies of the 'system builders' had shifted way beyond anything envisaged before and certainly compared to the British systems of a century before; with the approach of the era of monopoly, mass production and modernism, they had shifted to a much grander scope of operation.

(1990, 3–4)

The elaboration of standardised networked infrastructures thus allowed all aspects of industrial urban life to be extended and intensified. The widespread application after World War I of 'Fordist' notions of scientific management, rational organisation and the mass production of standardised goods was, in essence, also a paradigm based on fully exploiting the potential of new networked infrastructures to support industrial societies based on mass production, distribution and consumption. 'The new era was articulating a new organisational philosophy extolling the virtues of economies of scale in lowering unit costs and exploding mass consumption (especially in Germany and the USA). These principles implied throughput, centralised control, coordination and diversification' (Preston, 1990, 4). This, in turn, required the construction of national systems of interconnected highways, rail, communications and energy infrastructure, bringing the urban infrastructural 'islands' into a radically new era of regulated interconnection and extension.

Such transitions were also intrinsically bound up with changes in culture and philosophy. Notions of space and time, speed and culture, subjects and objects, technology and society, were gradually recomposed. Urban landscapes were transformed and modernised. 'The history of an ever-expanding landscape of light which we now take for granted cannot be ignored,' writes Nigel Thrift (1996a, 268). At the end of the nineteenth century electrically lit arcades, in particular, emerged as the 'dream spaces' of the modern city, fantastical spaces which so obsessed cultural theorists like Walter Benjamin (1969, 1979, 1999).

From the initial, general, picture of heterogeneous, partial networks, of poorly interconnected 'islands' of infrastructure and of extreme uneven development in the infrastructural capacities of different urban spaces emerged, over the period 1850–1960, single, integrated and standardised road, water, waste, energy and communications grids covering municipalities, cities, regions and even nations. These were legitimised through notions of ubiquity of access, modernisation and societal progress, all within the rubric of widening state power. By the 1940s in virtually all Western cities:

the technological networks promised by nineteenth century reformers were finally in place: sewers, water systems, tracks and electrical and telephone wires criss-crossed the city, providing services at a level that was inconceivable in the 1890s.

(Perry, 1995, 6)

Finally, such ideas were also closely woven into wider rationalities and ideals of the emerging urban planning movement during this period, based on the idea of rational, comprehensive planning, driving 'progress' towards unitary, coherent and emancipatory cities (Fischler, 1998). Both realms, in fact – infrastructure development and urban planning – were constructed as key elements of the broader project of modernity, as Enlightenment ideals of universal rationality, progress, justice, emancipation and reason were applied to all areas of social life (see Heynen, 1999). As Nan Ellin suggests, such ideals:

sought to discover that which is universal and eternal through the scientific method and human creativity, in order to dominate natural forces and thereby liberate people from the irrational and arbitrary ways of religion, superstition, and our own human nature. It was through reason that Enlightenment, the conceiving of infinite possibilities, would enable the emancipation of humanity to take place: emancipation from ignorance, poverty, insecurity and violence.

(1996, 6)

The new technologies surrounding networked urban infrastructures were thus seen to be mechanisms to control time through instigating waves of societal progress. Space, in the form of the physical city, was seen as an object to be rationally manipulated. First, this was to occur through the discourses of modern city planning. Second, the annihilating effects of space and time-transcending sewer, gas, water, road, railway, telegraphy and telephone networks were to allow the 'tyranny of space' to be overcome. Networked-based modernity thus promised the joys of perpetual transformation towards a scientifically rational and technologically intense urbanism. As Marshall Berman so tellingly described, to be modern was to 'find ourselves in an environment that promise[d] power, joy, growth, transformation of ourselves and the world – and at the same time, that threaten[ed] to destroy everything we ha[d], everything we kn[e]w, everything we [were]' (1983, 89).

Of course, experiences of actual infrastructure development in real cities during this broad period were much more diverse than this generalised portrait suggests. The symbolic importance of the modern ideals of integration and cohesion was also radically different from their effects in practice. Beneath the universalising rhetoric, modernising cities were always about rupture, contradiction and inequality. Extending metro, sewer, water, highway, energy and communications grids through the fabric of city spaces was always laden with social and political biases, highly uneven power struggles and cultural and historical specificities.

Nevertheless, the modern networked city, dominated by notions of order, coherence and rationality, through the harmonious planning of networked connections and urban space, became the very embodiment of the modern project throughout much of the urban world (Ellin, 1996; Gold, 1997). Images of mobility of speed, light and power, mediated by the fantastical new water, waste, subway, streetcar, electricity, gas, telegraph and telephone networks, were, as Raymond Williams (1973) and Nigel Thrift (1996a) have pointed out, crucial in underpinning the changing 'structures of feeling' surrounding modern urban life.

THE AIMS OF THIS CHAPTER

How did the ideals of 'progress' via planned, publicly regulated or monopolistic transport, telecommunications, energy and water networks emerge in the first place? And how did such notions become so closely wedded to the modern rationalities of urban planning, the elaboration of modern states, and practices and principles of modern urban consumption? This chapter explores these two questions. This is vital for the purposes of the book, for these questions set the context against which the parallel processes of splintering infrastructure networks and fragmenting cities are occurring.

In this chapter we piece together the construction of the modern ideal of the integrated, networked city. Not surprisingly, given that we are addressing all networked infrastructures in a wide variety of contexts, it is a huge and complex story. We can merely scratch its surface in this chapter. Along the way we shall need to embrace many disciplines, subdisciplines and subplots.

To make the story of the construction of the modern infrastructural ideal manageable we will explore each of the four essential 'pillars' upon which the whole edifice was constructed. First we focus on the ideologies of science, technology and the city that fed the belief that 'progress' and modernisation had to be achieved through standardised infrastructure monopolies. Second, we explore how the emerging discipline of modern urban planning developed to take the modern infrastructural ideal for granted as a central tenet. Third, we look at the ways in which the modern ideal became implicated in wider practices of home-based consumption, mediated by energy, water, transport and communications grids. Finally, we briefly explore how modern municipalities and nation states became so founded on the idea of providing public infrastructure monopolies within their respective territories. The chapter finishes by considering how the four pillars of the modern ideal were exported to the context of cities in the developing world.

BELIEVING IN MODERNITY: URBAN REFORM AND THE 'TECHNOLOGICAL SUBLIME'

Good roads and canals will shorten the distances, facilitate commercial and personal intercourse, and unite, by a still more intimate community of interests, the most remote quarters of the United States. No other single operation, with the power of Government, can more effectively tend to strengthen and perpetuate that Union which secures external independence, domestic peace and internal liberty.

(Albert Gallatin, Secretary to the US Treasury, 1808, quoted in Perry, 1995, 1)

The first 'pillar' underpinning the construction of the modern infrastructural ideal was a powerful set of ideological beliefs asserting the positive transformative powers of modern science and networked technologies (Mattelart, 1994). In 1994 Herbert Muschamp, a leading public official with the City of New York, captured perfectly the modern ideology surrounding urban infrastructure. Infrastructure networks, he argued, were nothing less than 'the connective tissue that knits people, places, social institutions and the natural environment into coherent urban relations'. They were, in other words, little less than 'the structural underpinnings of the public realm' (cited in Perry, 1995, 1).

Our first concern is to explore how such ideologies of networked infrastructure came to be accepted as the basis of planning, policy, development and management. How, in other words, did energy, water, transport, streets and communications grids become so closely associated with modern ideologies, allocating them apparent power to bind and connect cities, transforming them in the process? Here we shall stress only two of the most important elements of the broader process.

THE CITY AS AN ENGINEERED SYSTEM IN THE NINETEENTH CENTURY

First, the idea that the city was a single, objective entity became a dominant basis of engineering, planning and urban reform debates between 1850 and 1920. Not only this, but the growing corps of urban engineers sought to understand the growing industrial city as a systemic 'machine' that needed to be rationally organised as a unitary 'thing', using the latest scientific and technological practices available. Increasingly 'the metropolis was believed to be an inorganic and fabricated environment, the product of mathematics and the creation of the engineer' (Boyer, 1994, 116).

Through this process urban 'engineers became the paragons of the public works reform culture – embodying all that was efficient, technologically superior, and devoid of political corruption' (Perry, 1995, 9). For example, in that icon of urban modernity, nineteenth century Paris, 'the idea of the engineer began to assert itself in the city's arcades, buildings and streets. In fact the practice of engineering geography through the application of a new science and technology of space was elevated to a greater scale of operation at this time' (Kirsch, 1995, 549; see Prendergast, 1992).

Fed first by concern among the urban bourgeoisies over disease, death rates, and the poor health and potential unrest of working class populations, networked technologies, and scientific practices, became imbued by reformers and urban engineers with the moral power to bring sanitation, cleanliness, rationality and order to the troubled and apparently chaotic industrial metropolis (Boyer, 1987; Chatzis, 1999). Infrastructure networks, particularly water and sewerage, quickly became associated with curative powers able to 'cleanse' city spaces, so emancipating 'good' working class people from the risks of immorality (Felbinger, 1996, 11).

Distributing access to these networks thus became the *Leitmotif*, not just of the 'good' city but of modern urban civilisation itself. What Anthony Vidler terms a 'technical ideology' of the metropolis developed and prevailed (quoted in Gandy, 1998, 4). The urban reform movements of the eighteenth century, 'led by sanitarians, engineers, urban planners, and the growing urban middle class, equated the efficiency of infrastructural systems with the quality of the entire civilisation' (Felbinger, 1996, 11). The application of scientific and statistical techniques to the design and routinised operation of these systems added further to the mystique surrounding urban engineering as a 'rational' and 'value-free' movement (Chatzis, 1999). And the extending and deepening construction of ducts, wires, dams, water treatment plants, power stations, gas installations, railways, subways and sewers in the city became celebrated icons of modernity. Such constructions were highly visible foci of media and touristic attention – an 'urban dowry' of 'iconic landmarks that were prominently visual and present' (Kaika and Swyngedouw, 2000).

At the same time, however, such landmarks increasingly disrupted and fragmented many urban landscapes as modern urban identity became constructed around the full range of mechanical and electromechanical infrastructure networks. Along with the rolling out of gas and electric light, this brought 'the great period of what one might call urban visibility', in the sense that 'the modern metropolis was increasingly cut open, laid bare and illuminated' (Whiteman, 1990, 26). From the point of view of Berlin, Neumeyer (1990, 17) recalls that:

Like blades, the engineers' iron structures chopped up the body of the city, fragmenting the urban tissue. . . . Bridges, elevated railroad structures, gasometers, and other modern objects of unfamiliar shape became significant new elements in the traditional cityscape, taking on a disturbingly powerful and threatening presence. . . . The image of a metropolis was laid bare in the multilayered movement and omnipresent bridges that were a response to the need for connections between the isolated objects of the cut-open city.

ELECTRIFICATION AND THE URBAN 'TECHNOLOGICAL SUBLIME', 1880–1940

Networked infrastructures were not only a central focus of moral debates about urban reform and iconic constructions of urban modernity; they were also, second, essential supports for the wider application of systems-based engineering techniques to the whole modernisation of metropolitan life. As Rose puts it, from the point of view of the United States, 'beginning around 1880, gas, electricity, light, and heat comprised only one portion of a much larger picture of technological innovation and systems-building thinking taking place in virtually every city' (1995, 190–1).

In parallel, the industrial factory system most famously associated with Henry Ford started to replace small workshops. Trolley systems and track-based rail complemented then replaced horse-drawn transport. Industrial cities mushroomed in population and physical scale to unprecedented dimensions. In this frenzy of growth 'politicians, business executives, and experts in finance, engineering, and administration competed with one another for the opportunity to shape the landscapes and to govern these technological systems and the metropolis' (Rose, 1995, 190–1).

Ever wider and larger infrastructure networks, thus, came to symbolise the emerging technological 'sublime' of the modern industrial city, especially in the most dynamic industrial culture of all: the United States (Nye, 1994). As many small private and municipal networks grew, spread and became interlinked into standardised urban and regional systems (see, for example, Figure 2.1) these networks became the basis of blossoming discourses celebrating the excitement of electrification, electric communications and water and transport systems in the modern city (see Marvin, 1988). 'By the middle of the nineteenth century mechanical systems had become the central subject for some narratives and a part of the ideological underpinning for many others' (Nye, 1997, 181).

Bridges, electricity grids, telephone poles, subways and port constructions thus became imbued with cultures of the technological sublime – triumphal testaments to the socio-technical 'progress' at work across urban life (Nye, 1994). A pervasive age of technological optimism became concretised in grand technological visions for cities realised through integrated infrastructural and urban planning.

The many uses of electricity, in particular, 'seemed to ensure a brilliant future for civilisation' (Nye, 1994, 66). The largest industrial cities, where electricity generation and consumption were concentrated, were widely labelled 'electropolises' (Thrift, 1996a, 275). Competition to have the largest, most powerful and most extensive electricity infrastructure,

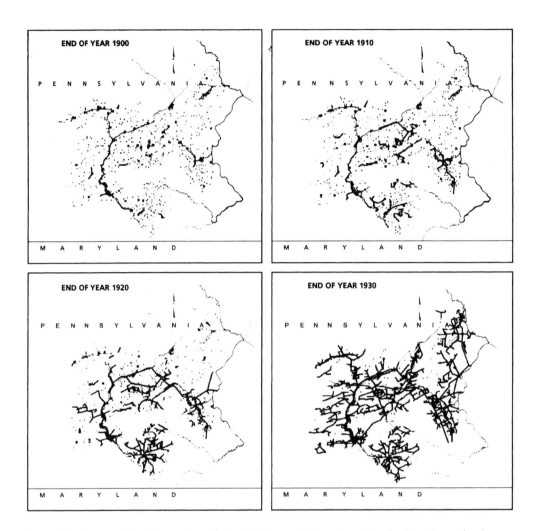

Figure 2.1 The growth and integration of electricity transmission systems in urbanising Pennsylvania, 1900–30. *Source*: Hughes (1983), 10–13

along with the brightest night-time cityscapes and streets, was intense. Ambitious municipalities wanted their cities to be a 'blaze of light', 'rearing out of the darkness of the surrounding, non-electrified regions. The coming of electric lighting . . . became the new status symbol of the elite. . . . Glittering city lights reinforced the magnetic appeal of America's urban centers' (Platt, 1988, 251). Standardising, extending and rolling out single, integrated electricity grids, from the uneven patchworks inherited from initial processes of entre-preneurialism, became a very metaphor for everything modern, exciting and transformative (Platt, 1988; Hall, 1998, 382).

Major industrial cities dominated all aspects of electrical innovation, technology and infrastructure construction (Hughes, 1983). Perhaps the most famous 'electropolis' of all, Berlin, for example, had over half of all German electrical employment between 1895 and

1925 (Hall, 1998, 382). Electrification straddled the dazzling cityscapes of consumption, the networked homes of affluent (and, increasingly, less affluent) consumers, and the emergence of an electrified urban economy (Rose, 1988).

City governments competed to develop the most awesome infrastructure networks. Showpiece projects, and the showmen 'system builders' (and they were virtually always men), were glorified through technologically trumphalist narratives (Hughes, 1983). Systems, and the men who built them, became harbingers of some naturalistic progress based on an essential and technologically determinist notion of how infrastructures related to cities. Highways, telephony and modern water systems were also often heralded as the very deliverers of benefits for all, promising an emancipatory future of linear, absolute progress (Figure 2.2). Modern appliances, and the vehicles to be used through these new networks, moreover, became laden with lustrous promises of modernity, and were increasingly portrayed by advertisers as being harbingers of a future of untold wealth and mobility (see Figure 2.3).

Based on the supportive notions of perspective vision and grand technological strategies, such discourses portrayed technologies as 'natural', inexorable and 'autonomous'. City-scapes, conversely, were increasingly portrayed as 'empty' Cartesian spaces – raw material for reworking through the new technological sublime as the systems 'blasted through topography' (Nye, 1997, 181).

Central to this period, then, was the emerging dominance of the notion of the city as an abstract object to be managed and controlled. The emerging perspectives of aerial photography and the perspective views of cities increasingly available from rising buildings, in turn, further contributed to the view that the spreading industrial city-region was a recognisable entity that needed to be managed as a unitary whole. In the United States, panoramic, electrically lit views from skyscrapers, writes David Nye, served to 'miniaturis[e] the city, making it into a pattern'. Thus 'the vast region visible from the top of the skyscraper appears intelligible, offering itself for decipherment like a huge hieroglyph' (1994, 105). According to him, 'attention was displaced from human beings and the apparent pettiness of their lives. Lifted up into the sky, the visitor was invited to see the city as a vast map and to call into existence a new relationship between the self and this concrete abstraction'.

As Box 2.1 demonstrates, a similar process of technological 'roll-out' and standardisation towards ubiquity accompanied the shift of that other great harbinger of modernity and progress in the nineteenth and early twentieth century city: the telephone. During this period the telephone shifted from the preserve of the social elite to become the essential interpersonal communications medium of the growing metropolis, operated as a standardised monopoly and integrated network geared to universal social and spatial access within the metropolis.

Figure 2.2 The modern idea of linear technological 'progress' via ever more elaborate and capable scientific, technological and infrastructural systems. *Source*: Smith (1995), from Compton (1952)

Figure 2.3 Electrical appliances and automobile tyres as harbingers of a dynamic, exciting urban future: advertisements for General Electric refrigerators (1931) and Goodrich tyres (1931). *Source*: Corn and Horrigan (1984), 42

URBAN PLANNING AND DEVELOPMENT: THE EMERGENCE OF THE UNITARY CITY IDEAL

The modern city, like the modern nation, was imagined as a space that should be unitary, coherent and ordered.

(Asu Aksoy and Kevin Robins, 1997, 26)

The second key 'pillar' supporting the elaboration of the modern infrastructural ideal was another important part of the broader project of modernity: theories and practices of modern urban planning. As Entrikin suggests, in most Western countries, between the mid nineteenth and mid twentieth centuries, state-backed urban planning was essential to maintain 'the integrity of the modernist project' (1989, 34). Its practices and associated architectural theories supported the 'rationalisation' of whole urban landscapes, backed by notions of rationality, science, technology, the celebration of machines, and ideas of 'modern' aesthetics (see Banham, 1980; McCarter, 1987). Urban planning helped to define the 'vision of the progressive force of modernity' through its attempts to impose systematically an 'abstract space' upon the complex social and lived spaces of the industrial metropolis (Entrikin, 1989, 34). It attempted to apply what Henri Lefebvre called 'the simple, regulated and methodical principle of coherent stability' to the spatial form and temporal rhythms of the massive, chaotic metropolis (Lefebvre, 1984, 238).

BOX 2.1 THE TELEPHONE AND THE METROPOLIS, 1880–1940

It is important to stress the often ignored importance of the telephone in the development of the modern metropolis (Fischer, 1992). The elaboration of standard telephone grids, from the small local systems that emerged in the nineteenth century, provided important infrastructural support to the booming industrial metropolis of the late nineteenth and early twentieth centuries (Abler, 1977; Gottman, 1977).

As with electricity, water and gas networks, the story of the telephone from the late 1880s to the mid twentieth century is one of the interconnection of small networks to allow the roll-out towards universal spatial coverage of the metropolis and urban system. Moyer (1977) recounts how in Boston, Massachusetts, the city where the telephone was first invented, phone provision moved from a situation where there were 174 different telephone tariffs around 1900 to a standard grid operated by the Bell company and closely interconnected across national and international space by 1930.

As 'natural monopolies' the operators of the consolidated systems emerged as giant industrial powerhouses which were closely regulated at the national and municipal levels to ensure spatial and social equality of tariffs and access. Moreover the definition of 'local call areas' around cities was crucial in emphasising the idea of the city as an integrated territory within which exchange and communication were to be maximised – in contrast to the higher tariffs inhibiting interaction with far-off spaces, especially those in other countries.

THE TELEPHONE AND THE 'SPIRALLING MASS OF BITS OF INFORMATION'

Widening access to the telephone beyond business cores and socioeconomic elites allowed it to emerge as the taken-for-granted means of mediating all aspects of big-city life (see Figure 2.4). Jean Gottman called this the 'spiralling mass of bits of information' that is characteristic of the modern metropolis. Thus the idea of the standardised telephone network, organised as a monopoly and planned in an integrated way to cover a city, region or country, became an essential component of the modern networked infrastructural ideal (Mattelart, 1994). As Jean Gottman suggested in 1977, 'the telephone has helped to make the city better, bigger, more efficient, more exciting, providing, when needed, "quasi-immediate verbal communication" between all elements of expanding metropolis at minimum cost' (1977, 312).

TELEPHONES AND URBAN CHANGE

The mediation of widening portions of urban life by the telephone facilitated many complex structural changes in urban form, enhancing the centrality of dominant business cores affluent users were used to cross-subsidise loss-making networks in poorer and rural areas. Such principles effectively drove the development of urban telephone networks in most Western countries from the 1920s and 1930s to the late 1970s. The ideal of a coherent, planned, unitary network, laid out to integrate the metropolis, sometimes still drives the development of state-of-the-art optic fibre infrastructures in areas which have been slow to open up to the recent global wave of liberalisation, as the example of Melbourne (Figure 2.5) demonstrates.

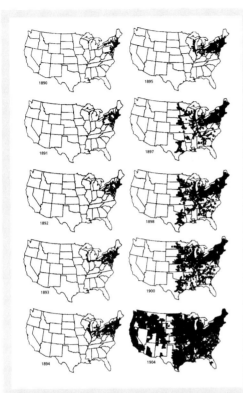

Figure 2.4 The diffusion of AT&T telephony in the United States towards ubiquitous, standardised access. *Source*: Abler (1977), 333

(which could now act at a distance with greater immediacy and power) whilst also supporting urban decentralisation through suburbs and exurbs. As Ron Abler suggests from the point of view of the United States, 'the nation's telephone services evolved with the nation's metropolitan regions in a highly complex and circular causal relationship in which cause and effect are impossible to discern ' (1977, 339).

As with other modern urban infrastructures, then, urban telephone networks were to be planned as singular, coherent wholes, delivering access to all spaces and users on a universal basis (usually under specific regulatory obligations). Profits from trunk routes and

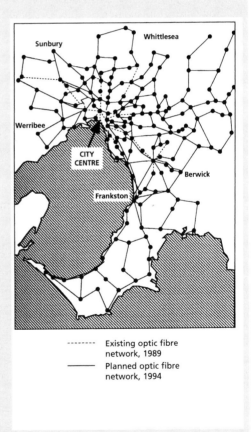

Figure 2.5 The continued influence of the modern infrastructural ideal on contemporary urban communications: the planned development of Melbourne's optic fibre network. *Source*: Newton (1991)

The late nineteenth and early twentieth centuries saw the crystallisation of planning movements in all Western countries, driven by a post-Enlightenment perspective which saw the city as 'a territory to be bounded, mapped, occupied and exploited [and] a population to be managed and perfected' (Donald, 1997, 182). The rationale of the movement was to 'design away the opacities, the confusions as well as the litter and detritus of a complex social world' (Robins, 1999, 46). Through its reliance on the perspectival and disciplinary gaze of the (usually male) urban planner, modern, and later modernist, planning sought 'the ideal of transparency where every functioning element of the city [was] supposedly open to inspection, the structure revealing itself to the inquisitive eye' (Wigley, 1996, 138; see Jay, 1988). In many cases the disdain of leading architects and planners for the inherited cityscape, especially during the modernist period, was such that the *tabula rasa* was 'the point of departure for the modernist city' – a strategy aimed at the complete elimination of the pre-modern cityscape (Gandelsonas, 1999, 27).

The practices through which the modern planning ideal of the unitary city was constructed, and how such approaches became predicated on the modern infrastructural ideal, were enormous and diverse. We can only highlight a series of five illustrative vignettes here.

THE EMERGENCE OF THE MODERN IDEAL OF URBAN COHESION, 1880–1940

First, we need to highlight the emergence of the very ideal of urban cohesion. In the face of apparently chaotic urbanisation in nineteenth century industrial cities, as well as the miseries faced by the working classes, health crises and uncoordinated infrastructures, urban planning was widely seen in this period as the means to realise technological progress. Social elites, in particular, increasingly saw modern urban planning as the mechanism to bring rational, expert-driven practices to the comprehensive reshaping of metropolitan life. Planning was to bring a whole new 'sociotechnical environment' to the city. It was to mould cities into the model of some 'new, efficient industrial apparatus' (Rabinow, 1994, 407).

An essential aspect of the drive for cohesion was the construction of the 'underground city' through the 'knitting together of necessary underground utilities' as well as above-ground street networks and transport and communication grids (Mumford, 1961, 478). Gradually, in keeping with broader prevailing ideologies of science and technology, urban planning doctrines built up universalising norms of access to gas, electricity, water, transport and communications. These became intrinsic to the new urban vision, as did standard practices for laying out and managing streets (Celik *et al.*, 1994).

All spaces of the modern city were thus to be integrated by ubiquitous, democratically accessible and homogeneous infrastructure grids, usually under public ownership or control (Chaoy, 1969). Such interventions, it was widely believed, would, in turn, help realise the social, economic and environmental benefits of mass production, distribution and consumption, integrated through the mediating powers of new infrastructure networks (Fillion, 1996). Above all, planning was seen as necessary to support a 'sense of cohesion' (ibid., 1939); it was an exercise to bring 'order to the fragmented form' of the industrial metropolis (Beauregard, 1989, 382).

STREETS AND SEWERS: HAUSSMANN'S 'REGULARISATION' PLANS FOR PARIS, 1853-1870

The straight, wide street with smooth traffic flow seems to lessen the distance and, as it were, put two points which before had seemed leagues apart to us in touch with each other.

(*Paris-Guide*, 1867, cited in Offner, 1999, 224)

Second, we need to explore Haussmann's Paris to understand the archaeology of the modern notion of comprehensive and integrated street and sewer systems. 'Streets,' write Celik *et al.*, have always provided 'the primary ingredient of urban existence. They provide a structure on which to weave the complex interactions of the architectural fabric and human organization' (1994, 1). As long as cities had existed, special metropolitan streets had been planned and laid out as sites of ritual, imperial symbolism, as conduits of communication, commerce and exchange, and as meeting spaces between more or less privileged citizens (Kostof, 1992). But the aqueducts, sewers and streets laid out between the times of Mesopotamia, ancient Egypt and imperial Rome and those of the mid nineteenth century industrial metropolis had tended to have only partial urban coverage (usually for rich users or districts) (see Hall, 1998, chapter 22; Soja, 2000).

What doctrines of modern urban planning added, however – starting in the nineteenth century – was the idea that comprehensive, integrated networks of streets could be laid across whole urban areas in a technocratic way, to bind the metropolis into a functioning 'machine' or 'organism'. This process saw a changing conceptualisation of the street which, by the end of the nineteenth century, started to tie in closely with growing motorisation and expanding demands for mobility and circulation. As Nann Ellin suggests, 'the street had been for walking to work or shops or for socialising. Now they were primarily for movement' (1997, 13). Such ideas of comprehensive, integrated provision of urban road systems, usually under the auspices of state bodies, remained a dominant doctrine of urban planning, even through to the efforts to surgically insert integrated urban highways through urban areas in the 1950s and 1960s.

One of the first and most celebrated total conceptions of the need to 'bind' a city through a comprehensive and integrated street system came with Haussmann's plans for the 'regularisation' of Paris between 1853 and 1870. It is the totality of Haussmann's scheme for the construction of massive highways and sewer networks through Paris that makes it nothing less than the archetype of the modern urban infrastructural ideal (see Figures 2.6 and 2.7). Haussmann's Paris was the cradle of the notion of 'regularising' the industrial city through combined land use planning and the construction of integrated infrastructure networks that were explicitly designed to foster free circulation both within and between cities (Offner, 1999). To Spiro Kostof:

Haussmann's treatment of Paris was in fact the first total conceptualisation of what we understand as the 'modern' city. It heralded a technocratically minded, comprehensive approach to town planning in which a rationalised circulatory network would once and for all sweep away [what was seen by the planners as] the dross of the community's promiscuous life through time.

(1994a, 11)

Figure 2.6 A crucial influence on the modern infrastructural ideal: Haussmann's 'regularising' strategy for Paris, 1853–70, showing the system of boulevards that was carved through the city. *Source*: Chaoy (1969), 50

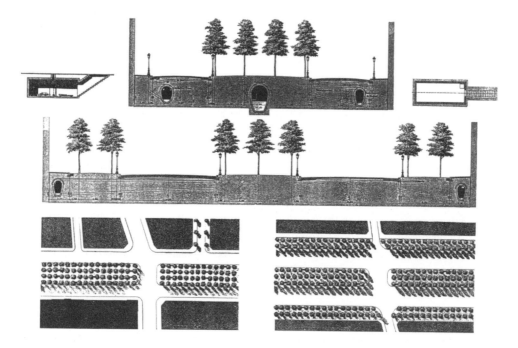

Figure 2.7 The boulevard as the organising framework for laying out gas lighting, water mains, drains and sewers in Haussmann's regularisation plans for Paris. *Source*: Kostof (1994b), 14

Haussmann specified his aims as to 'regularise the disordered city, to disclose its new order by means of a pure, schematic layout which [would] disentangle it from its dross, the sediment of past and present failures' (Haussmann, quoted in Chaoy, 1969, 16). The celebrated new networks, ploughed ruthlessly and with unerring accuracy through the social and physical fabric of the city, were designed 'to give unity to and transform the operative whole' of Paris (Haussmann, quoted in Chaoy 1969, 16).

Here, then, we have the social construction of the very idea that coordinated infrastructure networks and urban plans might meaningfully connect the dispersed parts of the modern industrial city into an 'organic' whole, thereby supporting its wider role as a dominant national and international metropolis (see Figure 2.6). To Haussmann the road network was the city's circulatory system; the rationally engineered sewer and cemetery systems were the waste disposal 'organs' of the metropolis; and green spaces were the city's 'respiratory' system. The street system, in fact, was the physical framework for the 'bundling' of buried water networks, lighting, drains and sewers – a situation so familiar today that we take it for granted (Figure 2.7; see Moss, 2000). In every sense, then, the networks were seen as coordinated allies in the effort to rationalise, systematise and control metropolitan space as a whole.

Haussmann dreamed of 'Paris as a whole [city which] could one day become a single organism quickened with a unique life' (Chaoy, 1969, 16). He aimed to 'cut a cross, north to south and east to west, through the centre of Paris, bringing the city's cardinal points into direct communication' (ibid., 26). The road and sewer networks were planned as a 'general circulatory system' with hierarchical tributaries linking the new plaza nodes (ibid., 16). Both sets of channels were celebrated as symbols of progress. His legacy was to 'open . . . up the whole of the city, for the first time in history, to all its inhabitants' who could experience Paris as 'a unified physical and human space' (Berman, 1983, 153; see Picon and Robert, 1999).

SEWERS, WATER AND THE 'DOMESTICATION' OF THE URBAN BODY, 1880–1940

Hygiene is the modern project's supreme act.

(Lahiji and Friedman, 1997, 7)

Surely the state is the sewer. Not just because it spews divine law from its ravenous mouth, but because it reigns as the law of cleanliness above its sewers.

(Laporte, 2000, 57)

In the past [waste] water simply flowed over the street. Then engineers brought water underground.

(Dieter Jacobi, cited in Moss, 2000, 63)

Which leads us neatly to our third vignette: the construction by nineteenth century reformers and social elites of the notion that comprehensive underground urban water and sewerage systems served to 'domesticate' and cleanse the unruly 'body' of the modern city (Kaika and Swyngedouw, 2000; Laporte, 2000).

'The city,' writes Erik Swyngedouw, 'cannot survive without capturing, transforming and transporting nature's water. The "metabolism of the city" depends on the incessant flow of water through its veins' (1995b, 390). The construction of systems to deliver such water supplies, however, is fraught with difficulties. Urban water systems necessarily 'demand some form of central control and a coordinated, combined and detailed division of labour' over long periods of time (Swyngedouw, 1995b, 390).

Again, Haussmann's regularisation plans, which also extended to water and sewerage, laid the foundations of the modern infrastructural ideal's treatment of the burgeoning water demands of the modern metropolis. With Haussmann's new Paris sewer system, as Matthew Gandy contends, the city's 'new boulevards and shopping arcades now had their subterranean counterpart beneath the city's streets. The transformation of Paris made urban space comprehensible and visible to the public, thereby dispelling much of the opacity and heterogeneity of the pre-modern city' (1998, 8). In an attempt to undermine the stench that characterised nineteenth century Paris, Haussmann, and other reformers, set in place many hygiene reforms with the central aim of the 'comprehensive "deodorization" of the urban environment. . . . Deodorization was to be applied to all areas of public and private space. . . . Thus emerged the fantasy of the odourless city, ideally sanitized to a zero degree of olfactory disturbance' (Prendergast, 1992, 79). From the experience of Paris, and many other modernising cities, 'the lessons of modern urbanism were clear. Water and sewer systems were a city's lifelines. As such they were too vital to be left to either the good intentions or the caprices of private enterprise alone' (Schultz and McShane, 1973, 395).

Thus we can also view Haussmann's schemes as the beginning of the broader project to 'domesticate' water as an agent in cleansing the city's 'circulatory' system (Swyngedouw, 1995b). The elaboration of extensive sewer and water systems, and the scientific discovery of bacteria, paved the way to a dramatic increase in the consumption of water which supported, in turn, the permanent washing of the urban 'body' and the privatisation of bodily hygiene (see Melosi, 2000; Lupton and Miller, 1992). On the eve of the twentieth century the Viennese architect, Adolf Loos, famously implored that 'increasing water usage is one of the most pressing tasks of culture' (cited in Lahiji and Friedman, 1997, 7).

Water thus became urbanised and commodified as another component in the infrastructural 'binding' of the city through standardised infrastructure services, accessible through single systems, access to which was initially limited to urban bourgeoisies (Melosi, 2000). Standardised water services allowed the emergence of a sanitised and deodorised public realm within the city

In France, for example, Guillerme recounts how, in the late nineteenth century, private water firms and municipalities struggled to find the finances and technologies to develop qualitatively new water systems to meet social demand (1988). Whilst still ridden with social inequality and bias, such networks eventually extended over whole urban areas as access to quality water and sewerage networks, at standard tariffs, became normalised as part of the modernising social world during the interwar period.

As with other urban infrastructure networks, then, the ideals of modern urban engineering stipulated that water networks needed to be rationally planned, systematically rolled out through the urban fabric and coherently integrated into an integrated and relatively standardised functioning whole. In Europe this transition was smoother than in the United States,

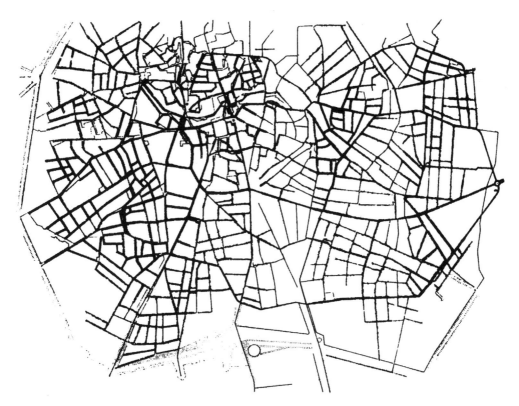

Figure 2.8 The water distribution network in Madrid, 1948: a fully integrated and ubiquitous system according to the demands of the modern infrastructural ideal. *Source*: Gavira (1995), 13

where the privatised provision of individual boreholes and cesspits persisted as the dominant model until the 1870s, despite many efforts to build systematic, public, modern water and sewer systems (Ogle, 1999; Melosi, 2000). Gavira (1995), discussing how the construction of a single monopolistic network occurred in Madrid, stresses the power of the idea of developing and mapping a systematic water and sewer network with maximum 'homogeneity and isotropy' (see Figure 2.8).

THE 'FOGGY GEOGRAPHIES' OF URBAN WATER SYSTEMS

Through such processes 'the urbanisation process itself [became] predicated on the mastering and engineering of nature's water' (Swyngedouw, 1995b, 21). But the enormous systems of reservoirs, channels, chambers and shafts through which metropolitan life was watered maintained a curiously invisible presence in the city, embedded, as they were, in deep subterranean passages, excavations and culverts. Theirs was a world of 'foggy geographies'; the 'extent of [these] nearly invisible system[s] was difficult to comprehend, extending to wider and wider fields and depths for sourcing increasingly scarce fresh water' (Reiser *et al.*,

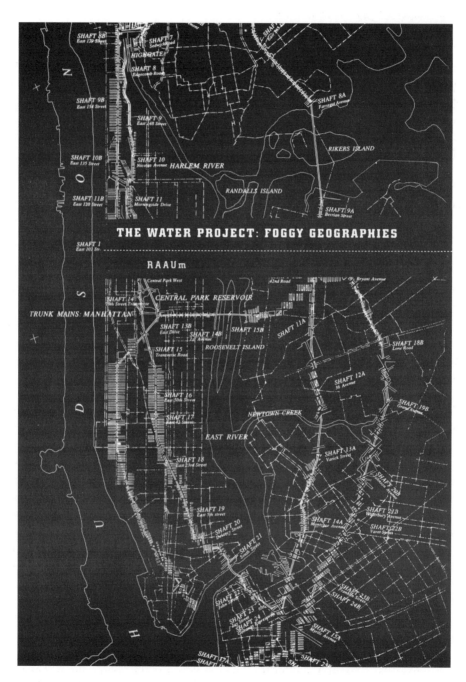

Figure 2.9 The 'foggy geographies' of urban water infrastructure: the example of Manhattan.
Source: Reiser *et al.* (1996), 73

1996; see Figure 2.9). Such linkages between water and the growing urban industrial metropolis remained complex and contested. The modern city relied on:

capturing and controlling ever-larger watersheds, water flows, water territories and an ever-changing, but immensely contested and socially significant (in terms of access and exclusion empowerment/ disempowerment), choreography of national laws, rules and engineering projects. Local, regional and national socio-natures are combined with engineering narratives, and speculation and global water and money flows.

(Swyngedouw, 1995b, 22)

INFRASTRUCTURE CRISES IN THE EXPANDING CITY: URBAN GRIDS, FUNCTIONAL ZONING AND URBAN INFRASTRUCTURE

Our fourth vignette concerns the ways in which Haussmann's idea of the 'regularisation' of the city through standardised, integrated networked infrastructures linked seamlessly with wider ideals then developing in the fast-emerging urban planning movement of the period. For Haussmann's ideas laid the essential foundations for the wider framework of unitary urban planning and urban beautification. They were quickly applied elsewhere, from Berlin, Cologne and Dresden to Barcelona, London and Chicago. Drawing from Haussmannian notions of the modern city, infrastructure planning was thus much more important to the development of the modern urban planning project than is generally realised (Fischler, 1992, 98). 'Planners grasped early on that different capitalists pursue different spatial investment strategies in an uncoordinated fashion' – including the rivalry between competing 'patch-works' of infrastructure networks (Beauregard, 1989, 382).

Efforts at technological standardisation and regulation were not new, however. As long ago as 1761 cities like London had been attempting to regularise the design and paving of street spaces and to systematise the delivery of street lighting and the naming of streets (Ogborn, 1999; Kostof, 1994a, 262). Standardised, paved, centrally managed and unitary street systems gradually started to emerge as the central, integrating network of the modern, public city (Southworth and Ben-Joseph, 1997). Street names in Stockholm, for example, were entirely changed and reorganised by the city government, 'in order to nominalize and control the city' (Whiteman, 1990, 28). A century later, Victorian London's extensive collective of private, gated streets came under public attack as symbols of elitist power. Intensive lobbying eventually led to their incorporation into the unitary public street system (Atkins, 1993).

But, even by the start of the twentieth century, much remained to be done to develop universal, standardised street and infrastructure networks. Often, peripheral areas had no access to the new technology services like sewers, potable water, telephones, gas and electricity networks, because private and municipal infrastructure providers targeted the more lucrative spaces at the centre. Christine Boyer shows how, after the turn of the century, dramatic urbanisation meant that 'the great needs of older and larger US cities were for the extension of gas, electricity, water, and sewer lines; the establishment of cheap and efficient streetcar systems, subways, bridges, and tunnels; and the construction of pubic buildings and private dwellings' (1994, 7). Of course, the laying out of integrated and open street systems provided the physical frameworks for the extension of these other services, as well as setting the legal and territorial boundaries of further urban growth, often on a speculative basis.

In much of North America the rectilinear grid became the norm for organising metropolitan expansion. The effort to use extensible grid patterns to define urban growth in a systematic and integrated way is best illustrated by the map used in 1811 in New York by the City Commissioners to lay the framework for expanding the city. The plan laid the foundations for a fivefold increase in the city's area, transforming 11,000 acres into 2,000 of the block patterns that still characterise Manhattan (Figure 2.10; see Pope, 1996).

For all their faults of repetitiveness and predictability, such grids forced openness on the urban form, overcoming the closure characteristic of premodern cities. 'The nineteenth century gridiron city [of North America can] be defined as an inherently open city,' suggests Albert Pope, because such open grids 'cut through and unite a sequence of scales connecting discrete urban artifacts to limitless space' (1996, 17). As gridded plans shaped development in North America, the bye-law streets of London and Cerda's extensions in Barcelona, Pope argues that 'every device of physical closure – the very idea of the closed urban system itself – was overwritten by subtle yet radical transformation of the urban grid' (Pope, 1996, 32; see Dupuy, 1991, 93). Many cities thus became more open and less bounded, their integrated street systems supporting the exploding demands for exchange, production and distribution that came with capitalist urban industrial development.

Functional zoning and the public development and planning of infrastructure networks were often developed together around the framework of the urban grid or street pattern. Clean, well maintained streets and functioning electricity, gas and water and sewer systems were thus an essential element in the strivings of early city planning for the rational, orderly and disciplined metropolis. In North America, for example, 'side by side with the creation of a disciplinary order and ceremonial harmony to the American City . . . , improvers gave heed to the creation of an infrastructural framework and regulatory land order' (Boyer, 1994, 7).

Indeed, without considering the infrastructure networks to connect the planned urban spaces, the wider trend in early twentieth century modern planning towards the functional separation of 'work', 'housing', 'leisure', 'transport' and 'administration' that was such an intrinsic element of all modern city plans could not have been sustained (Rabinow, 1994). The modern urban planning ideal, from the early 1900s to the 1960s and 1970s, was thus to integrate, either explicitly or implicitly, coherent networks of transport, energy, water and communications grids with the public spaces, and industrial zones, of the functionally planned physical city (see Lefebvre, 1984). This was epitomised by the modern functional urban planning principles drawn up by the Congrès Internationaux d'Architecture Moderne (CIAM) group in the 1920s and 1930s. 'In this way, the emerging political economy of industrial capitalism would be manifested in a planned built environment' (Beauregard, 1989, 382). As Matthew Gandy suggests, from the point of view of US city planning:

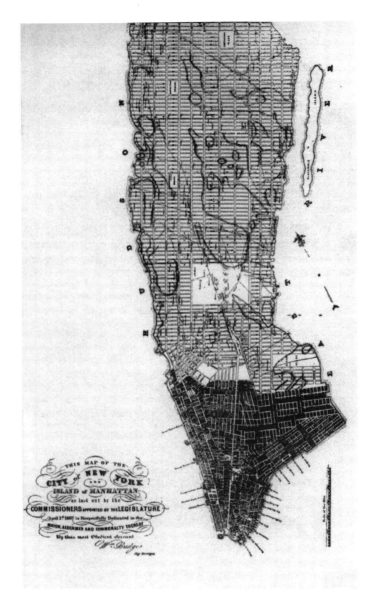

Figure 2.10 The 1811 commissioners' plan of Manhattan, which established the uniform, and infinitely extendible, rectilinear grid. *Source*: Pope (1996), 34

in the early decades of the twentieth century American modernism became closely bound up with newly emerging conceptions of urban planning within which there was to be a progressive move towards greater degrees of spatial rationalisation to accommodate new urban technologies and changing social and cultural aspirations.

(1998, 2)

BRINGING THE STRANDS TOGETHER: THE MODERN
UNITARY CITY IDEAL, 1880–1960

Finally, we need to explore how these four strands came together to support the broader notion of the ordered, unitary city, mediated by standard, ubiquitous infrastructure networks. Between the mid nineteenth century and the mid 1960s a dominant rhetoric of modern planning existed in the West which idealised the notion of the orderly, unitary city, tied together by a visible and non-visible web of standardised infrastructure grids. Through reason and democracy modern urban planning purported to 'produce a coordinated and functional urban form organised around collective goals' (Entrikin, 1989, 381).

Such an emphasis on integration and coordination led to stress on comprehensive city plans and standardised street standards enforced through the growing power of traffic engineering (Southworth and Ben-Joseph, 1997). These were 'the culmination of the modernist project', seeing the city as a 'synthetic whole. Henceforth, it was believed, all the contradictory tendencies of capitalist urban development could be resolved, into a unitary vision which stressed order and coherence' (Goodchild, 1990, 128).

Following Haussmann, cities were often seen metaphorically to be either 'machines' or 'organisms' whose functioning or metabolism rested on the appropriate connective systems. Rational plans 'would treat the city like a machine, to be planned as an engineer plans an industrial process, breaking it down into its essential functions (housing, work, recreation and traffic), Taylorizing and standardising them (in a Master Plan) as a totality' (King, 1998, 23). Urban space was seen to be 'an infinitely malleable matter, susceptible to the deigns of those who set themselves to reshape it' (Olalquiaga, 1994, 48).

In the process, of course, normative ideas of the appropriate uses of urban spaces became embodied into networks and spaces: the (male) public space of circulation, work and production separated functionally from the (female) domestic and suburban spaces of social reproduction and family rearing (Sandercock, 1998a). Fully elaborated energy, water, transport, street, highway and communications grids provided the 'connective tissue' of such a geometric vision.

As Mel Webber argued, 'In both the urban sciences and urban planning, the dominant conception of the metropolitan area sees each as a unitary place' (1964, 81). This he defined as a:

physically urbanized segment of land on which building and other physical equipment are closely spaced, and where people conduct activities that are typically more closely related to and dependent upon each other than they are to activities in other settlements.

(ibid.)

Innovations in infrastructure would help tie the '"synthetic" city, that is, the city of singular form invariant over time' together (Beauregard, 1989, 385).

Two key further developments were critical in taking the notion of the unitary, networked city to an unassailable, axiomatic position at the heart of modern urban planning practice. The first was the close linkage between integrated networked infrastructure and the idea of the comprehensive urban development plan (Box 2.2). The second was the close linkage between the modern infrastructural ideal and an influential range of utopian urban schemes that emerged during the first half of the twentieth century (see Box 2.3).

BOX 2.2   DEVELOPMENT PLANS AND INTEGRATED INFRASTRUCTURE

A key to the realisation of the unitary city ideal was the notion of the urban development plan which, when backed by sufficient weight of political and economic power, was cast as a means to steer the development of the land parcels and infrastructure networks, so that the whole city could emerge as a unitary, modern metropolis. 'For a century the plan was the centrepiece of modern city planning' (Neuman, 1998, 208). Zoning, together with the adoption of plans, was seen to be a means to rationalise and systematise the ways in which infrastructures related to the city.

One of the first comprehensive plans in North America, for example, the 1916 New York ordinance, was a 'vision for promoting the planned, controlled, orderly, sustained growth of the metropolitan region' through a 'vast, coordinated program of public improvements for the city's harbor, dock, streets, street railway, and raid transit facilities' (Hammack, 1982 ,188, quoted in Fischler, 1998, 179). Other North American master plans, notably those of the City Beautiful movement, epitomised the search for regular, geometric order, using the standardised, integrated street, water, sewerage, telephone and energy networks as integrated 'bundles' of networks for laying out and regularising the growing metropolis (see Figure 2.11).

In the interwar period Lewis Mumford wanted to extend such planning to the regional scale, using new infrastructures to support broad mechanisation, electrification

Figure 2.11 The Chicago plan of 1909: a master plan to impose civic order on an expanding metropolis. *Source*: Hall (1988), 177

and human-scale decentralisation. At the same time he sought to humanise the 'mega-technics' of city life by developing a more balanced urban system through coherent regional planning (Mumford, 1934, 1961; Corn and Horrigan, 1984). His vision helped inspire postwar efforts at extending the modern infrastructural ideal to the regional scale, notably with Patrick Abercrombie's well known strategy for Greater London.

As markets became increasingly national and international, corporations, too, lobbied for the regional integration of phone, electricity and water infrastructures. As Christine Boyer suggests, the issue arose: 'were these corporations to bargain with each fragmented governmental unit for adequate highway systems and sufficient land, water and electricity to parallel production needs?' (1994, 182).

BOX 2.3 INTEGRATED INFRASTRUCTURE AND MODERN URBAN UTOPIAS

The final source of urban planning's stress on integrated infrastructures came from a long line of influential modern urban utopianists, from Ebenezer Howard to Le Corbusier and Frank Lloyd Wright. All such visionaries were fascinated by the potential of the new network technologies to sustain radical shifts in the social order, embodied in the transformed physical landscapes of their preferred urban utopia. 'Modern technology,' they believed, 'had outstripped the anti-quated social order, and the result was chaos and strife. In their ideal cities, how-ever, technology would fulfil its proper role,' namely, integrating the planned metropolis into a functioning, efficient, harmonious and modern city designed according to their personal blueprints (Fishman, 1982, 13).

Thus, whilst nearly all utopian urbanists suggested a radical decentralisation of the industrial metropolis, it was to be a planned and ordered dispersal, using the new medi-ating capabilities of singular infrastructure networks to sustain the coherence of the new extended city. Tying into the prevailing ideologies of science, technology and the city, all utopianist urban visionaries of the first half of the twentieth century painted a picture whereby emancipatory progress for all could be achieved through combining the new powers of mediating urban life by the latest energy, water, transport and com-munications systems, integrated through a planned, modern urban landscape. All 'based their ideas on the technological innovations that inspired their age: the express train, the automobile, the telephone and radio' (Fishman, 1992, 13).

NETWORKED TECHNOLOGIES AND URBAN UTOPIAS: EBENEZER HOWARD, FRANK LLOYD WRIGHT AND LE CORBUSIER

The three most famous urban visionaries were Ebenezer Howard (with his garden cities), Frank Lloyd Wright (with his decen-tralised Broadacre City model), and Le Corbusier (with his *Ville Contemporaine*). All assumed that new networked infrastructures could be laid out to support an inherently harmonious urban and social order (Fishman, 1992, 13). Time and space were to be mastered by these new networks, allowing perfectly ordered and coherent urban land-scapes to be engineered.

In the late nineteenth century Howard wanted to set out integrated systems of suburban railways to sustain the planned decentralisation of older cities and the development of garden cities. In the 1920s Wright saw universal car ownership and an ever-extending grid of public highways as supporting a shift towards decentralised, self-sufficient living and the progressive abandonment of the 'obsolete' big cities. In Le Corbusier's *Ville Contemporaine* the old chaotic and disorderly city was to be swept away in the wake of rationalised urban landscapes with clean, modern intercon-nections. A utopian city of 3 million, it was inspired by the clean modernity of aero-planes and ocean liners. A testament to what he called the emerging 'machinery society' it was, in essence, a 'colossal and well ordered mechanism' (Corn and Horrigan, 1984, 38). Le Corbusier demanded that the integrative network architecture of the city had to function through new straight-line systems. 'The modern city,' he wrote, 'lives by the straight line, inevitably; for the construction of buildings, sewers and tunnels, highways, pavements, the circulation of traffic demands the straight line' (1929, 10).

La Ville Contemporaine, and other 'new city' plans like it, were predicated on seamless, rapid and futuristic infrastructure networks – 'multilevel circulation' systems based on massive integrated highways, airports atop railway stations, airship hangars, ubiqui-

tous (although invisible) energy, communications and water systems – providing limitless opportunities for exchange and integration (King, 1996, 49). Within the massive skyscrapers of the geometrically laid out central business districts was concentrated the 'apparatus for abolishing time and space, telephones, cables and wireless' (Le Corbusier, quoted in King, 1996, 51). Pedestrian flows, as in nearly all modernist schemes, were to be entirely separated from traffic within their own walkway systems.

LINEAR CITY UTOPIAS

The most extreme notions of the integrative power of new networked infrastructures, however, came from linear city ideas theorists, whose concepts of idealised cities involved strung-out linear corridors of housing and facilities that were closely bound together by integral infrastructure networks. One example was Edgar Chambers' (1910) 'Roadtown' – an endless corridor of housing laced by three levels of underground trams and rooftop roads (Hayden, 1981, 246).

Another example was the 'linear city' (La Ciudad Lineal) proposed by the Spanish Republican, Arturo Soria y Mata (1844–1920). As a theorist of communication Soria y Mata's ideal city is one of the best early examples of the use of high-quality, integrated networked infrastructures to reformulate the whole notion of the modern city (see Figure 2.12). 'The form of the city is, or must be, derived from the necessities of locomotion,' he wrote (quoted in Choay, 1969, 100). With a single all-encompassing 500 m wide strip housing integrative infrastructures of all kinds, Soria y Mata concluded that the imperative was to:

put in the center of this immense belt trains and trams, conduits for water, gas

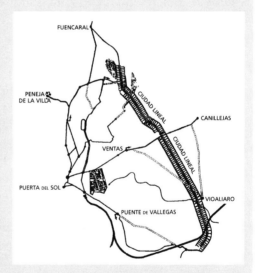

Figure 2.12 Soria y Mata's strategy for a linear city, 1882: an extreme form of using integrated infrastructure networks to structure urban growth. Source: Barnett (1986), 176

and electricity, reservoirs, gardens and, at intervals, buildings for different municipal services – and there would be resolved at once almost all the complex problems that are produced by the massive populations of our urban life.

(Quoted in Choay, 1969, 100)

CONSTRUCTING NETWORKED URBAN CONSUMPTION, 1920–60

For our third 'pillar' supporting the construction of the modern networked ideal we need to look at how the above ideologies of technology and planning practices broadly came together to support new types of mass production and consumption between 1920 and 1960, mediated by standard infrastructure networks in the city.

Whilst the emergence of modern forms of household-based consumption mediated by infrastructure networks can be traced back to the nineteenth century, it is in the first half of the twentieth century that such developments reached their greatest intensity. For, between the early 1900s and World War II, 'Fordist' systems of mass production, distribution and consumption were elaborated in most Western nations. Stressing automation, standardisation, economies of scale and a technical division of labour, the 'heartlands' of Fordism were the major metropolitan regions (Giannopoulos and Gillespie, 1993). Most important from our point of view, all aspects of Fordist social and economic life became predicated on access to the integrated energy, transport, water and communications grids so central to the modern planning ideals of the time.

NETWORKED INFRASTRUCTURE, FORDIST PRODUCTION AND THE CITY, 1920–60

> [In Fordism] it was incumbent upon communication to ensure the welding together of serial production and mass consumption, work and entertainment.
>
> (Mattelart, 1986, xiv)

The complex and territorially dispersed production systems of Fordism imposed considerable requirements upon urban infrastructure systems (see Table 2.1). With their high degree of territorial segregation and dispersed production, the variants of Fordist production developed would not have been possible without significant innovations in communication, transport and energy networks, organised as standardised monopolies across space (Warf, 1995). Differences between spaces needed to be ironed out in terms of the infrastructure available and technological standards. Standardised infrastructure grids needed to match the broader moves towards standardisation in production, consumption and, later, mass public housing within cities and national systems of cities (for the French case see Lucan, 1992). And central and local states needed to ensure that modern, integrated infrastructure networks were provided to underpin virtuous cycles of production and consumption within the Fordist model.

Fordist development was thus predicated upon a set of technologies which together rendered 'industrial location and the management of production itself largely independent of geographical distance' (Frobel *et al.*, 1980, 36). Their requirements were met by massive national and local state investment in standardised and dependable infrastructure services. Containerisation and the long-distance highway network facilitated freight movement.

Table 2.1 Fordist spatial organisation, transport and communications

Industrial organisation	Spatial organisation	Transport and communication patterns	Infrastructure requirements
• Economies of scale • Vertical integration • Hierarchical control • Mass production • Large firm-dominated • Mass consumption of standardised goods	• Spatial division of labour • Corporate control linked to urban hierarchy • Metropolitan labour markets • National and international production systems • Linkages maintained over long distances	• Long-distance product movement to assembly lines and markets • Long-distance movement of people to maintain corporate control • Vertical information flows • Commuting to large urban and suburban production sites	• Reliable long-distance goods transport • Reliable air and rail travel to maintain national corporate control • Reliable national and international telephone and telex • Urban and suburban mass transit for labour force • Goods and information networks predominantly national and international to connect core with periphery • Little need for local communication networks

Source: Capello and Gillespie (1993), 37.

Reliable voice communication permitted the long-distance control of spatially dispersed production. Transport, telecommunications and the roll-out of electricity networks meant that locations in peripheral regions and cities became economically feasible (Gillespie and Giannopoulos, 1993).

Increasing international connections between cities helped to meld an increasingly international urban system, further freeing up location choices for growing international corporations. As the interwar practices of US Fordism diffused into Europe during postwar reconstruction, a twenty-five-year boom followed, based on the virtuous linkage of mass production techniques, mass consumption and advertising based on the nuclear family household, Taylorist work organisation, collective wage bargaining, the hegemony of the large corporation, Keynesian demand management, the welfare state and the mass production of standardised housing (Giannopoulos and Gillespie, 1993, 36). Underpinning all, of course, was the elaboration of often taken-for-granted and increasingly standardised infrastructure grids to bind Fordist space economies, and the city-regions that made them up, into functioning entities.

As cities integrated more closely into national urban systems, so planning doctrines in turn contributed to the virtuous circles of growing mass consumption and production. As Fillion suggests, 'new infrastructures, zoning regulations favouring low density development, and financial support for single family housing transformed time–space relations in a way that was particularly propitious to the mass consumption of durable goods' (1996, 1939).

DOMESTIC CONSUMPTION AND MODERN NETWORKED URBANISM

In most Western cities between the 1920s and 1960s the above ideologies of technological progress, of the need for bright, modern, technologically advanced cities, wove together powerfully with planning debates about urban rationalisation. The result was to shape a remarkable growth of domestic mass consumption, mediated by the increasingly ubiquitous electricity, gas, telephony, broadcasting and transport grids. Both the transforming urban landscapes of the modern city and the fast-developing highway, telephone, television, energy and water networks underpinned new cultures of domestic mass consumption of a vast range of goods and services over unprecedented distances.

Distance was less and less a barrier to interaction, mobility and exchange as networked cities merged into networked urban systems. Consumers could live at ever-increasing distances from power stations, and highways, water grids, phone networks, television and radio broadcasting networks maintained the illusion of proximity. Electricity, for example, was now pushed down high-voltage cables across regions, nations and cities, to enter the networked home invisibly, giving 'the impression that [it] was a sourceless source, an absent presence' (Thrift, 1996a, 271).

Such mediated, dispersed consumption, in turn, fuelled the demand that sustained extending domains of Fordist mass production. As Figure 2.13 shows, we can, in fact, interpret the period from 1900 to 1960 as one in which household access to the 'bundle' of modern infrastructure networks diffused towards near ubiquity – an essential foundation of all aspects of economic, social and cultural life during the period (see Figure 2.13).

It is important, therefore, as our third pillar of the modern infrastructural ideal, to explore how this 'bundle' of integrated networked infrastructures served not only to underpin the elaboration of mass consumption in the modern city, but also to sustain normative conceptions of social and cultural behaviour. Most particularly, we need to understand how modern networked consumption was constructed to underpin starkly gendered divisions of urban work and space. Three particular aspects of this shift deserve closer scrutiny.

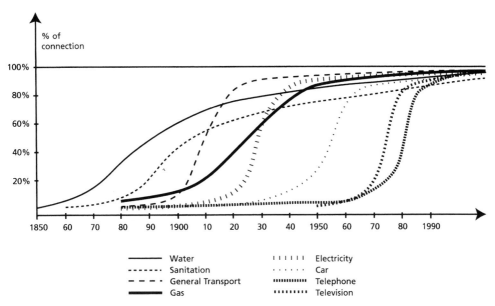

Figure 2.13 The expansion of household access to urban infrastructures in the West. *Source*: Dupuy (1991), 42

THE GENDERED CONSTRUCTION OF MASS, NETWORKED CONSUMPTION

We have learned that the way to break the vicious deadlock of low standard of living is to spend freely and waste creatively.

(US 'consumer engineer' Christine Frederick, 1929, cited in Lupton and Miller, 1992, 7)

First, there was the (profoundly gendered) construction of mass, home-based consumption. In the United States, and elsewhere, this was constructed as the modern urban ideal of 'cleanliness, comfort, convenience and economy of effort' and became closely linked with the efforts of infrastructure operators to generate domestic demand (Rose, 1995, 200). Such efforts, in turn, linked with wider strategies of national and federal governments to maximise the impacts of their widening infrastructure grids on mass production and consumption. In US cities like Kansas, for example, utility companies worked closely with consumer appliance manufacturers to target directly female 'housewives' to sell all sorts of appliances. They were 'most effective in selling irons or gas water heaters, especially in presentations to women. Public policy had encouraged [utility executives in Kansas] to set in motion a program of mass production and mass consumption of light and heat that placed gender at the center of the sale pitch' (Rose, 1995).

In this and many other cases a whole range of advertisers and advocates worked to build up consumers' demands for emerging integrated urban infrastructures, standing 'between generalized desire for comfort and convenience, on the one hand, and complex and little understood infrastructural systems, on the other' (ibid., 200).

UNEXPECTED happenings often detain the business man at his office.

With a Bell telephone on his desk and one in his home, he can reach his family in a moment. A few words relieve all anxiety.

The Bell telephone system is daily bringing comforting assurances to millions of people in all parts of the country by means of Local and Long Distance Service.

Are you a subscriber?

NAME OF ASSOCIATED COMPANY

Figure 2.14 The starkly gendered construction of the telephone: An AT&T advertisement of 1910. *Source*: AT&T archives; Fischer (1992), 158

At this time a whole range of supporting practices – advertising, market research and 'consumer engineering' – emerged aimed at orchestrating the construction, surveillance and stimulation of networked household consumption and 'housekeeping' (Lupton and Miller, 1992; Hayden, 1981). As we can see in Figure 2.14, which shows one of AT&T's adverts for the telephone from 1910, the starkly gendered representations used resonated strongly with the dualistic distinction between the (female) domestic private space and (male) public space that was so strongly embedded in modern planning doctrines about urban streets and landscapes (Weisman, 1994). Thus paternalistic and patriarchal images were often used, configuring 'housewives' as grateful recipients of new networks and appliances, supporting their responsibility for maintaining comfortable, modern family homes. 'The obligation of men', meanwhile, was seen to be to 'protect women from the hazards of industrializing cities' (Rose, 1995, 200).

This applied to gendered notions of telephony, television use, automobile access and use, water, bodily hygiene and the widening range of appliances linked with gas and electricity. In each case the social and cultural meanings of vast, extending, engineered systems became constructed within an endless plurality of domestic spaces in profoundly gendered ways. But householders were not just victims of broader social norms: they actively appropriated and shaped domestic appliances and their uses 'in a fashion that specifies the significance of a particular machine for the gender, income, habits, popular ideas, and contemporary ideologies of buyers during a specific time period' (Rose, 1995, 196). This was especially so in the construction of American suburban households as sites of intense consumption, disciplined bodily and kitchen hygiene, and the systematised and profligate disposal of wastes (see Lupton and Miller, 1992).

DECENTRALISATION, SUBURBANISATION AND MEGALOPOLITAN URBAN FORM

Second, there were dramatic transformations in the physical forms and landscapes of cities. As the American suburban ideal grew to dominate representations of mass consumption, so European and American cities themselves were decentralising physically. Such decentralisation was essentially underpinned by the achievements of the modern infrastructural ideal: reliable access across the metropolitan region to electricity, telecommunications, road and highway links, and water and sewerage networks to allow social, economic and cultural participation over ever greater geographical areas (Fishman, 1990).

The bundle of networked infrastructures that underpinned such changes supported the growth, particularly in North America, of whole urban landscapes made up of networks of controlled and interconnected environments: malls, offices, cars and homes. As we shall see in the rest of this book, such decentralised metropolitan forms in turn provide the perfect landscapes to later sustain parallel processes of infrastructural splintering and urban fragmentation.

In supporting processes of decentralisation, ideologies, planning practices and state policies interlaced with changing consumer behaviour to create immense megalopolitan urban landscapes. As Mark Rose suggests, from the point of view of the United States, the:

Federal and state government built highway networks that allowed Americans to live far from central cities. Government also financed construction of water and sewer systems extending into distant suburbs, all the while guaranteeing the mortgages of the residents. At the same time, electric and gas rates declined; engineers built larger and more efficient plants; regulators kept energy prices low, particularly for natural gas; and lengthy pipelines and electrical interchanges carried that energy throughout the continent.

(1995, 201)

THE EMERGENCE OF THE MODERN NETWORKED HOME

Finally, of course, the notion of the modern urban networked home quickly became reconstructed to support the wider dynamics of mass consumption mediated by bundles of integrated infrastructure networks (Lucan, 1992). As Lewis Mumford describes, the new networked household meant that the 'cost of a whole room was buried in the street, in the various mechanical utilities necessary for the house's functioning' (1961, 48). Rybczynski notes further that:

The main difference between, say, the house of one hundred years ago and one of today is that the latter contains a great deal of machinery. The contemporary house, as the French architect Le Corbusier remarked, has become a 'machine for living', that is, it has become an environment that is conditioned primarily by technology. Electricity, power pumps, motors, furnaces, air conditioners, toasters, and hair dryers. There are technologies for providing hot and cold water, and for getting rid of it. There are telephone systems and cable television systems; unseen waves carry radio and television signals. The house is also full of automated devices – relays and thermostats – which turn these machines on and off, regulate the heat and cold, or simply open the garage door. Remove technologies from the modern home and most would consider it uninhabitable. Cut off the power that fuels the machine for long enough and the dwelling must be evacuated.

(1983, 22–3)

Such trends were supported by the modern prophets' idealisation of technologically advanced homes. Le Corbusier's 'machine for living', for example – the basic element of his utopian vision of the modern city – was to be a space where 'drudgery [was] eliminated by machines' (Corn and Horrigan, 1984, 69). Buckminster Fuller, another architectural utopian, spoke of the home as a 'modular unit'. But the application to housing construction of the assembly line techniques of the Fordist factory, in both the private and the state sectors, allowed the fastest possible diffusion of the attainable notion of the house as a multiply networked space 'filled with durable consumer goods' (Corn and Horrigan, 1984, 79).

The mass diffusion of multiply networked homes, especially through expanding Western cities, and the spin-offs in construction and consumer goods industries, acted as a kind of 'super multiplier effect'. It did much to sustain the whole long boom of Fordist–Keynesian economic growth between 1950 and the end of the 1960s. Gershuny notes how an extending national electricity grid in Britain after World War II provided such a 'super multiplier' to all aspects of consumption, particularly stimulating mass consumption of vast ranges of 'specific new products for innovative household electrical products' in a remarkably short time (1983, 196). Consumer durables, linked with networked infrastructures, thus, in effect, became normalised, as essential supports to all aspects of 'normal' domestic life, sustaining an ever-extending web of 'commodity futurism' (Corn and Horrigan, 1984), often based on the idealisation of US styles of suburban life.

In the process, urban cultures stressing privatism, enclosure and self-sufficiency have become dominant in many cities across the world (Kostof, 1994a). As Nan Ellin suggests, while 'private transit (the automobile) had served to accelerate privatization during the first half of the twentieth century, widespread access to communications technologies – particularly television, VCRs, the mobile phone, the Internet and personal computer, cast new dimensions on it' (1996, 108). The withdrawal of consumption politics into the networked spaces of individualised households is thus a common trend driving developments in modern urban infrastructure.

GOVERNING THE INTEGRATED INFRASTRUCTURAL IDEAL: CITIES, INFRASTRUCTURE AND THE NATION STATE, 1900–60

From the eighteenth century on, every discussion of politics as the art of government of men [*sic*] necessarily involved a chapter or a series of chapters on urbanism, on collective facilities, on hygiene, and on private architecture.

(Foucault, 1984, 240)

The final 'pillar' supporting the construction and elaboration of the modern urban infrastructural ideal was provided by efforts by governments and states to support the shift to regulated, near universal access to infrastructure networks across cities, regions and nations.

Reviewing the development of infrastructure networks, it is clear that broad agreement was reached between the end of the nineteenth century and the late 1960s, especially across the Western world, about the need to roll out rapidly a relatively standardised set of technologies to the city and the wider, urbanising nation. The general view was that infrastructure networks needed to be delivered by social institutions based on private or public monopoly control (McGowan, 1999). Roads, utilities, water systems and telephony were generally seen to connect and mediate all aspects of modern production, distribution and consumption. Without public control of these grids, local operators would, as the National Civic Federation of US Utilities argued in 1907, 'be left to do as they please' (Simon, 1993, 35). From the initial private and public local utilities in the late eighteenth century, in fact, many efforts had been made by municipalities and states in the United States to fight against the vested interests of private capital in developing the single integrated public water, sewer and later energy systems for cities that were able to match the enormous pace of urbanisation at the time (Tarr, 1984).

In France the agents of the nation state ultimately began to consider the country's 'territory on the model of the city' – a space to be ordered, regulated and configured through managing the interplay of territory and infrastructure networks (Foucault, 1984, 241). The essential idea was that 'a state will be well organized when a system of policing as tight and efficient as that of the cities extends over the entire territory . . . What was discovered at that time was the idea of *society*' (ibid., 241–2, original emphasis).

THE 'QUANTUM LEAP TOWARDS UBIQUITY': INFRASTRUCTURE NETWORKS AND THE CONSTRUCTION OF THE MODERN NATION STATE AS A TERRITORIAL 'CONTAINER'

It is important, though, to locate the shift to nationally regulated, and often owned, infrastructure networks during this period. Between 1880 and 1950 modern nation states emerged as great territorial 'containers' with growing powers over many domains. First they 'captur[ed] politics, then economics, followed by cultural identity and finally the idea of society itself' (Taylor, 1994, 157).

The nation state emerged to treat 'the people of a state as a society, a cohesive social grouping that constituted a moral and practical social system' (Taylor, 1994, 156; see Urry, 2000b). Infrastructure networks, now widely seen through organic metaphors as the very 'connective tissue', 'nervous systems' or 'circulation systems' of the nation, became an essential focus of the power, legitimacy and territorial definition of the modern nation state. The laying out of contiguous, monopolistic infrastructure networks, starting with the legacies of the more localised and internationalised networks constructed up to the 1930s, was, in fact, tied very closely not just to the modern view of the state but to much deeper views of territorial scale and space (Jessop, 2000). As Neil Brenner suggests, during this period, 'scales were viewed as relatively stable, nested, geographical arenas inside which the production of space occurred' (1998c, 460).

Most important, though, the broader elaboration during this period of Keynesian models of state policy and demand management, to balance Fordist production and consumption practices, seemed predicated on a 'quantum leap towards ubiquity' in access to publicly regulated or controlled infrastructure networks (Sawnhey, 1992, 539). Fragmented 'islands' of incompatible and uneven infrastructure within and between cities became a source of much concern throughout the Western world.

Strategies such as the New Deal initiative in the United States, which did much to support extension towards national phone, electricity and highway grids, sought to use integrated public works programmes to 'bind' cities, regions and the nation whilst bringing social 'harmony', utilising new technologies and also creating much-needed employment (Gandy, 1998; Easterling, 1999a). Figure 2.15, which shows an advertisement from AT&T in 1915, captures perfectly the spirit of extending infrastructure networks universally to 'unite' the United States. State-led regional development initiatives in the United Kingdom, similarly, worked closely with public utilities to ensure infrastructure capacity was built in advance of demand to ensure 'national prosperity' (see Figure 2.16).

Taking control over the supply of networked infrastructure supplies to production, the territorial roll-out of networks over space, and the application of new services to modern consumption, were therefore essential components of the growth of the modern nation state itself. The shift towards near-ubiquitous access to infrastructures across the territory of a nation, in fact, had much to do, as Thrift (1990) found with transport and communications in Britain between 1730 and 1914, with 'the gradual melding of the country economically, socially and culturally' .

The democratisation and diffusion of infrastructure were therefore critical to the emergence of a national sense of 'cohesion'. As Neil Brenner suggests, infrastructure policies were the central way in which national states engaged in shaping capitalist territorial organisation, especially between 1890 and the 1930s (1998c, 469). Later, between the 1950s and 1970s, infrastructure strategies helped entrench the national approach to the production of scale that most suited Fordist industrial policies. In fact, as Brenner argues, 'throughout the twentieth century, the state has operated as a form of territorialization for capital, above all through the planning, production and regulation of large-scale infrastructural configurations that serve as "general conditions of production" . . . on differential geographic scales' (ibid.).

The Telephone Unites the Nation

AT this time, our country looms large on the world horizon as an example of the popular faith in the underlying principles of the republic.

We are truly one people in all that the forefathers, in their most exalted moments, meant by that phrase.

In making us a homogeneous people, the railroad, the telegraph and the telephone have been important factors. They have facilitated communication and intervisiting, bringing us closer together, giving us a better understanding and promoting more intimate relations.

The telephone has played its part as the situation has required. That it should have been planned for its present usefulness is as wonderful as

that the vision of the forefathers should have beheld the nation as it is today.

At first, the telephone was the voice of the community. As the population increased and its interests grew more varied, the larger task of the telephone was to connect the communities and keep all the people in touch, regardless of local conditions or distance.

The need that the service should be universal was just as great as that there should be a common language. This need defined the duty of the Bell System.

Inspired by this need and repeatedly aided by new inventions and improvements, the Bell System has become the welder of the nation. It has made the continent a community.

AMERICAN TELEPHONE AND TELEGRAPH COMPANY
AND ASSOCIATED COMPANIES

One Policy   *One System*   *Universal Service*

Figure 2.15 'The telephone unites the nation': the portrayal of AT&T's nationwide telephone system as supporting the construction of a powerful, homogeneous country. *Source*: Bell Telephone Canada; Fischer (1992), 163

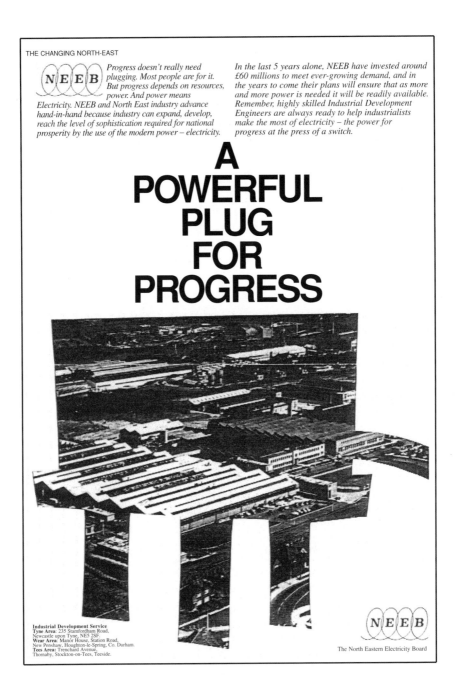

THE CHANGING NORTH-EAST

NEEB

Progress doesn't really need plugging. Most people are for it. But progress depends on resources, power. And power means Electricity. NEEB and North East industry advance hand-in-hand because industry can expand, develop, reach the level of sophistication required for national prosperity by the use of the modern power – electricity.

In the last 5 years alone, NEEB have invested around £60 millions to meet ever-growing demand, and in the years to come their plans will ensure that as more and more power is needed it will be readily available. Remember, highly skilled Industrial Development Engineers are always ready to help industrialists make the most of electricity – the power for progress at the press of a switch.

A POWERFUL PLUG FOR PROGRESS

Industrial Development Service
Tyne Area: 235 Stamfordham Road,
Newcastle upon Tyne, NE5 2SF.
Wear Area: Manor House, Station Road,
New Penshaw, Houghton-le-Spring, Co. Durham.
Tees Area: Trenchard Avenue,
Thornaby, Stockton-on-Tees, Teeside.

NEEB

The North Eastern Electricity Board

Figure 2.16 Keynesian regional infrastructure planning: North Eastern Electricity Board advertisement, 1967. *Source*: NEEB

Such state-organised investments in the productive basis of the territorial organisation of capitalism involved:

The construction of transportation systems such as highways, canals, ports, tunnels, bridges, railroads, airports and public transportation systems; the management of public utilities and energy resources such as gasoline, electricity and nuclear power, as well as water, sewerage, and waste disposal systems; . . . the maintenance of communications networks such as postal, telephone, and telecommunications systems; and the planning and construction of '*grands ensembles*' and other infrastructural configurations on urban–regional scales to coordinate the reproduction both of labour power and of capital.

(ibid.)

Ensuring nationally integrated infrastructure thus allowed the state to impose its own rationality on to the territorial scales, and social processes, within it. Large-scale Keynesian infrastructure projects allowed the nation state, in particular, to figure to an unprecedented degree 'in the promotion and spread of technological change' (Waites *et al.*, 1989, 27). The US nation state, for example, 'built roads, extended railways, organized electricity grids, put municipal sewerage systems into place' (ibid.). It also supported the emergence of the massive national Bell/AT&T telephone monopoly.

The rolling out of infrastructure networks, thus, helped to define the modernity and ideology of nation states in very direct ways. Consider the Nazis' *Autobahn* network, the electrification of the Ukraine and the Soviet Union (Figure 2.17), the New Deal regional projects of the Tennessee Valley and the national highway programme in the United States. Aimed directly to help pull the United States out of the Great Depression of the 1920s and 1930s, the New Deal allowed the federal government to construct 'a huge range of projects including roads, sewers, waterworks, multi-purpose dams, bridges, parks, docks, airports, hospitals, and other public buildings' (Tarr, 1984).

Later, following the Great Depression and World War II, the large-scale infrastructure projects of Fordist/Keynesian nation states further cemented their roles as geographical 'containers', supporting a tight 'fit' between urban development and planning and systems of national economic subsidies, grants, loans and public ownership. At this time, regional and local policies towards infrastructure were often little more than 'transmission belts' for national policies (Brenner, 1998c, 475).

LEGITIMISING INFRASTRUCTURE NETWORKS AS STANDARDISED MONOPOLIES

There were, of course, substantial national and local variations in the specific technological and social organisation of infrastructure providers. But these worked within a general and powerful consensus that networked infrastructures were characterised by three particular features that required a high degree of public involvement in the roll-out of the networks.

First, there was a broad consensus that the networks through which services were distributed were most effectively managed through 'natural monopolies'. Infrastructure

Figure 2.17 Electricity pylon as an icon of modernisation and collectivisation in postwar, communist Ukraine, then part of the Soviet Union. *Source*: Gold (1997), 101

networks, especially in the early stages of development, were seen to require significant capital outlay, recouped over a long period of time. Networks represented a form of 'embedded' or 'sunk' capital that realistically could not be dismantled or moved. Investments were often very 'lumpy' because new capacity must be created in large increments. There was therefore little scope for infrastructure to be a 'contestable' activity because of the high sunk costs involved in establishing a competitive network.

To Sleeman, a typical commentator on state utility policy in the United Kingdom in the early postwar period, public utilities were therefore not normal commodities to be bought and traded in markets. They were now considered 'essential to a civilised life' (1953, 3). The supply of services was unusual in that 'reliability and regularity are all-important' (ibid.). And, despite the efforts of some entrepreneurs to develop competing networks in some places, the huge physical costs of infrastructure networks meant that services were not generally competitive or 'transferable from the point of view of users' without the user moving physically to a new location. Utilities, roads and rail services (as well as crucial communications services such as the post), therefore, stood 'in a special relation to the State, and need[ed] to be publicly regulated to ensure reasonable charges and adequate services' (ibid.).

Taken together, these features usually meant that networks were treated as natural monopolies because infrastructure was an area of activity where several suppliers were less efficient than a single one. To Sleeman and many others, the natural monopoly status of infrastructure networks was 'usually an essential condition of efficient operation, that one undertaking should have a monopoly of the supply of service in any particular area' (1953, 13). A publicly regulated monopoly was able to benefit from economies of scale by developing one network whereas a fragmented industry was likely to lead to duplication of costs.

Rationalisation and the interconnection of local networks to create national 'grids' thus allowed economies of scale to be realised in all infrastructural domains. In electricity, for example, 'diversity of demand in an interconnected network can save a great deal of generating capacity for a given aggregate power of appliances' (Byatt, 1979, 96). Indeed, Britain's slow economic progress in the 1920s in integrating its many electricity suppliers meant that it still suffered from many incompatible local power suppliers. As Byatt notes, as a result, Britain was seen to suffer from a debilitating 'electrical backwardness'; 'electricity was already so well established as a symbol of progress and modernity, that the identification of Britain's industrial future with it was compelling' (1979, 93). Thomas Hughes (1983) notes that 'London was a backward metropolis' at this time, compared with its great rivals, the 'electropolis' cities of Berlin and New York. London had sixty-five electricity companies, forty-nine systems, thirty-two different voltages and seventy different pricing systems, all of which led to low consumption, poor levels of innovation and reliability, and few economies of scale and scope. Rationalising these, and many other local systems, into the national grid in the 1930s thus 'both brought down the price of electricity and gave an enormous boost to British industry' (Thrift, 1996a, 276) (see Figure 2.18).

In all networks, then, nation states thus faced a technological, economic and territorial imperative to meld standardised, efficient national or subnational networks from the myriad network patchworks that they inherited. And as Hall and Preston argued:

this, by definition, ha[d] to be achieved comprehensively and in a relatively short time; it require[d] large scale organization and massive capitalization; it also require[d] technical standardization (railway gauges; standard times; voltage; communication protocols; rules of the road; speed limits; driving tests; motor insurance; international air traffic control; computer reservation systems). We invariably find, therefore, that the infrastructural consequences of major innovations involve[d] the creation of very large-scale, vertically integrated enterprises, often the largest seen down to their day . . . requiring state provision, or at least, major state concessions.

(1988, 273)

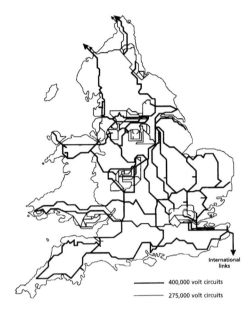

Figure 2.18 The national extension of the modern infrastructural ideal: the National electricity grid in England and Wales, 1994. *Source*: Thrift (1996a), 307

Second, infrastructure networks were largely considered to be 'public goods'. Public consumption goods were defined as having features that made them difficult for distribution within private markets. Three basic criteria defined 'pure' public goods (see Pinch, 1985). There was, first, the concept of 'joint supply' (or non-rivalrousness), meaning that if a service, such as the national defence or legal system, was supplied to one person it could also be supplied to all other persons at no extra cost.

The second concept, 'non-excludability', meant that once a supply had been built a user could not be prevented from consuming the service, including infrastructure such as rural roads. Third, there was the notion of 'non-rejectability'. This meant that once a service was supplied it must be equally consumed by all, even by those who do not wish to consume it. An example would be traffic control systems.

Finally, infrastructure services were seen to produce 'spillovers' or 'externality' effects that might negatively affect the environment or particular social groups. These were usually unpriced effects – such as the negative externality of fossil fuel emissions which contribute to global warming or the health impacts of not having access to clean water supplies and endangering health in other parts of the city. For this reason it was accepted that state regulation was needed to manage the negative externalities associated with infrastructure to ensure that wider economic, social and environmental objectives were met.

Because the development of infrastructure during this period was driven by the notion of trying to 'bind' the nation together socially and spatially, a set of practices were developed to ensure the rapid roll-out of standardised infrastructure at equal prices across national economic space. The institutions responsible for infrastructure provision, whether public or private, were given considerable autonomy and powers to ensure the rapid roll-out and integration of the national space economy through networks, especially into domestic and rural areas. Cross-subsidies from large to small users, and universal service obligations to ensure a minimum cost of connection, were developed to ensure rapid roll-out, standardisation and equalisation.

As nation states drove the integration of networked infrastructures, extremely powerful supply-oriented logics of network development emerged. Expansion of utility and infrastructure networks became intimately connected with the drive to improve national economic performance and the quality of life. Levels of energy consumption, connection to water and waste networks and levels of telephone ownership became surrogate indicators of levels of national economic performance. In the search for greater economies of scale the electricity industry built larger power stations and upgraded the national electricity transmission network. In the United Kingdom, in the space of twenty years, following the Second World War, electricity generating capacity multiplied seventeenfold (Reid and Allen, 1970, 9). Driven by the basic assumption that economic growth would generate new demands for utility and infrastructure services, network providers became locked into a logic of network management that focused on the supply of networked services. Major investments in national transport, energy and telecommunications services were made during this period in order to develop standardised systems of network supply.

EXPORTING THE MODERN IDEAL: THE EXPERIENCE OF DEVELOPING CITIES, 1850-1960

The modern infrastructural ideal evident in the cities of advanced Western economies had important implications for the style of infrastructure provision adopted in developing and colonial cities. It is important to understand how the modern integrated ideal was adapted and exported from cities in the dominant economies of the North to the colonial cities of Africa, South America and Asia (King, 1990). Several key questions emerge here. How successfully did the modern ideal translate into quite different social, economic and cultural contexts? How did unequal power relations between the colonial powers and colonial urban peripheries shape the style of urban infrastructure provision? And how was infrastructure linked with the developmental agendas of the colonial and postcolonial states in developing nations?

There are powerful resonances between the four main pillars of the integrated infrastructural ideal and the colonialist policies shaping the attempted roll-out of infrastructure networks in developing cities. To explore these, we would identify two broad phases of development in the style of infrastructure provision for developing cities: formal colonialism (1820s–1930s) and neocolonialism (1940s–1980s). Within each period there were enough parallels to enable us to talk meaningfully about the emergence of 'styles' of infrastructure provision in developing cities.

EXPORTING CUSTOMISED INFRASTRUCTURE: FORMAL COLONIALISM, 1820s–1930s

First, during the period of formal and direct control by core countries over their colonial 'dependencies', infrastructure was developed along a highly selective trajectory (see Yeoh, 1996). Investments in infrastructure were designed to meet two broad objectives. The first was to 'rationalise' the economies of colonies to create an 'open' structure heavily dependent on the export of primary products to the metropolitan core. Colonial economies also became increasingly important as a market for consumer, and later producer, goods from the core economies. The second objective was the creation of well serviced urban cores for colonial and local elites to organise production, exert political and administrative control, and mediate relations with the metropolitan core in the global North. During this period, infrastructure investment was explicitly geared to the needs of colonial interests. Networks were designed to minimise the risks and obligations of metropolitan power. Two forms of infrastructure development were related to these two broad objectives.

Metropolitan infrastructure, 'spatial apartheid' and the assertion of moral superiority

The first focused on the provision of networked infrastructures for the colonial metropolis. Here, urban infrastructure systems were a key part of the local creation of variants of the unitary city ideal. However, in colonial cities networks and plans largely focused on the needs of metropolitan and local elites (with the later, often unrealised, promise of later network extensions to the majority population). The Western ideal of a unitary, orderly city, laced by networked infrastructure, was thus effectively remodelled as a system of spatial apartheid (Balbo, 1993). Modern networks were laid out for the population; the 'natives' remained confined to premodern, non-networked and informal settlements beyond *cordons sanitaires* of walls and major boulevards.

In French colonial cities in North Africa like Fez and Algiers, for example, garden suburbs were laid out according to best practice but only for European settlers. The native towns – the Muslim-dominated medinas – were mostly left intact and generally neglected in terms of improvement in sanitation and services (Robinson, 1999, 161). As Balbo argues, this partial completion of modern infrastructure was a very deliberate attempt to symbolise the superiority of colonial power holders over colonised civilisations. The large avenues of the European city, he writes, with 'its modern services and infrastructures were to show very clearly on which side progress, wealth and power were situated' (1993, 25). Thus Western infrastructural and disciplinary concepts and practices were adapted and imposed in order to 'make non-western societies legible, ordered, and controllable' (Crang and Thrift, 2000, 10).

Along with the construction of 'colonial medicine' and 'sanitary science' to support the networked infrastructures of the colonisers, the existing infrastructural practices of indigenous populations tended to be denigrated as 'backward', 'disease-ridden' and full of 'latent poisons' (Yeoh, 1996, chapter 3). In Singapore, for example, 'the colonial medical and sanitary campaign' of constructing Western-style water and sewer systems, first for the colonial core of the city, 'not only served to legitimize imperial rule and to impart to it a gloss of munificence, an illusion of permanence, but was in itself an exercise of disciplinary power which penetrated the smallest details of everyday life' (Yeoh, 1996, 28; for a discussion of similar practices in South Africa see Minkley, n.d.).

Those 'majority' populations beyond the very limited reach of modern infrastructure networks, in traditional and informal settlements, were therefore rarely acknowledged as urban citizens within the discourse of urban planning, modernisation and colonialism. At best they were ignored; at worst they were labelled illegal and their settlements were torn down in the name of modernisation (still a widespread practice today) (Bhabha, 1994). To Balbo then:

the network city is the concretisation of the master planning approach to the idea of the unitarian city. Those who cannot afford to have their own w.c. or water tap and adopt other types of solution for their needs (oil lamps, street water vendors, foot travelling, pit latrines) are not acknowledged as citizens of the network city, even if they are the majority of the population.

(1993, 29)

COLONIAL HEADLINKS: INFRASTRUCTURE NETWORKS AND ECONOMIC EXPLOITATION

The second form of development was marked by the emergence of economic enclaves serviced with infrastructure. Colonial powers provided infrastructure networks, particularly communication systems such as rail and seaports, and international and regional telegraph and telephone cables, to incorporate selected areas of their dependencies into the world market, but on highly unequal terms. Usually this was done to support mineral exploitation, mines and plantations. Infrastructure was often explicitly designed to support the extraction of resources from productive enclaves whilst servicing the metropolitan elites in cities who organised production and maintained political control. Technically, local urban infrastructure tended to follow the same design and specifications as those of colonial powers – voltages, pressures, gauges, etc. – locking peripheries into particular trajectories of development and dependence on metropolitan powers for spares, maintenance and the capital equipment for major network extensions.

The creation of 'enclaves' either took place through direct external control or through a relationship with local elites (Cardosa and Faletto, 1979, 60). The objective was to incorporate local production processes, resources and labour into an economic system under strong external influence. The key economic function of this form of development was the growth of a node – a port or city – to serve as an infrastructural point of connection between local resources and international flows of raw materials and manufactured goods, a node through which metropolitan and colonial goods could flow. 'Colonial cities were hence planted as "headlinks" and designed to facilitate European capitalist penetration' (Yeoh, 1996, 18). Highly specialised infrastructures were developed with a powerful external orientation towards the export of resources to the Northern metropolitan core.

Take, for example, British railway construction in Africa. In this case the overarching rationale was to focus on what could be profitably exploited. As a result, the system reinforced a dendritic pattern of exchange. 'Communications were designed mainly to evacuate exports. There were few lateral or intercolonial links, and little attempt was made to use railways and roads as a stimulus to internal exchange' (Hopkins, 1973, 198). The routing of externally imposed infrastructure systems often seemed entirely arbitrary, for the links often bypassed important indigenous cities and trading centres.

MODERNISATION AND INFRASTRUCTURE: NEOCOLONIALISM, 1940s–1960s

In the period following World War II, increasing numbers of colonised territories obtained formal independence from the metropolitan nations of the North. This did not, however, signal a dramatic change in the style of infrastructure provision. Former colonies were usually locked into particular pathways of development through the technologies employed and powerful social and political links that maintained a high degree of continuity in infrastructural development.

THE EMERGENCE OF MODERNISATION AND INFRASTRUCTURAL 'TRICKLE DOWN'

However, what did change dramatically was the perceived role of infrastructure in urban and national development. Modernisation theory became the dominant development paradigm. It had important implications for infrastructural policy. Accepting that networks were already highly unevenly developed, the objective of policy became the acceleration of the Western pathway of urban infrastructural development. Modernisation theory suggested a model of infrastructural development initially focused on key cities and users. The assumption was that the benefits would 'trickle down' through the urban hierarchy and into rural areas, and that more marginal users would eventually be connected to the networks. The expansion of infrastructure networks began to be seen as the material representation of modernisation and the assertion of an embryonic national identity in the form of airports, four-lane highways and power stations that would sweep away the divisions of colonialism and the barriers of traditionalism (Bhabha, 1994).

It is interesting to review the relationship between cities and infrastructure implied within the paradigm of modernisation theory. The best-known exposition of modernisation theory was popularised by Walt Rostow (1960), who put forward a model of economic development comprising four successive stages from pre-industrial to post-industrial through which developing countries should pass. Rostow presented a limited conception of a single linear path to attain 'lift-off to self-sustaining growth'. This idea was universalising, technologically historicised, and oversimplistic but it became a powerful rhetorical device and was translated into spatial models of urban and regional development. John Friedmann (1966), for example, developed a four-stage model in which a single dynamic and modernising urban core expanded through the urban hierarchy and across the rural periphery, reducing urban–rural disparities and producing a homogeneous, fully integrated and modern development space. Many social scientists and policy makers assumed that the urbanisation process in less developed countries would pass through a progression of stages approximating the phases of city growth in Western urban history. This approach emphasised the similarities between urban growth in the Developing World and urbanisation in the advanced Western economies, implying that urbanisation was tied to economic growth and industrialisation.

According to modernisation theory the prime objective of policy was the modernisation of a region's economy through a form of industrialisation that closely followed the Western model of development. Former neocolonial powers 'sought to establish some of the essential infrastructure and facilities that the new independent states would require, and which had been ignored or neglected during colonial rule' (Simon, 1996, 36). The intention was to accelerate the newly independent Developing World through an industrial transition of rapid modernisation that would become evident in improved living conditions and standards. Clearly the role of water and energy infrastructure and reliable transport in ensuring the movement of energy, goods and people was central to this model of development. For instance, 'the infrastructure and public services utility industry in Brazil, during the developmentalist rule, evolved upon a model of huge networks, with growing territorial encompassment and functional complexity' (Schiffer, 1997).

Although initially concentrated in one or two urban centres for reasons of 'economic efficiency', these investments, it was widely hoped, would 'trickle down' through the urban hierarchy and rural areas of developing nations, replacing traditional lifestyles, modes of production and poverty in the process. In the early stages, massive infrastructural investment in the urban cores, usually capital cities, actually tended to drain the periphery, concentrating resources and skills in cities. But, it was argued that, once the diseconomies of urban growth outweighed the agglomeration benefits, the balance would shift in favour of trickle-down and the diffusion of modernity in all its infrastructural and cultural forms. These processes were seen as positive and beneficial and the spatial, economic and social inequalities as a necessary price to pay in the course of development. In any case inequalities were believed to be of limited duration, since industrialised development, when completed, would be characterised by a high degree of spatial homogeneity, therefore eliminating poverty in the periphery (see Box 2.4). It was perceived that Western models of infrastructural provision could be translated unproblematically into developing contexts.

IMPORT SUBSTITUTION INDUSTRIALISATION

The second approach to infrastructure development – import substitution industrialisation (ISI) – was closely linked with modernisation theory. The development of import-substituting industrialisation in many developing countries was designed to decrease dependence on foreign imports and help improve the balance of payments. But this required the availability of a small, highly trained, disciplined labour force and the existence of a highly concentrated consumer market. In this context cities were viewed as generative catalysts of development,

BOX 2.4 A MODERNISATION APPROACH TO THE DEVELOPMENT OF TRANSPORT NETWORKS IN DEVELOPING COUNTRIES

The model by Taffe *et al.*, published in 1963, provides a good example of the modernisation approach to the development of transport networks in a Developing World context (see Figure 2.19). Based on historical research in West Africa, it built a six-stage approach to the development of transport networks. Although this echoed Rostow's stages of growth Taffe *et al.* (1963) argued that not all countries would experience all the stages in a sequential form.

In stage one of the model a series of trading posts and ports are separately established along the coast to serve as points of exchange for exports and imports. In the second stage, trading routes are extended into the interior from the coast to make contact with new people and resources. Stage three is characterised by the development of extensions of the hinterlands through the construction of lateral feeders from the trunk links, but each port has its own distinct network and regional system. In the fourth stage these local and regional systems become interconnected.

As new economic resources – minerals and plantations – are exploited, new administrative and transport interchanges develop as intermediate urban centres. However, in stage five the creation of an increasingly integrated network reduces the need for so many ports. Trade is therefore increasingly concentrated in one or two large ports connected either with the largest cities or with the most favourable locations and facilities. Most of the small local ports decline and the level of connectivity with others increases rapidly. The final, sixth, stage echoes the recent development of rail networks in the North, with the closure of

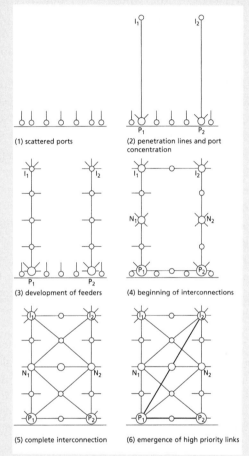

Figure 2.19 A modernisation approach to the development of transport networks in developing countries: the model of Taffe *et al.* (1963)

branch and low-volume lines in favour of the enhancement of a limited number of high-volume and priority links between the principal centres.

The modernisation approach, however, presents the development of networks in an unproblematic and neutral manner that assumes this style of development is positive for the host country. No account is taken of the fundamentally unequal power relations between colonists and colonised. Colonial infrastructure also reflected the priorities of the metropolitan power, often with little regard for the indigenous population. Traditional modes of transport were often undermined by new rail and road networks, substantially disadvantaging many important indigenous towns that were economically and administratively bypassed.

'the foci of modernisation and dynamism because they served as conduits for information to developing societies and as loci for innovation, opportunity, and political transformation' (Smith, 1996, 5). The shift towards capital-intensive export substitution had important implications for urbanisation and infrastructure development. As Robbers suggested:

The concentration of middle and high-income populations in a few urban *entrées* makes investments in capital-intensive consumer goods attractive. These industries are located in, or close to, centres of population and contribute to the attraction of large cities for rural migrants. Improvements in urban infrastructure such as roads, lighting, sanitation and housing are part of the dynamic of this industrialisation.

(1978, 81)

ISI required an internal market of consumers clustered together in urban areas who were wealthy enough to purchase final products. Consequently, urban infrastructure policies in this context reflected 'strategies for facilitating and subsidising the profit-making activities of the administrative elite and their partners, particularly transnational enterprises' (Smith, 1996, 67). Nation states therefore developed policies, including patterns of urbanisation and infrastructure development, that promoted the interests of this alliance by concentrating infrastructure in the capital cities or ports, providing for the material needs of the local elites but also subsidising the lifestyles of employees of transnational corporations. Like modernisation theory, ISI tended to downplay poverty and unemployment problems, seeing them as largely transitional. Problems such as massive rural to urban migration were seen largely in demographic terms that could be muted and controlled by planning designed to 're-equilibrate the system'. However, there was insufficient institutional capacity to manage and plan cities while the problem of low quality and even non-connection to infrastructure services became an increasing issue for poor and marginal users.

CONCLUSIONS: NETWORKING MULTIPLE MODERNITIES

We can see from the wide (but necessary) breadth of the above discussion, and the complex range of aspects covered by its four 'pillars', that the modern urban infrastructural ideal was a complex and multifaceted construction. More properly, it was a complex and multifaceted, yet diverse, set of constructions. The ideal embraced many intersections between ideology, technology, geography and culture; between politics, history, philosophy and society.

Discussion above has been able to give only a small number of examples of how the modern infrastructural ideal was elaborated in different places in different ways. We have found, nevertheless, that variations of the modern infrastructural ideal were an essential component of the elaboration of modern nation states and urban planning movements. They were central to the construction of modern notions of time and space and of gendered and racialised constructions of the 'urban'. They were invoked in the wider extension of modern consumption and the development of space economies. They helped symbolically and materially to support the construction of national identities, welfare states, and technocratic urban and infrastructural professions. And they were adapted to the very different contexts of developing cities, where infrastructural configurations were central in structuring power relations between colonised and colonisers.

Inevitably, then, exploring the archaeology and origins of the modern ideal of urban infrastructures has forced us to be unusually, even athletically, interdisciplinary. We have had to bring together many usually separate debates: from Geography, Planning, Architecture, Urban Studies and Cultural Theory, from Science and Technology Studies, and from debates about consumption, culture, governance and the state. We have also had to draw on the stories of a wide range of cities: developed and developing, colonial and postcolonial.

So intrinsic are infrastructures and urbanity to 'modernity' that such stories are inevitably a good part of the story of modern society itself. Indeed, so woven are the (unusually implicit and unspoken) tenets and axioms of modern ideals of integrated infrastructure into the fabric of modern civilisation that commentators across many disciplines have been extremely reluctant to recognise their current demise. As Kaika and Swyngedouw (2000) argue, once they were completed within the framework of the modern infrastructural ideal, 'the networks became buried underground, invisible, banalised, and relegated to an apparently marginal, subterranean urban world'. It is only now, through processes of splintering urbanism, that infrastructure networks are being reproblematised and (unevenly) brought back into view as major foci of debate, renegotiation and reconstruction within contemporary cities.

This is not to imply, however, that the modern infrastructural ideal was ever perfectly 'realised' or that it was a universal and uniform 'thing'. Always heterogeneous and dynamic, efforts materially or discursively to construct ubiquitous, normalised and standardised infrastructure networks emerged in a myriad of different ways covering different networks, spaces, cities and times. What we have explored here, then, is not some overarching story but the complex interplay and emergence of *multiple* modernities (Eisenstadt, 2000). Whilst the range of stories and experiences in these domains was broad, we can see from this chapter that there was a notable 'fit' between the diverse resonances and constituent forces of the modern ideal such as to allow us scope to generalise in the way that we have.

Having pieced together some of the stories surrounding the emergence of the modern infrastructural ideal, we are now in a position to begin to understand the significance of the contemporary shift towards the wholesale and widespread unravelling of that ideal. The rest of this book will be concerned with exploring in detail how the parallel and interdependent dynamics of infrastructural splintering and urban fragmentation are becoming manifest in a wide range of contexts across the world. One immediate challenge arises at this point, however. We need to understand exactly why the modern integrated ideal is unravelling. It is to this question that we turn in the next chapter.

3 THE COLLAPSE OF THE INTEGRATED IDEAL

The modern networked city in crisis

Plate 6 Dedicated highways and transit routes linking the Sheraton Hotel and Paris Charles De Gaulle airport. *Photograph*: Andreu (1997)

PROCESSES UNDERMINING THE MODERN INFRASTRUCTURAL IDEAL

> We know nothing of vast multiplicity – we cannot come to grips with it – not as architects, planners, or anybody else.
>
> (Aldo van Eyck, quoted in Holston, 1998, 27)

From the late 1960s a series of powerful critiques of the assumptions underlying the modern urban infrastructural ideal, and its results in practice, emerged. Most of the central tenets of the ideal – the need for public or private infrastructure monopolies, for singular and standardised technological grids across territories, for the 'binding' of cities into supposedly 'coherent' entities – became deeply problematic and difficult to defend. Twenty years later, as privatisation, liberalisation, globalisation and the application of new technologies weave across the planet, fewer and fewer networks, in fewer and fewer cities, regions and countries, continue to develop in isolation from these critiques, along the 'pure' lines of the modern integrated ideal.

A set of wider societal and technical shifts has created the context for this rapid transformation in the logics underpinning urban infrastructure. The long capitalist boom from the 1950s to 1970s – the period where the modern infrastructural ideal reached virtually hegemonic dominance in most Western cities – has collapsed through complex processes of economic and technological restructuring. Profound economic, political and cultural shifts have surrounded the emergence of an intensively, but unevenly, interconnected global capitalist economy, society and culture (whether one captures such trends with phrases such as 'post-Fordism', 'postmodernism' or the emergence of a new Kondratiev 'long wave' of economic development – see Amin, 1994; Leonard, 1997).

The Keynesian welfare states that were so central to applying the modern ideal in Western nations in its latter stages have everywhere experienced deep fiscal and legitimacy crises. In the Developing World, meanwhile, notions of publicly building modern infrastructure to stimulate economic modernisation have been under severe strain, with many states and cities selling off their infrastructural assets to private transnational firms. In addition, Soviet and East European communism has collapsed, opening up its territories – with their standardised, centralised and often obsolescent infrastructure networks – to be incorporated unevenly into global capitalist divisions of labour and flows of capital, information and technology. In most contexts, it seems, politically neoliberal critiques of the 'inefficiencies' of centralised public control and ownership have fuelled a widespread wave of infrastructural liberalisation and privatisation which is still accelerating.

Above all, we have seen a dramatic, 'global assertion of the moral superiority of individual choice compared to the "tyranny" of collective decision making' (Leonard, 1997, 4). 'Market forces' and liberalised models of infrastructural competition are widely attested to deserve hegemonic status as modes of distributing many types of goods and services previously considered to be 'public'. To a considerable degree, this is being driven by the hugely influential lobbying of private transnational firms. On the one hand, such firms are keen to (re)commodify public goods for profit. On the other, they are keen to benefit from the global archipelagoes of high-quality infrastructural investment that are necessary to sustain their own internationalising divisions of labour, production and consumption.

However, the shift away from the modern infrastructural ideal has been about much more than geopolitical and economic shifts. Powerful social and cultural critiques have also exposed its inadequacies. In particular, feminist, anti-racist, postcolonialist and environmentalist critiques have dramatically exposed the social, gender and environmental biases inherent within the various elaborations of the modern infrastructural ideal. Finally, notions of urban planning and the city have also experienced radical overhaul, with the demise of the idea that it is either possible or desirable comprehensively and rigidly to plan 'order' and 'rationality' into the form, structure and life of cities.

These changes are, in turn, embedded within wider transitions. Within the current contexts of economic volatility, a proliferation of boundary-transcending environmental risks, cultural and ethic migrations, and the decay of the notion that capitalist or communist states can steer social change, 'the very idea of controllability, certainty or security collapses' (Beck, 1999, 2).

Addressing these complex and diverse shifts in detail would require a whole library rather than one relatively brief chapter. Moreover, attempting to explore shifts away from the modern ideal, across the full range of urban infrastructures, and right across the planet, is clearly an impossible task. In this chapter we seek, rather, to present a sweeping, but selective, perspective. To do this, we explore the interlocking critiques and challenges that, we argue, have so effectively undermined the modern infrastructural ideal – in the process demolishing the four 'pillars' upon which it was constructed.

We identify, in particular, five interrelated shifts that have worked together to render the modern infrastructural ideal severely problematic. These we label: the urban infrastructure 'crisis'; changing political economies of urban infrastructure development; the collapse of the modern notion of comprehensive urban planning; the physical growth and extension of metropolitan regions; and the challenge of social movements and critiques.

THE URBAN INFRASTRUCTURE 'CRISIS'

The first broad force undermining the modern infrastructural ideal has been a perceived 'crisis' in the infrastructural underpinnings of urban life across the developed, developing and post-communist worlds, especially since the late 1960s.

In the period since the late 1960s debates about the deterioration of infrastructure services, and their implications for the economic and environmental development of cities, have periodically lifted networked infrastructures out of the domain of the technical engineers. This occurred especially at the end of the 1970s, when the promised delights of urban modernity that surrounded the modern infrastructural ideal started to seem decidedly ironic, given the widespread physical collapse of urban infrastructure networks.

In older industrial cities of the North, especially, urban residents became much more 'aware of visible signs of decay, notably in the form of potholes, breaks in water mains, and bridge closings. The media have picked up on these visible signs and dramatised the worst cases in articles and news stories' (Petersen, 1984, 180). Central to these debates is the concern that the physical deterioration of infrastructure, the lack of spending on new facilities, and a huge backlog in maintenance and rehabilitation, actually threaten to slow and even reverse economic growth in cities. As we see in Box 3.1, the infrastructure crisis in US cities is particularly emblematic of this wider deterioration.

In developing countries, meanwhile, the debate is focusing on the lack of any networked infrastructure across wide swathes of cities, and the inability of providers even to keep pace with rapid demographic growth and urbanisation that are creating huge unmet demand for services (Potter and Lloyd-Evans, 1998).

BOX 3.1   THE OBSOLESCENCE AND PHYSICAL DECAY OF URBAN INFRASTRUCTURE: THE EXAMPLE OF THE UNITED STATES

The most emblematic and widely publicised of urban infrastructure crises occurred in US cities in the late 1970s and early 1980s. Although the experience is specific to one particular context, it did broadly mirror a developing discourse around cities with declining infrastructural assets in advanced Western economies. It is useful for our purposes, therefore, to explore the US infrastructural crises of between 1965 and the early 1980s in a little more detail.

The key element in the development of the crisis debate was the notion that urban infrastructure was rapidly deteriorating and becoming physically obsolescent. Widely publicised examples of crumbling bridges, worn-out roads, poor-quality sewage treatment, dirty water, and inadequate energy and telecommunications infrastructures all became dramatic symptoms of a widening collective sense of urban infrastructure crisis. In 1995, for example, Perry found that 40 per cent of the nation's 600,000 bridges could be classed as deficient and that two-thirds of the country's water treatment facilities were substandard (1995, 3). Slightly earlier, the 1982 US national urban policy report had expressed concern about the 'signs of erosion in the condition and performance of the urban spatial plant, especially in the oldest urban areas' (US Congress, 1984, 11).

Thus, almost as soon as the modern urban infrastructural ideal had become hegemonic in shaping policies for urban infrastructure in US cities, the very urban infrastructure networks that had been developed under the ideal appeared to be in a state of crisis. There was increasing evidence that much of the infrastructure in older cities had reached or passed its design life. The main problem was insufficient maintenance or rehabilitation, caused by the combination of massively extended infrastructure grids built up through the modern ideal and a collapse of fiscal capacity at local, federal and national levels to support these infrastructure grids. The result of the collapse of cross-subsidies within and between cities was growing inequality between the richer and poorer parts of cities to finance maintenance and new infrastructure projects.

In 1989 the US Department of Transportation estimated that $50 billion were required to repair the nation's 240,000 bridges, $315 billion was needed to repair its highways and that national spending on new infrastructure had fallen from 2.3 per cent of GNP in 1963 to 1 per cent in 1989 (Reich, 1992, 254). The crisis in water and sanitation was just as severe.

NEW YORK'S INFRASTRUCTURE CRISES

For instance, New York had one of the oldest and most extensive water distribution systems in the United States. An estimate of water distribution replacement needs was made in *Rebuilding During the 1980s: New York City's Capital Requirements for the Next Decade*. This report concluded that 'approximately $12.45 billion should be spent over the next decade to replace 39 per cent of the city's water mains (2,404 miles) and 8 per cent of the valves that are at least sixty years old and have exceeded their design life'. Such a programme would be extremely disruptive to traffic and it was not even clear whether such a costly programme would solve the city's chronic water supply problems. Instead, the State of New York, in co-operation with the city, requested the army Corps of Engineers to investigate replacement water mains needs.

THE CAUSES OF THE US INFRASTRUCTURE CRISIS

At one level, the infrastructure crisis had an apparently simple cause: state and local expenditure on infrastructure was on an erratic path downward as a result of a widespread fiscal crisis at all levels of the US state. Expenditure was approaching the point at which it could barely maintain investment in net infrastructure assets. In fact, with the extra demands caused by the completion of interstate highways, net disinvestment was actually occurring (Petersen, 1984, 111). Although overall public expenditure on infrastructure quadrupled in 1960–84 to $40 billion (1984 prices) this actually represented a decline when measured as a fraction of GNP. Capital spending declined from 2.3 per cent of GNP in 1960 to 1.15 per cent in 1984. The picture looked very different when considering expenditure on infrastructure maintenance and operations, which had increased from $800 million in 1960 to $50 billion in 1984 (1984 prices), surpassing capital spending after 1977. Operating expenditure had remained fairly steady at 1.4 per cent of GNP over this period. While overall expenditure had increased, the composition of spending had changed dramatically:

in the 1960s highway spending predominated, accounting for 60 per cent of all spending. In the 1970s highway spending began to fall and waste water treatment and water supply projects began to increase. By 1980 the interstate highway system was 97 per cent complete, and highway spending dropped to 45 per cent of all spending; spending on waste water treatment and water supply continued to grow, and mass transit doubled. The three programs accounted for 41 per cent of all government infrastructure spending.

(US Congress, 1984, 47)

CHANGING POLITICAL ECONOMIES OF URBAN INFRASTRUCTURE DEVELOPMENT AND GOVERNANCE

The second broad set of forces undermining the modern infrastructural ideal surrounds the wholesale political economic shift in processes of urban and infrastructural development that has occurred since the early 1970s. The classic territorial 'containers' of nation states and national markets are being tied together into integrated regional blocs supporting integrated flows of investment, capital and technology. Moreover, fiscal crises are forcing virtually all types of nation state – advanced industrial, newly industrialising, developmental and post-communist – to explore transferring some or all of their infrastructure operations to private operators, in the search for the 'one off' spoils of privatisation (Martin, 1999).

The privatisation of infrastructure has been most widespread in the Anglo-American, post-communist and Developing worlds (Offner, 2000). Perhaps the most famous and extreme example has been the United Kingdom's programme of wholesale infrastructural privatisation since 1984 (see Table 3.1). But infrastructural privatisation is a growing trend in all types of nation state, as the global pressures of the IMF, World Bank, WTO and regional trade blocs are forcing the colonisation of public infrastructure by global finance capital and a widespread retreat from collectivised, integrated and 'bundled' ways of managing urban infrastructure (Schiller, 1999a; Clark, 1999).

Within the emerging internationalised capitalist political economy, transnational corporations – of which there were over 37,000 in the early 1990s – dominate trade, invest-ment patterns, technological innovation and the reshaping of systems for the provision of infrastructure networks. Operators of national infrastructure monopolies geared to rolling out networks coherently over national and regional spaces are increasingly striving to piece together the global–local transport, communications and energy grids that most 'fit' the demands of transnationals, lucrative consumers and the investment strategies of large institutions (Offner, 2000).

Within such a massive, complex and diverse transition, we would like to emphasise only three aspects here. These are: the retreat of state-backed, collectivised forms of urban

Table 3.1 The programme of wholesale infrastructure privatisation in the United Kingdom, 1984–91

Privatisation	Total shares offered	General public allocation (%)	Institutional allocation (%)
British Telecom (1984)	3 012 000 000	38.5	61.5
British Telecom (1991)	1 597 500 000	65.7	34.3
British Telecom (1993)	1 311 500 000	55.8	44.2
British Gas (1986)	4 025 500 000	62.0	38.0
British Airport Authority (1987)	500 000 000	46.0	54.0
Regional water companies (1989)	2 185 000 000	44.0	56.0
Regional electricity companies (1990)	2 311 614 000	50.6	49.4
Northern Ireland Electricity (1991)	164 600 000	67.0	33.0
Scottish Electricity (1991)	1 198 259 000	58.0	42.0
RailTrack (1995–97)	500 000 000	42.0	58.0
GENCO (1991)	1 253 565 706	49.4	51.6
GENCO (1995)	759 691 098	51.3	48.7
British Energy (1991)	700 000 000	42.6	57.4

Source: Martin (1999), 271.

infrastructure provision; the rising imperative of 'competing' locally through particular configurations of urban infrastructure; and the widespread retreat of the idea that networked services are 'public' services that should be available to all at standard tariffs.

NEOLIBERALISM AND THE RETREAT OF THE STATE: FROM 'PUBLIC WORKS' MONOPOLIES TO INTERNATIONAL INFRASTRUCTURAL CAPITAL

First, supranational, national and local governments are easing restrictions on private entry into previously monopolistic infrastructure markets in many economic and urban contexts around the world. This is allowing many new, customised infrastructure networks to be overlaid within, through, above and below the monopolistic legacies of modern infrastructural planning and development.

Encouraging liberalised competition is once again a growing approach to national infrastructure regulation. On the supply side, powerful and transnational alliances and mergers between network operators in telecommunications, energy, water and transport are rapidly growing as newly private or entrepreneurial infrastructure firms attempt to position themselves favourably within dominant and emerging markets (see, for example, Summerton, 1999; Curwen, 1999). Whilst such internationalisation is far from new – before the era of national infrastructure 'champions', intense patterns of cross-national ownership existed in energy and telecommunications – it is now intensifying to a level never seen before (McGowan, 1999). Consolidation deals to create larger utility power companies in the electricity sector alone amounted to US$50 billion in 1998 across the world (Rider, 1999). In telecommunications and water the figures were much higher still.

INFRASTRUCTURE DEVELOPMENT AND FINANCIAL MARKETS

Stock market flotations and the speculative effects of globalising financial markets fuel much of the frenzied process of internationalisation that is occurring today in global infrastructure capital (Hirsch, 2000). In a widening range of cases, 'in urban infrastructure and development . . . the state has displaced its responsibility for financing and provision to the financial sector' (Clark, 1999, 242). Private infrastructure firms need to attract investment from pension funds, institutional investors and private shareholders, which have an extraordinarily diverse choice of investment options (equities, property, securities, etc.). But, as Poole argues, 'the world's financial markets are awash with private capital looking for economically sound infrastructure projects to invest in. Several multibillion dollar infrastructure funds have already been assembled' (1998, 7). The remarkable investment in international and national telecommunications grids, by alliances of private telecom, media, entertainment and Internet firms, is especially noteworthy, as it makes it very difficult for nation states to direct infrastructure development in this crucial field (Everard, 2000; Sassen, 1999).

FROM LARGE-SCALE INTEGRATED INVESTMENTS TO PROJECT-BY-PROJECT RISK ASSESSMENTS

Given the long-term and risky nature of infrastructural investment, investors from financial markets are likely to be reluctant to invest in large-scale, comprehensive and 'bundled' networks unless there are ways to guarantee certain rates of return (Clark, 1999). Usually, such investors will tend to demand a project-by-project risk assessment, identifying individual revenue and profitability streams for particular infrastructural developments, within tight definitions of accounting that minimise social or geographical cross-subsidies (Schiller, 1999a). Thus we could argue that the supplementation of state forms of collectivised infrastructure development that supported the modern ideal with privatised regimes that need to attract international finance capital seems very likely to support the splintering of integrated and 'bundled' networks into a myriad of individually financed and managed infrastructure projects. As Gordon Clarke suggests, 'one result of scrutiny has been a shift away from long-term investment relationships to project-by-project assessments ruled by the law of contract' (1999, 257).

This growing crossover between private finance capital and infrastructural development thus increasingly works to 'unbundle' the more or less coherent and integrated infrastructure networks that were the legacy of the modern infrastructural ideal. Such unbundling can happen organisationally, sociotechnically and geographically. From the point of view of infrastructure privatisation in Brazil, for example, Sueli Ramos Schiffer observes that private capital is often attracted only by the low-risk elements of infrastructure networks that can be 'splintered' off from the whole and directly managed for private profit:

the functional and territorial unbundling of infrastructure networks is necessary to make the private operation of public utilities feasible. Besides the desirable doctrinaire appeal to competition, the unbundling of complex unitary networks is a precondition for schemes of project finance based strictly on each project's risk.

(1997, 19)

Such a transition seems very likely to exacerbate the uneven development of urban infrastructure. Investment seems likely to focus on low-risk, lucrative projects with short-term, demonstrable profitability. Networks supporting more socially and economically marginal parts of cities are likely to experience increasing underinvestment, neglect and marginalisation. If such a logic works to shape infrastructural investment and disinvestment over a long period, fewer and fewer material connections will work to integrate the diverse social and economic circuits of cities. Such stark increases in the unevenness within cities seem especially likely when pension funds have systematically taken over the financing function, as in parts of the Anglo-American world. 'With the advent of pension fund capitalism,' suggests Gordon Clarke:

Urban structure will be increasingly an investment good managed with respect to the interests of pension funds and their beneficiaries. . . . It is likely that the urban fabric of Anglo-American societies [will] be systematically discounted by underinvestment over the coming generations with selective private investment replacing comprehensive investment by the state. There need be no connection between the goals of funds' investment strategies and the economic and social coherence of urban society.

(1999, 258)

But, to understand the broader significance of this shift, we need to look beyond the urban scale to the wider role that infrastructure capital is playing within the mushrooming political economies of international finance as a whole. Here it becomes clear that the increasing incursion of global finance capital into infrastructure is unleashing a frenzied process of alliance formation, mergers and acquisitions across the planet. In the interests of profitability and speculative growth, newly privatised national and regional monopolies are diversifying into new territories and sectors. Foreign infrastructure companies are acquiring networks in each other's countries. New entrants are taking on incumbent companies. And infrastructure capital is diversifying into, and making alliances with, other sectors (retailing, financial services, home entertainment, media, insurance, etc.). The 'map' of alliances in the global telecommunications industry, the sector that is rationalising most quickly on a global scale, is especially complex and fast-moving (see Figure 3.1; Curwen, 1999).

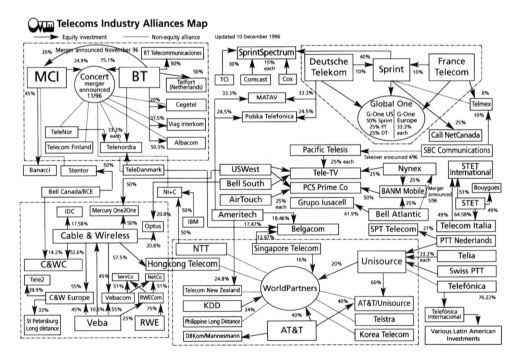

Figure 3.1 Telecommunications industry alliances, 1997. The map is already considerably out of date, for example MCI severed relations with BT on being taken over by WorldCom in 1998. *Source*: Winsbury (1997), 29

NATIONAL AND TRANSNATIONAL RE-REGULATION

But nation states do not tend to simply abandon infrastructure to private capital; they must find new ways of regulating. Many regulatory institutions, increasingly transnational themselves, are, in turn, starting to develop mandates to support the splintering of infrastructure networks within and between national borders. Neoliberal orthodoxy, supported by global economic institutions like the IMF, World Bank, World Trade Organisation, G8 and some regional trading blocs, is looking towards constructing the new infrastructural grids to support a global system of 'free trade' and foreign direct investment (FDI) by transnational corporations (TNCs).

In the Developing World, 'structural adjustment' programmes are now being combined with the efforts of national states to gain global financial credibility to encourage the growth of state sell-offs in infrastructure (Harris and Fabricius, 1996). Financial constraints on governments, rapid population growth, the qualitative infrastructural demands of foreign investors and elite residents, the perceived 'inefficiencies' of state-owned enterprises and general dissatisfaction with the supposed inflexibilities of centralised infrastructure planning have all supported policies of privatising and liberalising infrastructure markets.

But the crisis in the developmental state has also been substantially induced by national debt repayments and the strategies of the IMF to force ideologically driven, neoliberal 'structural adjustment' packages on to nation states. Such packages often 'give *no option*' but to privatise, liberalise and sometimes 'dismantle the public sector' as a whole (Hoogvelt, 1997, 169, original emphasis). The implication is that 'any hint of Keynesian notions of national economic management [is to be consigned] to the dustbin of history' (ibid.). As Western infrastructure firms have sought to acquire the infrastructure networks subject to privatisation, so global–local linkages and flows of capital, technology, infrastructure specialists and information have intensified further.

In post-communist states, finally, the highly standardised and extensive but often poor-quality infrastructure networks rolled out under the communist versions of the modern infrastructural ideal have often been decapitalised and rapidly decayed. In response, state-owned infrastructure monopolies have been widely privatised and transnational infrastructure firms have been invited in to lay new networks which often tend to 'bypass' the technically obsolescent legacies of the old systems (Enyedi, 1996; Berlage, 1997; Marcuse, 1996). In many cases small 'islands' of new and relatively functional telecommunications, water, power and road infrastructure networks, patched together to serve defensive enclaves or massive new development and consumption zones, are rising out of the decaying, comprehensively planned networks rolled out with such resolute determination during the communist era of centralised economic and urban planning (see Castells, 1998; Herrschel, 1998).

ECONOMIC INTEGRATION, URBAN COMPETITION, AND THE IMPERATIVES OF GLOBAL–LOCAL CONNECTIVITY

Our second point follows directly from the first. In virtually all contexts, it seems, the development agencies representing cities and regions are now struggling 'entrepreneurially' to develop the networked infrastructures that they think will lure foreign and tourist-related investment, and so position their cities favourably within divisions of labour which are being constructed across international borders with unprecedented precision (Hall and Hubbard, 1998; Knight and Gappert, 1989). Thus, single, monopolistic infrastructure grids start giving way to multiple, separate circuits of infrastructure which are customised to the needs of different (usually powerful) users and spaces. 'If there is a technical trend,' writes Paul Trenor, 'it is to division: separate infrastructures become more feasible' (1997).

Premium infrastructure networks and transnational connectivity

Increasingly, as we shall see in the rest of this book, such infrastructures straddle and interconnect urban, interurban and international scales, aligning with dominant vectors of global–local flow rather than the modern ideal of intra-urban connectivity. The economic development of spaces emerges as a 'war-like power struggle of deterritorialisation and reterritorialisation' within an internationalising capitalist system (Thrift, 1996a, 285).

As a result, infrastructure development increasingly centres on seamlessly interconnecting highly valued local spaces and global networks to support new vectors of flow and interaction between highly valued spaces and users locked into highly sophisticated international divisions of labour (Peck, 1996). As any examination of hub airports, logistics zones, global telecommunications grids and international energy connections will now demonstrate, seamless global–local technological and organisational connection of infrastructure networks is now the central emphasis, geared to the logistical and exchange demands of foreign direct investors, tourist spaces or socioeconomically affluent groups (Schiller, 1999a; Castells, 1996).

Through these processes, the global–local connections of cities and spaces are increasingly scrutinised by agents of global capital and other mobile investors, as they search to locate in areas with maximum infrastructural capabilities, lowest costs, and maximum flexibility and mobility potential. As an example, a frenzy of alliances, acquisitions and joint ventures in the fast-growing world of telecommunications and digital media is now under way (McGowan, 1999). These promise 'one-stop shop' service and distance-independent tariffs on a global scale for corporate clients in highly valued locations but effectively bypass less profitable or more marginal spaces and users (Schiller, 1999a; Noam and Wolfson, 1997; Mosco, 1999b).

'Not only is this new industrial and spatial scenario built on the assumption that long-distance, reliable transport and communications networks are implemented,' write Giannopoulos and Gillespie (1993, 51), 'but it rests on the idea that these networks have to be "integrated networks", both geographically and technologically.' Thus infrastructure networks can simultaneously be 'unbundled' locally whilst being integrated internationally. This fundamentally challenges the modern notion that a 'city' or 'nation' necessarily has territorial coherence in its own right, as a spatial container for economic activity which is somehow 'naturally' separate from surrounding spaces (Virilio, 1991).

The growing interconnections between national and urban economies, and the widespread crisis of confidence in Keynesian approaches to infrastructural development, have thus significantly undermined the notion that infrastructure networks should be publicly planned and laid out as single monopolies to add 'coherence' to urban and national territories. Spatially:

> networks which are confined to national [or urban] territories, whether for the movement of material or people, will be of limited use. Segmented markets need flexibility and responsiveness. New markets and competitive strategies require a highly flexible transport system, able to cope with frequent movements of small quantities, rather than with the predictable, less frequent, larger volumes of transported goods that characterised Fordism.
>
> (Giannopoulos and Gillespie, 1993, 46–51)

Customising networks for premium spaces

Development processes in developed, developing and post-communist cities are thus starting to emphasise the provision of customised, networked spaces within cities and regions. Frank Peck (1996) has shown that transnational corporations engaged in decisions over foreign direct investment now encourage public and private infrastructure developers and urban agencies to develop highways, ports, logistics centres and telecommunications, energy and water services that are specifically tailored to their requirements. There is a general trend 'towards the customisation of public infrastructure on behalf of private firms' (Peck, 1996, 36). Linked with this is the widespread 're-infrastructuring' of redundant spaces in the form of trade zones, enterprise zones, export processing areas, teleport spaces, logistics platforms, world trade centres, media zones, tourist enclaves and the like.

To Peck, the combination of profit-hungry private infrastructure firms and entrepreneurial city authorities provides an irresistible support towards focusing investment on fewer spaces and networks that most meet the perceived needs of global transnational capital (see Jessop, 1998). He writes that 'this pressure to comply with very precise [infrastructural] requirements means that the creation of customised space can be a vital factor in levering in new inward investment' (Peck, 1996, 329).

In this changing context, the strategic priorities of dominant economic players now emphasise flexibility, responsiveness, technical specialisation and, above all, the ability precisely to control and monitor processes of logistical flows across space and time (Hepworth and Ducatel, 1992). In an increasingly liberalised global market place for infrastructure services, the providers of infrastructure services are having to rethink their styles of planning and management to tie in with the demands of their most lucrative corporate customers (for 'one stop' corporate telecommunications services, transparent logistical management, cost-effective and uninterruptible energy and water services, and high-quality back-up and value-added service support) (Schiller, 1999a). In the process, infrastructure services become much more than the laying of the hardware within and between spaces; value added now centres on customising services to corporate needs on a seamless, multilocational basis. In these changing competitive conditions, infrastructure providers must either develop strategies to define and expand their markets or become obsolete, bypassed or taken over.

REMOVING NETWORKS FROM THE WELFARE STATE: THE DEVELOPMENT OF INFRASTRUCTURAL CONSUMERISM

Finally, the changing political economies of cities and infrastructure have often amounted to a decollectivisation of energy, water, waste, transport and telecommunications services, and the reduction of their status as quasi-public goods to be consumed by all, at similar, generalised, tariffs. Instead, infrastructure services are being remodelled and recommodified to be distributed within more or less regulated markets between liberalised, competing providers. 'Throughout Western countries, it seems now self-evident that the role of the state as the provider of a wide range of public services, rooted in the promise of dramatically evening up the life chances of individuals and populations, is coming to an end' (Leonard, 1997, 1).

As Saunders argues, it seems possible from current standpoints that 'collective consumption is proving to be not a permanent feature of advanced capitalism but a historically specific phenomenon' – a 'holding operation' between old and new forms of market provision (Saunders and Harris, 1994, 211). This is supporting the fragmentation of the production of infrastructural goods and services as firms struggle to engage with the instruments of market research and advertising, as well as the individualising capabilities of information technology, to carve lucrative niches for themselves within volatile contexts (Clarke and Bradford, 1998, 874).

SOCIAL LANDSCAPES OF CONNECTION: NEW EXTREMES OF INEQUALITY

Infrastructure networks, thus, can no longer be dismissed as immanent, universal and homo-geneous grids; as local public goods which can remain the arcane and technical preserve of the civil engineer. Market-based and consumerist logics are increasingly being imprinted on to such networks. The assumptions that underpinned the public, monopolistic provision of infrastructure services are increasingly being challenged. Across the advanced industrial world, utility infrastructures are now the focus of radical reregulation. Public, monopolistic models of regulation and ownership are being challenged by waves of privatisation and liberalisation. Generally, this 'means a loss of the redistributive, social role implied by such public monopolies' (Little, 1995, 9).

Such a shift is 'imposing an ethos of individual choice which belies the role of consumption in the systemic reproduction of capitalism' (Clarke and Bradford, 1998, 874). This infrastructural 'choice', however, often tends to be limited to certain social and spatial groups within the city. The ability to access competing providers is usually highly dependent on wealth, location, skills and how lucrative one is to serve.

In some parts of the city, then – perhaps those that used to pay above-cost rates for services so that cross-subsidies could support poorer districts – the splintering of networks that come with imposing a logic of consumer choice on previously public infrastructures is likely to lead to considerable variety, choice and improved service. Such groups will be actively seduced into 'premium' markets for the most capable road, energy, telecommunications and water networks.

In poorer parts of cities, however, large parts of the population seem likely to be forced to remain with incumbent monopolies, as they are not lucrative enough to attract competition and the seductive attention of new, risk-averse, market entrants. Such groups and areas are likely to remain highly vulnerable to the efforts of states to shift from the mass, collective organisation of social infrastructure to a dwindling 'safety net' covering only the needs of the most desperate for fuel, communications and water.

The danger, of course, is that the consumerism and individualism of the new debates on social access to infrastructure will undermine the position of the poor, who often tended to benefit most from universal service obligations and cross-subsidies inherent in the approaches of the modern infrastructural ideal. Markets for advanced infrastructural services seem likely to fail such people, possibly even excluding them from access to very basic and essential infrastructures in the process. Sophie Body-Gendrot wonders whether, even in Western Europe, where public service principles and welfare states have been most comprehensively elaborated, the restructuring of welfare states means that:

we are now observing the exhaustion of a model for state-provided protection against hardship. . . . National societies seem to be disarticulating in a strange movement of demodernisation. . . . In an era of globalization, the processes of disintegration, disempowerment, social invalidation, marginalization – whatever terms one wishes to use – fracture post-industrial cities . . . into a myriad of patterns.

(2000, xx)

THE COLLAPSE OF THE COMPREHENSIVE IDEAL IN URBAN PLANNING

The concept city is decaying.

(De Certeau, 1984, 95)

Which brings us neatly on to the third set of forces which have undermined the modern infrastructural ideal: the related collapse of notions of comprehensive and 'rational' urban planning first built up by Haussmann over a century before. The technocratic and comprehensive styles of urban planning most closely allied with the rolling out of the modern infrastructural ideal have also found it difficult to survive the shift to an increasingly globalised political economy driven by liberalised flows of capital, technology and information. It has also lost much of its legitimacy in Western nations as a result of being undermined by powerful 'postmodern' social and cultural critiques.

As a result, urban planning now tends to centre on projects rather than comprehensive and strategic plans; on getting other agencies to deliver required urban services or infrastructures; and on pragmatic attempts to address perceived local problems rather than utopian or visionary frameworks for re-engineering metropolitan regions according to idealised blueprints or desired urban forms. This shift, along with the withdrawal through privatisation and liberalisation of many infrastructure networks from even the peripheral orbit of public sector planning, has significantly contributed to the onset of splintered models of infrastructural development. Overall, as David Harvey notes, there has been a decisive break 'with large-scale, metropolitan-wide, technologically rational and efficient urban *plans*' (1989, 66, original emphasis).

CRITIQUES OF INTEGRATED URBAN AND INFRASTRUCTURE PLANNING

During the period of the urban infrastructure crises, and the rapid restructuring of the public institutions forged to implement the modern infrastructural ideal, a range of powerful critiques effectively destroyed the idea of comprehensive urban planning. Increasingly vociferous economic sectors and firms argued that such plans were inflexible, unwieldy, and failed to deliver infrastructure networks able to meet their increasingly sophisticated locational and technological demands (Fillion, 1996).

Many planners, in turn, have themselves developed a 'growing skepticism towards large-scale infrastructure projects' and urban 'renewal' schemes, especially in the West (Fillion, 1996, 1640). Increasing social resistance to major highway, port, rail, road, airport and construction projects has faced them. In a smaller-scale version of the collapse of the modern nation state, planners, and their local state employers, found themselves increasingly unable actually to *control* or orchestrate the development of their territories in any meaningful manner. Leonie Sandercock writes that now 'the local state is less comfortable exerting control over its territory in terms of who is investing and what kinds of investments are being made in local development' (1998a, 28).

MODERN PLANNING'S PARADIGM CRISES OF INFLEXIBILITY

Modern urban planning also tended to lose confidence in its core notions of 'progress', technical rationality and benefits for all as environmental and social movements lambasted its underlying assumptions. The static, orderly models of cities at the root of the modern urban ideal found it impossible to cope with the turmoil of social, economic and cultural change between the 1960s and 1980s. As Polo puts it:

the predominance of flows, deformations and dimensional and dynamic heterogeneity within the urban structure of advanced capitalism puts into question the static spatiality, homogeneity and constancy of urban form in time that once characterised urban structures and planning methods.

(1994, 29)

Sandercock points out that 'there are processes of socio-cultural change that have been reshaping cities and regions over the past 20 years in ways not dreamed of in the Chicago model of the rational, orderly, homogeneous city' (1998a, 27). The arrogance, unrecognised bias and relentless homogeneity of the planned city forms dreamt up by modernist planners were especially lambasted by a wide range of social movements. As a result, as Fillion argues, 'the postmodern fragmentation of values and the proliferation of conflicting interest groups' tend to 'undermine the political consensus required to carry out large-scale planning projects' (1996, 1640).

Urban planning has thus lost its ability to conceive of a 'public interest' against which to justify the on-going reorganisation and rationalisation of urban space through modern infrastructure and urban planning (Gandy, 1998). As Paul Knox puts it, in many cases urban planning has tended to become:

fragmented, pragmatically tuned to economic and political constraints and oriented toward stability rather than being committed to change through comprehensive plans. . . . It became increasingly geared to the needs of producers and the wants of consumers and less concerned with overarching notions of rationality or criteria of public good. The outcome has been a disorganised approach that has led to a collage of highly differentiated spaces and settings.

(1993b, 12)

The traditional tools of modern urban planning – development plans and zoning ordinances – have thus in many cases become more and more discredited. Modern planning's analytical techniques – gravity modelling, regression analyses, cluster analyses, cost–benefit analyses – have been widely criticised for providing the obfuscating jargon through which essentially political decisions could be represented as somehow 'technical' and value-free (see Box 3.2).

BOX 3.2 PARADIGM CHALLENGES TO INFRASTRUCTURE PLANNING: THE CASE OF URBAN TRANSPORT

As a result of wider shifts in urban planning and its social context, it was becoming increasingly untenable by the late 1970s to maintain that infrastructure networks were simply technical, engineered systems existing somehow separate from society which operate to 'impact' on society. The methodological and analytical tools underlying urban infrastructure planning were similarly under question. The collapse of the notion that civil engineers could roll out integrated infrastructures rationally to meet perceived needs, whilst abstracted from the social and political worlds of their city, has been especially important. But, as Chatzis suggests, such processes have been profoundly destabilising to the professional cultures that rested on the axioms of rationality. 'Technicians, deprived of the objectivity of the standardised formula,' he writes, 'often find themselves in a difficult situation concerning the evaluation and justification of their actions ' (1992, 12).

Whilst utilities and telecommunications operators are having to explore new ways of planning infrastructural development and investment, the crisis is most visible in transport. For example, the whole subfield of regional transport modelling and regional science, built up since the 1950s and 1960s to apply quantitative methods to understanding spatial interaction within and between cities, is undergoing severe criticism. To Barney Warf, for example, the spatial analysis models long used to support transport planning are 'a profoundly ahistorical' approach which resort to 'static analyses that reify time' (1995, 187).

PROBLEMS WITH MODERN TRANSPORT PLANNING

In their efforts to uncover transcendental 'laws' of aggregate transport behaviour, such technocratic and essentialist models have tended to squeeze the whole gamut of human life into crude, quantitative, deterministic, mechanistic equations based on the notion that the social world is analogous to Newton's mechanistic 'billiard ball' universe. Such models as the gravity model – the basis of countless transport and infrastructure plans in the postwar years – followed Newton in treating space and time as absolute, essential objects. It reduced the complex social world to overarching geometric and morphological laws. And it relied on essentialist technological determinism of the simplest kind in extrapolating and forecasting into the future. In effect, 'history became little more than a static objectified data source, and all that marched through it were abstracted spatial processes purged of social meaning' (Duncan, 1979, 1). At the root of the problem was the false assumption that mechanistic cause-and-effect models of social behaviour were possible (when Physics had actually ditched such modes of the behaviour of subatomic particles forty years before in favour of probabilistic, indeterminate and relativistic thinking).

To Warf, whilst there are important shifts towards complexity theories and probabilistic modelling within such approaches, regional science, and the whole edifice of econometric infrastructure modelling based on it, 'remains enraptured with that desolate, asocial, self-centred individual, *homo economicus*. Even behavioural approaches have essentially replaced deterministic approaches with stochastic and probabilistic ones' (Warf, 1995, 190).

The field of transport planning is an especially powerful example of the collapse of the notion that infrastructure is simply a technical, engineered system. It shows especially clearly how such notions effectively collapse in the maelstrom of the crisis of technical rationality, the undermining of mechanistic thinking, the highly embarrassing modelling failures of the past, and the broader context of socio-technological and political–economic change discussed above. Tim Marshall (1997) points out that transport planners have virtually given up forecasting the future, not only because past efforts were so embarrassingly inaccurate, but because of transport 'planners' sense of being adrift in a confusing and uncontrollable flux' (31). He cites recent attempts in the United Kingdom at foresighting (rather than forecasting) technological shifts in transport which confronted the essential problem that 'Technology Foresight is something of a contradiction in terms, because the future development of the transport sector is determined as much by economic, social, political and environmental factors as it is by the availability of technology' (Technology Foresight Panel on Transport, 1995, 59, quoted in Marshall).

SIGNS OF A PARADIGM SHIFT

Guy and Marvin (1999) point out that this paradigm shift involves a shift from simply promulgating extra road space to offering integrated packages of private and public transport solutions and land use changes; from using positivist analytical models like the gravity model towards more social concepts of real social accessibility; and towards a managerial engagement with users as diverse groups of social beings rather than a hands-off treatment of people as homogeneous, atomistic, economic men, behaving in aggregate according to mathematical formulas.

THE LIMITATIONS OF PLANNING'S BINARY DUALISMS

Moreover, modern urban planning's attempt to establish order within the urban fabric was exposed as the imposition of static geometries based on binary distinctions: centre and periphery, core and fringe, inside and outside – categories which are 'no longer adequate to describe the urban conglomerate' of today in all its complexity and dynamism (Sandercock, 1998b). In fact, 'in the new city formal principles of composition miss their target. Its morphology instead unfolds from a system of relations between different, sometimes contradictory forces, no longer as an absolute but in reference to other structures' (Angélil and Klingmann, 1999, 24).

The engineering-dominated ethos of the modern city has also found it especially hard to accommodate public political and social challenges for more democratic ways of organising urban infrastructure development. As Matthew Gandy suggests, there emerged a:

conflict between the centralised engineering dominated ethos behind infrastructural development and growing demands for greater public participation in urban policy making. Urban planning faced the disintegration of the kind of putative 'public interest' which had sustained the kind of urban renewal through large scale investment in infrastructure. Planners themselves increasingly recognised that 'the ideal of master planning' was illusionary and began exploring ways of bolstering legitimacy through wider public consultation.

(1998, 13)

SCRUTINISING THE ACHIEVEMENTS OF COMPREHENSIVE URBAN INFRASTRUCTURE PLANNING

It is important not to lose sight of [modern planning's] utopian ideals but . . . these were achieved by less than democratic means. . . . In effect – and in isolation – it has often proven elitist and alienating.

(Ley and Mills, 1993, 265–6)

Not only were the axioms of modernist planning systematically dismantled; its achievements came under increasingly close scrutiny, too. To King, this scrutiny of the effects of the modern infrastructure ideal, and of modern urban planning more widely, revealed modern city planning to be little less than 'a supreme delusion' (1996, 1). Rather than 'unify' and 'bind' cities by distributing benefits to all – its alleged purpose – it became increasingly clear that practices of modern city planning had tended, in practice, bureaucratically to perpetuate inequalities across space. They had often worked, moreover, 'to the disadvantage of women and of all those able to be labelled as "others" – ethnic minorities, the handicapped, the unsmart' (King, 1996, 3).

Rather than support urban 'coherence', 'order' and 'cohesion', as argued in so many modern infrastructure plans since the days of Haussmann, it was increasingly realised that such planning had, more often than not, supported social turmoil through the on-going fracturing and restructuring of urban space, especially within the spaces of the weak or marginal (see Box 3.3). From the 1960s onwards, books like *The Death and Life of Great American Cities* (Jacobs, 1961) and *The Evangelistic Bureaucrat* (Davies, 1972) helped to stimulate enormous popular protest against the whole edifice of modernist planning and architecture.

BOX 3.3 THE INTERNAL CONTRADICTIONS OF COMPREHENSIVE URBAN HIGHWAY PROGRAMMES: URBAN 'COHERENCE' VERSUS URBAN 'FRAGMENTATION'

The urban highway, that *Leitmotif* of virtually all modern urban plans between the 1920s and 1960s, was particularly associated with promising urban 'cohesion' whilst delivering fracturing and fragmentation (Dear, 1999, 110). Multilane highways, cutting through the urban fabric to 'integrate' urban regions through tunnels, cuttings and elevated sections, were 'seen as a marvel of the modern metropolis' (Gandy, 1998, 1). Designed as totalising systems for smoothly integrating their limited entry points, major highways epitomised the standardised, centrally imposed design of segregated transport systems that was so central to modernist planning and traffic engineering. As with the mass-produced housing of the postwar era, the ethos was 'the more neutral, uniform, and quantifiable, the more bankable' (Easterling, 1999a, 3).

'YOU HAVE TO HACK YOUR WAY THROUGH WITH A MEAT AXE': THE CASE OF ROBERT MOSES IN NEW YORK

Marshall Berman (1983) famously discussed perhaps the best-known example of the use of highway networks to force an industrial metropolis into some form of 'modern' integration through highway construction: Robert Moses' plans in New York. Over a period of about thirty years, Moses used bond issues to finance an ever-expanding web of public highways and works projects across New York, a programme linked with the federal New Deal and highway programmes (see McShane, 1988). Aimed at maximising regional markets, productivity and ease of circulation, Moses wanted the highways to unify the city, creating an 'integrated car-oriented urban form' (Gandy, 1998, 6). The resulting highways ploughed through disadvantaged neighbourhoods like Brooklyn in the process. But Moses' modernising zeal was unabashed. He declared, in his defence, that 'when you operate in an overbuilt metropolis you have to hack your way through with a meat axe' (Robert Moses, quoted in Berman, 1983, 307).

Echoing Haussmann, Moses wanted to 'create a system in perpetual motion' (ibid., 306). He imagined a 'utopian new city of unified flow whose lifeblood was the automobile'. City spaces were thus 'conceived principally as obstructions to the flow of traffic' (Berman, 1982, 307). In that period, Berman remembers, 'to oppose his bridges, tunnels, expressways, housing developments, power dams, stadia, cultural centers, was – or so it seemed – to oppose modernity itself' (Berman, 1982, 293).

THE SOCIAL BIASES OF HIGHWAY DESIGN

But the parkways were, according to Langdon Winner (1980), carefully configured to meet the needs of increasingly affluent suburban commuters whilst excluding the poor and black inner city populations (see Bayor, 1988). In one of the most infamous examples of building social bias into technology, Winner alleges, the Long Island State Park Commission engineers designed the parkways and highway bridges on Long Island to be only 9 ft at the kerb – 2 ft lower than the height of buses – thus guaranteeing that the roads would remain permanently for car use only (McShane, 1988, 81; although see Joerges, 1999a, for a refutation of this relatively simple interpretation).

One of the professed aims of a co-ordinated, national system of urban highways in the United States was to bind cities and systems of cities into modern coherent wholes. After all, such strategies drew their inspiration from Haussman's regularisation plans in Paris in the nineteenth century. In effect, though, they contributed not only to the enormous urban sprawl of the 1980s and 1990s, but to a retreat from the notion of the open, interconnected city (Pope, 1996). For they have served to fragment the traditional North American cityscape based on the urban grid, leading to a series of closed, hierarchical 'ladder'-style urban highways rather than an open urban street system.

FREEWAYS AS A 'REVERSAL' OF GRIDDED SPACE

'With the introduction of the freeway,' writes Alexander Pope, 'the continuities of gridded space are thrown into a fantastic reversal. [It] eliminates choice by enforcing a strict hierarchical movement along a primary route of transportation, [so] dramatically coarsening the urban fabric' (1996, 109). Pope believes that such 'ladders' have supported a broader trend towards xenophobic enclaves. Through such processes, 'historically open urban centres degenerate into tourist sites. Main Street becomes a festival marketplace, the campus becomes a theme park, traditional suburbs become xenophobic enclaves, and one is (still) left wondering how it ever happened' (ibid., 96). To Pope, serendipitous interconnections and relationships – the essence of the urban – are severely limited by the new 'laddered' urban landscapes. The erosion of grids has also supported the centripetal shift of cities to ever wider regional fields where closed-off places are increasingly defined by rapid interconnections elsewhere by limited-access highways.

THE CRISIS IN PLANNING AND THE RETREAT
FROM THE UNITARY CITY IDEAL

Most especially, it was increasingly argued that the ideals of modern, unitary city planning had required the 'total subordination of the individual to the collective industrial machine' (Ravetz, 1980, 345). In its use of single, idealised representations of urban space, modern urban planning had thus wielded crude social power based on arrogant, undifferentiated notions of identity and need that were little more than laughable in the context of the fragmenting identity politics of the postmodern city.

Urban planning is, in short, facing a 'paradigm crisis' as its classical foundations are exposed as anachronistic, dangerous and intellectually spurious (Sandercock, 1998a, b). Ezquiga notes that 'the nature of the planning crisis is twofold: the bankruptcy of its epistemological fundamentals as a discipline, but also a crisis in the culture that has considered the urban plan as the sole, holistic expression of the public interest' (1998, 7).

The idea of the comprehensive urban plan, as guarantor of some single, orderly 'progress' offering 'benefits for all' through the laying out of urban activities and their connective infrastructures, has been the major casualty here. The classic urban planning tradition of equating order with equilibrium and disorder with disequilibrium can have little place within the volatile and complex dynamics of the postmodern metropolis. Such approaches have been shown to be reductionist and naively functionalist; in their endless search for a perfect equilibrium through controlling 'urban morphology and building typologies, the passage of time stands still, dead, petrified' (Solà-Morales, 1998). Miles argues that:

the urban plan is a representation of space which enables idealised conceptions of the city. From it are derived the methodologies of planning and design which depend on the reduction of realities to geometries. And, through the representation of space in urban planning, users become an undifferentiated public ascribed the same disempowered role as women, slaves and strangers in classical Athens.

(1997, 131)

ACCEPTING CITIES AS 'ARCHIPELAGOES
OF ENCLAVES'

In a supreme irony, urban planning faced the realisation that it had often had precisely the reverse effects to those it aspired to and coveted (at least rhetorically). As Aksoy and Robins (1997, 26) point out, in the case of Istanbul, in assuming that orderly plans could make cities orderly, and in imagining that urban space could ever be truly unitary and coherent, urban planning had often directly supported the development of enclaves and the social fragmentation of urban space.

MODERNIST PLANNING INSTRUMENTS AND THE TURN TO URBAN FORTIFICATION

In practice, then, the instruments of modernist planning, ostensibly developed to support coherence and urban egalitarianism, have often been appropriated to support fragmentation and social exclusion. This has especially been the case as economic change, polarising income levels, motorisation for the well-off, the privatisation of public services, rising crime levels and the growth of corporate and consumer 'mega-projects' have rewrought the fabric of metropolitan life (see Olds, 1995). As Teresa Caldeira argues, from the point of view of Brazil:

ironically, the instruments of modernist planning, with little adaptation, become perfect instruments to produce inequality. . . . Streets only for vehicular traffic, the absence of sidewalks, the enclosure and internalisation of shopping areas, and spatial voids isolating large sculptural buildings and rich residential areas are great instruments of generating and maintaining social separations. . . . Contemporary fortified enclaves use basically modernist instruments of planning with some notable adaptations.

(1996, 317)

ACCEPTING PLANNING'S 'OTHERS'

Modern urban planning, it was also increasingly realised, had neglected many voices, in its 'mainstream' depiction of the modernist planner as an omniscient, benevolent (inevitably male) 'hero', taming the wild chaos of the disorderly metropolis. The views of women, minority ethnic groups, indigenous people, disabled people, gay men and women, older people and children were largely ignored (Sandercock, 1998b). Modern urban planning had often therefore ignored the essentially patriarchal, racist, disablist, socially divisive and colonialist assumptions woven into its master plans and utopian visions, being even less concerned when such assumptions were imprinted on to cities and city life (Sandercock, 1998b; Weisman, 1994). Urban highway networks, for example, which purported to deliver 'access for all' and add 'coherence' to cities, were often found to destroy communities, undermine interactions in places, and worsen social and gender unevenness in access to transport (see Box 3.3).

The result of this coming together of social critiques and uncomfortable self-realisations was a widening sense of the failure of comprehensive, modern plans for cities and networked urban infrastructure. This, in turn, resulted in the growing politicisation of urban space, a consequence of the challenge that has been posed to the modernist vision (Aksoy and Robins, 1997, 33). As King suggests, as the twentieth century progressed, 'modernity [increasingly] present[ed] itself as fragmentation itself, and the nineteenth century was the age when its opposite, in the unitary dreams of the Enlightenment, finally disintegrate[d]' (1996, 44).

Very quickly the linkage of large-scale infrastructure projects with urban 'progress' and improvement was replaced by representations that portrayed modern infrastructure grids, especially highway networks, as destroyers of valuable social and urban environments. 'Aspects of infrastructural investment which had previously been conceived as integral to urban revitalization,' suggests Gandy, 'had now become directly implicated in post-war urban decline and the destruction of city life' (1998, 13).

All these intertwined realisations forced urban planning to retreat systematically from the notion of comprehensive urban and infrastructural planning, effectively ditching the idea that the development of cities could be somehow orchestrated and shaped as a whole. Virtually all planning concepts today agree, at least implicitly, that 'the primary matter of importance is no longer an integral approach, but the cheerful acceptance of regions as an archipelago of enclaves' (Bosma and Hellinga, 1998, 16).

Increasingly, then, planners are forced to accept that their cities are 'collages of fragmented spaces' defined by multiple identities, aspirations, life worlds and socioeconomic and time–space circuits (Fillion, 1996, 1640). Imposing some simplistic notion of order or representation on such places is not only a power-laden act, but it is an arrogant act which privileges the 'technical' knowledge of the 'expert' over all other forms of knowledge, experience and opinion (Healey, 1997). And, suggests David Harvey, 'since the metropolis is impossible to command except in bits and pieces, urban *design* (and note that postmodernists design rather than plan) simply aims to be sensitive to vernacular traditions, local histories, particular wants, needs, and fancies' (1989, 66, original emphasis).

URBAN PLANNING, ENCLAVE CONSTRUCTION, AND GLOBAL ECONOMIC INTEGRATION

Key works like Oscar Newman's *Defensible Space* (1972) helped further to encourage planners to reject large-scale integrated modernist city schemes for smaller-scale, inward-looking projects designed to address people's 'need' for enclosed personal spaces. 'Defending our own turf' increasingly became sanctioned by certain strands of planning theory (Luymes, 1997, 201). Waterhouse criticises the role played by 'a new generation of planners in extending, then partitioning, the urban realm' (1996, 311). More positively, though, many planning agencies sought to undermine the modern ideal's insistence on rigidly enforced technocratic standards for street layouts, allowing a whole range of shared streets and traffic calming measures to be introduced at the grass-roots level (Southworth and Ben-Joseph, 1997).

As part of a wider professional paradigm crisis, however, the worry is that formal planning processes, whilst not entirely marginalised, seem increasingly driven by the entrepreneurial imperatives of making specific spaces 'competitive' within the metropolis, where 'competitiveness' needs to be seen as 'an essentially contested, inherently relational and politically controversial concept' (Jessop, 1998, 81). The result, suggests Bob Beauregard, is that, in the United States at least, 'at present, the physical city exists within planning as a series of unconnected fragments rather than as a practical and theoretical synthesis of planning thought and action' (1989, 382). Leonie Sandercock even worries that there is a serious risk that 'the profession of planning is becoming increasingly irrelevant except in its role of facilitating global economic integration' (1998a, 2).

Certainly, as neoliberal agendas drive changes in urban governance and policy in a growing range of cities, the 'collages of fragmented spaces' that make up many cities are, in turn, becoming subject to widening arrays of urban governance agencies, special economic development and enterprise zones, 'partnership' organisations and single-purpose economic development bodies (see Foster, 1996; Nunn and Schoedel, 1997). Many Western cities now exhibit a myriad of small, special-purpose zones, from cultural, heritage or leisure and sports districts to Business Improvement Districts, Enterprise Zones, shopping malls, financial districts, cultural quarters, affluent enclaves, etc. Such spaces increasingly tend to be developed, organised and managed by property-led development bodies, urban marketing organisations and special infrastructure developers (Boyer, 1996).

Public and private sector planners supporting the construction of the new edifices of urban development thus articulate a very different strategy and set of ideals from those which aspired to use modern plans to bring order and coherence to a whole city space. The logic is now for planners to fight for the best possible networked infrastructures for their specialised district, in partnership with (often privatised and internationalised) network operators, rather than striving to orchestrate how networks roll out through the city as a whole (Nunn and Schoedel, 1997). As Sandercock argues, 'to fast track many of these mega projects governments have short circuited established planning processes and removed these developments from public scrutiny and democratic politics, creating such entities as "special exemptions" and the like' (1998a, 28). As a result:

the planners working for the state on the kinds of mega-development projects that have become common to this round of capital accumulation (the Docklands in London, Darling Harbour in Sydney, the casino and Docklands projects in Melbourne, New York's World Trade Center, Mitterand's *Grands Projets* in Paris . . .) have a hard time convincing anyone of their 'critical distance' or objectivity.

(Ibid.)

Thus not only is the physical and technical fabric of urban infrastructure splintering in many urban regions; in many cities, the fabric of urban governance and planning is, too (Fillion, 1996, 1640). Increasingly, the nexus between fragmented urban governance and splintering infrastructure networks is strengthening as special-purpose governance agencies try to get actively involved in customising networked infrastructures – highways, telecommunications, energy, water and waste – to the precise needs of the targeted users of the space of the city they are responsible for developing.

Within high-value spaces colonised by users that are intensively international and even global in their operations – the global financial and research and development clusters, the logistics hubs around airports, the advanced port districts, the global media spaces – attention now centres powerfully on equipping city districts with the best possible infrastructure to link the local into global matrices of flow through powerful connections elsewhere (Castells, 1996). The emphasis is on developing sophisticated new infrastructures – teleports, 'smart' ports, global airport hubs, special rail links to downtown cores, direct links to transoceanic fibre networks – that effectively bypass the relatively homogeneous street, energy, transport and communications grids that are the legacy of the modern infrastructural ideal.

Inevitably, with the growing emphasis on global–local rather than intra-urban connections, the notion that the level of the city *per se* is the most appropriate scale at which to manage and articulate infrastructure is transcended. Contemporary planning practices further support the notion of buildings and zones as terminals on grids of global connection – hubs on international networks which effectively radiate spokes to other spaces through time–space 'tunnel effects' between them (Graham and Marvin, 1996, 58–9). The idea that cities necessarily work to enclose their contents, providing infrastructural coherence in the process, becomes yet further undermined (Virilio, 1991).

THE RESILIENCE OF THE URBAN AS A POLITICAL ARENA

And yet the idea of the unified city as a political jurisdiction has often managed to maintain a powerful discursive hold on urban politics (see, for example, Bagnasco and Le Galès, 2000). Despite the fragmenting logic of global–local exchanges, in certain cases, cities remain powerful, perhaps *increasingly* powerful, political actors, and urban political actions strive to address the ambivalent roles of urban places within contemporary social change (Body-Gendrot, 2000). Complex political coalitions and 'regimes' articulate how national and international political processes relate to the particular development of individual cities (see Jonas and Wilson, 1999; Hall and Hubbard, 1998). Even with liberalisation and privatisation, complex intergovernmental relations, deep institutional histories and diverse regulatory approaches shape how the regulation and development of networked infrastructures is constructed locally (Lorrain, 2000; Lorrain and Stoker, 1997). This ensures a wide variety of particular ways in which the political action of city agencies and politics intersects with that of nation states and international governance bodies, to shape the reconfiguration of infrastructure networks.

Cities therefore remain as places where significant and potent power exists to plan and act to address the complex and ambivalent position of place within globalising vectors of flow (Amin and Graham, 1998a; Healey, 1997; Bagnasco and Le Galès, 2000). What is needed, however, as we suggest later in this book, are new ways of thinking about networked urban infrastructure, and new ways of imagining the territorial politics of networked urbanism, to best address contemporary challenges.

NEW URBAN LANDSCAPES: PHYSICAL DECENTRALISATION, MOTORISATION AND THE POLYNUCLEATED CITY

> The city's reign over our senses, our moods, our very ways of being is outmoded. The suburban metropolis has superseded the city. . . . The city's long shadow fades in the dappled light of the suburban metropolis.
>
> (Lerup, 2000, 85)

Discussions by urban planners and urbanists about the changing nature of urban landscapes and development processes lead us to the fourth set of processes that have undermined the modern infrastructural ideal: the physical spread of cities and the widespread shift from traditional core-dominated cities to polycentric and extended urban regions. For global trends towards urban decentralisation and the growth of the urban 'periphery' seem further to undermine the whole project of using integrated infrastructure networks coherently to bind cities as a whole (Woodroffe *et al.*, 1994; Keil, 1994).

URBAN INFRASTRUCTURE AND THE RISE OF THE URBAN PERIPHERY

> The future is peri-urban. We must stop considering the hinterland as an indescribable horror, as an illegitimate and residual part of the city.
>
> (Guido Martinotti, 1997, cited in Foot, 2000, 7)

There is little doubt that current urban trends, based on the growing mediation of urban life by highly capable infrastructure networks, are 'contributing to the mono-centric city eroding, fragmenting, and metamorphosing into a poly-centric metropolis' (Woodroffe *et al.*, 1994, 6). Such a transition exposes strange urban landscapes where the marginal can be central; centrality can be on the urban margin; and the 'urban' expands far into spaces previously considered as 'countryside'.

Urban peripheries and the 'liquefaction' of urban structure

Many cityscapes now reveal what Deyan Sudjic has called a 'single urban soup' (1995, 30). Within them complex patchworks of growth and decline, concentration and decentralisation, poverty and extreme wealth are juxtaposed. Whilst downtowns may maintain their dominance of some high-level service functions, back offices, corporate plazas, research and development and university campuses, malls, airport and logistics zones, and retail, leisure and residential spaces spread further and further around the metropolitan core. In many cities, complex social and cultural 'turfs' driven by international migration, add further levels of complexity and polynucleation. 'The contemporary city, exposed to the instability of late-capitalist production, cannot maintain the rigidities of an organic structure that articulates urban events within a global structure' (Polo, 1994).

Instead of the ordered, hierarchical and cohesive structure of the modern city (always an oversimplification), we increasingly encounter what Polo terms the '"liquefaction" of the urban structure' (1994, 26). This unleashes 'a discontinuous, unarticulated, urban growth' of polycentric, intensively (but highly unevenly) networked urban regions (ibid.). In many sprawling megalopolises, in both developed and developing nations, Bernard Tschumi observes an apparently 'scaleless juxtaposition of highways, shopping centres, high-rise buildings and small houses' (1996, 41). In the 'generic' landscapes of the spreading urban periphery, car-oriented buildings, no longer defined by systems of walls, are increasingly 'lone objects', 'one element within a rhythmic succession of space and matter, voids and solids' (Neumeyer, 1990, 19). In the process of peripheralisation, Woodroffe *et al.* argue that the modern 'binary metropolis [of core and periphery] explodes into a regional carpet' of fragmented communities, zones and spaces. Each is an enclave in its own right with its own boundaries and edge conditions (1994, 8).

THE CHALLENGE TO CLASSICAL NOTIONS OF THE CITY IN URBAN GEOGRAPHY AND URBANISM

In short, the idealised structures of classical urbanism and urban geography – always dramatic oversimplifications – are increasingly at odds with the forms and landscapes of most contemporary cities. 'The textbook geometry of sectors and zones has become increasingly difficult to discern in the landscapes, social economies, and bid-rent patters of cities' (Knox, 1993b, 1). To Richard Skeates the growth of megacities and transnational urban corridors means that the very distinctiveness of a place called the 'city' is now threatened. He believes that:

we can no longer use the term city in the way it has been used to describe an entity which, however big and bloated, is still recognisable as a limited and bounded structure which occupies a specific space. In its place we are left with the urban: neither city in the classical sense of the word, nor country, but an all-devouring monster that is engulfing both city and country and in so doing effectively collapsing the old distinction.

(1997, 6; see Castells, 1999c)

Overall, then, the expanding periphery seems increasingly to drive the development dynamics of many metropolitan regions. In the United States, for example, suburbs accounted for 60–85 per cent of new construction, new investment and new jobs in the last thirty years of the twentieth century. Suburban office stock rose 300 per cent during the 1980s, accounting for over 70 per cent of commercial office space in cities like Atlanta, Dallas and Detroit (Dunham-Jones, 1999, 3). Roger Keil asks, 'Have we reached the era of the outer city, and does the real urbanism of the waning century really happen on the edge?' (1994, 131).

Urban planning is poorly equipped to coordinate or control development across the regionally extending and polycentric metropolis. The architects Koolhaas and Mau (1994) famously argued that planners and urbanists, with their obsession with tidiness and order within the confines of the traditional city, were, in effect, doing little but 'making sandcastles'. But, with the traditional city swamped by the polycentric urban region, urbanists and planners can now do little but 'swim in the sea that swept them away'.

ECLIPSING THE URBAN CORE: GROWTH BEYOND THE LIMITS OF THE MODERN IDEAL

The spectacular growth of urban peripheries tends geographically to eclipse or even isolate the networked urban cores that were the legacy of the modern infrastructural ideal (Jackson, 1985). Such legacies, whilst often geographically far from the new centres of economic and demographic growth around urban peripheries, tend in any case to be physically or technologically obsolescent, requiring significant and costly retrofitting. To the municipal jurisdictions in old urban cores, securing the financial and technological expertise to update infrastructure, whilst also facing the fiscal and social crises surrounding deindustrialisation and social polarisation, poses enormous challenges.

In the United States, for example, the physical expansion and economic restructuring of US cities has meant that much of their infrastructural inheritance is technologically obsolescent and physically decaying – a key element of the broader urban infrastructure crisis discussed above (Lewis, 1997). Simply put, the legacies of modern infrastructure planning failed to support the changing geographies and economies of the nation's cities. The Port Authority of New York and New Jersey identified in a 1979 survey that:

Much of the infrastructure in our nation's larger cities was put in place during the nineteenth and early part of the twentieth century. The average life expectancy of these systems was considered to be somewhere between fifty and seventy-five years. Upgrading and modernisation of these systems probably should have started during the 1930s, but the nation was in the midst of a great depression.

(Quoted in Perry, 1995, 18)

THE 'TERRITORIAL ADAPTER' AND THE 'ROAD PLACE': MOTORISATION AND THE URBAN EFFECTS OF CAR CULTURE

Traffic circulation is the organization of universal isolation.

(Kotamyi and Vaneigem, 1961, 17)

It is a profoundly modern idea that we can enter a flow, be carried along with it, and exit again effortlessly, unscathed.

(Mau, 1999, 204)

Of course, the extension and growth of urban peripheries is also intimately bound up with the mass diffusion of the automobile and the increasing dominance of car culture within virtually all contemporary urban contexts. This process itself has been intimately linked with the construction of a whole system of supportive infrastructure, from highways to service stations, to drive-through fast food centres and out-of-town malls and auto-access leisure and retail complexes (Lewis, 1997; Dupuy, 1995).

The landscapes of many contemporary cities have been powerfully shaped by the standardised laying out of circulation and storage systems for automobiles, under the powerful supervision of professional corps of traffic engineers. In the United States, for example, most cities now devote over half their entire land area to the car; in extreme cases of automobile dependence like Los Angeles, the fraction reaches almost two-thirds (Southworth and Ben-Joseph, 1997, 5). 'Endemic to sprawl, the square footage required for the parking lot typically matches or even exceeds that required for the building. . . . Employees' office cubicles are smaller . . . than the space allotted for their car' (Dunham-Jones, 1999, 6).

In many cities virtually all urban uses are now constructed in articulation to the dominant and space-hungry technological systems that surround the car. In a sense, cars, and the enormously complex sociotechnical 'hybrid' infrastructures that support them, work as 'territorial adapters' for the decentralised, polynucleated metropolis (Dupuy, 1995; see Urry, 1999). Along with plug sockets, mobile phones, Internet terminals, and television transmissions, cars work to adapt the 'traveller' to the multiplying and multiscalar territories and spaces of the extending, global, urban world. They do this through a widening and interconnected array of roads of various sizes that were often laid out according to the strict formulas and protocols of traffic engineering (see Easterling, 1999a; Southworth and Ben-Joseph, 1997; Urry, 1999).

CAR CULTURE AS 'COERCED FLEXIBILITY'

Thirty-four man-years are spent per day commuting on the freeways of Houston. Yet the ride heals, soothes, and eases the jump cuts between home and work, between nature and culture, between byways and freeway, between his and hers.

(*Sic.*; Lerup, 2000, 19)

On the one hand, then, as John Urry writes, 'automobility is a source of freedom, the "freedom of the road". Its flexibility enables the car-driver to travel at speed, at any time in any direction, along the complex road systems of Western societies that link together most houses, workplaces and leisure sites' (1999, 12). Automobiles and their associated roads and support infrastructures of service stations, communications and information services have thus become a fully integrated and extended system (Dupuy, 1995, 4). This allows flexibilities of time–space movement that public transport systems, geared to traditional monocentric cities, simply cannot match.

In reality, however, the flexibility of the car is a 'coerced flexibility' in the sense that the extended, polycentric cities that automobility supports entail an ever-increasing spatial separation of uses (Urry, 1999, 13–14). This, in turn, necessitates more and more use of the car to bring the distanciated and fragmented time–space 'bundles' of the metropolis into some manageable articulation. For this reason, and very importantly, 'mass mobility does not generate mass accessibility', despite the widespread depiction in advertising and the media of automobiles as harbingers of unproblematic liberation (ibid.).

A central paradox of processes of splintering urbanism is that the extension of standardised highways and roads across and beyond the metropolitan region – ostensibly to support metropolitan integration – has tended in practice to support the partitioning and fragmentation of urban space. Three interrelated processes can be identified surrounding the motorisation of metropolitan life which directly work to support the processes of splintering urbanism.

THE ROAD AND THE 'ROAD PLACE'

First, motorisation has supported a shift in the use of streets from multi-use meeting and transit spaces to single-use spaces which do nothing but vector car flow or house parked or gridlocked stationary vehicles (Southworth and Ben-Joseph, 1997). This role is not to be underestimated. Such is the dominance of car-based mobility in most cities that car-oriented roads and highways are about more than flow; they are complex social spaces in their own right. 'Today the road transcends its function as connector and becomes both a threshold and a place,' writes Alex Wall. 'If the space of the car is sometimes an office, a home, a place of courtship, then the roadscape becomes the place where we live' (1994, 10).

It follows that 'the "road place", rather than a residual or leftover place . . . should be seen as a contiguous realm, the site of smooth and fluid interchange between freeway, arterial road or city street and adjoining land' (Wall, 1996, 161). But the road place is also a highly personalised and cocooned space, riddled with clashing personal territorialities, fears of incursion, and the real and perceived dangers of crime and conflict. For the car has powerfully supported the notion that private enclosure is equated with personal freedom:

the car represents freedom because: it is privacy, a place to be alone, as a mobile apartment, a place where your children can misbehave without embarrassing you and themselves in public, a place for sexual activity; you can pick and choose your companions; the car waits for you; waiting at a bus stop is far less comfortable than sitting in traffic jams.

(Hamilton and Hoyle, 1999, 29)

The modernist zeal to eradicate traditional street patterns has often also served to undermine the social patterns of life that were associated with those traditional urban forms. Social life has often been internalised, encouraging the gradual privatisation of social relations (Holston, 1998, 45). Such privatisation, in turn, has tended to allow 'greater control over access to space, and that control almost invariably has served to stratify the public that uses it. The empty "no man's spaces" and privatised interiors that tend to result contradict modernism's declared intention to revitalise the urban public sphere, rendering it more egalitarian' (sic. ibid.).

MOTORISATION AND URBAN FRAGMENTATION

Second, the widespread shift to highways and automobiles as the dominant transport system of polynuclear cities has strongly supported the broader shift towards urban physical and social fragmentation and separation. Highways and motorisation have contributed to a coarsening, widening and stretching of the urban fabric. To Haug 'the private car, together with the running down of public transport, carves up the towns no less effectively than saturation bombing, and creates distances that can no longer be crossed without the car' (1986, 54, quoted in Clarke and Bradford, 1998, 874). To Woodroffe *et al.*, in 'transforming from main street to freeway culture, North American cities have most clearly demonstrated the paradigmatic qualities of fragmentation' (1994, 7).

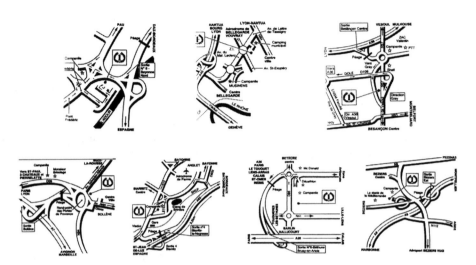

Figure 3.2   Defining urban location by topological connection: 'first class' hotels on French highway networks. *Source*: Dupuy (1995), 27

As strictly hierarchical highway systems with highly limited access points have been superimposed on more fine-grained street structures the possibilities of interaction have been radically altered. Increasingly, as the automobile and road–highway systems become normalised as the dominant mode of personal transport in many cities, the location of an urban space is defined by its topological relation (or lack of relation) to highway access points within the regional, national or international urban field (see Figure 3.2).

Automobiles and highways have thus tended to support the horizontal segregation of uses within the extending metropolitan region. As Calthorpe suggests, 'the car is now the defining technology of our built environment. And, more importantly, it allows the ultimate segregation of our culture: land uses which separate old from young, home from job, and rich from poor, and owner from renter' (1993, 21). In many countries highway construction has also tended to support and entrench patterns of ethnic segregation (see, for example, Bayor, 1988).

These new horizontal segregations, reflecting the chaotic nature of polynuclear urbanism, tend to be difficult to interpret and understand, resisting all the easy distinctions and categories of classical urban and infrastructure planning and analysis. In this new condition, social spaces may be less visible and not necessarily comprehensible. Further, they may no longer be static or embedded in the physical fabric of the city in any simple or comprehensible way (Wall, 1994, 10).

RESIDUALISING THE STREET: CAR CULTURE AND THE TURN INWARD IN URBAN DESIGN

Finally, urban spaces are being increasingly reconfigured towards dominant car-using users as inward-looking 'islands' or 'enclaves', surrounded by the physical highways, connections and services to support motor access, parking and use (see Box 3.3). In a growing number of cities, traditional streets, laid out under the influence of the early stages of the modern infrastructural ideal and before, have often been marginalised by the growth of highways as places of danger, fear and mistrust.

Complex causal loops are clearly at work here. They defy easy analysis and should make us wary of simple technological determinism. But connections between motorisation (and the associated individualisation and decollectivisation of movement within cities), a shift towards private, inward-looking urban developments, the demise of the 'public' nature of many streets and the associated rise in fear of crime can be identified. Kerry Hamilton and Susan Hoyle, for example, argue that:

society's notion of what the street is for has been profoundly altered by car-culture. One of the first effects of the car's take-over of a city street is to denude it of people. This creates a physical as well as a conceptual space on the street, a space which is filled (or is perceived to be filled) by crime, insanity, and other social deviation – and by more cars. Thus a downward spiral is created of experience and expectation of what it is to be on the street (except in a car). People come to resist venturing into public space: they seek bolts and entry phones on the door, moats around the house, gates on the street. . . . We are not suggesting a simple causal model here. . . . What we do want to stress, however, is that there is a connection.

(1999, 35)

Thus urban highway programmes, whilst often justified in the language of 'urban cohesion' reminiscent of the modern infrastructural ideal, have actually tended to support fragmentation and fracturing of the fine-grained urban fabric. The case of New York is typical (see Box 3.3).

NETWORKED INFRASTRUCTURES AND THE 'INFINITE CITY'

Urban development processes now centre on combining flexible bundles of tailored infrastructure networks to suit a myriad of needs of organisations, buildings and clusters, within extending urban regions which may or may not resemble the modern notion of a 'city'. Right across the planet, the traditional urban cores that were subject to the modern infrastructural ideal have been swamped by widening, polynuclear, urban regions.

As the expanding cities emerge as archipelagoes of economic, social, ethnic and cultural enclaves, so the use of networks to link beyond the city with international infrastructures supporting intense flow, exchange and mobility, managed through unprecedented logistical precision, becomes critical. 'The landscape currently being punched out [of cities] is characterised by a major reorganisation of production and distribution, mostly linked to changes in logistical processes (such as just-in-time production)' (Storper and Walker, 1989, 205).

As Richard Skeates suggests, our world of exploding megacities, of extending urban regions, of expanding urban corridors and of transnational 'network cities' thus raises the prospect of the 'infinite city' (1997, 6). To him, we are moving from the modern world of the discrete 'city' to an 'all-pervasive and ever-present urbanisation. . . . The urban is becoming the overwhelmingly predominant way in which the world is experienced by the majority of people whether they live in the city or not.'

In such sprawling and 'supermodern' urban landscapes, notions of networked infrastructures configuring 'centre' and 'periphery' become much less clear. An 'amorphous homogeneity' (ibid.) of the urban emerges which, in rapidly growing 'megacities' like São Paulo or the Pearl River Delta, seems to eclipse completely any traditional, modern notion of 'cityness' at all (Castells, 1996).

New, sprawling Special Economic Zones like Shenzen, for example, are massive archipelagoes loosely laced together by huge new infrastructure networks – private tolled superhighways, airports, optic fibre grids, new water and energy networks, ports and logistics centres. In such places position and centrality are configured less by geographical location with respect to 'downtown' than by the conditions of buildings and places with respect to global–local networked infrastructures like international airports, high-speed rail and port terminals, transglobal optic fibre links, broadband Internet 'pipes' and satellite terminals, and, increasingly, links with distant water and energy reserves.

THE LOGISTICAL SWITCHING CENTRES OF 'SUPERMODERNITY'

Urban sites thus emerge as staging posts on internationally organised flows of goods, people, signs, images, commodities and information – all mediated by physical urban infrastructures (Lash and Urry, 1994). Managing exploding traffic flows – in transport, telecommunications and media, energy and water – becomes the central focus of infrastructure development, requiring new 'megastructures' like high-speed rail networks, global airport hubs, transglobal optic fibre grids and international water and energy networks. Global property capital increasingly centres on constructing such edifices to mobility; global infrastructure capital increasingly focuses on linking them up into (highly uneven) international grids of exchange via sophisticated infrastructure networks. Separation becomes not so much a function of distance as of time, the time it takes to exchange with and relate to far-off places via infrastructure (Woodroffe *et al.*, 1994; Urry, 1999).

Mark Augé (1995) terms such sites 'non-places' – sites that are wholly constructed and controlled to support the mobility of global commodities, signs and transnational travellers. In these 'generic cities', intense concentrations of networked infrastructure are laid down to construct 'supermodern' spaces of flow that are, arguably, the dominant urban landscapes of the late twentieth century (Koolhaas and Mau, 1994; see Augé, 1995). Whilst the construction of ever more heavily networked technopoles, research and development parks, edge cities, peripheral malls, airport cities, leisure complexes and exurban housing areas creates what Skeates terms 'an overabundance of space', the disjointed and instrumental nature of such new networked spaces means that 'there is simultaneously a lack of place' in the exploding urban peripheries (1997, 9).

NEW 'STRUCTURES OF FEELING': THE CHALLENGE OF SOCIAL AND CULTURAL CHANGE AND SOCIAL MOVEMENTS

The fifth and final challenge to the modern infrastructural ideal has been thrown down by processes of social change and an interrelated array of social and political movements. Social theorists capture these interrelated social and cultural shifts within the concept of the shift from the modern to the postmodern urban condition (Harvey, 1989; Ellin, 1996; Dear, 1999). Such transitions have been complex and varied and their interpretation has been fiercely contested. A key element, however, has been the ways in which a wide range of new social movements have brought resistance to bear on the technical and ideological assumptions that underpinned the establishment and propagation of the modern infrastructural ideal.

In combination with the political-economic, planning and urban shifts outlined above, a diverse range of social and cultural movements have contested the very notion that any single, coherent notion of the networked city, with its single public realm, is either possible or desirable. No universal, essential subject, common to all humanity, is possible. No all-encompassing struggle for emancipation, whether based on 'progress' or on new infrastructural technology, can ever be attained. The starting point for any real process of development must be recognition of the cultural politics of difference which ground the diverse and multiplying social worlds of contemporary cities (Fincher and Jacobs, 1998).

The social democratic welfarist consensus in Western nations in the postwar period has emerged from such critiques as not a 'consensus' at all. Rather, in many cases, it is cast as a patriarchal, racist construction based on privileging bureaucratic, masculine and technical rationalities over all other forms of social being. Legitimised by scientific and technological practices, such a state-backed bureaucratic construction worked relentlessly to impose its vision of order on city and society, treating humans as objects in the process in both the capitalist and the socialist worlds. In this context, as Nigel Thrift argues, 'the importance of new generations and new social groups can be seen in the way that matters of gender, sexuality and race have been taken up and have led to much greater attention being given to borders, transgression, third cultures, and other motifs of the new structure of feeling' (1996a, 260).

The cultural, social and political struggle to move away from the totalising and reductionist orders of the modern infrastructural ideal is wide and diverse. It touches on the whole complex of contemporary cultural and environmental politics and theory: social movements revolving around the quest for environmental 'sustainability' as well as cultural movements asserting the rights of women, gay people, the disabled, ethnic minorities, aboriginal groups, religious and spiritual minorities, the young, and older people (see Sandercock, 1998b). Such struggles encompass Western, postcolonial, developing and post-communist cities in different ways. In the brief space available here we can only stress the social movements that were instrumental in bringing down the modern infrastructural ideal: the feminist movement, postcolonial critiques of the modern infrastructural ideal in developing cities, and the environmental and appropriate technology movements. We add to this a brief, broader discussion of how the fragmentation of cultural politics has also served to undermine the modern ideal.

FEMINIST CRITIQUES OF THE MODERN INFRASTRUCTURAL IDEAL

> Feminists, in particular, have been able to demonstrate the existence of deeply embedded impulses to domination and gender inequality often lurking beneath the surface of socialist rhetoric of equity and liberation.
>
> (Leonard, 1997, xi)

First, one of the most powerful social challenges to the modern infrastructural project came from the feminist movement. It increasingly became clear that urban infrastructure, as developed within the rubric of modern urban planning and design, had provided physical services that were central in structuring social relations between men and women in very particular and biased ways (see Kirkup and Smith, 1992; Weisman, 1994). As Leonie Sandercock (1998a, 52) suggests, 'the spatial order of the modern industrial city came to be seen as a profoundly patriarchal spatial order; that is, an arrangement of space in which the domination of men over women was written into the architecture, urban design, and form of the city'.

THE MODERN IDEAL AS A 'POEM OF MALE DESIRES'

Following the first feminist critiques of the late 1960s, it quickly became apparent that the 'coherence' and order that were such a central aspiration of modern urban plans and infrastructure strategies were *imposed* (see Hayden, 1981). To a large extent, the practices and ideologies of modern city planning were, as Barbara Hooper puts it, 'a poem of male desires' (1998, 227). They tended to privilege overwhelmingly masculine, technocratic and perspectival ways of seeing the 'city' and its inhabitants. They gave the masculine gaze of the 'rational' planner and engineer all-encompassing power to structure urban space according to notions of instrumental rationality which systematically marginalised women as 'other' within the form and process of urban life. And they worked within a wider system of orderly separation where the structure and functioning of cities were driven by a reductionist and repressive system of binary divisions:

> public space separated from private, moving vehicles separated from pedestrians, recreation and housing separated from work, underground from above-ground, poor from rich, respectable from dangerous, sick from well, dead from living, women from men.
>
> (Hooper, 1998, 239)

Hooper argues that female bodies, in particular, tended to be cast as a threat to such masculinised notions of social and urban order (ibid.). The fantasy of the straight line within both Haussmann's and Corbusier's strategies for regularising cities and urban infrastructure, for example, was, in a sense, directed against the 'disorder' of the body, and more particularly against (what were seen as) 'the dangerous curves and excesses of the female body' (Hooper, 1998, 230). Women were configured through such practices as domesticated 'home makers' or 'housewives' whose 'natural' domain was the private, newly networked household spaces of the suburbs (Weisman, 1994, 72). Women remaining on the street – for example, homeless women or prostitutes – tended to be rendered invisible, or were assumed to be uncontrollable, pathological, dangerous or sexually threatening. In a sense, then, both planners and doctors were, to paraphrase Foucault, 'specialists in space' – one of the body, one of the city (1977, 150). Both were 'men of reason whose medical and medicalised attentions will restore the logic of borders – the logic of function and system – that is order and health' (Hooper, 1998, 236).

THE GENDERING OF INFRASTRUCTURE NETWORKS

The networks laid out to service the new networked households of the increasingly polycentric city – electricity, gas, streets, water and the telephone – were also constructed in gendered ways to reinforce this patriarchal stereotyping in the partitioning of bodies in urban space. Other networks – notably the car and highway system – tended to be constructed overwhelmingly as masculine systems geared to the male breadwinner's commute to work. All other identities found it hard to escape the relentless construction of dominant patriarchal social, spatial and technical forms based on simplistic, binary divisions.

Feminist movements, linked closely with other critiques from black, lesbian and disabled groups, quickly sought to expose the inherent biases and disciplinary goals within modern urban and infrastructural planning. In particular, the binary separation of 'public' and 'private' spheres, each supported by differently configured networked infrastructures, was widely criticised, as were the efforts of planners and infrastructure companies to maintain the resulting gendered notions of urban space – the 'home', the 'suburb', the 'street', the 'city'.

As Box 3.4 shows, the suburban, networked household, in particular, came to be seen by feminists as 'the prominent locus of confinement. Its street – the milieu of quiet, discreet, respectful surveillance – and its neighbourhood becomes the new panopticon. Women become their own keepers, and the new bourgeois suburbia produces and reproduces the ideology of self-surveillance' (King, 1996, 39). Similarly, as Box 3.5 shows, the application of the modern infrastructural ideal embedded profoundly gendered landscapes of movement and mobility within cities.

BOX 3.4 FEMINIST CRITIQUES OF THE MODERN NETWORKED HOUSEHOLD

One of the first achievements of feminist critiques of the modern infrastructural ideal was to expose the gendered construction of urban infrastructure networks – particularly electricity. Most central here was the contestation of the notion that electrification had displaced the need for housework. Despite the widespread assumption that electric 'labour-saving devices' had brought housewives into an age of leisure, research on womens' time budgets between 1920 and 1960 indicated that the hours spent on housework had changed very little. For instance:

> Despite massive technological changes in the home, such as running water, gas and electric cookers, central heating, washing machines, refrigerators, . . . studies show that household work in industrialised countries still accounts for approximately half of total working time.
>
> (Wajcman, 1991, 238)

GENDER DIMENSIONS OF HOUSEHOLD MECHANISATION

The development of the four other household infrastructure systems – water, gas, transport and the telephone – was equally ambivalent. While they could be used to transform housework and dramatically increased housewives' productivity, they did not reduce the necessity for time-consuming labour. Urban infrastructure had thus failed to ease or eliminate household tasks (R. Cowan, 1997). This was because mechanisation had:

- Given rise to a set of new tasks that were just as time-consuming, if not as physically demanding, as the tasks they displaced. Middle-class housewives, rather than servants, were expected to do all the housework.
- Been accompanied by rising expectations of the housewives' role – generating even more work. The germ theory of disease and the domestic science movement led to exacting new standards of housework (Lupton and Miller, 1992).
- Only a limited effect because it took place in the context of a privatised single family home. This context reinforced the sexual division of labour and hindered the reallocation of work. Millions of American women cooked supper each night in millions of separate homes on millions of separate stoves. Reinforcement of gender relations was also reflected in the construction of women as users of new technology whilst men were constructed as the 'fixers' of such technology.

The atomised logic of the suburban networked household meant that the dominant feature of the use of infrastructure in the home was that energy and expertise were devoted to the mechanisation of housework in individual households rather than through collectivised or socialised efforts. During the early part of the century there were a number of alternatives, including commercial services, the establishment of alternative communities, co-operatives and the invention of different types of machinery (Hayden, 1981).

EXPLAINING THE TRIUMPH OF INDIVIDUALISED HOUSEHOLDS

Yet the privatised and individualised household triumphed. There were a variety of reasons for this:

- Women embraced new technologies because they made possible a higher material standard of living for an unchanged expenditure of housewives' time.
- A more powerful factor may have been the value ascribed to the 'privacy' and 'autonomy' of the family over technical efficiency and community interest when making decisions about the expenditure of limited funds.
- Major contradictions underlay the rationalisation and mechanisation of domestic life. Individual forces were constrained by powerful structural forces. For instance, the 1930s provision of municipal wash houses, laundries and communal areas was not always seen as progressive but was associated with poor-quality sanitation, shared water supply and squalid housing. In any case the alternative to single family houses were extremely limited. 'State policy in the area of housing and town planning played a key role in promoting privatism' (Wajcman, 1991, 249). At the same time, the manufacturers of appliances and electricity supply networks played a central role in shaping domestic technology and their diffusion to build loads for their networks and profits for manufacturers.

The importance of domestic infrastructure, therefore:

lies in its location at the interface of public and private worlds. The fact that men in the public sphere of industry, invention and commerce design and produce technology for use by women in the private domestic sphere reflects and embodies a complex web of patriarchal and capitalist relations . . . By refusing to take technologies for granted we help to make visible the relations of structural inequality that give rise to them.

(Wajcman, 1991, 254)

BOX 3.5   TRANSPORT AND THE GENDERING OF THE MODERN CITY

Feminist critiques also examined the wider built environment as a biased technological construct. The structuring and form of modern cities, they argued, worked to reinforce the sexual division of labour as architecture and urban planning structured separation between men and women, private and public, home and paid employment, consumption and production, suburb and city (Weisman, 1994).

Although it was clear that people did not live according to these simple dichotomies, the widespread belief in them profoundly influenced women's lives. The design of the modern networked city was stamped with wider normative assumptions about social and economic relations.

Feminist critiques of transport planning and policy thus emerged, arguing that communications systems and the car actually worked to restrict women's mobility and reinforce their confinement to the home and immediate locality. Wajcman (1981, 126), for example, suggested that:

> Women's and men's daily lives trace very different patterns of time, space and movement, and the modern city is predicated on a mode of transport that reflects and is organised around men's interests, activities and desires, to the detriment of women.
>
> (Wajcman, 1981, 126)

During the development of the modern city the car had been widely expected to emerge as the future of urban transport. Many of the land use planning and transport planning styles of the 1950s and 1960s had centred on planning for roads and elaborate highway, freeway and motorway systems (see Box 3.3). However, new road construction had generated more traffic which, in turn, led to a response based on the construction of more roads to eliminate congestion. This set in motion a vicious circle that has only recently been questioned. Economic, largely masculine, vested interests from car manufacturers, road construction companies, oil companies and property developers all shared in accelerating the development of the motorised city. But these developments affected different social groups, and especially women, in different ways:

The assumption of car ownership discriminates against the poor and working class in general. Older women and single mothers are amongst the poorest groups in society and have been left literally stranded in, or outside of, cities designed around the motor car. Although the automobile did not create suburbia, it certainly expanded and accelerated this process. The promotion of mass motor-car ownership has tended to exacerbate a greater dispersal of residential settlement often without any other mode of transport to service such areas.

(Wajcman, 1981, 129)

Despite women's relatively low mobility, their travel needs were expanding as increasing numbers of them entered paid employment and the location of health, shopping and educational services became more decentralised. Women were more reliant on off-peak and unreliable public transport. Women's journeys were more complex and multipurpose, owing to their multiple roles as mothers, domestic workers, social agents and paid employees. This meant that they did more journeys of shorter duration, requiring time-consuming and expensive changes. Because much of the public transport laid out by modern infrastructural planning was designed around the needs of full-time, largely male, workers commuting to a city centre, services rarely met women's needs, limiting them to a more spatially restricted job market.

'POSTCOLONIAL' CRITIQUES OF THE APPLICATION OF THE MODERN IDEAL IN DEVELOPING CITIES

Second, from the early 1970s, postcolonial critiques in developing cities began to question the usefulness of modernisation and import substitution theories as the basis of infrastructural policy in developing cities (see Balbo, 1993).

THE FAILURES OF INFRASTRUCTURAL 'TRICKLE DOWN'

The empirical evidence did not support the theory that infrastructural investment in urban cores, targeted at the needs of local and international elites, 'trickled down' through the rapidly growing spaces of developing cities. Instead there was increasing recognition that the processes of city growth were fundamentally different in developing cities, particularly in the stark urban–rural differences in wealth levels, the intense concentration of limited resources in capital and primary cities, and the very high economic disparities between the main bulk of the population and small political and cultural elites within dominant cities (Balbo, 1993). It was increasingly realised that colonial infrastructure policies, and broader ideologies of modernist urban planning, had supported a powerful fragmentation in developing cities between minority elites in their well networked enclaves and the poorly served majority. Informal economies, fast-growing squatter settlements – in many cases the majority of a city's population – had simply been ignored by the adapted notion of the modern infrastructural ideal in developing countries.

THE BIASES OF MODERNISATION THEORY

These features were dramatically reflected within, and reinforced by, infrastructure networks as developed under modernisation theory. Such networks tended to be heavily concentrated in urban areas, most highly developed in quality and coverage in primary cities and, within cities, largely configured to exclude the poor by providing formal infrastructure services only to wealthy elites. It was increasingly obvious that infrastructure networks in developing cities were 'outcomes not of smooth, natural processes of innovation and diffusion but of political and economic battles. In most cases, these contests [were] rather one-sided, and the resultant transportation and communication systems [were] much more likely to facilitate the interests and needs of wealthy people' (D. Smith, 1996, 147; see Box 3.6). By channelling scarce resources into highly uneven networked connections, it emerged that:

BOX 3.6 THE POLITICAL ECONOMY OF TRANSPORT NETWORKS IN DEVELOPING COUNTRIES

Political economy perspectives argue that developing countries are held in unequal and exploitative relationships with metropolitan cores in the wealthy North. Capitalist development in colonial peripheries is a distorted or subordinate form of development that will lack the potential or dynamism to develop any autonomy. Economic development in the periphery is geared to supporting accumulation in the North. Any industrialisation in the South therefore tends to be of the enclave type, with little potential to support development outside urban centres. Transport technologies form the key bridge between the core and the periphery. While the technologies themselves may be neutral they were developed in the North under particular social conditions and utilised to serve those interests.

In the context of transport, Slater (1975) shows how colonial transport networks reflected colonialists' perceptions of the strategic importance and economic value of colonial territories. Transport networks provided the material sinews and shaped the extent to which different regions were incorporated into the global economy. In Tanzania, for example, the territorial space economy was divided into three zones by the 1930s. First, the coastal belt comprising major towns and cash crop-producing regions was most directly linked to the transport network. It was surrounded by a second zone that supplied food and services, while a third and most peripheral zone served as a source of migrant labour from declining subsistence economies. This process cut across pre-existing modes of production and transport routes. Slater illustrates how this was a simultaneous process of internal disintegration and external reintegration.

At independence, colonial states faced massive problems as they inherited infrastructure networks designed to serve metropolitan rather than local needs. Many newly independent states won power on an agenda designed to meet local needs and distribute development more equitably. In Tanzania the transport infrastructure was a major hurdle to the new development trajectory. Physically embedded networks could not simply be rerouted to link bypassed regions. While foreign aid supported the construction of trunk and secondary roads in some regions, little change occurred in the inherited rail system of three unlinked parallel lines running east–west to the main ports. Virtually all railways built in Africa since independence have replicated the colonial pattern of linking enclaves with the nearest port.

The provision of specific types of infrastructure (where and when to build highways, ports, rapid transit, 'smart-wired' buildings, and so on) had become implicitly urban policy. The construction or selective upgrading of particular roads, railroads and ports to accommodate the evacuation of particular export products, and the disproportionate expenditure of scarce national resources on telecommunications and highway grids in capital cities, [were] bound to skew city growth and settlement patterns. . . . They [were] the result of policy making decisions that usually reflect[ed] dominant class interests.

(Ibid., 147)

As Box 3.7 demonstrates in the case of water infrastructure, powerful elites in developing cities tended to use modernisation theory to orchestrate the development of urban policies that reflected their own narrow interests. Surpluses that were extracted under colonial rule often accrued directly to the national capital in the form of taxation and trade duties. A large part of the surplus usually went into public works disproportionately benefiting the national capital. These public works were often directed towards conspicuous infrastructure investments designed to create and symbolise a 'modern' aura in the capital city – airports, highways and skyscrapers. Such policies exacerbated inequality of access to infrastructure. Numerous urban residents, particularly with the explosive growth of informal settlements to house rural in-migrants, lacked access even to the most basic services. From the late 1970s there was increasing recognition that modernisation theory was effectively unable to deliver affordable and reliable infrastructure services to new users in rapidly growing developing cities.

BOX 3.7 THE SOCIAL AND GEOGRAPHIC DUALISATION OF URBAN INFRASTRUCTURE IN DEVELOPING CITIES: THE EXAMPLE OF WATER

The example of water services illustrates how hundreds of millions of users have been excluded from access to modern infrastructure networks in developing cities, even though the development of such networks has often been legitimised by notions of ubiquity and universal roll-out adapted from the modern infrastructural ideal. The modern logic has thus come in for increasing criticism from development agencies, environmental organisations, international aid agencies, nongovernmental organisations and local social movements.

THE GLOBAL WATER AND SANITATION CRISIS

Initially the development of networks tended to keep pace with the accelerating process of urbanisation (see Goubert, 1989). But nation states have found it extremely difficult to keep pace with increasing urban demand for water (see Gilbert, 1992; Serageldin 1994). The UN Habitat programme estimated that by the year 2000 an urban population of 450 million people would be deprived of urban water services and a further 720 million would lack urban sanitation. Domestic underinvestment and chronic dependence on external capital resulted in the systematic exclusion of the new urban poor from easy and cheap access to potable water (see Banes *et al.*, 1996; Bhatia and Falkenmark, 1993; Black, 1995; Gilbert, 1994; Swyngedouw, 1995a).

Although there are major variations in the quality of service provision between develop-ing countries and cities, the variations between different areas within the same city are usually even more marked (see Table 3.2). A common pattern of water provision has emerged (see Banes *et al.*, 1996; Black, 1996; World Bank, 1994). The upper and middle income groups are usually well served, while delivery to the less affluent areas of the city is often very poor. For instance, in Lima over 90 per cent of the top 10 per cent income group have direct connections to water and sewerage, compared with an approximately 60 per cent rate of connection for the bottom 10 per cent of income groups (Glewwe and Hall, 1992, 30). The level of connection is also linked with the age of the settlement, the poor living in older-established areas tending to have access to most services while those in the newest are often less well provided for.

OPTIONS FOR NON-SERVED LOW-INCOME SPACES

Communities without access to the formal water network are more likely to buy water from private vendors, and are prone to have more interruptions of supply, to have more polluted water, and not to be connected to sanitation services. Residents in marginal communities typically spend between 10 per cent and 40 per cent of their income on water services (see Black, 1996, 6). Paradoxically, these users also pay much higher prices for

Table 3.2 Dual circuits of water provision

Formal	Features	Informal
Public monopoly	Provider	Private water traders
Higher socioeconomic groups connected	Network	Marginal socioeconomic groups unconnected
Subsidised prices	Costs	Expensive
Higher quality and reliability	Quality	Lower quality and disruption
Slowly enrols new users	New users	Large numbers of new users enrolled
Low cost to users but low returns to water provider	Consequences	High health, environmental, social and economic costs to users

Source: Marvin and Laurie (1999), 343.

water, with charges up to ten, twenty and even 400 times higher than those paid by domestic users of the public utility (see Black, 1996; Petrei, 1989; Swyngedouw, 1995a). In many cities a form of negative redistribution operates because the unconnected poor pay very high prices for water from private traders because the public system cannot deliver services in a comprehensive way.

The deficit and underpricing of water results in a massive transfer of income from marginal users to the middle and upper-class consumers and to commerce and industry. A study of five Latin American countries found that wealthy users always benefit disproportionately from subsidies for water and sanitation services (Petrei, 1989). Utilities face increasing difficulty in extending networks: the cost of new supply is often two or three times more than that of existing supply (IBRD, 1993, 37) and it is estimated that complete coverage of the water network will require US$5 billion investment and sewerage US$7 billion over the decade 2000–10 (Idelovitch and Ringskog 1995, v).

Increasing evidence from studies of water resource management in developing cities indicates that many elements of the current logic of provision are based on dual circuits of supply (see Montgomery, 1988). The consequences of each circuit are very well known and documented (see Black, 1996; Banes et al., 1996; World Bank, 1994). In the formal sector, users benefit from low prices and lax recovery of charges. There are few incentives to restrict consumption while the water provider fails to recover the costs, and there is insufficient finance to extend the network to new users. In the informal sector, users suffer serious economic, health, social and environmental costs associated with uncertain, low-quality and expensive water supply, while the total charges paid by such users are often greater than the revenue collected by the formal provider. Many of the social, economic, health, social equity and political conflicts over urban water supply can be traced to these dual circuits of differential access to urban water resources.

ENVIRONMENTAL, ENERGY, AND INTERMEDIATE TECHNOLOGY MOVEMENTS

The third social movement which powerfully undermined the modern infrastructural ideal, the environmental movement, began to have an important role in shaping infrastructure development in the 1970s. The 1973 energy crisis brought into sharper focus the connections between energy supply, infrastructure networks and the vulnerability of societies dependent on external energy sources. This led to the rapid expansion of research and development into the diversification of energy supplies.

REASSERTING THE POWER OF HUMAN CHOICE: FROM 'HARD' TO 'SOFT' ENERGY PATHS

Central to this debate was the notion that energy futures were not technologically determined but open to human choice. Amory Lovins's (1977) seminal work systematically articulated the view that urban industrial societies needed to make a transition from the dominant 'hard' energy strategy to alternative 'soft' energy strategies. Distortions in the market such as massive government subsidies, and lack of knowledge about alternatives, prevented the emergence of a softer 'pathway'. (An 'energy pathway' refers not only to the direction of energy use but the institutional, social, economic and cultural factors that make up an energy system in a society.) Lovins argued that different paths have different social impacts: a 'hard' path will have disruptive impacts while a 'soft' pathway will be relatively benign. Lovins portrayed the dominant hard energy path as involving capital-intensive systems with complex, large-scale, centralised and resource-depleting technologies that alienated human beings, generated inequality and damaged the environment.

There were a number of further elements to the environmental critique of the conventional approach to urban infrastructure provision that surrounded the modern ideal. In the energy sector, environmental and tax payers were increasingly critical of the large tariff rises needed to support supply-oriented investment in large nuclear and coal-fired power stations. Under pressure, state regulators forced utilities to consider equally the value of both supply- and demand-oriented measures to meet energy needs. In response, power companies postponed or even cancelled new power stations and instead invested in energy efficiency and conservation measures to push peak demand for power downwards. Increasing resistance to highly disruptive urban highway projects also eventually forced government agencies to consider alternative options, rerouting highways and increasing investment in mass transit.

The US case is illustrative of the major policy turn-rounds prompted by such environmental critiques. Between 1973 and 1977, for example, federal funding for mass transit grew fourfold, to $1.3 billion, with a large increase in the level of public subsidy to services. In the waste and water sector, environmental pressure caused the US government to improve the environmental performance of networks. Between 1967 and 1977 federal expenditure on sewer systems increased from $150 million to $4.1 billion. By 1977 over 30 per cent of federal aid for cities was for expenditure on sewer systems. New environmental concerns have therefore had quite dramatic implications for the rehabilitation and style of urban infrastructure management. The conventional supply logic that traditionally ignored environmental concerns, or operated only to minimal standards, was dramatically challenged. Infrastructure providers had to consider alternatives to the conventional supply-oriented approach in both the transport and energy sectors, and massive investment in sewerage and water treatment was required to bring systems up to new environmental standards.

CRITIQUES OF 'BIG' TECHNOLOGY

Environmental and energy critiques were also backed up by a broader attack on the large-scale centralised technologies surrounding the modern infrastructural ideal, led by the appropriate or intermediate technology movement (Willoughby, 1990). The impetus for the development of this movement came from a response to the growth of modern 'technological' society as a whole (Nelkins, 1979; Yearly, 1992). In particular, a groundswell of dissent started to challenge the dominant notion that technology was inherently progressive. Responses to, and perceptions of, technology became increasingly controversial and vehement.

There were a number of dimensions to these critiques. The 'big' technoscience of nuclear weapons and energy led many to question the neutrality and supposed benevolence of technology. The rapid growth of environmentalism was linked with the realisation that technology had insidious effects on health and the environment. It was increasingly obvious that harmful impacts could arise from the unforeseen side effects of technology. And a growing number of writers argued that the dynamics and imperatives of capitalist and state technological development transcended the control of individuals and communities (Willoughby, 1990).

Intermediate technology was initially formulated as a response to the problems of the modern integrated infrastructural ideal in cities in the Developing World. Later the concept was developed in the advanced economies of the North when new energy technologies and energy conservation measures were linked with the development of employment and cooperatives. In particular, critiques focused on the problems of technological dependence, technology transfer and technology underdevelopment. The main issue has been the inappropriateness of applying the infrastructural models developed in the context of Western cities to the very different socioeconomic and cultural contexts in developing cities. New approaches to infrastructure development, based on Schumacher's (1973) famous dictum 'Small is beautiful,' were developed, based on notions of grass-roots-scale technology, minimising capital costs and technological complexity, and maximising the degree to which communities could support their own infrastructural needs.

TECHNOLOGICAL FETISHISM AND THE ASCENDANCY OF SYMBOLIC CULTURAL POLITICS

Finally, it is important to locate this broad range of social movements and social critiques of the modern infrastructural ideal within the wider processes of social and cultural change which also served to render the modern infrastructural project increasingly problematic.

Currently, the 'postmodern' turn in social theory and social and cultural change increasingly stresses the aesthetics of cultural production as the defining aspect of cultural identity (Dear, 1999; Soja, 2000). Both the production and the consumption of cities, and the production and consumption of infrastructure products and services, are increasingly dominated by the language of commodity aesthetics and semiotics (Knox, 1993b, 16). The relatively standardised 'mass society' of Fordist production, distribution and consumption has fractured into a massive pluralisation of practices, enjoyments, tastes and needs.

The differentiation of lifestyles is supported by combinations of specialised urban neighbourhoods and spaces, and customised infrastructure networks. The cultural politics of cities which result are increasingly driven by the construction of places, commodities and services as signs defining social identity. From the Fordist period, where 'consumption was a means of social integration', we have shifted to a world where 'consumption is viewed increasingly as a means of asserting distinctiveness within mass society' (Knox, 1993b, 20). Within increasingly differentiated, mobile, and socially and culturally diverse cities, such consumption is now tied to socioeconomic status, ethnicity and 'race', lifestyle and interests, and profession as means to construct and differentiate identities.

INFRASTRUCTURE NETWORKS AND LIFESTYLE DIFFERENTIATION

Thus, in many cases, standardised and 'black-boxed' infrastructural services – such as the classic 'black' bakelite telephone, the ubiquitous public electricity supplier, the mass-broadcast television signal, the public water monopoly and the standardised 'Fordist' motor car on the public street or road – have become hard to maintain. The aesthetic and stylistic forms of 'high-tech' infrastructures and appliances, in particular – laptop computers, personal digital assistants but also computerised cars, phones, television sets, houses and domestic appliances – have become deeply fetishised as symbols of power, status, mobility and worth. The fetishism of technology – that is, the celebration of its surface appeal whilst ignoring or covering up the broader social relations that produce and surround it – 'seem to be inherent in the very notion of high-tech' (Rutsky, 1999, 129). As Nan Ellin (1996, 12) suggests:

in this new period of late capitalism, cultural production assumes a centrality and significance never previously attained. Image, appearance and surface effect dominate forms of cultural production in which the distinction between original and simulacra dissolves, nostalgia and kitsch supersede realism and naturalism in art, and cultural aesthetics dominate everyday life.

(Ellin, 1996, 12)

CONCLUSIONS

In this chapter we have explored and exposed the complex range of pressures behind the collapse of the modern urban infrastructural ideal. This exploration has necessarily been multifaceted and diverse; such pressures are far from simple or straightforward. The collapse of the modern ideal has been found to be inextricably bound up with the changing political economies of capitalist urbanisation, with transitions in cultural politics, with changing practices and ideals of urban planning, with radical restructurings in all types of nation state, with transformations in the form and structure of cities, and with transitions in the politics and sociologies of technology.

What is left, it seems, is a logic which threatens to support the separating circuits of exchange and interaction, through the interlacing of customised networks and specialised spaces within and between cities. Notions of 'public' spaces and realms where the mixing of differing social, political and cultural groups is actively encouraged are becoming increasingly problematic. Ideas of the unitary nature or 'wholeness' of both cities and urban infrastructure networks are unravelling. Territorially driven policies and politics that try to use networked infrastructures to redistribute between the circuits and spaces of the metropolis are starting to appear at odds with the wider transformations under way. When backed by the ascendancy of neoliberal policies an infrastructural individualism threatens to emerge, bearing all the divisive and polarising hallmarks of wider transformations in 'public' and 'welfare' services. As Leonard suggests:

welfare as a function of the state under capitalism, postmodern critiques argue, epitomizes that contradiction between domination and emancipation which has been historically an invariable feature of modernity. As the political agenda of the Right comes to dominate the discourse on welfare, its empowering and caring side loses ground to a brutal individualism intent on the further degradation of the poor.

(1997, xii)

And yet, in keeping with the wider ambivalence that runs through 'postmodern' critiques of urban and technological change, it is difficult to criticise some of the more positive aspects of the collapse of the modern ideal. For example, can we possibly argue against the idea that spaces and networks should be splintered from the homogeneous, stultifying, gendered and dominating logics of modernist urban and infrastructure planning? Are we to regret the reduced power of the all-seeing planning 'expert' who systematically imbues urban structures and infrastructure networks with masculinised, partial ideologies whilst ignoring all other forms of knowledge and experience? Can we seriously challenge the notion that network market places cannot, in some cases at least, deliver appropriate services in a more diverse and flexible manner than homogeneous public monopolies? And would we argue against those groups most marginalised by modern planning – the disabled, ethnic minorities, women – gaining the ability to negotiate energy, transport and communications services within markets that might be specifically tailored to their needs?

Clearly, the processes surrounding the collapse of the modern ideal are profoundly ambivalent. The risks of urban fragmentation and the reduction in the degree to which large cities bring together and interconnect their diversifying socioeconomic circuits need to be weighed against the benefits that come from recognising and addressing the needs of urban diversity. With such ambivalence in mind, it is time for us to look in more detail at the precise mechanisms and processes through which infrastructure networks can become splintered and cities fragmented. It is to these mechanisms that we turn in the next chapter.

Plate 7 'Power up. Have you switched on to electricity competition?' An advertisement in *Woman's Journal*, 1998, 89

UNBUNDLING INFRASTRUCTURE AND THE RECONFIGURATION OF CITIES

In the last chapter we explored the many forces that are contributing to the demise of the modern ideal of monopolistic, standardised and integrated networked infrastructures. Our broad analysis of the processes that are undermining the modern infrastructural ideal raises a series of questions. How are integrated infrastructure networks being restructured in practice? What are the implications of these transformations for urban change in developed, developing, post-communist and newly industrialised cities? Is it possible to develop a broad perspective on the various practices of splintering urbanism which can be applied to all networked infrastructures, and all types of city, without falling into the trap of oversimplification, technological determinism or ethnocentrism?

This chapter explores how processes of splintering urbanism operate in practice. It is divided into two parts. In the first we analyse in detail how processes of infrastructural 'unbundling' work in practice. We closely examine the implications of privatisation, liberalisation and the application of new technologies for integrated urban infrastructures. The transitions are varied and complex; we illustrate how the unbundling of networked infrastructures actually encompasses a wide variety of shifts. But we argue that the current unbundling of networks is to a considerable extent an overarching, universal process that is, at the same time, diverse and multifaceted. As such, the unbundling of infrastructure networks defies easy generalisation. It varies considerably between particular political, economic, territorial and social contexts. And it is bound up with complex processes of urban change across the world.

The second section constructs an exploratory typology of how unbundled networks are involved in the simultaneous reshaping of social and spatial relations in cities. Here we develop an understanding of how real cities are being fragmented in parallel with processes of infrastructural unbundling. Central to this is an understanding of how unbundled infrastructure networks more intensively and actively connect valued places, while at the same time progressively withdrawing and disengaging from less valued places. Our typology attempts to examine how this shift works its way through different cities in varied contexts across the globe.

PRACTICES OF INFRASTRUCTURAL UNBUNDLING

Within the last two decades there has been a paradigmatic shift across all networked infrastructure sectors based on the movement from integrated to unbundled urban networks. As Curien argues:

what is now becoming quite clear is that the era of integrated monopolies operating infrastructure and providing the totality of services is over. The 'industrial structure' of future networks is a complex one, where a dominant firm, the former monopoly, maintains its monopoly over some segments of the market, is open to competition in others and does not intervene at all in segments where there are other providers.

(1997, 51)

Working together, the economic liberalisation of infrastructure and the development of new technologies have made possible an entirely new infrastructural landscape that radically challenges established assumptions that have underpinned the relations between integrated networks and cities. In this section we provide a guide to the complex processes of infrastructural unbundling across different networks and within particular national contexts. We focus on the processes through which integrated infrastructure networks become segmented (see Guy *et al.*, 1997). We review the marketability of different segments of unbundled networks. Finally, we map the multiple pathways towards unbundled infrastructure.

NEW TECHNOLOGIES AND SEGMENTED INFRASTRUCTURE

We need, first, to understand the ways in which new technological innovations have supported network unbundling. For it is clear that broader institutional shifts surrounding infrastructural liberalisation have been paralleled by a set of technological innovations that provide the ability to challenge the conventional technical assumption that infrastructure had to be provided by public monopolies. Although a number of these innovations have existed for some time, it was only in the 1990s that the applications were exploited to provide support for the measures that effectively unbundle integrated infrastructure.

It is important to stress that liberalisation, combined with new technology, creates great flexibility in the styles of unbundling that can be applied to the integrated modern infrastructures created through the modern ideal. In particular, we suggest that new technologies have:

- *Helped to reduce the conditions for natural monopolies in infrastructure supply.* In the telecommunications sector, new technologies such as cable, radio, microwave and satellite have reduced the cost of providing both long-distance and local telecom networks, effectively undermining economies of scale and the cost barriers to new entrants. These innovations are paralleled by the development of decentralised technologies in the power and water sector, including gas turbines, renewable energy systems and smaller-scale water treatment and waste disposal techniques. The lower costs, high efficiency and flexibility of new and decentralised technologies facilitate the entry of new competitors into increasingly contestable infrastructure markets.
- *Enabled the unbundling of integrated networks and operations.* These technologies facilitate the division of integrated networks into monopolistic and non-monopolistic segments that are contestable by new entrants. Complex control, monitoring and data management systems deal with all the millions of transactions involved in segmenting

power generation, the transmission network and the local distribution system in the electricity sector. Information and communications technologies (ICT) allow new entrants to participate in the contestable segments of unbundled networks – power generation and the supply of power to users – while providing access to those parts of the network that still have monopoly status – power transmission and the local distribution network. These technologies also facilitate the segmentation of gas, water, transport and telecommunication networks whilst effectively maintaining the appearance of an integrated network. The networks are effectively split into the monopolistic and competitive elements, with ICT seamlessly managing transactions between millions of users and multiple suppliers in contestable markets.

- *Increased the range and quality of infrastructure services.* Value-added services in the telecommunications sector, such as high-speed data transmission and mobile communications, are the most dynamic sources of demand and are based on new digital transmission and processing technologies. The containerisation of freight, together with complex logistics systems, permits cost-effective, high-speed and high-quality transfer of freight across multiple transport modes. Power companies use specially tailored distribution networks and back-up systems to offer extremely reliable and high-quality energy supplies to industrial and commercial users. In the domestic sector, premium users are increasingly offered differentiated packages of enhanced and high-quality utility services – these include green power, special maintenance and security services, and services tailored to their specific needs (e.g. swimming pool or greenhouse heating). Information and communication technologies thus challenge the traditional notion that infrastructure services are standardised and homogeneous products. Instead they become increasingly complex and differentiated products offered to a much more fragmented market of users (Graham, 1997).

- *Expanded options for the management of demand for infrastructure services.* In the transport sector, new electronic road pricing technologies permit the introduction of user charges that can be differentiated to reflect the impact of vehicle loads and the level of congestion, so internalising the social costs of pollution. In the power and water sectors, small-scale demand management technologies can improve the efficiency and promote the conservation of energy and water on networks that have economic or environmental limitations on the expansion of supply options.

These technologies therefore challenge the assumption that relations between users and providers are standardised and homogeneous. Instead the interface becomes much more complex: users may be enrolled by producers to shift the timing and level of demand; new pricing technologies can send real-time economic signals to shift consumption patterns; and new intermediaries such as logistics specialists, energy and water conservation agencies, and property developers, increasingly manage relations between users and network providers.

- *Supported the development of appropriate technologies which have facilitated low-cost infrastructure supply options.* Intermediate water and waste technologies have lower costs than conventional, heavily engineered supply options, making them potentially much more affordable to low-income communities, especially in developing nations. Smaller-scale irrigation, energy production, water treatment and waste disposal technologies can permit lower-cost alternatives and facilitate community involvement in the planning, installation, maintenance and administration of infrastructure networks. These users traditionally rely on informal systems of infrastructure provision – private water vendors, charcoal sellers and private transport. Yet, with technical and institutional support, such users can become enrolled on to community infrastructure networks, challenging the traditional assumption that new users have to connect to formal public or private infrastructure networks.

STYLES OF NETWORK SEGMENTATION AND UNBUNDLING

New technical capabilities can therefore actively facilitate the unbundling of infrastructure networks. Central to the notion of unbundled networks is the concept of 'segmenting' integrated infrastructure into different network elements and service packages. Segmentation involves detaching activities and functions that were previously integrated within monopolies and opening them to different forms of competition. Segmentation challenges the assumption that one infrastructure monopoly can provide a service at a lower cost than two or more competitive providers can. Economic liberalisation has also challenged the idea that there are strong technical and institutional reasons why infrastructure networks should be provided on a monopoly basis by the public sector. Many infrastructure suppliers usually provide a range of services, only some of which can genuinely be defined as 'natural monopolies'. The process of unbundling, therefore, attempts to separate the natural monopoly segments of a network and then promote new entrants and competition in segments that are potentially competitive.

Network unbundling can take three different forms: vertical segmentation, horizontal segmentation or virtual segmentation. These are not mutually exclusive, however; they can coexist within a single unbundled infrastructure network.

VERTICAL SEGMENTATION

This involves the division of vertically integrated infrastructure networks. In the power sector, it usually encompasses separating the generation, transmission and distribution of power into a number of different segments open to competitive entry, while some segments retain their natural monopoly status.

As Figure 4.1 shows, in the unbundled electricity sector, generation is usually open to independent power producers, using a range of new, small renewable and conventional power production technologies, who enter the market and compete with incumbent generators. To facilitate competition in the distribution of power the transmission and distribution parts of the networks are also often separated. The transmission network tends to remain a natural monopoly in either the public or the private sector, within a national or regional regulatory framework structured to allow reasonable access and connection to new power generators and energy suppliers. The local distribution network is also likely to remain a monopoly, while virtual network unbundling allows new suppliers to compete for the right to supply users over a single distribution network. It would not be economic to run more than one distribution network to customers' premises.

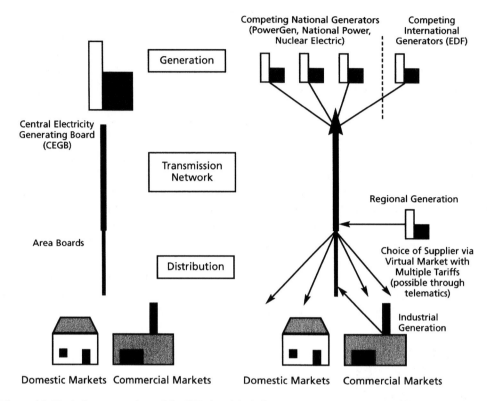

Figure 4.1 Vertical segmentation of the UK electricity infrastructure. *Source*: Guy *et al.* (1997), 200

In the gas sector, resources, treatment, transmission pipelines and local distribution networks can also be owned and operated by different entities. Networks that were previously considered to be integrated can be effectively taken apart and unbundled into different elements to stimulate new entrants and competition in non-monopoly segments of the networks.

Horizontal segmentation

Here, activities are separated by markets, either geographically or by service categories. Geographical unbundling into a number of producers allows direct competition within a single region. Alternatively, the creation of regional monopolies can facilitate performance comparisons between different providers and more efficient regulatory monitoring. For instance, with privatisation, the UK water sector was divided into a number of regional monopolies, with the regulatory agency using a range of comparators to monitor the efficiency of the industry. The telecommunications sector lends itself to horizontal unbundling when the operation of cellular and satellite-based services is typically separated from the provision of traditional landline-based services.

VIRTUAL SEGMENTATION

Virtual segmentation cannot always be sharply distinguished from vertical or horizontal unbundling. Virtual unbundling refers to the development of new competitive services that can be effectively superimposed on the monopolistic elements of unbundled infrastructure networks. Although the local gas, electricity and telecommunications distribution network is effectively a natural monopoly, new entrants gain access to individual users over these networks because of the ability of new information infrastructure to handle all the millions of transactions between users and multiple suppliers.

In the transport sector, virtual network unbundling allows new entrants to offer a range of additional services, such as traffic information and route guidance systems, that can be layered on top of the existing monopoly highway networks. This creates an enhanced set of premium service options. Within the United States the conventional freeways can now be effectively segmented into a series of different lanes offering a range of mobility services: privately funded and tolled express lanes, high-occupancy vehicle (HOV) lanes, lanes for commercial trucks that can all coexist with conventional public-access freeways (see Figure 4.2). New information and surveillance technologies can control access and police these differentiated lanes. Virtual network unbundling thus provides the technological capacity effectively to segment and splinter established integrated infrastructure networks.

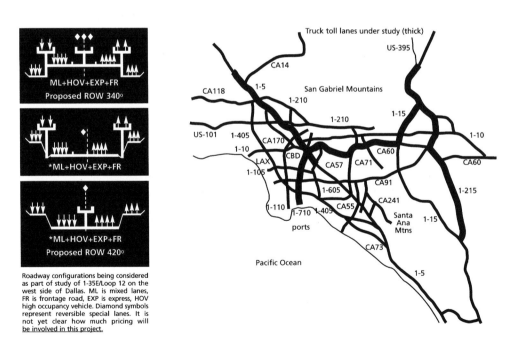

Figure 4.2 Virtual segmentation of the US highway infrastructure. *Source: Toll Roads Newsletter*, March 1999, 8, 14

NEW TECHNOLOGIES AND INFRASTRUCTURAL UNBUNDLING: A SUMMARY

New technologies have challenged many of the traditional assumptions that infrastructure networks needed to be provided by public or private monopolies. There are significant new potentials to unbundle integrated networks using a number of different styles of segmentation. Yet the distinctions between these three forms of unbundling are not necessarily easy to make. Within a particular infrastructure sector, various forms of network unbundling are likely to coexist in parallel. For instance, in the UK telecommunications sector, new entrants can provide value-added services over their own parallel networks or use local distribution networks owned by incumbent operators (see Figure 4.3). In such cases, vertical unbundling between the management of distribution networks and the supply of services is needed to allow competition between horizontally separated service providers. Virtual network unbundling allows these new entrants to use the incumbent's distribution network because telematic systems manage information and transaction flows between users and multiple suppliers. With this wide range of options how do policy makers identify elements of a network that are provided as a monopoly and those that are contestable?

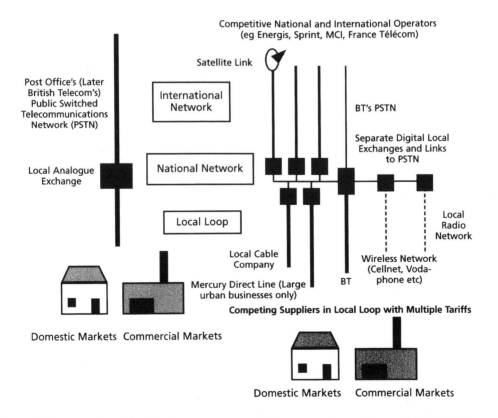

Figure 4.3 Segmentation of the UK telecommunications infrastructure. *Source*: Guy *et al.* (1997), 201

THE MARKETABILITY OF UNBUNDLED INFRASTRUCTURE

The second key area we need to address is the ways in which economic, technical, social and political constraints on the segmentation of networks work to limit the degree to which the integrated infrastructure networks of the modern ideal can be totally segmented within infrastructural markets. Across national contexts and infrastructure sectors there is wide variation in the potential for, and acceptability of, the different options for segmenting networks.

The key to understanding the break-up of integrated infrastructure is to explore how the marketability of the various segmented elements of a network is contested and defined. In what follows we develop a conceptual framework that builds an understanding of the marketability of different segments of infrastructure networks. This framework explains the process through which infrastructure is unbundled and presents a guide to the process of infrastructural unbundling, across different sectors, identifying where competitive conditions could apply or be approximated and contexts where they do not apply. The framework is based on an understanding of the four characteristics that influence the 'marketability' of infrastructure:

- *The character of the service*. The degree to which an infrastructure service is a jointly consumed 'public good' or individually consumed as a 'private good'.
- *The conditions of production of the infrastructure service*. The extent to which an infrastructure service is 'contestable' and open to new entrants who can compete with incumbents.
- *The environmental externalities and social objectives of service provision*. The extent to which the costs and benefits of an infrastructure affect other persons than those directly involved in consuming the service.
- *The character of users' demand for infrastructure services*. The degree of consumer access to information about supply alternatives and the existence of substitutes for particular kinds of services.

In what follows we will discuss each of these in turn.

RIVALRY AND EXCLUDABILITY: THE CHARACTERISTICS OF THE INFRASTRUCTURE SERVICE

Turning first to infrastructure service characteristics, Figure 4.4 presents a matrix that attempts to assess the potential for introducing competition into various types of infrastructure. Each infrastructure sector contains activities which can be unbundled or segmented, each of which is shaped by two key concepts: 'rivalrousness' and 'excludability'. *Purely private goods* are usually consumed by one person at a time. These goods, such as food and consumer durables, are highly rivalrous in consumption.

At the opposite extreme are *purely public goods*. These have low rivalry because consumption by one individual does not lessen availability to others. Such goods are said to be jointly

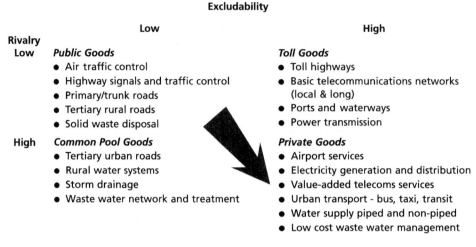

Excludability

	Low	**High**
Rivalry		
Low	**Public Goods**	**Toll Goods**
	• Air traffic control	• Toll highways
	• Highway signals and traffic control	• Basic telecommunications networks (local & long)
	• Primary/trunk roads	• Ports and waterways
	• Tertiary rural roads	• Power transmission
	• Solid waste disposal	
High	**Common Pool Goods**	**Private Goods**
	• Tertiary urban roads	• Airport services
	• Rural water systems	• Electricity generation and distribution
	• Storm drainage	• Value-added telecoms services
	• Waste water network and treatment	• Urban transport - bus, taxi, transit
		• Water supply piped and non-piped
		• Low cost waste water management

Notes: Rivalry refers to the impact that consumption by additional users has on the consumption opportunities of all users. Low rivalry (nonrival) occurs when consumption by one user does not reduce availability to other users - such goods are jointly consumed. High rivalry refers to consumption by one user that then imposes identifiable costs on other users - such goods are privately consumed. Excludability refers to the feasibility of control access to a good - high excludability means it is easy and non-costly to prevent users from consuming the good.

Figure 4.4 Infrastructure potential for competition. *Source*: Israel (1992), 52

consumed. For example, highway use by one vehicle does not usually affect use by other vehicles. But there are circumstances when a highway starts to become congested as additional users impose additional costs on existing users.

The second criterion is the concept of 'excludability', which can be defined as the feasibility of controlling access to infrastructure. Usually, individual consumers can be excluded from transactions involving purely private goods. Such exclusion is usually not feasible, or very costly, in the case of public goods.

Between the two extremes of purely private and public goods are *toll* and *common pool* goods. *Toll* goods are characterised by high levels of excludability but a low level of rivalrousness. For instance, it is possible to control access to a piped sewage system, but consumption by one user does not usually lessen its availability to others. *Common pool* goods are rivalrous in consumption but have low feasibility of excluding individual users – examples would include small rural roads and access to storm drainage. There is an increasing tendency for infrastructures that were seen as purely public goods, such as urban roads, to be viewed as private or tollable because the technology now exists to restrict access through road pricing technologies. Particular judgements in these cases are powerfully shaped by assumptions about the type of technology and the costs of exclusion.

CONDITIONS OF PRODUCTION OF INFRASTRUCTURE SERVICES

The efficiency of the private market depends on the existence of effective competition. This does, however, preclude services that are delivered by *natural monopolies* – transmission networks, local distribution networks and rural roads.

Natural monopoly conditions occur where there are high economies of scale, which implies that the unit cost of supplying a service will be minimised when the market is served by a single, rather than multiple, supplier. A natural monopoly has a high degree of market power that it can exploit by increasing prices in excess of marginal cost. The lack of competition may blunt the incentive to increase efficiency while high levels of market power are a barrier to new entrants. New technologies may reduce economies of scale and decrease the capital investment needed to enter the market.

However, the main deterrent to competition in infrastructure provision is the magnitude of *sunk costs* in the event of exit from the market. The costs are sunk to the extent that they cannot be recovered for other uses – this is generally the case with specialised equipment and fixed installations such as roads and sewers. When the production of a service requires no sunk costs it is referred to as *contestable* and the threat of entry is usually considered to provide similar market discipline to an incumbent monopoly. But there may still be practical barriers to entry imposed by other policies or shortage of financing.

A concept which is related but distinct from natural monopoly is the degree of coordination needed for an infrastructure network to function effectively. Because of the interlocking networks involved in infrastructure systems, and the complexity of the resources that flow along networks, their management must follow a number of minimum rules and regulations. Consequently, regulation needs to ensure that there is a degree of formal co-ordination in the planning of investment, technical operation and setting of minimum standards for connections between networks.

EXTERNALITIES AND SOCIAL OBJECTIVES

Externalities occur where the benefits and costs of producing or consuming goods affect persons other than those involved in the transaction. In the infrastructure sector *negative* externalities include air, noise, water and land pollution from motor vehicles and electricity production. *Positive* externalities include the public health benefits of access to water and sanitation infrastructure. Many infrastructure activities also generate *network* externalities where all users benefit when a new user joins the network because of the ability to communicate with more people. Similarly, broader social and political objectives, valued by the wider social community, such as universal access regardless of location to a minimum level of service, have wider social benefits. Regulation needs to ensure that externalities and social obligations are managed to meet wider societal objectives.

CHARACTERISTICS OF DEMAND AND SERVICE USE

Finally, there are five features of infrastructure demand that suggest an enlarged set of requirements for consumers to obtain satisfaction from infrastructure supply. These will have important implications for regulatory policies and the form of privatisation:

- *The existence of substitutes for infrastructure.* In some cases, acceptable and affordable alternatives can substitute for the services provided by the incumbent supplier, weakening their market power. For instance, consumers may turn to private electricity generation even at a higher cost because of the unreliability of public power sources. Households may resort to private water vendors when the quality and/or reliability of the public system fails to provide an adequate service. Specialised communication networks for high-volume business users can bypass congested public telephone networks. Yet many consumers, especially low-income users, are financially or physically restricted in their access to substitutes. Regulation may need to ensure that competition from substitutes can discipline an incumbent monopoly provider.
- *The price elasticity of demand* for infrastructure services. For certain infrastructure services in certain minimum quantities – such as water and energy – demand is virtually inelastic because the service would have to be consumed at almost any price. Water is an essential service that cannot be substituted for and must be consumed in certain minimum qualities to sustain life. Beyond this particular case, the price elasticity of infrastructure varies among different groups of users. There is a recognition that a minimum level of public transport, electricity and the basic telephone demonstrate inelastic demand as development increases. This is because they are necessities if people are to gain access to jobs, health care, etc. Where demand is price-inelastic, especially for low-income groups, suppliers must be regulated on welfare grounds.
- *Consumers' access to information about infrastructure services.* Market-based provision assumes that consumers have access to information about the cost and quality of infrastructure services. But, in a number of sectors, the quality and cost of a service are difficult to assess, e.g. water or energy quality, the safety of different urban transport modes and the availability of services at a particular point in time. All these factors mean that it is difficult for consumers to make informed choices about the services they want to use. Regulators may need to develop indicators of service quality to make the performance of services more transparent to increase the marketability of competing services.
- *The pattern of demand for infrastructure.* Demand for many infrastructure services, especially power, voice telecommunications, water supply and urban transport, is not distributed evenly over time but shows distinct peak and off-peak periods. Such services cannot simply be stored, so the capacity of the system must be designed to meet peak, rather than average, demand in real time. The supplier must therefore meet the high cost of providing excess capacity in slack periods to meet higher demand during peaks. Infrastructure networks are often configured to meet peak demand that only occurs once or twice a day at different times during the year. The construction of additional capacity to meet peaks can be delayed or postponed by using pricing mechanisms to shift consumption from peak to off-peak periods. Regulation may need to ensure sufficient capacity is provided to meet peak demand or provide incentives to accelerate the development of demand management activities.
- *The diversity of users' infrastructure needs.* The modern conception of infrastructure as homogeneous products provided by a standardised production process, through monopolistic providers, is no longer relevant. User needs in the water, power, transport and telecommunications sectors are becoming much more diverse, owing to rapid

changes in production processes and users' lifestyles. Industrial firms, using sensitive and electronically mediated production processes, are increasingly reliant on extremely reliable and high-quality electricity sources that minimise voltage 'spikes' and 'troughs' to avoid interference with sensitive equipment. The financial services sector depends on extremely reliable telecommunications networks with stand-by systems designed to allow seamless operations if the primary network is disrupted. Domestic users are now interested in value-added utility services such as green electricity produced by renewable sources. Regulation may need to ensure that the structure of supply has to become much more diverse to reflect the differing infrastructural needs of different users.

THE MARKETABILITY OF UNBUNDLED INFRASTRUCTURE: A SUMMARY

Table 4.1 summarises this assessment of the marketability of infrastructure. There are three key conclusions here. First, the provision of networked infrastructure networks, especially the primary or trunk elements of the network – the transmission of gas, electricity, water, sewerage and highways – tends to exhibit the characteristics of public goods. That is, they have high sunk costs and are most effectively provided by natural monopolies. Second, the operation of the networks often does not now entail large sunk costs of equipment and is usually contestable. Finally, the infrastructure sector is highly heterogeneous, both within and between sectors, in the level of marketability assigned to specific activities and different segments of the networks.

INSTITUTIONAL PATHWAYS TO UNBUNDLED INFRASTRUCTURE

The above framework has shown that there is considerable economic and technological potential to reassign infrastructural activities between the public and private sectors. However, it also demonstrates the need for a third area of discussion to explore the different pathways to unbundled infrastructure networks and the ways in which they reflect the broader social, political and economic feasibility of segmenting networks in quite different national and political contexts.

The transition from integrated to unbundled networks is often captured within a single concept – privatisation or deregulation – but the process is actually much more complex and 'messy'. The transition from integrated infrastructure can follow a series of pathways characterised by significant variation across infrastructure sectors and nation states. When the marketability of different infrastructural activities has been assessed within particular contexts the responsibility for different services and segments can potentially be reassigned between the public and private sectors.

The range of alternatives, as illustrated in Table 4.2, is very broad. Options range along a continuum between completely public and completely private sector participation. The range

Table 4.1 The marketability of infrastructure

	Service character		Production		Co-ordination requirements	Externalities and social		Summary
	Rivalness	Excludability	Economies of scale	Sunkness of costs		Externalities	Social objectives	Potential for competition
Telecommunications						Network effects	USO, national integration	
Local networks	Private	High	Low	Low	High			Medium
Long-distance value added	Private	High	Low	Low	High			High
Energy						Pollution	USO	
Generation	Private	High	Medium	Medium	High			Low
Transmission	Private	High	High	Medium	High			Medium
Distribution	Private	High	Medium	Medium	High			High
Water						Allocation and public health	USO	
Urban piped	Private	High	High	High	High			Medium
Non-piped	Private	High	Low	Low	Low			High
Sanitation						Pollution and health	USO	
Urban piped sewage and treatment	Toll	Medium	Medium	High	High			High
On site	High	Private	Low	Low/medium	Low			High
Transport								
Urban bus and taxi	Private	High	Low	Low	Low	Pollution	Affordability	High
Urban transit	Private	High	Medium	High	High	Land use	Affordability	Medium
Urban roads	Common/ private	Low/high	Low	High	High	Land use, environment	Integration	Medium
Port and airport facilities	Toll	High	Medium	High	High	Noise safety	Integration	Low
Port and airport services	Private	High	Low	Low	High	Noise safety	Integration	High
Rail network and stations	Toll	High	Medium	High	High	Network effects	Remote access	Medium
Rail services	Private	High	Low	Low	High			Medium

Source: Israel (1992), 115.

of options does not extend from government monopoly to unfettered markets. Rather, it involves a wider series of different options. Yet, while this typology provides a useful indication of the range of options for involving the private sector in the provision of infrastructure, it does not provide an assessment of the importance of network segmentation. We therefore need to move from overly simplistic typologies that only focus on the level of private sector participation in the provision of networks. A more useful conceptualisation would be able to recognise the diversity of institutional forms that can surround network unbundling. Our central concern in what follows is not with privatisation *per se*. Rather, we are concerned with those urban contexts in which infrastructure networks become unbundled, bringing new parallel contestable networks to coexist with the incumbent network.

Figure 4.5 provides a more detailed conceptualisation of the institutional alternatives for infrastructure provision. This transcends the limitations of a simple comparative framework based on degree of public and private sector involvement. There are three key features of this model.

Table 4.2 Infrastructure: degree of public/private responsibility

Role of market incentives	Range of responsibility	Institutional options	Description
	Public sector		
Low	••••••••	Government department	Service provided by civil servants and accounts in government budget
	•••••••	Parasatal	An organisation owned and controlled by the state
	••••••	Service contract	Contracting out services to the private sector for fixed period and fee
	•••••	Management contract	Private sector manages publicly owned infrastructure for fee or performance-related fee
	••••	Leasing	Private sector operates a public facility for a fixed period but does not provide fixed assets
	•••	Concessions (BOT)	Private sector leases an asset for an extended period – investment reverts to public sector
	••	Communal arrangements	Users cooperatively plan, build, maintain and manage infrastructure
High	•	Private entrepreneurship	Ownership by private sector either through transfer of assets from public sector or new entry
	Public sector		

Source: Kessides (1983a), 18.

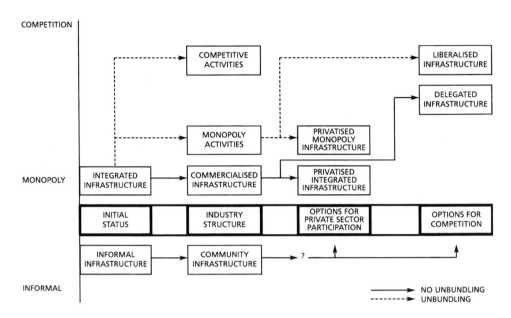

Figure 4.5 Pathways to unbundled networks. *Source*: based on World Bank (1994), 56

First, the new model is able to map a much more complex set of pathways towards unbundled infrastructure than the usual polarities of 'private' and 'public' provision. In many contexts the transition away from integrated networks is a complex and highly contested process and at least in the early stage the move is not to competitive markets. There may actually be a number of intermediate stages.

Second, the model illustrates that shifts in the social organisation of infrastructure services can occur, incorporate more private sector participation, but not require the unbundling of networks. Private sector participation cannot simply be equated with unbundling and the competitive provision of infrastructure. Yet privatisation may be able to increase accountability through the discipline of budgetary constraints and the signal sent through government regulation.

Finally, two of the options do create the context for competitive conditions:

- *Delegated infrastructure*. Governments create the conditions through which firms compete for the right to supply an entire market of users through leases or concessions.
- *Liberalised infrastructure*. Multiple providers compete directly with each to provide services within a regulatory framework.

It is worth briefly examining the key features of these six different pathways. In what follows we do this, whilst selecting examples from different infrastructure and national contexts to illustrate the shifts involved.

0 INTEGRATED INFRASTRUCTURE: GOVERNMENT DEPARTMENT

Option 0 is the *status quo*, with integrated infrastructure still being provided by a government department or a publicly owned utility. This option is increasingly untenable with international regulatory agencies placing increasing pressure on developing countries to reform public infrastructure and with regional trade agreements forcing governments in the advanced economies to introduce private sector participation and liberalise integrated infrastructure networks. There are a range of different options for restructuring infrastructure provision, with the selection being shaped by the particular socioeconomic and institutional context within which the transition away from integrated networks is being developed.

1 COMMERCIALISED INFRASTRUCTURE: PUBLIC OWNERSHIP AND OPERATION

Option 1 does not require the unbundling or segmentation of the networks. Instead, an integrated infrastructure provided by a government department or a public utility is transformed into an independent but publicly owned corporatised parastatal organisation designed to run on commercial lines. Networks are still owned by the state, limited subsidies from government may fund capital expenditure, but the objective is that costs should be covered by revenues. The public sector will continue to have the primary responsibility for the provision of infrastructure networks in many countries and sectors.

In low-income countries weak private sector and management capabilities will improve only slowly. Even with a dynamic private sector, roads and major public works are likely to remain in the public domain. National governments may also decide for strategic, regulatory and political reasons to retain public sector responsibility for infrastructure provision. In many contexts, however, especially low-income countries, governments are under pressure from international aid agencies and the World Bank to improve the performance of publicly owned infrastructure organisations because more effective performance by the public sector will actively facilitate future private sector participation.

2 DELEGATED INFRASTRUCTURE: PUBLIC OWNERSHIP AND PRIVATE OPERATION

In option 2 the integrated network can remain bundled or be segmented into contestable and monopoly elements. Delegated infrastructure involves suppliers competing for *the right to supply a market*. Governments create market conditions by offering leases or concessions for either integrated or monopoly elements of the networks. These can range from specific contracts to longer-term concessions that require operation, maintenance and facility expansion. Competition occurs before the contract is signed and when it expires and is due for renewal. Therefore the government creates competitive conditions through leases or concessions and firms compete for the right to supply the entire market (see Box 4.1). The

BOX 4.1 EXAMPLES OF DELEGATED INFRASTRUCTURE

THE FRENCH WATER SECTOR

The ownership of infrastructural assets in France remains with municipalities, which contract-out operations and maintenance to private firms under management contracts, lease contracts or concessions. Under lease contracts and concessions, fees are established through negotiation or on the basis of competitive bidding, with adjustments for inflation according to an agreed formula. Tariff levels are monitored by the Ministry of the Economy and Finance, and environmental standards, and are enforced at the *département* level of government. Competition for the market is the key to efficiency. The water supply market of each municipality is contestable, and the existence of two major firms, Lyonnais des Eaux and Compagnie Général des Eaux, also permits the use of benchmark indicators.

However, the separation of ownership and operations could distort investment incentives. One mechanism that helps is full cost recovery through water tariffs: since both owner and operator recover costs from the same source, they have the incentive to collaborate in making a sound investment decision. The length of concessions may also result in better asset management, but it reduces the potential for competition.

The French system has been the model for several developing countries. The approach can be adapted to a variety of situations. If properly designed, for example with compensation linked to the volume of water delivered to consumers, such arrangements can transfer all the commercial risk of operation and maintenance to the private sector. Actual ownership of the assets may not matter very much because the utility owns the right to a stream of revenues.

WATER SUPPLY IN GUINEA

Before 1989 Guinea had one of the least developed water supply sectors in West Africa. Then the sector was restructured: ownership of the urban water supply infrastructure and responsibility for sector planning and investment were transferred to a new autonomous water authority, SONEG. A new company, SEEG, was created to operate and maintain the facilities; it was a joint venture, 49 per cent owned by the government and 51 per cent by a private foreign consortium.

Under the ten-year lease signed with SONEG, SEEG operates and maintains the system at its own risk, with its remuneration based on user charges actually collected and new connections. The company benefits from improvements it achieves in the collection ratio and the reduction of operating costs and unaccounted-for water. It has incentives to seek adequate tariffs and to invest carefully based on realistic demand forecasts, since it has responsibility for capital finance. Water tariffs have increased from US$0.12 per cubic metre in 1989 to about US$0.90 in 1995. Despite higher tariffs, the collection ratio of private customers has increased dramatically from 20 per cent to around 85 per cent in 1995. Peak technical efficiency of the network and service coverage have also improved.

commitments in contracts can then provide an alternative to independent regulation. The use of leases and concessions has become increasingly common in the infrastructure sector in both developed and low-income countries. They are particularly common in the transport sector for large fixed facilities such as ports and toll roads and have been very common in the water sector. There are, as Table 4.3 demonstrates, a wide variety of arrangements possible

Table 4.3 Contractual arrangements for water supply

Contract	Applications	Incentives	Examples
Service	Meter reading, billing and collection, and maintenance of private connections	Permits competition among multiple providers, each with short and specific contracts	A public water company, EMOS, in Santiago, Chile, encouraged employees to leave the company in 1977 and compete for service contracts for tasks previously performed internally – resulting in large productivity gains
Management	Operation and maintenance of the water supply system or major subsystem	Contract renewed every one to three years, and remuneration based on physical parameters, such as volume of water produced and improvement in collection rates	Electricity and water company of Guinea-Bissau (EAGB); contract awarded to Electricité de France, with about 75% of the remuneration and a possible additional 25% based on performance
Lease	Extended operational contract	Contract bidding, with contract duration of about ten years; provider assumes operational risk	Water supply in Guinea owned by state enterprise (SONEG) and leased to operating company (SEEG) from 1989 for ten years; achieved large increases in bill collection
Concession	All features of the lease contract, plus financing of some fixed assets	Contract bidding, with contract period up to thirty years; provider assumes operational and investment risk	Côte d'Ivoire's urban water supply concession went to SODECI, a consortium of Ivoirien and French companies; SODECI receives no operating subsidies and all investments are self-financed

Source: Triche (1990).

within these agreements. Using this option, the public sector can delegate the operation of infrastructure and responsibility for new investment to the private sector without dismantling existing institutions or creating new regulatory frameworks.

3 PRIVATISED INFRASTRUCTURE: PRIVATE OWNERSHIP AND OPERATION

Option 3 can also apply to integrated and unbundled elements of the network. In both cases, monopolies continue to exist. No competitive element is involved in this transition but it is assumed that the transfer from public to private ownership will produce efficiency gains. This option involves the private ownership and operation of infrastructure, usually in contexts when commercial and political risks are low and there is high potential for securing revenue from user charges. A system of regulation is needed to ensure that private monopolies do not exploit market power and that providers seek increased efficiency gains.

Although privatisation has a long history it is comparatively new in the infrastructure sector. Privatised utilities usually undergo major corporate restructuring. The benefits of privatisation are ambiguous. Many studies show short-term benefits to producers, consumers and employees, but longer-term effects have not been demonstrated. Infrastructure privatisations are often linked with a requirement to undertake certain minimum investments and roll-out obligations contained with service conditions imposed on companies. Governments transfer monopolies to private ownership with a regulatory mechanism to reward performance and efficiency gains. This may lead to improvements in economic efficiency but is not a sufficient condition for improved economic performance because privatisation simply transfers assets out of the public sector.

Privatisation has spread rapidly through developed and developing cities. In the Western world the United Kingdom has led in the scale of its privatisation of infrastructure, while many developed and middle-income nations have privatised telecommunications, power and, increasingly, their water sectors. In such processes, public infrastructure undergoes major corporate restructuring. But there is often considerable controversy about the requirement to undertaken minimum levels of investment, roll-out obligations and universal service requirements.

Table 4.4 illustrates that, between 1988 and 1993 the global value of infrastructure privatisation was over US$30 billion. Telecommunications and power are the most dynamic sectors and over 60 per cent of all infrastructure privatisation by value has taken place in Latin America. In the advanced economies the value of infrastructure privatisation is likely to be in the hundreds of billions. Yet, in all these contexts, privatisation does not necessarily create contestable infrastructure markets. Without unbundling, integrated infrastructure networks can be privatised as integrated privatised monopolies. If the networks are unbundled the monopolistic elements of the trunk networks can also be privatised. But, in both cases, privatisation does not inevitably lead to competition in the provision of networked services.

Table 4.4a Value of infrastructure privatisations in developing countries, 1988–93, by subsector (US$ billion)

Subsector	1988	1989	1990	1991	1992	Total, 1988–92	No. of countries
Telecommunications	325	212	4,036	5,743	1,504	11,821	14
Power generation	106	2,100	20	248	1,689	4,164	9
Power distribution	0	0	0	98	1,037	1,135	2
Gas distribution	0	0	0	0	1,906	1,906	2
Railroads	0	0	0	110	217	327	1
Road infrastructure	0	0	250	0	0	250	1
Ports	0	0	0	0	7	7	2
Water	0	0	0	0	175	175	2
Total	431	2,312	4,307	6,200	6,535	19,785	15
Closely related privatisations							
Airlines	367	42	775	168	1,461	2,813	14
Shipping	0	0	0	135	1	136	2
Road transport	0	0	0	1	12	13	3
Total developing country privatisations	2,587	5,188	8,618	22,049	23,187	61,629	25

Note: Countries undertaking infrastructure privatisations:

1988. Power: Mexico; telecom: Belize, Chile, Jamaica, Turkey; airlines: Argentina, Mexico.

1989. Power: Korea; telecom: Chile, Jamaica; airlines: Chile.

1990. Power: Malaysia, Turkey; telecom: Argentina, Belize, Chile, Jamaica, Malaysia, Mexico, Poland; roads: Argentina; airlines: Argentina, Brazil, Mexico, Pakistan.

1991. Power generation: Chile, Hungary; power distribution: Philippines; roads: Argentina; telecom: Argentina, Barbados, Belize, Hungary, Jamaica, Mexico, Peru, Philippines, Venezuela; airlines: Honduras, Hungary, Panama, Turkey, Venezuela; shipping: Malaysia; road transport: Togo.

1992. Power generation: Argentina, Belize, Malaysia, Poland; power distribution: Argentina, Philippines; gas distribution: Argentina, Turkey; telecom: Argentina, Estonia, Malaysia, Turkey; railroads: Argentina; ports: Colombia, Pakistan; water: Argentina, Malaysia; airlines: Czechoslovakia, Hungary, Malaysia, Mexico, Panama, Philippines, Thailand; shipping: Sri Lanka; road transport: China, Peru.

Table 4.4b Value of infrastructure privatisations
in developing countries, 1988–93, by region
(US$ billion)

Region	Value
Africa	0.1
Asia	7.4
Latin America	22.5
Eastern Europe and Central Asia	2.0
Total	32.0

Note: Number of divestitures 267.

4 LIBERALISED INFRASTRUCTURE: PRIVATE COMPETITORS

Liberalised infrastructure requires unbundling and the segmentation of contestable elements of the network to allow multiple providers to compete directly for users within the context of a regulatory framework. The shift towards the liberalisation of infrastructure restructures both the size and the structure of the industry. This is a most dramatic shift in the infrastructure sector where a combination of technological and regulatory change is making competition possible. Central to the process is the unbundling of networks by detaching activities previously undertaken by monopolies and opening them to various forms of competition. Unbundling of infrastructure is moving at a brisk pace as governments isolate the natural monopoly elements of networks and encourage new entrants in segments that are competitive. In the energy sector energy supply, generation, transmission and distribution can now be owned and operated by different companies. While the physical distribution network will retain natural monopoly characteristics, the alternative suppliers can gain access to the network on an equal basis. Box 4.2 illustrates the claims made for liberalised infrastructure policies in the United States while Box 4.3 provides an example of liberalisation in the tele-communications sector of a developing country.

5 COMMUNITY INFRASTRUCTURE: USER PROVISION WITH POLICY SUPPORT

There are a number of problems with market-based approaches to infrastructure provision that cannot be addressed in the above options. In many developing cities, for instance, no formal infrastructure exists and there are few incentives to extend services, especially to low-income users. In these contexts, coordination across and between infrastructure sectors may not receive enough attention. In those urban areas where there is no infrastructure, the amount and type of infrastructure must be planned, investments made, and infrastructure

BOX 4.2 LIBERALISED INFRASTRUCTURE IN THE UNITED STATES

With its long history of private infrastructure provision, the United States exemplifies the changes in regulatory goals and implementation of the ensuing cycles in regulatory policy. In the late nineteenth century, and well into the early part of the twentieth, much competition prevailed, especially in electric power and telecommunications.

An early instance of economic regulation, the Interstate Commerce Act of 1887 was concerned with monopoly power in railway operations. The bounds of economic regulation were extended gradually but, especially during the Great Depression and the 1930s, to virtually all infrastructure sectors and to other areas of public interest (for example, creating service obligations and information disclosure requirements).

Delivery of infrastructure thus came to be based on a particular social compact. The service provider was typically granted exclusive rights to specific markets, and, in return, the government took on the public responsibility of ensuring that service obligations were fulfilled at reasonable and just prices. Inflationary pressures in the early 1970s caused regulators to intervene even more heavily in the operations of service providers. Health, safety and environmental regulation also gained momentum around that time.

Public dissatisfaction with regulatory outcomes resulted in a move to reduce economic regulation in many sectors in the late 1970s and 1980s. According to one estimate, 17 per cent of the US gross national product (GNP) in 1977 was produced by fully regulated industries. By 1988 the proportion had declined to 6.6 per cent as large parts of the transport, communications, energy and financial sectors were freed from economic regulation. Greater operational freedom, and competitive threats, stimulated service providers to adopt new marketing, technological and organisational practices. Evidence from the United States points to substantial economic gains from deregulation.

provided and maintained. Private entities are not likely to make investments in roads and water, since user charges that fully cover costs are not always feasible. Private sector participation in the provision of infrastructure is not, therefore, likely to be an option open to governments.

An alternative approach is a shift towards the devolution of infrastructure planning and management, with a higher degree of user and community involvement. As Box 4.4 shows, the aim is to use appropriate technologies, local organisation and installation, and lower technical standards, to create quite literally a self-built infrastructure in cities that have been abandoned by formal networks and are not attractive to private sector participation. A high degree of support, usually delivered through development organisations, is required to develop these bottom-up models of infrastructure provision. Once the network is built and operated with a significant consumer base it may be possible for such informal networks to be incorporated into the formal system of infrastructure provision. At that stage there may be the potential for private sector participation in the provision, management or ownership of the network.

BOX 4.3 LIBERALISATION IN THE GHANA CELLULAR TELEPHONE MARKET

Ghana's telecommunications sector is substantially underdeveloped. In the 1980s there were on average twenty-three telephones per 1,000 people and levels of connection were actually declining. Over half the country's districts were entirely without a telephone service and effectively cut off from the capital, Accra. In 1994 30 per cent of local calls could not be completed, international calls had only a 15 per cent completion rate and it was estimated that every line in the country experienced at least one or more faults per year.

The response to these problems was to encourage private sector participation and create a competitive market in cellular tele-

phones. In 1990 six private companies were licensed to operate cellular telephone services and the market is open to other operators who wish to enter. Competition between the cellular companies to provide local telephone services has had the effect of expanding the network and lowering the prices of mobile telephony. The cost of adding cellular users is lower: about US$1,000 per customer versus US$3,600 per customer for the hard wire service operated by Ghana Telecom. Services may therefore be expanded to geographical areas that contain at least half the country's population and the regional capitals over the next few years.

BOX 4.4 COMMUNITY INFRASTRUCTURE: WATER SUPPLY

World Bank-funded projects in Brazil provide a useful example of how demand-oriented planning of low-cost water and sanitation requires considerable adjustment by the formal institutions of government, engineering and external donors. In Brazil the water and sanitation programme for low-income urban populations is investing US$100 million to provide water and sanitation infrastructure to about 800,000 people in low-income areas in eleven cities.

Participation must be tailored to the population. Since 1997 the programme has taken a variety of approaches to involve beneficiaries and the design of projects. In one approach, leaders of community organisations

are consulted on basic choices and the details are worked out with the actual beneficiaries. In another approach, agreement is reached between design engineers and beneficiaries directly in consultation with community leaders and organisations. In both these models, conflicts of interest between the water companies and community-based organisations are resolved through negotiation, with the project design consultant acting as facilitator. These approaches have dramatically lowered investment costs and increased the sense of project ownership among communities. The process has directly affected the kind of engineering advice adopted.

6 Informal infrastructure: self-help and private provision

The final option is a pathway that continues to allow the development of informal infrastructure services. Basically, those users who cannot afford to obtain access to formal systems are forced to search for informal and usually unregulated alternatives. These informal options are very costly for low-income households, especially in the water sector, where there is overwhelming evidence that households pay significantly higher tariffs from private water vendors, shown to be between 20 per cent and 2,000 per cent higher than normal public tariffs. Without reform of the state-owned infrastructure networks in these contexts, significant costs are imposed on households and businesses. There are, however, major difficulties in sustaining institutional reform, attracting private sector interest and maintaining wider social equity objectives. The options for reform are likely to be limited without significant external financial support.

Transitions to unbundled infrastructure

The pathways followed in an infrastructure sector are also powerfully shaped by particular national contexts of transition and the particular characteristics of the infrastructure network. These include the coverage, effectiveness and demand for existing infrastructure; the institutional capacity for the development of commercial and competitive infrastructure: management and technical capabilities; the development of a regulatory environment that facilitates private activity; and the private sector's interests and capacity. Table 4.5 compares the coverage and quality of infrastructure in five types of national economy, and here we review the relative importance of different pathways from integrated infrastructure among each country bloc:

- *Low-income developing nations.* In low-income countries infrastructure is characterised by low coverage and poor performance, and a high rate of population growth means that demand for new services is growing rapidly. The challenge is to improve the performance of existing networks and expand networks to meet unmet and growing levels of demand. In such countries the technical capability, enabling environment and private sector interest in unbundled infrastructure are often very low. Table 4.6 illustrates that the dominant pathway away from integrated infrastructure in Africa has tended to focus almost exclusively upon the commercialisation and devolution of infrastructure networks, with few examples of privatisation or liberalisation.
- *Post-communist states.* The transition economies of former communist countries are usually characterised by high levels of infrastructure coverage and reasonably good, although rapidly deteriorating, technical performance. The central challenge is to reorientate infrastructure supply to meet changing demands brought about by economic restructuring. While technical capability can be high, the enabling environment for private activity is still developing. The middle-income reforming economies have reasonable coverage but weak performance, especially in maintenance. The key challenge is to improve efficiency to support higher economic growth. While the capacity to privatise

Table 4.5 Country infrastructure: coverage and performance

Indicator	Low-income economies	Transition economies	Middle-income reforming economies	High-growth economies	OECD economies
Coverage of infrastructure					
Main lines per thousand persons	3	95	73	122	475
Households with access to safe water (%)	47	95	76	86	99
Households with electricity (%)	21	85	62	61	98
Performance of infrastructure					
Diesel locomotives unavailable (%)	55	27	36	26	16
Unaccounted-for water (%)	35	28	37	39	13
Paved roads not in good condition (%)	59	50	63	46	15
Power system losses (%)	22	14	17	13	7
Basic indicators					
GNP per capita, 1991 (US$)	293	2,042	1,941	3,145	20,535
GNP per capita average annual growth rate, 1980–91 (%)	-0.2	1	-0.6	5	2
Population average annual growth rate, 1980–91 (%)					
Urban	6	1	3	4	1
Total	3	0.3	2	2	0.5

Source: World Bank (1994), 12.

Table 4.6 Infrastructure transitions in Africa

Management contract	Lease	Concession/BOOT	Demonopolise BOO	Divestiture
Water				
Gabon	CAR			
Gambia	Côte d'Ivoire			
Mali	Guinea			
	South Africa			
Electricity				
Gabon	Côte d'Ivoire		Côte d'Ivoire	
Gambia	Guinea		Mozambique	
Ghana				
Guinea				
Guinea-Bissau				
Mali				
Rwanda				
Sierra Leone				
Telecoms				
Benin		Guinea-Bissau	Burundi	Sudan
Botswana			Ghana	
Guinea			Guinea	
Madagascar			Madagascar	
			Mauritius	
			Namibia	
			Nigeria	
			South Africa	
			Tanzania	
			Uganda	
			Zaire	
Railways				
Cameroon	Burkina Faso			
Tanzania	Côte d'Ivoire			
Togo	Gabon			
	Zaire			
Airports				
Togo	Benin	Cameroon		
	Gabon	Mali		
	Guinea			
	Madagascar			
	Mauritania			
Ports				
Sierra Leone	Cameroon	Mali	South Africa	
	Mozambique	Mozambique		

Source: Kerf and Smith (1996), 19.

infrastructure is well established, low rates of economic growth restrict private sector involvement.

• *High-income developed economies* have comparatively good coverage and performance. The key challenge is to meet the demands of economic growth and increasing urban population growth. The OECD countries have the highest level of coverage and the most effective infrastructure performance. Population growth rates are low, so the main requirement is to reorientate the supply of infrastructure to meet new demands brought about by economic and technological change, especially in telecommunications. In such contexts there is usually strong technical capacity, a supportive regulatory environment and strong interest from the private sector; all institutional pathways are open but with strong emphasis on the development of private and liberalised options. We saw in Box 4.2 that nearly all the major US infrastructure sectors have been liberalised, with electricity currently undergoing rapid unbundling.

SECTORAL VARIATIONS IN UNBUNDLING OPTIONS

Finally, the institutional pathways are not simply determined by the particular technical, economic, institutional and social characteristics of a national state. We have seen that there are important variations in the marketability of different infrastructure networks, therefore the institutional options are also likely to vary by infrastructure. Table 4.7 identifies the pathways that are open to different types of infrastructure networks.

TELECOMMUNICATIONS

The marketability of unbundled *telecommunication* services is high – particularly for the provision of long-distance and value-added services. Growth in demand for telecommunication services, coupled with technological innovation such as wireless and mobile, which have lowered the cost of network construction, have boosted the potential for competitive entry even at the local level. This sector is most often characterised by liberalised provision, with a mix of public and private service providers using a range of telecommunication technologies and offering different services tailored to meet different user needs.

ELECTRICITY AND GAS

In the *electricity* sector, generation and distribution can be separated from transmission and operated under concession or by privatised and liberalised provision. The national power transmission system tends to retain elements of a natural monopoly and can be either publicly or privately operated, providing access to the network is regulated. *Gas* can also be competitively supplied in many countries. Usually, natural gas production is vertically integrated, often with petroleum production, under public ownership. Transition from integrated supply requires segmentation of the networks to permit competitive production – through delegation or private ownership. The main regulatory issue is to ensure competitive access of producers to the transmission and distribution network. The UK gas sector has also introduced competition into local distribution.

Table 4.7 Pathways applicable to infrastructure networks

Form

Sector	Government department (national or sub-national)	Parastatal/ Public utility	Service contract	Management contract	Leasing (affermage)	Concession BOT	Regulated entrepreneur cooperative	Unregulated Entrepreneurs	Unregulated Cooperative communal
	(Investment plus O&M)	(Investment plus O&M)	(O&M only)	(O&M only)			(Investment plus O&M)	(Investment plus O&M)	
Power	Overall sectoral planning and policy making	Generation						Small scale	Communal systems
		Transmission system						Enclave	
		Distribution system							
Telecoms		Local transmission, switching system				Long distance / Transmission		Terminal equipment / Service extension	Communal systems
Transport Railways		Passenger and freight rail						Enclave	
Urban transport		Commuter rapid transit					Urban bus	Taxi, van interurban bus	
Road freight									
Roads		Primary, secondary, urban roads				Toll road		Enclave	Rural and local roads
Airport/airlines		General-use airports						Enclave and airlines	
Ports and waterways		General-use ports						Enclave and shipping	
Water and sanitation Water supply		Pipe water trunk and distribution						Enclave / Shallow well; vendor	Communal systems
Sewerage		Conventional sewerage and treatment						Intermediate sewerage	

Source: Kessides (1993a), 20.

WATER, WASTE AND SEWERAGE

Activities involving *water and waste* all have strong environmental links that make them less marketable than telecoms and power. But imaginative ways are being developed to facilitate competition in different elements of the networks or for the right to supply the market. The main pressure is for urban piped water and sewage networks to be provided by municipal or public enterprises that are run on commercial principles. The pressures on governments are to commercialise aspects of the public networks, to improve management and efficiency, and to ensure commercial operation, either through delegation or privatisation. Public regulation is necessary to ensure access for low-income customers and to protect health and the quality of water.

Sewage services can also be dealt with in a similar fashion. Low-income countries also have to consider another option for users not connected to the formal network. This includes the development of community-based options to be financed, constructed and managed by users to provide a minimum level of service. Finally, the transport sector creates the potential for a complex mix of options involving different forms of public and private provision.

INSTITUTIONAL PATHWAYS TO UNBUNDLED INFRASTRUCTURE: A SUMMARY

In this section we have outlined the six main pathways from integrated to unbundled networks. These lead us to two key conclusions. First, only a limited number of options require the unbundling of infrastructure and then only two options create competition. Delegated networks stimulate competition for the market, while liberalised infrastructure stimulates competition in infrastructure markets. Second, there is a huge degree of diversity in the pathways away from integrated networks that can be followed in a particular national context and for a particular infrastructure network.

UNBUNDLING INFRASTRUCTURE AND URBAN TRANSFORMATIONS

With our analysis of detailed practices of infrastructure unbundling now complete, to round off this chapter's discussion of practices of splintering urbanism we now need to return to the central concern of the book: how the transitions away from integrated networked infrastructures towards unbundled networks are involved in reconfigurations of social and spatial relations within and between cities.

In this section we develop a broad conceptual framework that can help us address this question in the remainder of the book. We will continue to look across water, energy and communications networks in both advanced and developing countries, focusing especially on the urban issues in contexts where networks have been unbundled. We primarily focus on those options that segment networks into monopoly and contestable elements. All transitions from the integrated ideal discussed above will involve a parallel reshaping of the sociotechnical relations and geographic landscapes of cities. In what follows, however, we focus on the most advanced and dramatic of these transitions: the creation of parallel or substitute networks that build multiple infrastructures in cities. We believe that the key to developing an understanding of the complex connections between unbundling and urban restructuring is to explore the uneven nature of three forms of what we call 'infrastructural bypass' (represented diagrammatically in Figure 4.6). These we define as:

- *Local bypass.* The physical development of a parallel infrastructure network that effectively connects valued users and places while simultaneously bypassing non-valued users and places within a city.
- *Glocal bypass.* The material development of a network that is configured to support interaction between local valued users and spaces and global circuits of infrastructural exchange.
- *Virtual network bypass.* The use of new information and communication technologies that support and facilitate the distribution of competitive infrastructure services over a single physically integrated network inherited from the modern ideal.

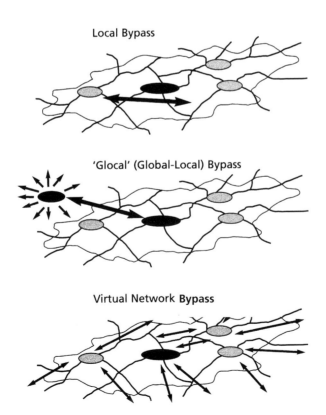

Figure 4.6 Diagrammatic representation of the three types of infrastructural bypass

We need to recognise that these categories are not mutually exclusive. Different styles of bypass usually coexist within the same network and city. For instance, local bypass usually relies on virtual network bypass to identify elements of the network, users and places in the city that should be targeted or ignored by new infrastructure networks, while facilitating connections between the monopolistic and competitive elements of different networks. Our typology should therefore be regarded as an ideal type through which we can build an understanding of how unbundled infrastructures have a central role in the reconfiguration, that is, the rebundling, of users and places within and between contemporary cities. We examine the three different forms of urban reconfiguration associated with unbundled infrastructure in more detail below.

LOCAL BYPASS

Our first type of infrastructural bypass, what we term 'local bypass', refers to the development of new parallel or substitute infrastructure networks that facilitate the development of contestable markets in infrastructure services within a city. These can take the form of multiple telecommunications networks (basic telephony, satellite, cable and mobile), the development of substitute energy sources (electricity, gas, district heating networks or local generation) and the creation of parallel or substitute telecommunications connections ('fat' Internet 'pipes') or highways, streets (private skywalks and tunnels as an alternative to the public street network). In each case a private company or public–private partnership can build an alternative or substitute network to the publicly provided infrastructure system. Such networks can be new, physically distinct material systems retrofitted through the city and/or they can be composed of the contestable segments of unbundled integrated infrastructure. These new networks are not ubiquitous or universal; they tend to be configured to connect only selected users and places while simultaneously bypassing others. Those users and places who are effectively bypassed by the new parallel networks then have to rely on the remaining public network or on informal mechanisms for infrastructure services. There are three main styles of local bypass.

SUPERIMPOSITION OF A PARALLEL INFRASTRUCTURE NETWORK

The first style of local bypass is linked with the provision of *entirely new infrastructure networks*. Local bypass often involves the construction of an entirely new infrastructure that is retrofitted through the cityscape and operates in parallel with existing public networks. The objective here is to develop a new parallel infrastructure network that provides valued zones and users with a high-quality and enlarged range of services that transcend the perceived problems – unreliability, poor quality, congestion, high costs and lack of choice – often associated with monopoly infrastructure networks. The parallel networks can be retrofitted within the existing city and are, therefore, often associated with initiatives designed to renew parts of the city.

Examples of the superimposition of parallel infrastructure networks are proliferating in all sectors.

- In the *energy* sector, steam networks or specialist air conditioning networks may be fitted in parallel with public energy supplies to provide more choice and cheaper supplies for commercial users. An example is the Citizen energy and steam distribution network in the city of London, which uses an abandoned pipe network to distribute heat to large users in competition with incumbent utilities and new entrants.
- In many North American city centres (for example Houston, Atlanta, Minneapolis and Montreal) massive *skywalk pedestrian systems* have been constructed to link malls, corporate office centres and entertainment complexes whilst bypassing the traditional municipal street system. Major new malls can also be considered as efforts to privatise and enclose pedestrian spaces, so bypassing local, public controls.
- New private *transport* systems are also being built in parallel with congested public networks to provide increased speed, certainty and reduced travel time for users who are prepared to pay. In Bangkok a network of private toll highways has been superimposed on top of the heavily congested public highway network. In North America and Australia there are a number of examples of new privately funded highway networks that have been built as alternatives to the public network. They include Highway 407, which runs for nearly 50 km in one of the most congested highway corridors near Toronto, and California Highway 91, located in the space between the lanes on the state highway with differential tariffs depending on the time of travel. There are at least another twenty-five proposals for privately funded networks to be developed in parallel with congested public highway networks.
- Even in *water and waste*, the least contestable of networks, recent UK policy has allowed large users to bypass local suppliers through pipelines and arrange what are known as 'inset agreements' with adjacent companies. Contestable water networks are not restricted to the commercial sector. Again in the United Kingdom, executive housing market-leading builders are constructing homes that are much less reliant on the incumbent utility's water and waste networks. In the first initiative in the north west of England 123 homes have been constructed with the Waterwise system, which recycles waste water that is routed back into the toilet cistern for reflushing, with the remaining water discharged to a nearby river. Water consumption is reduced by 30 per cent, and the housing does not have to be connected to the incumbent utility's waste network. The system enables house builders to develop executive housing estates in areas that would conventionally lack the necessary water and waste infrastructure to support major development projects.

SEGMENTATION OF AN EXISTING INFRASTRUCTURE NETWORK

The second form of local bypass involves the physical segmentation of existing infrastructure systems as elements of the network are seceded from the public sector to alternative providers and are then rebundled with the production of new zones within existing cities. This process can take a number of different forms. In the water and power sectors specialist users are provided with quality water, waste disposal or energy supplies to meet rigorous production needs; these are usually linked with special tariff reductions to attract users into these customised infrastructure zones. Although the network looks very similar it may be managed quite differently in different parts of the city to meet the particular needs of commercial users. In the telecommunications sector incumbents may respond to new entrants by segmenting their networks in areas of intense competition to respond to tariff reductions and the provision of value-added services – targeted at particular users under threat from cherry-picking.

Examples of the superimposition of the segmentation of existing infrastructure networks include:

- The emergence of special bodies for managing highly valued portions of the *city street* systems, such as the Business Improvement Districts (BIDs), of which there are over 1,800 in the United States; the so-called 'Ring of Steel' cordon which has been established to control entry into the financial district of London; and the recent spate of electronic road pricing (ERP) schemes in cities such as Singapore and Tromsø.
- The growing separation of *highway* lanes on existing public highway systems exclusive to particular types of users – high occupancy vehicle (HOV) lanes, special lanes for trucks, and toll lanes.
- The construction of new 'premium' *rail* networks between valued spaces that bypass older, slower networks. Here we would include the whole movement towards interurban fast rail and *trains à grande vitesse* (TGV), as well as premium links between major airports and city centres such as the Heathrow Express rail link between central London and Heathrow airport.
- The creation of special zones in the *water and waste* sector, designed to have highly reliable or cheaper infrastructure services that sit within, yet transcend, the incumbent's network.
- The construction of tailored zones in the *energy* sector. For example, the Baglan Energy Park in south Wales is based on a public–private partnership to attract manufacturing industries back to the area. A 500 MW gas-fired combined heat and power plant is being built to provide steam and power to factories on the site. The developers are hoping to attract new users to a 1,000 acre development site with the incentive of power costs that are 30 per cent lower than conventional industrial tariffs, and the power plant may eventually provide power and heat for off-site users.

COPING WITH THE WITHDRAWAL OF INFRASTRUCTURE

The final style of local bypass focuses on the coping strategies of those users who are least valued and who inhabit the places that are effectively bypassed by infrastructure networks. These local users and places tend to have three forms of bypass imposed upon them and respond with different types of coping strategy.

- First, there are those marginal users who have *never been connected* to an infrastructure network. The logic of local bypass may mean that they are even less likely to be connected now than they were to the formal networks that characterised the period of integrated infrastructural development. Such users rely on a whole set of coping strategies to obtain access to informal systems of infrastructure provision – private water vendors, non-networked energy supplies, walking or private transport systems. These tend to be more expensive, less reliable, more hazardous and of lower quality than public networks. Where users are provided with technical and institutional assistance they may be able to organise community-based networks based on local construction, maintenance and operation of networks.

- The second category are those users who are *left tenuously connected to highly unreliable public networks* as new investment is focused on new, parallel, segmented or more reliable networks for the more valued users. Such users must either turn to informal networks or develop other coping strategies. In Africa, for example, many small businesses have taken to boring their own wells and buying their own generators to provide electricity during the frequent power cuts. These additional costs are imposed on businesses already struggling to compete.

- The final category are those users who *have infrastructure networks withdrawn from them* as they are not profitable and new investment is focused elsewhere on valued enclaves. There are a range of different forms of withdrawal. For instance, infrastructure services could become more expensive to small users and the service may become less reliable. Marginal and low-income users are often asked to pay in advance, to manage problems of debt and disconnection, as is the case with the United Kingdom's 'prepayment' utility meters. Alternatively, public transport may be withdrawn from enclaves to prevent inappropriate users from gaining access to retail or shopping centres, even whole towns.

GLOCAL BYPASS

The second type of infrastructural bypass, what we call 'glocal bypass', involves the construction of new, materially distinct networks that are configured to support interaction between local valued users and places and global circuits of infrastructural exchange. Although this style is often closely linked with local bypass and virtual network bypass, the main feature is the reorientation of networks to connect local segments of cities to other valued segments in different parts of the globe. Glocal networks effectively bypass local infrastructure networks, as they are based on the physical construction of new networks or the segmentation of existing networks. Glocal bypass often coexists with major physical planning schemes that link the customisation of places targeted at meeting the needs of global capital and foreign direct investment (FDI) with the infrastructural networks that make location in low-cost production zones possible. The infrastructure is provided either by private firms specifically for global companies or by the local providers who develop their network to provide infrastructure targeted at this valued segment. In this context there may be insufficient investment to meet the infrastructural needs of the local population, as investment is targeted at privileged users in an effort to attract foreign capital investment. Two forms of glocal bypass can be identified.

New and segmented glocal infrastructure networks

The distinction between the construction of new physical networks and the segmentation of existing infrastructure networks is less easy to make in cases of glocal bypass. Nearly all the examples of glocal bypass fundamentally involve the construction of new networks and the segmentation of the existing public network. Additionally, there are also often close links between glocal and local bypass. For instance, the Heathrow Express rail link is clearly a form of local bypass providing a parallel and much faster private link to the public Underground system. But it is also about providing a premium local link to the global airline hub of Heathrow airport.

There are a wide range of examples of glocal bypass across different infrastructure sectors:

- In the *water and waste* sectors specialist tourist and industrial zones are often provided with water treatment and supply networks that are designed to meet international rather than local standards. These systems usually operate only within particular enclaves and, although not physically connected internationally, the whole rationale is to build an infrastructure that is tailored to meet the exacting needs of foreign direct investment.
- In the *energy* sector, glocal infrastructure supports international trading of power to support the needs of large commercial and industrial users. The French electricity utility EDF claims that it will supply business users on all their sites in 'Europe and beyond'. Access to glocal energy infrastructure is often linked with the development of special zones for commercial and industrial users who are able to import power over new and segmented networks to gain access to better-quality and more reliable power sources.
- In the *telecommunications* sector there are numerous examples of glocal infrastructure networks. Multiple providers offer private fibre optic networks that are configured to bypass local networks and interconnect sites on global corporate networks seamlessly and reliably. These networks are highly selective; they tend to be limited to the top fifty business and finance cities and are configured to meet the needs of the largest corporate users. In specialist 'back office zones' in the Caribbean and Ireland, meanwhile, specialist telecom operators offer multiple networks to allow the insurance, retail and financial service sectors to export routine administrative functions from low-wage enclaves. In addition, a wide range of private Internet 'pipes' are being deployed to bypass the constraints of old Internet trunks so that content delivery networks can be operated which enable the high-speed delivery of media and e-commerce services to selected affluent markets by the major media conglomerates (Tseng, 2000).
- This style of infrastructure development is also echoed in the *transport* sector, where the

TGV, the Channel tunnel, logistics hubs, and airports are designed to provide specialist high-quality, fast and reliable networks that connect the most valued users and places in cities with similar locations internationally.

BUILDING RESILIENT GLOCAL INFRASTRUCTURE

The second type of glocal bypass concerns the development of 'zero defect' or extremely resilient infrastructural enclaves (Pawley, 1997). It is associated with premium users creating multiple alternatives and points of access through different infrastructure at the level of specific premises rather than zones. Such systems are designed to provide extreme reliability and alternative supplies if an existing network is threatened by natural disaster, terrorism, technical collapse or any form of risk that might disrupt the operation of infrastructure services. In cases where many users are totally reliant on infrastructure for the effective and profitable operation of their business, particular attention is focused on the resilience of infrastructure services. In this context, selected users and zones are provided with multiple networks, usually based on the construction of new systems, parallel systems, the segmentation of existing networks and, finally, back-up provision of devolved, decentralised or even autonomous infrastructure services.

Resilient glocal networks themselves are likely to be characterised by a high degree of surveillance to prevent interference and attack. But, in the event of failure, these users are likely to have a number of back-up systems that build up the basic redundancy offered by the public network. Specialist infrastructure, communications and logistics companies are likely to have access to multiple and redundant systems to provide back-up in emergency situations. In worst case scenarios, individual users are likely to have their own water supply and treatment, an emergency energy supply, specialist mobile or satellite communications and travel plans to cope with the breakdown of transport services. Such services are likely to become a more important part of infrastructural unbundling as recognition increases of the threat of disruption from natural disasters such as earthquakes in San Francisco, ice storms in Canada and the breakdown of power networks in Auckland, New Zealand. The cost of disruption is measured in billions of dollars; therefore users are likely to demand much more infrastructural asset resilience.

VIRTUAL NETWORK BYPASS

Our final type of infrastructural bypass, what we term 'virtual network bypass', entails the use of new technologies to support and facilitate the distribution of customised infrastructure services over a single physically integrated network inherited from the modern ideal. This has important implications for cities and the global competitiveness of networks. Virtual network bypass supports both local and glocal bypass – it allows the segmentation and splintering of a single physically integrated infrastructure that can effectively differentiate between different types of users, allowing new entrants to gain access to the most lucrative users. Virtual network bypass, therefore, is based on a more subtle set of processes than those which support other forms of bypass. Three, in particular, can be identified.

SEGMENTING MASS MARKETS

The first process involves the segmentation of infrastructure markets. The mass infrastructure markets of the integrated ideal required very little segmentation between different types of user because services were provided on a monopolistic and homogeneous basis. But the unbundling of networks forces providers to segment what were previously mass markets into different elements, in order to identify the most profitable and valuable users to target for new network and service provision. Geographical information systems (GISs), geodemographic and new marketing techniques are being imported from the retail and financial services sectors to segment utility markets. Billing data, housing information, payment methods and service use are being combined with other forms of data to build much more detailed profiles of users.

This information is used in a number of different ways by infrastructure providers. It enables utilities to identify the most and least profitable users, to identify users who may purchase value-added services, to build customer retention and loyalty schemes, to plan the roll-out of new parallel infrastructure networks, and to provide a range of value-added and enhanced services to selected users. New information and marketing techniques are utilised to segment the mass markets of the integrated era, allowing new and incumbent utilities to develop detailed social and spatial profiles of the most and least valued users. These can be used to plan the construction of new physical networks and support the development of strategies designed to effect disengagement from the least valued users.

SEGMENTATION OF INTEGRATED NETWORKS

The second form of virtual bypass involves the segmentation of integrated networks into contestable and monopoly elements. Telecommunication and ICT systems actively facilitate the unbundling of integrated networks, allowing new entrants to construct parallel segments of the networks that are contestable and to interconnect with monopolistic segments of the network. In the water sector these systems facilitate the development of contestable water supply and waste treatment by remotely monitoring the levels of water consumption, network use and waste disposal over the incumbent's network. In the power sector, integrated networks can be segmented into generation, transmission and distribution. ICT technologies facilitate flows of power and transactions between the contestable and monopolistic elements of the networks. Multiple local telecommunications networks can be constructed within a city with complex control systems ensuring that calls are routed seamlessly over mobile, cable and telecommunication networks, despite their separate technologies and ownerships. Road networks can be segmented with access control and pricing technologies monitoring and charging users for access to different parts of the network.

Taken together, then, this form of virtual bypass allows the monitoring and control of access to complex and fragmented networks. It also supports the complex pricing formulas involved in user networks owned by multiple users, manage complex pathways of resource flows over many different elements of the network. Effectively these technologies are able to support a 'seamless' infrastructure service, even though the network over which the services flow can be in multiple ownership, and flow over a number of different pathways and technologies.

Virtual network competition

The final form of virtual bypass involves laying entire ICT infrastructures over a physically integrated infrastructure to support competition between incumbents and new entrants for users whilst providing specialist customised services to selected users. The key to these systems is the development of 'intelligent' infrastructural terminals: smart meters, road pricing, and the specialist technologies of tolling and electronic route guidance and driver information systems. These technologies allow users to differentiate between specialist services provided to different groups of users with all the relations mediated through terminal technologies, even though the same physical infrastructure network is used to distribute the service:

- In the *water and power* sectors smart card prepayment technologies allow providers effectively to disengage from the least valued users, as all relations are mediated through prepayment cards. These users usually pay higher tariffs than more valued users. They also effectively self-disconnect themselves from networks when they cannot afford to charge their smart card.
- In the *telecommunications* sector similar technologies allow domestic users to gain access to alternative national and international telecommunication operators over the incumbent's local network.

Users simply purchase a brand, which then provides heat, light and movement. All transactions take place over virtually integrated infrastructure networks. Customers simply select the service level, speed, time, cost and enhanced features for an infrastructure service that intermediaries will bundle together from a range of different producers so that they are seamlessly delivered to users.

CONCLUSIONS

URBAN INFRASTRUCTURE: TOWARDS NEW PARADIGMS

This chapter has provided a broad descriptive and conceptual framework of the social and technical practices through which integrated infrastructure networks can be actively segmented and unbundled. We have examined these processes in some detail in order to illustrate the complexity of the unbundling process, the diversity of pathways through which integrated networks can be splintered, and the complex variety of ways in which monopolistic and contestable elements are selectively reassembled. Our discussion leads to five conclusions.

First, we argue that the processes of infrastructural unbundling analysed above are likely to reshape dramatically relations between cities and their networked infrastructures. The selective reassembly of segmented network elements effectively leads to the selective rebuilding of different sets of social and spatial relationships. This is a multiscalar process which operates both within and between cities. Generally, it involves intensifying the connections between most valued users and places while simultaneously weakening the connections with least valued users and places. Multiple providers of contestable infrastructure networks, new entrants and incumbents, build new and segment existing networks that, quite literally, bypass the least valued users, districts and cities, leaving such places to the remnants of monopoly networks while providing higher-quality, more resilient and less costly infrastructure networks for the most valued users. These processes of reassembly usually take place in parallel with new styles of urban planning: the creation of zones and enclaves for users provided with specialised infrastructure services.

Second, it is essential to recognise that the infrastructural unbundling is not a simple process but one that can take a wide range of different styles and institutional forms. The transitions away from integrated towards unbundled networks are extremely complex, 'messy' and time-consuming. In many developing, and some advanced, economies the initial shift has been towards commercialised, devolved infrastructure or privatised monopolies that do not necessarily involve unbundling. But the logic of the new paradigm would suggest that these may be intermediate stages of infrastructural restructuring that represent pathways towards unbundled infrastructure.

Third, it is not easy to characterise the trajectories of individual cities along pathways to segmentation and unbundling in a simple way. Within any particular city a variety of different pathways towards or away from integrated networks are likely to coexist across the full range of infrastructure networks. Although the cities of the advanced economies are likely to exhibit more fully liberalised networks, there are also clearly significant opportunities for privatisation and liberalisation in many cities in the developing world.

Fourth, it is also crucial to stress that the changing infrastructural landscapes of cities are much more complex than the simple displacement of the old by the new. The evidence seems to point to a much more complex set of pathways towards unbundled networks that are moving at a range of speeds across different infrastructural sectors and national contexts. Mapping the social organisation of any city's infrastructural assets in this new context would undoubtedly reveal that, across different networks, quite different pathways away from the integrated ideal actually coexist. Some networks will be more liberalised while others still have to be unbundled. A key challenge for researchers and policy makers is to map the dramatic reconfigurations that are taking place in the infrastructural assets of cities.

Finally, while infrastructure networks are undergoing a period of dramatic restructuring, they have certainly not all been unbundled. Many countries simply do not have the capacity to segment networks. They lack the private sector interest in providing services. And they cannot overcome social and political resistance to segmentation as strategic and political forces work to limit the loss of government control of infrastructural assets. Yet the defining feature of the emerging paradigm is that the pathways we have identified lead logically, but not inevitably, towards the eventual unbundling and liberalisation of infrastructure networks where ever more segmented elements of the networks can be made marketable and can be contested by new entrants. The logic of international trade liberalisation agreements and the conditions tied to international aid in many developing countries increases the pressure on national governments to move away from government monopolies that provide integrated infrastructure and towards institutional models that actively facilitate greater private sector participation in the provision of services. Each of the pathways away from integrated networks effectively seeks to create contexts in which networks can be unbundled and infrastructure markets made contestable.

The framework constructed in this chapter has focused on describing and identifying pathways from integrated networks that involve infrastructural unbundling. In particular we wanted to show that this very specific form of network transition is involved in the dramatic reshaping of cities. We showed that the development of substitute and new parallel infrastructure networks, either physically material networks or existing monopoly networks opened up to competition, has significant implications for different types of users and the organisation of movement, resources and information flows within and between cities. Central to this process was the notion of bypass. The development of alternative, parallel and substitute networks allows both new and incumbent network providers to make choices about the users and places that are connected or not connected.

In this chapter we have been forced by the very complexity of the shifts in infrastructure networks to consider the unbundling of networks and fragmentation of cities as largely separate processes, in order to build a conceptual understanding of the different pathways away from integrated networks and cities. At the same time, we have emphasised the importance of building a relational understanding of the linkages between shifts in the style of infrastructure provision and the restructuring of sociospatial relations in cities. Cities and infrastructure are mutually constructed and highly interdependent.

In the next chapter we therefore aim to develop a theoretical understanding of the complex interdependences between infrastructural unbundling and urban fragmentation by drawing insights from social, technological and urban theories.

5 THE CITY AS SOCIOTECHNICAL PROCESS

Theorising splintering urbanism

Plate 8 The Pudong redevelopment area, Shanghai, China.
Photograph: Stephen Graham

Formerly the dominant forces were separation and specialisation, the struggle for clarity, and the reduction of the world to calculable proportions; now we talk about simultaneity, multiplicity, uncertainty, chaos theory, networks, hubs and nodal points, interaction, the hybrid, ambivalence, schizophrenia, space of flows, cyborgs, and so on.

(van Toorn, 1999, 90)

The last chapter outlined in detail the parallel processes that surround the unbundling of infrastructure networks and the sociospatial fragmentation of cities that, we argue, are so comprehensively undermining the legacies of the modern infrastructural ideal in many cities around the world. In this chapter we turn from the description of these trends to their explanation. Our task now is to engage with contemporary social, technological and urban theories that help us to understand further why parallel processes of infrastructural unbundling and urban splintering are under way.

We seek help in this task from four particular strands of theory, each of which helps us in different ways to explore further how current urban trends across the world relate with the current revolution in infrastructure networks surrounding globalisation, privatisation, liberalisation and the application of new technologies. These are the theories of large technical systems (LTSs), actor network theory (ANT), theories of the changing political economies of capitalist infrastructure, and what we call 'relational' theories of contemporary cities.

These four perspectives are especially useful for our purposes. All four suggest that it is possible to develop a theoretical perspective which can be broadly applied to all networked infrastructures, and all types of city, without falling into the trap of oversimplification, technological determinism or ethno-centricism. They support the notion that parallel trends towards unbundling networks and splintering urbanism exist. And they help support the building of an explanation of these trends which does not overly separate treatment of the 'city' and the 'urban' from 'technology' and 'infrastructure'.

This chapter uses these four strands of theory to develop a theoretical synthesis through which we can go on to explore and understand the parallel, yet diverse, transformations of cities and urban infrastructures across the globe in Part Two. This chapter therefore provides the final element of empirical, theoretical and conceptual foundation in Part One. We aim to do this in a *parallel* way which does not overprivilege cities over infrastructure, or infrastructure over cities. In fact, we want to avoid the problem that has beset so much literature in the field, that of *separating* the ontological status of 'infrastructure' or 'technology' from the 'urban' or the 'city', as though they stand in a state of simple distinction.

We recognise, rather, that much of the 'urban' is infrastructure; that most infrastructure actually constitutes the very physical and sociotechnical fabric of cities; and that cities and infrastructure are seamlessly coproduced, and co-evolve, together within contemporary society. This problem resonates with the wider problem of separating 'technology' and 'society' into an easy, modernist, dualism with clear boundaries and simple causal links between them. As Kirsch suggests, 'the point is that whilst technology is a thoroughly social construction, society is a technological construction as well' (1995, 531) . Thus, by implication, we need to remember that infrastructures are thoroughly social constructions whilst cities are also infrastructural constructions.

OPENING 'BLACK BOXES': THE DESTABILISATION OF LARGE TECHNICAL SYSTEMS

With such points in mind, we can turn to our first perspective: the so-called 'large technical systems' approach (see Mayntz and Hughes, 1988; Summerton, 1994a; Coutard, 1999). The notion of infrastructure networks as large technical systems derives from the pioneering work of Thomas Hughes (1983) on the evolution of electricity networks in early twentieth century Britain, the United States and Germany. Using an explicitly sociological and historical perspective, Hughes demonstrated how the linkage of technical apparatus into widening, engineered infrastructure networks involved complex economic, political and social negotiations. By linking venture capital with engineering, innovation and organisation building, Hughes showed how what he called 'system building' entrepreneurs struggled to impose *systemic* qualities on their infrastructures, through a particular technical style, in often difficult and usually volatile circumstances.

BUILDING LARGE TECHNICAL SYSTEMS AND THE IMAGE OF TECHNOLOGICAL PERMANENCE

Through such system-building processes the diverse, local systems set up by initial entrepreneurs gradually merged into the standardised, national, widely accessible large technical systems that became central to the modern infrastructural ideal, especially in the West. Such systems were gradually 'rolled out' on an increasingly monopolistic basis to cover whole national and regional territories; Hughes's thinking was that 'LTSs can be managed only through growth' (Offner, 1999, 230) (see, for example, Figure 2.1). Through such growth, large technical systems gradually diffused to become taken for granted and 'normalised' as essential, but largely invisible, supports of modern urban life. In the language of LTS research, such infrastructures became 'black-boxed' by their users, who often had no other functional alternative to relying on the large technical system, whether it was water, sewerage, electricity, the telephone or the automobile system.

Thus, in general, Hughes argued, large technical systems tend to be characterised by an initial growth phase, an accelerated growth phase, a stabilisation phase (where the system became black-boxed and taken for granted), and (sometimes) a decline phase (where newer infrastructure systems came in to substitute for it, as, for example, with the postwar decline of mass transit through automobile use in many North American cities) (Gökalp, 1992, 58).

The large technical system perspective therefore helps demonstrate how systemic changes appear in the whole technological fabric of society, as interrelated clusters of innovations sometimes cohere into large technical systems through processes of social, political and institutional agency and entrepreneurship. Through complex interactions between technology 'push' and demand 'pull' factors, capitalist society has come to rely on a whole interconnected web of primary large technical systems: the automobile transport system, telecommunications, media networks, gas and electricity systems, water, waste and sewerage, and air transport.

In addition, an extending assembly of secondary large technical systems – such as scientific research systems or electronic finance, currency exchange, state bureaucracy systems and organ transplant systems – have also emerged. These rely, in turn, on primary large technical systems to operate (Mayntz, 1995). Large technical system research suggests that a fully sociotechnical perspective on the development of cities and infrastructure networks is necessary: neither technically determinist nor socially voluntarist approaches are satisfactory (Gökalp, 1992).

THE INTERDEPENDENCE OF LARGE TECHNICAL SYSTEMS

Invariably, large technical systems work in combination and are subtly interdependent. The latest fuel pump at a typical US filling station is a classic example. It allows users to insert a credit or ATM card to pay for the transaction. The pump is, in effect, a machine that seamlessly integrates the operations of (at least) four separate large technical systems: the highway/ automobile system, the computer and telecommunications system, the banks' financial, credit card and ATM chequing system, and the global oil production and supply system.

Such a perspective implies strongly that infrastructure networks tend to accrue in society on an incremental basis, creating ever denser and more elaborate systems, strung out over wider and wider distances. Together these interconnected systems are much more important in social development than individual innovations (Mayntz, 1995; Beniger, 1986). Whilst some substitution may take place (air for rail, teleworking for commuting, car for metro, gas for coal, debit card for cash, etc.), it is less common for older networks to disappear altogether. Instead they tend to become taken for granted, ubiquitous, standardised and 'black-boxed' – an essential prop to society and the economy which few take much notice of until failures or collapses.

THE BANALITY OF TECHNOLOGICAL ARTEFACTS AND THE DESTABILISATION OF LARGE TECHNICAL SYSTEMS

Telephones, electric plug sockets, water taps, flushing toilets and cars are thus so utterly ubiquitous in advanced industrial societies that these apparently banal artefacts give no hint to the average user of the huge technical systems that invisibly sustain them. Few venture to understand the inner workings of the technology or the giant lattices of connection and flow that link these network access points seamlessly to distant elsewheres.

The entire technological systems are black-boxed. 'Black boxes are therefore settled items whose users and colleagues (human and non-human) act in ways which are unchallenging to the technology' (Hinchcliffe, 1996, 665). However, the shift towards unbundled networks forces many users to rethink this idea of permanence and stability. But, at the same time, our tendency to take large technical systems for granted still means that, for most of us, 'technical systems [now] conjure up images of stability and permanence' (Summerton, 1994a, 1). This is because of their historical evolution from small, fragmented, specialised systems to integrated, often (quasi) universal, and technologically standardised ones that can be regarded as 'functional subsystems of society as a whole' (Mayntz, 1995).

Because of the apparent permanence of black-boxed large technical systems, infrastructure networks thus retain powerful images of stability. Often they are regarded as 'symbols of the complexity, ubiquity and the embodied power of modern technology' (Summerton, 1994a, 1). This explains why Urban Studies, for example, still often uses a language such as 'public infrastructure' or 'public works' that traps these networks within a historically specific period, so utterly failing to acknowledge radical shifts in the social organisation of the sectors. Urban Studies appears to have difficulty acknowledging the intrinsically dynamic nature of network changes. It, too, has, in effect, tended to 'black-box' networks like electricity as permanent, ubiquitous and banal underpinnings of urban life that do not really warrant contemporary attention.

LARGE TECHNICAL SYSTEMS AS 'PRECARIOUS ACHIEVEMENTS'

Recent analyses, however, have shown that infrastructure networks, despite the veneer of permanence, stability and ubiquity built up by the modern ideal, are never structures that are given in the order of things. Instead of being static material artefacts to be relied on without much thought, they are, in effect, processes that have to be worked towards. The dynamic achievement of a functioning energy, communications, water or transport network requires constant effort to keep the system functioning. It is easy to overplay the degree to which infrastructure networks necessarily 'mature' to become socially ignored and 'embedded'. Jane Summerton writes that:

we sometimes seem to view mature Large Technical Systems as invulnerable, embodying more and more power over time and developing along a path whose basic direction is as foreseeable as it is impossible to detour. [But] systems are more vulnerable, less stable and less predictable in their various phases than most of us tend to think.

(1994b, 56)

Infrastructure networks are, in short, precarious achievements. The links between nodes do not last by themselves; they need constant support and maintenance. When the heterogeneous elements are coupled and interact according to their assigned roles, allowing the intended effects to be expected with high reliability, the network is described as stable and closed. Take the 'large technical system' of the automobile and its related highways and service infrastructure, for example:

the techno-structure of automobile traffic is a striking example of this stability: the strongly knit relations between automobile manufacturers and suppliers, the close intertwining of transport and taxation policy, the long-lasting tradition of motor-car engineering and the mass myth and mass practice of automobilism. Each of these relations guarantees the continuation of a technological trajectory, although the automobile traffic system has been deeply shaken by the crisis of oil supply, air pollution, and urban traffic jams. This close coupling of things, people, and signs and its continuous production by routines are the social base of the technological momentum and the myth of technics-out-of-control.

(Ranmert, 1997, 186)

PROBLEMATISING AND 'UNBLACK-BOXING' LARGE TECHNICAL SYSTEMS

This precarious nature of large technical systems means that they can be reconfigured and 'unblack-boxed' as a result of major social, political, economic or organisational upheaval – such as those associated with the current transformations surrounding the unbundling of networks (Summerton, 1994a, b). Taken-for-granted assumptions about large technical systems – such as the idea that they are monopolies requiring universal service regulations – can thus be challenged. Periods of volatility may emerge. Through such volatility, entrepreneurial forces may strive to reach new periods of stability, through new linkages of the 'social' and 'technical' into large technical systems across distance (see Coutard, 1999).

Large technical system research has thus begun to acknowledge that the large, centralised, publicly owned and controlled systems that have been the object of its attentions thus far now seem to be out of keeping with contemporary societal shifts towards globalisation, liberalisation, privatisation and general scepticism about centralised bureaucracies and 'big technology' (Summerton, 1994a; Coutard, 1999). Renata Mayntz, for example, notes the recent:

deregulation of LTSs that are organised as monopolies. [This implies] vertical deconcentration, or uncoupling, [as has] happened in the case of telecommunications, the railways and electricity networks. The identity of network owner, system operator, service provider, and supplier of the user interface, e.g. telephones, has been broken up; these different functional parts now tend to be separated and transformed into independently operating profit centres.

(1995, 7)

THE FAILURE TO DEVELOP TERRITORIAL ANALYSES OF LARGE TECHNICAL SYSTEMS

Unfortunately for the purposes of this book, the historical and supply-side focus of LTS research, and the overwhelming concern with 'system builders', mean that it has been slow to explore fully contemporary infrastructural transformations. Some useful work on the current reconfiguration of large technical systems has, however, emerged, especially from the work of Jane Summerton. Her edited book (1994a) offers case studies of the reconfiguration of large technical systems, from organ transplant systems, global corporate telematics networks, railways and electricity to the Internet and the emergence of the notion of the smart highway (ibid.; see also Coutard, 1999). As well as emphasising supply-side trans-formations, Summerton has attempted to focus more on the changing orientation of the *users* of large technical systems, particularly the growing assertion of diversified user demands within previously homogeneous energy markets – in effect, opening the 'black box' of domestic energy consumption by allowing users to choose from many different styles of provision: 'green' renewable energy, 'red' energy provided through social housing and left-wing municipalities, or special packages designed to meet their particular needs, say, for swimming pool or greenhouse heating (Summerton, 1995).

Its origins in sociology and history also mean that this strand of research has tended to neglect the ways in which large technical systems become embroiled in the production and reconfiguration of urban spaces and places. Research has also almost completely neglected the territorial foundations and implications of large technical systems, and the necessary relations between infrastructure and the governance of the cities, regions and spaces that they pass through. Whilst there are now signs of a long overdue 'reterritorialisation' of large technical system research (Joerges, 1999b), Olivier Coutard has lamented that:

apart from some specific studies, the territorial dimension of LTS has been ignored. The relation between LTS and territories seems to boil down to the assumption that system development goes with (more or less impeded) spatial expansion – which need not always be true. This seems to leave a number of issues unexplored: the urban issues related to LTS, the impact of LTS on the organi-sation of territories (local, regional, global), [and] the legitimacy of LTS in relation to political spaces.

(1996, 47; see Joerges, 1999b; Offner, 1999)

ACTOR NETWORKS AND CYBORG URBANISATION

Our second perspective stems from the recent work of actor network theory (ANT). Anchored in the work of Michel Callon (1986, 1991) and Bruno Latour (1993), a range of researchers, including those writing of the proliferation of blended human–technological 'cybernetic organisms', or 'cyborgs' (Haraway, 1991), have argued for a highly contingent, relational perspective on the subtle linkages between technologies and social worlds. By 'relational' we mean focusing in detail on the ways in which relations are constructed which entail both social and technical connections across time and space. Such a perspective necessarily emphasises the social nature of infrastructural and technological innovations and the active constitutive roles of technologies as well as people.

Actor network theory emphasises how particular social situations and human actors enrol pieces of technology, machines, as well as documents, texts and money, into actor networks, configured across space and time (see Law and Hassard, 1999). Its central message is that 'modern societies cannot be described without recognising them as having a fibrous, thread-like, wiry, stringy, ropy, capillary character that is never captured by the notions of levels, layers, territories, spheres, categories, structure, systems' (Latour, 1997, 2).

BREAKING DOWN *A PRIORI* DISTINCTIONS BETWEEN THE 'TECHNICAL' AND THE 'SOCIAL'

Through such an approach, actor network theory abandons any *a priori* distinctions between the 'social' and the 'technological'. Rather, contemporary life is seen to be made up of complex and heterogeneous *assemblies* of both social and technological actors, strung out across time and space and linked through processes of human and technological agency (Murdoch, 1995). Drawing on Thomas Hughes's (1983) idea that society is a 'seamless web' of sociotechnical constructions, actor network theory 'eschews the modern's language of purity, of wrapped packages and firm boundaries in favour of an emphasis on connection, interdependence, mutuality and . . . flux' (Bingham, 1996, 644).

The networked infrastructures or 'technical networks' which are the focus of this book are only 'one possible final and stabilised state of an actor network' (Latour, 1997, 8). Nevertheless, the ANT perspective is useful to an understanding of infrastructure networks, as it attempts to develop a fully relational understanding of 'how all sorts of bits and pieces, bodies, machines, and buildings, as well as texts, are associated together in attempts to build order' (Bingham, 1996, 32). Absolute spaces and times, and simple separations between society and technology, are meaningless here. Agency is a purely relational process. Technologies have contingent and diverse effects only through the ways they become linked into specific social contexts by human and technological agency. What Pile and Thrift call a 'vivid, moving, contingent and open-ended cosmology' emerges (1996, 37). The boundaries between humans and machines become ever more blurred, permeable, interpenetrating and 'cyborgian'. And 'nothing *means* outside of its relations: it makes no sense to talk of a machine in general than it does to talk of a "human in general"' (Bingham, 1996, 17, original emphasis). To Nigel Thrift a key conclusion of the approach is that:

no technology is ever found working in splendid isolation as though it is the central node in the social universe. It is linked – by the social purposes to which it is put – to humans and other technologies of different kinds. It is linked to a chain of different activities involving other technologies. And it is heavily contextualised. Thus the telephone, say, at someone's place of work had (and has) different meanings from the telephone in, say, their bedroom, and is often used in quite different ways.

(1996a, 1468)

Thus the development of networked infrastructures within and between cities boils down to the linkage of massive, heterogeneous arrays of technological elements and actors, configured across multiple spaces and times (see Picon, 1998). This is a profoundly difficult process requiring continuing effort to sustain relations which are 'necessarily *both* social and technical' (Akrich, 1992, 206, original emphasis).

Once infrastructure networks are successfully built, 'unconnected localities' can be linked through what Latour calls 'provisionally commensurable connections' (1997, 2). Infrastructure networks, then, are vast collectivities of social and technical actors blended together as sociotechnical hybrids that support the construction of multiple materialities and space–times.

INFRASTRUCTURE NETWORKS, REMOTE CONTROL, AND TECHNOLOGICAL ORDERING

Actor network theory also stresses that successfully linking arrays of social and technical actors over distance requires continuous effort, even within mature, black-boxed infrastructure networks. Connections are always perilous and fragile; never-ending effort is required to sustain them. For example, the growing ability of telecommunications to support action at a distance and by remote control does not negate the need for the human actors who use them to struggle to enrol passive technological agents into their efforts to attain meaningful remote control. 'Stories of remote control tend to tell of the sheer amount of work that needs to be performed before any sort of ordering through space becomes possible' (Bingham, 1996, 27). In telecommunications such:

heterogeneous work involving programmers, silicon chips, international transmission protocols, users, telephones, institutions, computer languages, modems, lawyers, fibre-optic cables, and governments, to name but a few, has had to be done to create envelopes stable enough to carry [electronic information].

(Bingham, 1996, 31)

Actor network theory thus also undermines the notion that we can simply and unproblematically generalise a single material 'thing' called an infrastructure network, just as it challenges the idea that we can simply generalise a city. Instead, sociotechnical worlds emerge as a continuing cacophony of endless flux and fragile, multiple interdependences. It follows – again using the example of telecommunications – that there is no single, unified 'cyberspace'. Rather, there are multiple, heterogeneous networks and 'cyberspaces' surrounding the Internet, and many other infrastructures, within which telecommunications and information technologies become closely linked with human actors, and with other technologies, into systems of sociotechnical relations linked across space and time. As Nick Bingham again argues, 'the real illusion is that cyberspace as a singular exists at all', rather than as an enormously varied skein of networks straddling and linking different space–times (1996, 32; see Latour, 1993, 120).

Bruno Latour captures the complex inclusionary/exclusionary nature of technical networks in words that have deep resonance with our arguments about unbundling networks and fragmenting cities. To him infrastructure networks:

are composed of particular places, aligned by a series of branchings that cross other places and require other branchings in order to spread. Between the lines of the network there is, strictly speaking, nothing at all: no train, no telephone, no intake pipe, no television sets. Technological networks, as the name suggests, are networks thrown over spaces, and they retain only a few scattered elements of those spaces. They are connected lines, not surfaces. They are by no means comprehensive, global or systematic, even through they embrace surfaces without covering them, and extend a very long way.

(Latour, 1993, 117–18)

NETWORKED SPACES OF CONNECTION AND THE 'INFINITY' OF DISCONNECTION

Actor network theory and related 'cyborgian' perspectives on the city offer a fully relational perspective which has important implications for the ways in which we conceptualise the links between cities, unbundling infrastructure networks, and space and time (see Box 5.1). For actor network theory suggests that, rather than simply being space- and time-transcending technologies, infrastructure networks actually act as technological networks within which new spaces and times, and new forms of human interaction, control and organisation are continually (re)constructed. As Bruno Latour argues:

BOX 5.1 'CYBORG' READINGS OF CONTEMPORARY URBANISM

Actor network and cyborgian theories offer powerful new insights for exploring the complex sociotechnical negotiations that have surrounded the mass application of 'machinic ensembles' and infrastructure networks through contemporary capitalist society. They help us to collapse unhelpful and overly dualised distinctions between the social and technical, the city and infrastructure, the local and global. And they underline that 'attempts to find the boundary of any practice – where one ends and the other begins – are increasingly artificial' (Mau, 1999, 203). Eric Swyngedouw suggests that:

the production of the city as a cyborg . . . opens up a new arena for thinking and acting on the city: an arena that is neither local nor global, but that weaves a network that is always simultaneously deeply localised and extends its reach

over a certain scale, a certain spatial surface.

(1995a, 22)

Such cyborgian conceptions of urbanisation demonstrate the *immanence* of relations between modern urban society, and urban subjects, and the technological infrastructures that support them (Picon, 1998). Processes of urban life are now intimately constituted through extended webs of technologies; these define the ways in which the city reconstructs nature, processes resources and serves as a crucible of social relations and the production of services and commodities. 'Cyborgs, then, are "networked" entities; they do not exist simply as autonomous individual subjects, but through connections and affinities, including their connections to technology. Indeed, cyborgs are never entirely separate from technology' (Rutsky, 1999, 148).

CYBORG LIAISONS AND THE CITY

At least for more affluent city dwellers, then, the human body now enters into a myriad of increasingly intimate, continuous and seamless 'cyborg' liaisons with cars, entertainment systems, infrastructure, energy, water and waste networks, electronic finance, communications, home management and air-conditioning systems, and constructed and digital spaces. Such liaisons serve to extend the influence of the human actor across many geographical scales and territories through its daily physical and virtual mobilities (Lupton, 1999).

Contemporary urban life, thus, can be conceptualised as a palimpsest of socio-technical relations and processes. However, Timothy Luke claims that it is only the *collapse* of the sociotechnical relations – rare in the West, common in war zones and but part of everyday life in many developing cities – that fully demonstrate how bound up modern urban life is with many superimposed webs of urban infrastructure networks (1997, 1368). He argues that:

> without the agriculture machine, the hous-ing machine, the oil machine, the water machine, the electrical machine, the media machine, or the clothing machine, most cyborganised humans cannot survive or thrive, because these concretions of machinic ensembles generate their basic environment. Hence cyborgs, like us, are endlessly fascinated by mechanic break-downs, which would cause disruptions in, or denials of access to, their mega-technical sources of being. Beirut in the fifteen years war, Sarajevo in the two year long siege, or the Road Warriors' travels in post-megamachine Australia [in the *Mad Max* films] are all dark revelations – in fact and fiction – of what once were highly evolved cyborg beings, struggling to survive decyborgised societies without all the life support systems of humachinic megatechnics.
>
> (Luke, 1997, 1368)

Most of the difficulties we have in understanding science and technology proceed from our belief that space and time exist independently as an unshakable frame of reference *inside which* events and place would occur. This belief makes it impossible to understand how different spaces and different times may be produced *inside the networks* built to mobilise, cumulate and recombine the world.

(1987, 228)

Actor network theory usefully collapses taken-for-granted notions of 'far' and 'close' by revealing that elements which seem at first to be physically close 'when disconnected [by networks] may be infinitely remote if their connections are analysed' (Latour, 1997, 3). Conversely, 'elements which would appear as infinitely distant may be close when their connections are brought back into the picture' (ibid.). Latour continues, 'it may be that the telephone has spread everywhere, but we still know that we can die right next to a phone line if we aren't plugged into an outlet and a receiver. The sewer system may be comprehensive, but nothing guarantees that the tissue I drop in my bedroom will end up there' (1993, 115). 'I can,' he suggests:

be one metre away from someone in the next telephone booth, and be nevertheless more closely connected to my mother 6,000 miles away . . . a gas pipe may lie in the ground close to a cable television glass fibre and near by a sewage pipe, and each of them will nevertheless continuously ignore the parallel world lying around them.

(1997, 3)

Technological networks, despite the modern rhetoric of universality, are thus always specific and contingent in linking one place with another. They 'always represent geographies of enablement and constraint' (Law and Bijker, 1992, 301). They continually link the local and non-local in intimate relational, and reciprocal, connections. And they always support unevenness in inclusion and exclusion, differentiation, and presence and absence – but in ways that challenge simplistic ideas about how geographical space links with these phenomena.

THE USEFULNESS OF ACTOR NETWORK AND CYBORGIAN PERSPECTIVES

So – and this is of crucial importance – unbundled infrastructure networks and fragmented cities emerge as *two sides of the same overarching societal process*. But the logic of unbundling networks actually merely *accentuates* the ways in which infrastructure networks both reflect and support relational differentiation between and within places, by tying them into multiple and diverse spatiotemporalities which challenge traditional assumptions that social and technological relations can simply be understood by adopting binary dualisms like local–global, close–far, social–technical or in city–out of city.

In the broad shift from the single, coherent infrastructure networks laid out during the modern ideal to the competing, overlaid patchworks of unbundled networks now emerging, we would argue that the geographies of topological connection and exclusion, as manifest within contemporary cities, are becoming much more complex and uneven. Single geographies where networks bind spaces and cities are giving way to multiple, overlaid and customised grids that unevenly connect parts of cities together and to intensifying interactions elsewhere. The fragmenting social and economic geographies of contemporary urban areas are simultaneously being structured by the emergence of multiple, superimposed infrastructure grids whose complex combinations of connections and disconnections constitute a major reconfiguration of the 'power geometries' of the city (Massey, 1992; Amin and Graham, 1998b).

As Bruno Latour suggests, infrastructure networks and their access points thus tend to 'warp', 'stretch' and 'compress' the natural and social spaces and times of our daily lives, based on who is enrolled into the networks and who remains physically or technologically disconnected (1997, 3). As the access points and topological geographies of urban networks become more and more unevenly customised to particular users, spaces and zones in the contemporary metropolis, such stark processes of relational incorporation or exclusion become closely and subtly woven into the contemporary urban fabric.

In Latour's schema it is therefore a mistake to develop binary oppositions of 'local' and 'global' infrastructure networks. Rather than one network being 'bigger' than another it is simply longer or more intensely connected. In this sense a network must always remain continuously local, as it inevitably touches down in particular places. The task is to trace how a node becomes strategic through the number of connections it commands and how it loses importance when the connections are lost.

Actor network and 'cyborgian' readings of contemporary urban and infrastructural change are thus of considerable value to this book. For they suggest a clear theoretical argument which may help account for the parallel trends towards unbundling networks and urban fragmentation. It is that current shifts in cities and infrastructure allow the heterogeneous engineering of technologies and infrastructural artefacts to be done in a much more customised manner than in the past. Users no longer necessarily need to be drawn into the comprehensive, standardised urban infrastructure grids, as in the modern infrastructural ideal. Relatively homogeneous associations, through standardised grids, are replaced by a great diversity of increasingly customised or specialised associations with and through network technologies. Network associations in 'actor networks' between users, artefacts, finance and texts can be done on a much more customised basis, in particularly through 'enrolling' the powers of information technologies as 'selective agents'.

Of course 'action at a distance' and 'technological ordering' are still complex and difficult, requiring continuous effort. But successfully enrolling the control capabilities of information technology as technological agents can make the ordering of specialised sociotechnical networks a great deal more possible than previously. This can be done, moreover, without losing the ability to connect from tailored infrastructures like smart highways, customised electricity supplies, global telematics networks or global airline links to planetary grids as a whole. Social power, thus, is wielded through heterogeneous associations of technical infrastructure networks and specialised urban places, linked in parallel.

Sociotechnical constructions increasingly resemble specialised, heterogeneous arrays laid out within and through technical networks. Sometimes these interconnect; at other times they run parallel and in isolation from each other. It is here, most often within the unbundled networked infrastructures of the contemporary city, that we see complex combinations of 'infinite' disconnection, physically cheek by jowl with customised arrays of highly capable global–local infrastructure.

CHANGING POLITICAL ECONOMIES OF CAPITALIST URBAN INFRASTRUCTURE

Capitalism as a mode of production has necessarily targeted the breaking down of spatial barriers and the acceleration of the turnover time as fundamental to its agenda of relentless capital accumulation.

(Harvey, 1996)

Our third perspective revolves around analyses of the geographical or spatial political economies of contemporary capitalism. The starting point of this perspective is that the ability to transcend space and time constraints selectively, by building or using transport, telecommunications, energy and water infrastructures, is central to the economic and geographical development of modern capitalism.

The production of infrastructure networks, and the financial, engineering and governance practices that support them, are therefore necessarily embedded within the broader power relations of global capitalism. 'Infrastructure networks of all types are deeply embedded and implicated in the process of production, reproduction, and legitimation in a functioning capitalist economy' (Hodge, 1990, 87).

Taking this idea as his starting point, Eric Swyngedouw goes one step further. He suggests that 'changes in mobility and communication infrastructure and patterns are not neutral processes in the light of given or changing technological–logistical conditions and capabilities. Rather, they are necessary elements in the struggle for maintaining, changing or consolidating social power' (1993, 305). Mobility, he continues, is 'one of the arenas in which the struggle for control and power is fought'. An important strategic weapon of the powerful in this struggle 'is the ideology of progress and the legitimising scientific discourse of scientists and engineers' that so dominated the construction of the modern urban infrastructural ideal discussed in Chapter 2 (Swyngedouw, 1993, 324).

But how, exactly? Critical geography, in exposing the spatial political economies of contemporary societies, has done most to reveal fully the subtle power relations surrounding the parallel production of infrastructure networks and new urban landscapes. In what follows we will examine the seven key arguments of spatial political economy approaches to understanding the contemporary transition of capitalist urban infrastructure. The first three encompass the core arguments of the perspective; the remaining four address its contribution to understanding contemporary processes of urban fragmentation and infrastructural unbundling.

CENTRAL ARGUMENTS OF THE POLITICAL ECONOMY PERSPECTIVE

Spatial infrastructure is embedded within capitalism, and complex terrains of winners and losers inevitably accompany urban infrastructural change. But the spatial political economy perspective to urban infrastructure makes three further central points.

INFRASTRUCTURE AS FIXED SUPPORTS FOR THE SPACE–TIME MOBILITIES OF CAPITALISM

First, the production of infrastructure networks to transcend space and time barriers simultaneously requires those infrastructure networks to be geographically fixed in space. David Harvey (1985), drawing on Marx (1976, 1978), argues that the changing configuration of infrastructure networks represents the development of new solutions to the basic tensions inherent in capitalism between what he calls 'fixity' and the need for 'motion', mobility and the global circulation of information, money, capital, services, labour and commodities. Even 'mobile' infrastructures like wireless telephony require fixed infrastructure or satellites to be constructed to cover geographical space.

To understand the tensions between fixity and mobility one must explore the essential geopolitical economy of capitalism in a little more detail. To Harvey (1985) capitalism is inevitably expansionary. Because it is driven by the search for new profits and 'capital accumulation', the aggregate effect of capitalism is to expand into new geographical markets, as well as continually to restructure the development of old ones, through new products and new technologies. This means that widely dispersed sites of production, consumption and exchange need to be integrated and coordinated, to support processes of production, distribution, consumption and social reproduction. In order to achieve this, space needs to be 'commanded' and controlled, particularly by large, dominant, transnational firms.

The theoretical 'goal' of global capitalism, and the key to maximum profits with minimum risk, is therefore *perfect mobility* for labour and goods (through transport), capital (through financial markets) and water, energy and information products (through infrastructure networks). Erik Swyngedouw calls this 'the desire to produce a spaceless world' (1993, 313).

But in a fundamentally spatial and geographical world, where infrastructure networks must be fixed and embedded in space, this goal is impossible. Both the new spatial structures that are the bases of production and consumption – cities, industrial areas, consumption zones, etc. – and the new infrastructure and transport networks needed to support this mobility are inevitably *fixed and embedded in produced space* (Swyngedouw, 1993). New infrastructure networks 'have to be immobilised in space, in order to facilitate greater movement for the remainder' (Harvey, 1985, 149). As Swyngedouw continues:

a railway, a motorway or communication line, for example, all liberate actions from place and reduce the friction associated with distance and other space-sensitive barriers. However, such transportation and communication organisation can only liberate activities from their embeddedness in space by producing new territorial configurations, by harnessing the social process in a new geography of places and connecting flows.

(1993, 306)

Produced infrastructure networks within, between and below the fabric of cities thus literally underpin the territorial configurations of global capitalism. To Swyngedouw:

geography is actively *produced* in a well-defined and relatively immobile physical infrastructural and social way. . . . The production of territorial organisation, a combination of economic, infrastructural and institutional–regulatory practices, is a historical product which simultaneously defines, shapes and transforms social relationships and daily practices.

(1993, 310)

In contemporary processes of capitalist globalisation, spatiality, power and infrastructure networks are therefore intimately connected. 'Some barriers are torn down; others are maintained. New spaces are created (for global trade, or new *favelas* [shanty towns]); others are destroyed (the spaces of more integrated national economies, and those of small-scale agriculture)' (Massey, 1995, 140).

An essentially relational theory of what Swyngedouw (1993, 308) terms the 'reshuffling of spatial relations' in capitalism thus emerges, as networked infrastructures support new configurations within and between places, in keeping with the wider geopolitical logics of capitalism.

Infrastructure networks thus become central to the reproduction and development of capitalism because they link multiple spaces and times together. This is a necessity because of 'the fundamental diversity and heterogeneity in time and space' in our contemporary globalising world (Dupuy, 1988, 5). As Jean-Marc Offner puts it, rather than somehow causing 'territory to disappear', as in the immaterial dream of so many IT utopianists, 'it is precisely *the fact that a multitude of places exist* which creates the need for exchange based on infrastructure networks' (1996, 26, emphasis added).

THE DIFFERING TIME–SPACE CAPABILITIES OF INFRASTRUCTURE NETWORKS

Second, infrastructure networks obviously differ in their ability to transcend space and time barriers. Telecommunications, transport, energy and water networks clearly vary dramatically in the degree to which they are fixed and embedded in space, in the degree to which the mobility they support is retarded by the frictional effects of overcoming distance and in the scale at which mobility is required. This is shown schematically in Table 5.1.

Table 5.1 Understanding the embeddedness and time–space capacity of different infrastructure networks

Network	Fixity/embeddedness in space	Frictional effects of distance	Scale of operation
Water/waste	Very high	Very high	Local/regional
Transport	Very high	Very high	Local–global
Energy	Medium	Medium	Local–international
Telecoms	High to low	Very low	Local to global

INFRASTRUCTURE NETWORKS AS LOCALLY DEPENDENT CAPITAL

Third, because they necessarily have to be produced in space, utilities, transport and telecom grids are therefore also classic examples of what Cox and Mair (1988) call *locally dependent capital*. They are, quite literally, materially embedded in territory. Gas, electricity, water, waste, transport and telecommunication networks inevitably require the investment over relatively long periods of large amounts of capital within territorially bounded areas. The capital is quite literally *sunk* into the vast (sometimes hidden) lattices of ductwork, pipes, wires, cables, roads, ports, airports and railways that lie beneath, within and above modern cities and fill the corridors between them. Even when it is not so – as with radio-based telecommunications networks – the antenna infrastructure is still rooted in largely immobile patterns across particular territories and often depends on the conventional phone system for transmission.

The local dependence of infrastructure capital makes it expensive, uncertain and risky to develop – especially for the profit-seeking firms that increasingly control it. This inflexibility means that sunk infrastructures go on to present problems later in further 'rounds' of restructuring as the continuous dynamism of capitalism plays out. Crises emerge where older infrastructure networks, which are embedded in space, become *barriers* to later rounds of capitalist accumulation. To David Harvey:

the tension between fixity and mobility erupts into generalised crises, when the landscape shaped in relation to a certain phase of development . . . becomes a barrier to further accumulation. The landscape must then be reshaped around new transport and communications systems and physical infrastructures, new centres and styles of production and consumption, new agglomerations of labour power and modified social infrastructures.

(1993, 7)

It follows that the production and reconfiguration of space are intrinsic to the development of infrastructure networks within and between cities. One needs to maintain a linked, relational perspective on how broad societal shifts are constituted through the production of *both* cities and urban landscapes *and* the infrastructure networks that are sunk into the territory within and between them.

UNDERSTANDING TRANSITIONS IN CONTEMPORARY URBAN INFRASTRUCTURE

With these essential starting points established we can now explore what spatial political economy perspectives have to say about contemporary transitions in urban infrastructure. Here we can separate the recent contribution of political economy perspectives into four further arguments.

THE SOCIAL BIASES OF INFRASTRUCTURAL CHANGE: 'TIME–SPACE COMPRESSION' AND 'PERSONAL EXTENSIBILITY'

First, it is clear that, at a general level, the new capabilities of infrastructure networks are underpinning what is called 'time–space compression' – the uneven collapse or reduction of time and space barriers (Harvey, 1989). This applies, in particular, to telecommunications and fast transport networks like airlines and high-speed trains. But it also applies to widening grids of long-distance energy and water flow.

In the process, the spaces within and between cities become 'wired' and gridded with highly uneven networks of ever more capable infrastructures. The technologies of infrastructure increasingly become the very 'organisation principle to everyday life', supporting 'ever-accelerating geographies of production, exchange, and consumption' (Kirsch, 1995, 541). As Harvey suggests, we have, as a result, 'learned to cope with an overwhelming sense of compression of our spatial and temporal worlds' (1989, 240).

But extreme care needs to be taken over the definition of 'we'. For it must be remembered that the linked production of fixed infrastructure networks and urban spaces is an exercise infused with struggles between groups, firms and institutions that possess highly uneven social, economic, environmental and cultural power. Such struggles tend to reflect and reinforce the wider reproduction of capitalist social relations. Far from being somehow neutral or intrinsically beneficial, urban infrastructure networks embody power relations and reflect highly uneven political-economic struggles between firms, state and public sector organisations and wider social agents, which are intrinsic to the spatial geopolitics of capitalism as a whole (Samarajiva and Shields, 1990).

The uneven and partial integration of territories via infrastructure networks, and the current era of infrastructural unbundling, are therefore dynamic social and political processes filled with ambivalence, conflict and multiply contested perspectives (Graham, 1994). Complex terrains of winners and losers, woven into the wider social and spatial inequalities of capitalist urbanism, are an inevitable feature of the continual uneven construction, and reconstruction, of infrastructure networks between and within cities. Moreover, the conceived technological logics of powerful firms and their infrastructural suppliers tend to dominate the lived, emotional spaces of human life (Lefebvre, 1984; Kirsch, 1995, 545).

Time–space compression via infrastructural development is therefore necessarily very uneven and partial. Powerful interests may have ever more power over space and time. But, counter to the widespread ideology that people benefit universally from such processes, the space–time constraints on poor or marginal communities may actually increase, both relatively and absolutely, because their position with respect to accessing new infrastructures may become even more marginal. As Erik Swyngedouw writes:

the changed mobility, and hence power patterns, associated with the installation of new mobility commodities and infrastructure may negatively affect the control over place of some while extending the control and power of others. . . . Being trapped in particular places and subject to processes of restructuring and depreciation undermines the control and command of spatially imprisoned individuals and social groups, while this very restructuring and depreciation is organised by those Cyborg men and women whose ability to command place is predicated upon their power and ability to move over hyperspace.

(1993, 322)

But it is also vital to stress the complexity of such processes. 'The spaces of exclusion and inclusion are always heterogeneous and subject to change' (Hinchcliffe, 1996, 674). A person's or group's ability to construct their identity, and to exercise social and economic power, thus derives very strongly from the degree to which they can mediate their lives with infrastructure and so extend their influence over space. To draw one last time on Swyngedouw's analysis, the key point here is that, in contemporary society:

cashing in on cultural capital assets is related to an individual's capacity to construct a multi-scaled and multiple identity; an identity which comes about in and through the command of space and the capacity to move across space. In other words, social power cannot any longer (if it ever could) be disconnected from the power or ability to move quickly over space. The necessary resources to minimise time–space distances and the unquestioned commodification of time–space compressing processes accentuate social, economic and cultural inequality.

(1993, 323)

Places are thus infused with complex 'power geometries' based on their highly uneven interconnection with the full range of infrastructural means to overcome time and space barriers (Massey, 1993). 'Equidistant destinations within the city may take radically different lengths of time and effort to reach, and different people, for a variety of reasons, may have quite different problems in reaching the same destination from the same origin' (Hamilton and Hoyle, 1999, 18). The presence of a road, telecommunication line, energy or water conduit dictates that exchanges must flow along their lines of connection. Equally, such networks can divide spaces and deter exchange in transverse ways. Whilst their axes and end points may powerfully connect parts of cities together, 'dramatically reducing those distances between people who happen to be connected along such routes', roadscapes tend to be physically very divisive in a lateral sense (Urry, 1999, 16). 'The shops may be in full view across the road from the place where you live, but if there is a three-lane dual carriageway in between, and the nearest footbridge is half a mile away, the shops are pretty inaccessible to you' (Hamilton and Hoyle, 1999, 20).

Along a similar line of argument, Paul Adams (1995) offers the concept of 'personal extensibility' to capture how a subject's use of networked infrastructures, to access or connect with (more or less) distant spaces and times, may allow them to extend their domination of excluded groups, and so support the production of divided spaces and cities. 'One person's (or group's) time–space compression,' he writes, 'may depend on another person's (or group's) persistent inability to access distant places.' As Adams suggests, 'the variation of extensibility according to race, class, age, gender, (dis)ability, and other socially significant categories binds micro-scale biographies to certain macro-level societal processes' (1995, 268; see Massey, 1993, 66). Whilst Adams considers only information technology and telecommunications, these ideas may also apply to transport or the use of water and energy infrastructures to access resources and reserves far from the point of consumption.

Thus, within contemporary cities, forms of 'super-inclusion', based on the intensive use of information technology and other networks to access far-off places, emerge for socio-economically affluent groups (Thrift, 1996c). Such groups may, at the same time, work to secure cocooned, fortified, urban (often now walled) enclosures, from which their intense access to personal and corporate transport and ICT networks allow them global extensibility. 'Those who already have more power within the power geometry can wall themselves in' (Massey, 1995, 104).

Meanwhile, however, a short distance away, in the interstitial urban zones, there are often 'off-line spaces' (Aurigi and Graham, 1997) or 'lag-time places' (Boyer, 1996, 20). In these often forgotten places time and space remain profoundly real, perhaps increasing, constraints on social life, (say) because of welfare and labour market restructuring or the withdrawal of bank branches or public transport services.

It is easy, in short, to overemphasise the mobility of people and things in simple, all-encompassing assumptions about place transcendence and 'globalisation' which often conveniently ignore the fragmenting reality of many urban spaces (Thrift, 1996a, 304; see Massey, 1993).

'STRATEGIC LOCALISM' AND THE 'GLOCAL SCALAR FIX': UNBUNDLING MODERN NOTIONS OF TERRITORIAL GOVERNANCE, INFRASTRUCTURE AND SPATIAL SCALE

The second contemporary point stressed by spatial political economists addresses the effects of globalisation on national infrastructural monopolies. In the contemporary world, as we began to discuss in Chapters 3 and 4, the changing geometries of infrastructural power tend to be bound up with internationalisation, liberalisation, privatisation and the application of new information technologies. Such transformations mean that the modern nation state, in particular, is tending to leak its power to continuous and largely invisible circuits of fast or instantaneous financial and infrastructurally mediated flow.

Some argue that the nation state is actually becoming fragmented or 'hollowed out' in the face of globalisation and the growth of transnational governance institutions (Taylor, 1994). Coherent national economies – always a deeply rhetorical construction – no longer exist. The territoriality of politicians survives despite the collapse of the notion of the national state as a social or wealth container (ibid.).

As we saw in the last chapter, such processes tend to support the active construction of highly capable and customised networked infrastructures for highly valued spaces whilst the remaining portions of national territories often become neglected. In response, nation states, whilst being far from redundant, are often becoming, in a sense, 'spatially selective' (Jones, 1997). They are constructing experimental models of urban planning and infrastructure provision for building local microgeographies within strategically significant regions whilst withdrawing policies geared to mass integration and redistribution (ibid.). Jones terms this a policy of 'strategic localism' (ibid., 852).

Contemporary shifts therefore suggest that preoccupation with the national scale of produced infrastructure, characteristic of the period of the modern infrastructural ideal between 1880 and 1960, is but one, historically specific, orientation towards geographical scale. In fact, it is increasingly realised that, like places, spatial scales themselves are socially constructed. As Neil Brenner argues, 'spatial scales can no longer be conceived as pre-given or natural arenas of social interaction, but are increasingly viewed as *historical* products – at once spatially constructed and politically contested' (1998b, 460). In fact, he suggests that current shifts require us to rethink taken-for-granted assumptions about how spatial scales and territories relate to economic and social life. 'One of the most daunting methodological challenges' facing contemporary social science, he believes (1998b, 28), is:

to rethink the role of spatial scale as a boundary, arena, hierarchy of social relations in an age of intensified capitalist globalization. The current round of globalization calls into question inherited Euclidean, Cartesian, and Newtonian conceptions of spatial scales as neutral or stable platforms for social relations, conceived as containers of different geographical sizes.

In such a view the 'state' or the 'city' should therefore not be allocated some natural, given coherence in ordering social life. Neither the local nor the global is pre-eminent in the construction of contemporary cities. Cities, rather, are bound up in a dynamic continuum of global–local or 'glocalised' interactions. Multiple geographical scales now intersect in potentially highly conflicting ways within the landscapes and sociotechnical fabric of cities

(Brenner, 1998b, 438). What Henri Lefebvre called a 'generalized explosion of spaces' occurs as new spatial and temporal 'fixes' are established and reworked between and within social scales via continually shifting mixtures of territorialisation and deterritorialisation (1979, 289–90, cited in Brenner, 2000, 361). The:

scales of capital accumulation, state territorial power, urbanization, societal networks and politico-cultural identities are being continually transformed, disarticulated and recombined in ways that severely undermine this pervasive naturalization of the national scale of social relations.

(Brenner, 1998b, 28)

Using this perspective, Brenner goes on to argue that the period since the mid 1970s has witnessed a progressive *denationalisation* of capitalist territorial organisation and the widening production of a new set of dominant social scales, oriented towards intense global–local interaction, especially for powerful interests (1998b, 473). Infrastructure developments by national states and subnational governments have provided the most significant supporters of the construction of such 'glocal' scales. In a sense, then, 'globalisation':

emerges when the expansion of capital accumulation becomes intrinsically premised upon the construction of large-scale territorial infrastructures, a second nature of socially produced spatial configurations such as railways, highways, ports, canals, airports, informational networks and state institutions that enable capital to circulate at ever-faster turnover times.

(2000, 7)

As a result, in a shift driven by the rising power of transnationals, global financial integration, the collapse of Fordist and Keynesian styles of policy, and growing competition between spaces, older national states had all, by the mid 1980s, 'substantially re-scaled their internal institutional hierarchies in order to play increasingly entrepreneurial roles in producing geographic infrastructures for a new round of capitalist accumulation' (1998b, 29).

In the infrastructural arena this has meant a widespread shift to privatisation, liberalisation, the opening up of public infrastructure to private investment, and increasing the freedom of private capital to develop limited, customised infrastructures in specific spaces, without worrying about the need to cross-subsidise networks in less favoured zones. In France and the United Kingdom, for example, the vast bulk of investment in telecommunications following liberalisation has been within the Île-de-France and M25 regions respectively (Graham, 1999).

Neil Brenner calls these new approaches to infrastructural development 'glocal scalar fixes' (1998b, 476). To him, they differ radically from the styles of infrastructure development that characterised the latter stages of the modern ideal. The Fordist–Keynesian practices that characterised the modern ideal were geared, at least discursively, to 'homogenising spatial practices on a national scale'. In contrast, 'a key result of these processes of state re-scaling has been to intensify capital's uneven geographical development' (ibid.). As Susan Christopherson has suggested, 'with the withdrawal of national "equalizing" investment, the privatisation of previously public investment and the concentration of public spending in some types of localities, public investment programmes have deepened trends toward uneven spatial development' (1992, 284).

THE 'PAY PER' REVOLUTION: NEW TECHNOLOGIES AND THE UNRAVELLING OF 'NATURAL MONOPOLIES'

The penultimate contemporary argument of political economy perspectives is that, as part of such restructuring and rescaling, the redistributive role of infrastructure networks in modern welfare states is under severe strain. In particular, the rationale for cross-subsidy and redistribution is becoming more threadbare. 'The old welfare state, that once-seeming jewel in the crown of moral progress in democratic societies, is finished, although it has taken twenty years for the gradual erosion to reach the present point of collapse' (Leonard, 1997, 22).

As we saw in Chapter 4, the control capabilities of information technologies, in particular, are allowing modern infrastructure grids to be 'unbundled' and (re)commodified. Information technologies can help turn 'dead' infrastructures, streets and networks, based on mechanical and electromechanical technologies, into 'smart' infrastructures, managed by computers linked over telecommunication links. As a result, 'technology has further challenged the notion of the natural monopoly' (Rosston and Teece, 1997, 5).

Paul Virilio has discussed current trends towards the replacement of bureaucratic and electromechanical systems of control by smart electronic systems for controlling and monitoring access to goods and services in cities. To him, such trends mean the evolution of complex, subtle forms of real-time control, within which 'the rites of passage [in cities] are no longer intermittent – they have become immanent', in that they consist of cybernetic, intelligent systems that are subtly woven deep into the fabric of cities and regions (1987, 16; see Graham, 1997).

This means that the traditional characteristics of public infrastructure networks that made them 'natural monopolies' can in many cases be eroded or avoided completely, allowing infrastructure goods and services to be delivered as commodities to be bought and sold in capitalist infrastructure markets. Thus the high capital cost of building new or competitive infrastructure networks (say a new highway or telecommunications network) can be offset against a predictable revenue stream that comes from users being charged market rates to use that system (even if the network runs parallel to older, public networks). Economies of scale become less important because, again, customised infrastructures can tap into specialist market niches. The 'public good' nature of infrastructure networks and services can also be undermined, because new technologies of access control make it possible to control who has access to the network and the precise flows that go over it with unprecedented accuracy and ease. Good examples are the shift from public highways to electronic road pricing and the shift from national public television broadcasting to 'pay per view' charging.

'The essence of what is happening,' suggests Vincent Mosco, is that 'new technology makes it possible to *measure* and *monitor* more and more' of our consumption and use of infrastructure networks and services in real time (1988, 5, original emphasis). This monitoring can be cybernetically linked with building up precise records of consumption, behaviour and tastes, so fuelling further rounds of direct marketing and infrastructure planning in a cybernetic and iterative cycle. In addition, the predictable revenue streams that 'pay per' arrangements offer can, in turn, mean that specific, targeted infrastructure investments can be undertaken which guarantee a substantial profit stream based on monitored and charged access and use.

Infrastructure services can thus be developed as pay-per-use commodities rather than natural monopolies (Mosco, 1988).

Finally, of course, as we saw in the last chapter, information technology and telecommunications can enable what we call virtual network competition to occur over previously standard and monopolistic infrastructures – as with the emerging electricity markets in North America and Europe. Here the enormous transactional complexity of infrastructure markets is supported by the monitoring and tracing ability of electronic and computerised controls. As with the case of telephone or energy liberalisation, numerous service providers can now use the monitoring and surveillance capabilities of information technology to use the old networks to access specialised, targeted markets, in any part of a city or region, on a tailored, controlled basis. At the same time, they can avoid the problems of 'free riders' or of servicing low-profit or no-profit users or districts. Thus virtual network competition, as a form of network unbundling, may be just as closely linked with the social and spatial fragmentation of users within cities as the 'hard' construction of new, bypassing transport, streets or telecommunication links. The process of transformation is simply less visible.

FROM HOMOGENISING INFRASTRUCTURES TO 'HUBS', 'SPOKES' AND 'TUNNEL EFFECTS'

> Space-time no longer corresponds to Euclidean space. Distance is no longer the relevant variable in assessing accessibility. Connectivity (being in relation to) is added to, even imposed upon, contiguity (being next to).
>
> (Offner, 2000, 172)

Which brings us to the fourth and final key issue that emerges in our discussion of spatial political economy: imagining the urban geographies of unbundled infrastructures, 'strategic localism', 'glocal scalar fixes' and (re)commodified infrastructures in a little more detail. Clearly the shift from relatively homogenising and hierarchically organised infrastructures, oriented to the urban and national scale, to infrastructures configured for glocal interaction makes the maintenance of infrastructure networks that were (more or less evenly) laid out over urban and national spaces to 'bind' them more and more problematic (Offner, 2000).

But what happens to such homogenising networks? We believe that they increasingly become punctured and ruptured; in our language, they are unbundled and splintered, ushering in new geopolitical and geoeconomic logics based on the highly uneven warping of time and space in highly localised and valued places. Advanced telecommunications and fast transport, in particular, are being used to link producers, distributors and consumers across distance in radically new ways. 'More complex geographical arrangements' are emerging, write Beckouche and Veltz (1988). In these 'the production–distribution system can fight it out in space using the length of the infrastructure and communication networks on a national, even planetary level'. Guiseppe Dematteis (1994, 18) notes 'the passage from a functional organisation [within cities] in which the centres are graded with a multi-level hierarchy (as in the models of the economic geographers Christaller and Lösch) to interconnected networks organised on the basis of the corresponding complementarities of the nodes and the synergies produced'.

To some, these trends mean that the old territorial identity of the city economy, as the heart of its hinterland, has been totally lost; instead 'the city is divided into as many fragments as the networks which traverse it' (Dematteis, 1988). This new emerging type of unbundled infrastructure logic is shown schematically in Figure 5.1. Telecommunications, fast transport networks (and, to a lesser extent, customised energy and water services) now interconnect cities into systems of 'hubs' and 'spokes' across wide distances.

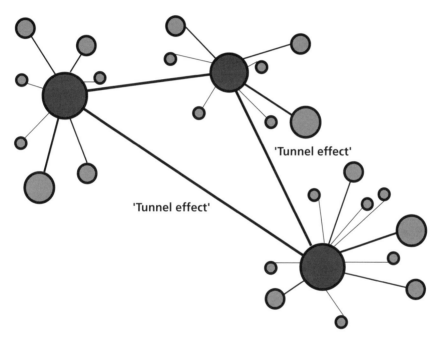

Figure 5.1 The logic of unbundled infrastructures: a schematic representation of 'hub and spoke' infrastructure networks which use 'tunnel effects' to traverse non-valued territory. *Source*: Graham and Marvin (1996), 59

Instead of standardised infrastructure networks operating more or less homogeneously to 'bind' a city, there is a set of so-called 'tunnel effects'. These are caused by the uneven 'warping' of time and space barriers by the advanced infrastructure networks, targeted on valued parts of the metropolis and drawing them into intense interaction with each other. Andreu believes that 'in our search for maximum speed, roads have been turned into tunnels. But this tunnelling effect is not only confined to roads. Present in all modes of transportation today, tunnelling isolates us from reality and cuts us off from the intelligible world. This is even true with trains and airplanes' (1998, 59).

Often these unbundled infrastructures link nodes together into networks whilst using such tunnel effects to exclude and bypass much of the intervening spaces, excluding them, in turn, from accessing the networks. Good examples of such 'tunnel effects' can be found in the advanced telecom systems that link New York, London and Tokyo in a single global 'virtual' financial market place, the global 'hub and spoke' arrangement of airline networks and airports, and the fast train or TGV networks that link up the major European cities whilst excluding smaller intervening centres from access.

Of course, we need to be careful not to overstress the uniformity, speed or simplicity of this transition from national, homogeneous infrastructures to 'glocal scalar fixes', unbundled networks and hubs, spokes and tunnel effects. Many national, regional and local infrastructural monopolies, which throw their networks across Euclidean definitions of territory, continue to define jurisdictions and political legitimacy. There is much inertia in developing and regulating infrastructures, with old practices continuing and many billions of sunk investment built into the urban fabric of every city. And many cases of privatisation do strive to keep the integrity of urban and national infrastructure networks intact through using franchising and detailed regulation of the whole territory (Offner, 2000).

But the key point is that there are fewer and fewer spaces where pressure for liberalisation and/or privatisation is not allowing new private infrastructural competitors to begin assailing the coherent urban networks left over from modern infrastructure planning. Crises of both corporatist welfare states and interventionist and developmental states have apparently been wholesale. 'It [has now] become more and more clear that there are limits to [the state's] ability to pursue a "structural policy" which can promote socio-technological processes or modernisation' (Hirsch, 1991, 20). The national policy imperative is now to liberalise and privatise infrastructure networks to support 'profitability of the national location for an increasingly international capital' (ibid., 21).

FROM 'UNIPLEX' TO 'MULTIPLEX' CITIES: RELATIONAL THEORIES OF CONTEMPORARY URBANISM

What can be studied is always a relationship or an infinite regress of relationships. Never a 'thing'.

(Bateson, 1978, 249, cited in Star, 1999, 379)

Which brings us to our final theoretical perspective: a broad swathe of recent theoretical writing about cities and social change that, like actor network theory and the more recent political economic work analysed above, can broadly be described as 'relational' (Amin and Graham, 1998a, 1999; Allen *et al.*, 1999). Such perspectives imply that 'infrastructure is a fundamentally relational concept, becoming real infrastructure [only] in relation to organized practices' (Star, 1999, 380). Whilst relational writings on contemporary cities are diverse and notoriously difficult to define precisely, they rest on two essential ideas about the interplay between cities, technologies and infrastructure networks which have critical importance for this book.

REJECTING SIMPLISTIC IDEAS OF SPACE, TIME AND PLACE

Place itself is no fixed thing. It has no steadfast essence. [The challenge is] to find place *at work*, part of something ongoing and dynamic, an ingredient *in something else*.

(Casey, 1998, 286, original emphasis)

First, relational urban and social theorists reject any idea that space, place and time have any essential, predefined or fixed meaning. Instead, relational urbanists suggest that 'both space and place are constituted out of spatialised social relations' working in practice over time (Allen *et al.*, 1999). Places are worked out through social action in dynamic, specific ways that resist easy generalisation and constantly change over time.

Relational urban theories thus help us to understand the ways in which infrastructure networks serve to mediate and construct our diverse experiences of time and space. On the one hand, space, rather than being some universal, Euclidean plane, is seen to be a multiple, socially constructed and increasingly fragmented phenomenon. Time, on the other, is also multiple and constructed – a contingent phenomenon, rather than some abstract and universal social 'container' for events.

Time and space are thus both socially constructed together in all sorts of diverse ways within and through the contemporary metropolis, often via the uneven use of and connection to networked infrastructures that selectively help construct new social times and spaces (Casey, 1998). As Nigel Thrift puts it, drawing on his long-standing work on time geography, 'time is a multiple phenomenon; many times are working themselves out simultaneously in resonant interaction with each other' (1996c, 2; see Thrift *et al.*, 1978). For example, stock exchanges, linked instantaneously with distant points on the planet, through billion-dollar-a-day electronic transactions, are often physically surrounded by homeless people living on the street – people whose times and spaces remain highly local and relatively unconnected with far-off places and time.

Place, thus, is a diverse *social process* rather than, as so often imagined in the utopian diagrams of modern urban planners and infrastructure engineers, a simple, bounded piece of 'Euclidean' territory to be designed and 'rationally' controlled, through physical plans and the configuration of infrastructure networks. Places are not contiguous zones on two-dimensional maps. They are not, suggests Doreen Massey, 'areas with boundaries around' (1993, 66). Rather, they are 'articulated moments in networks of social relations and understandings' (ibid.). It is how these 'articulated moments' in the diverse circuits and space–times of urban life come (and do not come) together within a place that shapes the dynamic nature of that place. David Harvey argues that it is 'cogredience' that matters most in making a place. He defines this as 'the way in which multiple processes flow together to construct a single consistent, coherent, though multi-faceted time–space system' (1996, 260–1).

THE CITY AS A 'GEARBOX FULL OF SPEEDS': THE IMPOSSIBILITY OF 'UNITARY' OR 'COHERENT' CITIES

> It is misleading to speak of '*the* city' as if it were a whole, organic entity.
>
> (Marcuse and van Kempen, 2000, 264, original emphasis)

The second point, which follows from the first, is that such a dynamic and multiple conception of place should make us highly sceptical about the notions within the modern infrastructural ideal suggesting some essential or necessary order, coherence or unitary quality to cities as 'things'. If cities are social processes, rather than 'things', and if such social processes are the 'messy' and dynamic results of on-going 'cogredience' between myriads of spaces and times, ideas of unitary coherence and order in cities cease to make much sense, other than as crude efforts to wield social power.

As Joe Painter puts it, 'cities are not unitary, cohesive or integrated . . . , any coherence that does emerge [in cities] will be unstable, fleeting, and probably unintended and un-reproducible' (1999, 13). Cities and urban life, especially in today's heterogeneous, culturally mixed and polynuclear metropolitan areas, therefore need to be considered as 'multiplex' rather than 'uniplex' phenomena (Amin and Graham, 1998a; Graham and Healey, 1999). Figure 5.2 captures a UK perspective on the broad shift from the classical 'unitary' or 'uniplex' metropolis to today's 'multiplex' extended urban regions, which bring together many social relations and experiences of space and time in an inevitably uneasy cogredience. The 'ruptures, deformations and dissonances' within the complex relations of the contemporary city actually 'constitute spatiality itself' (Leong, 1998, 203).

Importantly for our understanding of processes of infrastructural unbundling and urban splintering, such an approach again implies that geographical proximity within cities is no guarantee of meaningful relations or connections. Relational links within and between cities are far too multiple and complex, and far too mediated by local and glocal networked infrastructure, to obey any naive geographical laws implying that far-off people and places do not relate whilst close-up ones do.

The spaces of contemporary cities thus resist easy categorisations of function and interaction; time patterns in the city are being stretched and reconfigured beyond the rigid routines of work, commuting and home time characteristic of the classic industrial metropolis. 'The city is a gearbox full of speeds' (Wark, 1998, 3). The contemporary urban fabric, as both the product and the site of multiplying and diverse relational networks, offers stark contradictions and huge tensions, which sometimes connect and sometimes do not. Time and space within cities present a 'multiple foldable diversity' (Crang and Thrift, 2000, 21).

As Joe Painter continues, 'urban space is radically discontinuous; metaphorically, the urban fabric is "torn" or "ragged" because of the non-integration between relational networks' (1999, 25). But important connections do remain among the disconnections; the spatialities of urban life still matter. So, whilst 'the urban–regional multiplex has become, more so than ever before, a fragmented kaleidoscope of apparently disjointed spaces and places, a collage

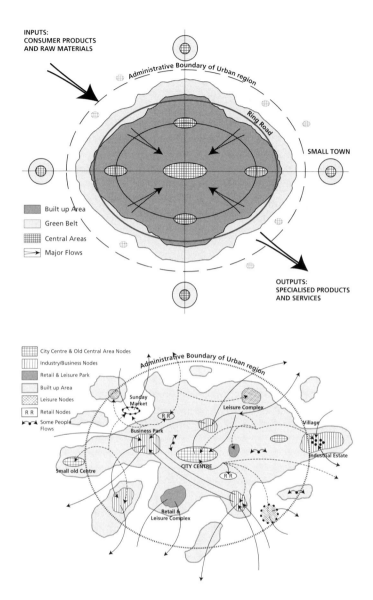

Figure 5.2 The shift in the United Kingdom from (*upper*) the uniplex metropolis to (*lower*) the multiplex urban region, as envisaged by Patsy Healey (personal communication, 1999)

of images, signs, functions, and activities', connections do remain between these 'in a myriad of ways' (Swyngedouw, 1998, 117). As Timothy Luke suggests:

given these larger structural trends, the concrete realities of place, expressed in terms of a sociocultural context of spatial location, gradually is being displaced by the tangible imaginary of flow, understood in terms of operational access to, or process through, zones of informational operation. The latter is not entirely disrupting the former; rather, they are coexisting together.

(1994, 613)

The result is that it is no longer easy (if it ever was) to separate a 'city' from the 'outside world'. (Parts of) cities are now so integrated into the outside world that we have to be very precise in identifying cities as separate entities at all. The urban fabric is (unevenly) porous and 'exposed' to the penetrating effects of infrastructure networks connecting it with multiple space–times around the globe (Virilio, 1991). Relations between people, firms, institutions, communities and buildings on the global scale, mediated by the sorts of 'glocal' infrastructure just discussed, may in many cases be more significant than their relations with urban activities or spaces that are physically adjacent.

The new paradoxes of connection and disconnection in contemporary cities, along with the collapse of the modern infrastructural ideal, therefore have major implications for how we think about both territoriality and temporality – the defining domains of human life (Amin and Graham, 1998b). The ordering power of Euclidean notions of space and Newtonian ideas of time that were so central to the modern infrastructural ideal, and the notion of urban coherence they were used to support, have collapsed in on themselves. To Peter Emberley, for example, the shift to global network societies, based particularly on instant electronic communications and uneven, customised infrastructures, is:

indicative of a significant alteration and ha[s] created an environment where the bases of our moral and political terms – the Euclidean notions of space (enclosure and exclusion) and time (succession and duration) which constituted the human experience of sequentiality, causality, continuity – have lost their ordering power. . . . Everything is fused with everything else; all is one, full and yet void in a perpetual movement of flow, of fragmented space, time, and objects. If there is coherence and integration, it comes not from socially imposed power, but from the circulation of power within the technological grid, of interlocking, interdependent agencies, practices and knowledges.

(1989, 745–58)

To authors like Ezechieli (1998) and Castells (1996, 1998, 1999b) global interconnections between highly valued spaces, via extremely capable infrastructure networks, are being combined with strengthening investment in security, access control, gates, walls, CCTV and the paradoxical reinforcement of local boundaries to movement and interaction within the city. Carlo Ezechieli writes that 'the glassy ramparts of corporate buildings, the brick and stone bastions of the temples of consumption, the aristocratic and exclusive stockades of gated communities, represent the new hard-edge boundaries of cities' (1998, 1). At the same time, 'increasingly advanced infrastructural connections compress space and time, reshape territorial patterns and challenge boundaries' (ibid.).

In Boxes 5.2 and 5.3 we examine in detail the authors who have perhaps done most to develop relational understandings of contemporary urban life: Martin Pawley, Manuel Castells, Mike Davis and Christine Boyer.

EVALUATING THE FOUR PERSPECTIVES

This chapter has sought to provide the final element of the book's conceptual perspective. It has done so by undertaking a long and complex conceptual journey across the four theoretical perspectives which are most suited to helping us understand the parallel processes of

BOX 5.2 RELATIONAL URBAN THEORY (I) 'TERMINAL ARCHITECTURES' AND 'ZERO DEFECT' ENCLOSURES

Our first example of how 'relational' thinking might help us understand contemporary urban and infrastructural change concerns the importance of infrastructural connections between a point in urban space and distant circuits of exchange. For the absolute reliability and quality of such infrastructural connectivity are becoming increasingly important to powerful international actors as corporate activity becomes driven by incessant, controlled flow over physical and electronic networks (Pawley, 1997). This idea resonates strongly with Brenner's (1998c) notion of 'glocal scalar fixes', discussed above.

The security of global–local network connections for the most powerful and highly valued spaces in cities is thus now reaching paramount importance. Buildings can increasingly be viewed as terminals articulating vast infrastructure connections with distant elsewheres. Such connections, moreover, are unevenly developed within, beneath and above the fabric of contemporary cities, a largely invisible panoply of hardware, technology and infrastructure that subtly weaves beneath the visual fabric of buildings and streets. Premium infrastructural connections, tailored to the needs of corporate users in highly valued spaces, start to become the dominant logic of (unbundled) network development. Consider, for example:

- The extreme sensitivity of global financial services, where *one day's* lost trading due to an optic fibre grid going 'down' – say through a misplaced shovel in a roadworks trench – may cost £1.4 billion sterling (Pawley 1997, 179).
- The way in which computer failure in one part of a global airline, energy or telecommunication network may throw the entire international system into costly, not to say dangerous, disarray. This was shown by the failure of the computerised baggage and freight handling system after the opening of Hong Kong's new airport, which led to utter chaos in the whole air freight and baggage movement system across the Asia Pacific region, costing hundreds of millions of dollars.
- How terrorist attacks or electricity outages in city centres can quickly lead to near economic collapse (as in the case of the IRA bombs in Docklands and the City of London, and the collapse of electricity in Auckland for five weeks during the winter of 1997–98).
- The way in which interruptions of sophisticated and 'just in time' logistical flows can disrupt the operation of strings of plants and transport networks on a planetary basis. Hence infrastructure industries increasingly emphasise the sophistication of 'back-up' services – unstaffed computer centres located in heavily protected bunker-style buildings waiting to be used at a minute's notice.

NETWORKED CONNECTIONS MEET BOUNDARY CONTROL

It is no wonder that boundary control, security and surveillance of the most vulnerable network hubs are starting to take on military proportions. Multiple, high-quality supplies of optic fibre loops, electricity and water become constructed and superimposed in case of a single failure, adding 'resilience' to a building's networked connections elsewhere.

'TERMINAL ARCHITECTURES'

Through such processes the central 'post-modern' combination emerges of intensely networked spaces and what Martin Pawley (1997) terms 'terminal architectures'. At the same time, such heavily protected architectures – stock exchange trading floors and computer centres, logistics hubs, airports, back offices – seem to establish highly controlled relationships with their surrounding urban districts. In such 'stealth buildings' or 'zero defect enclosures' 'the cost of the computer-controlled mechanical and electronic equipment . . . can often exceed their construction costs' (Pawley, 1997, 184).

Infrastructure networks become a means of securing spaces from surrounding cities whilst at the same time tying their inner workings intensively into global vectors of flow and interconnection. As the relational connections of buildings elsewhere through customised infrastructure start to mediate ever greater proportions of economic interaction, topological positions on infrastructure grids start to define the location of spaces (numbered autoroute stops, computer addresses, airline destinations).

Writing about the new financial dealing houses subtly woven into the fabric of 'world cities' like London, Pawley (1997, 194) argues that:

these blank-walled buildings are visible manifestations of the abstract, invisible, digital network that now links all the European Community countries and their neighbours through container ports, airports and railway stations, automated freezer stores, sealed warehouses, vast truck parks, and transient dormitories of mobile homes.

Pawley (1997, 171) even contends that the articulation of urban spaces through such 'terminal architectures' is now of considerably greater importance than the articulation of networks with the surrounding 'city' (which, in any case, is being drawn into incoherent and transplanetary urban fields). To him 'the act of building can be better understood, and valued,' he writes, 'as the provision of "terminals" for the systems and networks that sustain modern life, rather than the creation of cultural monuments' (1997, 97).

To Pawley, it follows that urbanity is being redefined as 'an instantaneously timed, infinitely apertured, omnidirectional phenomenon' which collapses any notion of urban coherence or the urban planning of coherent infrastructure networks (1997, 171). The challenge for architecture and urbanism, according to this argument, is to acknowledge rather than camouflage the intense infrastructural networking of contemporary built space. Such a practice – what Pawley calls 'posturban terminal architecture' – 'would involve a dramatic shift from the worship of the culture of enclosure to the belated recognition that the hidden networks that provide us with transport, energy, nutrients and information are the real riches of the modern world' (1998, 9).

BOX 5.3   RELATIONAL URBAN THEORY (II) CONNECTION PARADOXES IN CONTEMPORARY URBAN LANDSCAPES

Our second example of relational urban theory concerns the way in which cities can be made up of apparently paradoxical combinations of intense connections with far-off places and profound disconnection between adjacent ones. As Wall suggests, there seem to be strong contradictions between highly designed and planned enclaves of contemporary development and wider urban spaces within the extending landscapes of contemporary cities. 'As we move through the contemporary city,' he writes:

the contradictory and opposing conditions that are the result of its dispersal become immediately apparent. On the one hand, the city is composed of highly planned private ensembles of buildings, often large commercial or residential developments that skilled consultants in many fields took years to put together; on the other hand, large areas of the city appear to be uncared for, forming an entropic landscape.

(1994, 8)

It is clear that intense global connections for people, institutions, buildings and firms within cities are not universal or general within the diverse spaces and times of cities. Rather, they tend to be customised spatially, financially and technologically towards the more powerful and economically valuable spaces and users within and between cities. Often such powerful connections are relationally combined with intense local disconnections, between the emerging metropolis' 'archipelagoes of enclaves' – a geometry which strikingly reflects the polarising social fabric of the contemporary urban world (in the developed, developing and post-communist contexts).

The three commentators who have done most to develop a relational understanding of contemporary urban restructuring are Manuel Castells, Mike Davis and Christine Boyer.

MANUEL CASTELLS'S 'NETWORK SOCIETY' THESIS

To Manuel Castells (1996, 1997a, 1998) it is the remarkable growth in the application of information technologies and telecommunications that is the prime supporter of the shift towards an integrated, global, 'network society'. The uneven architecture and 'variable geometry' of this society are necessarily about the application of these technologies to help splinter and fragment urban space. This is because all spaces are being drawn into an integrated logic based on globalisation, asym-metrical power and the highly differentiated articulation of people and places with what he calls the 'space of flows' – the incessantly mobile, technologically mediated flows of finance, capital, information and media that dominate contemporary capitalist societies.

Castells argues that the shift to such a network society means that socioeconomically affluent groups – what he calls 'producers of high value (based on information labour)' – are everywhere being drawn into powerful articulation with global communications infra-structures. These are tailored to their needs in the urban districts and spaces where they live, work and play. Elite spaces in all cities become 'superconnected' via high-capacity tailored infrastructures, through which they articulate with the global 'space of flows' from a position of considerable power (Offner, 1996).

On the other hand, however, Castells believes that virtually all nations and cities are being penetrated by social and geographical logics of disconnection, as 'redundant', devalued labour that is of no functional use to the logic of the network society – typically manual labour in *barrios*, ghettoes and *favelas* – becomes more and more distanced from formal circuits of social and economic life (not just in the city but globally). Such groups face the collapse of employment prospects and worsening poverty.

MIKE DAVIS'S VISION OF SOCIO-SPATIAL 'APARTHEID' IN THE US METROPOLIS

To Mike Davis the sprawling, networked megalopolises of the United States increas-ingly surround the decrepit infrastructures of the declining urban cores, where crime, alienation and unemployment concentrate (1992, 6). In Davis's scenario, drawing on Los Angeles experience and the darker portrayals of cyberpunk science fiction, the concentration of new networked infrastructures into the valued peripheries of cities, backed up by private capital, repressive policing and special infrastructure measures, leads to the collapse of the coherence of the old city as global capital and wealthier social groups flee. Global eco-nomic convergence may, according to this argument, parallel local social divergence. As Roger Keil (1994, 131) suggests:

what is converging is the space of global capital whereas communities in different locales are diverging. Is the only counter-force to the convergence of global capitalist interests the tribalist fragmen-tation of these diverging communities: guarded and fenced off from one another, crammed in between the barriers of high speed traffic and humming to the deafen-ing sound of electronic highways?

Increasingly, in many urban landscapes, the emphasis is on retreat into the corporate, domestic, consumption or transport cocoons of the postmodern city whilst using highly capable networks – particularly highways, telecommunications, television – physically to extend one's actions to link into the wider social worlds within and beyond the urban region. Such trends threaten the 'public' nature of the legacies of modern infrastructural planning, especially the street. They also tend to support the withdrawal into new mediated forms of experience, as recent debates about cyberspace and the city suggest (see Robins, 1996; Graham and Marvin, 1996).

CHRISTINE BOYER'S 'FIGURED' AND 'DISFIGURED' CITIES

Our final relational urban theorist, Christine Boyer (1995, 82), also addresses US cities. She argues that contemporary restructuring trends are supporting the divergence of what she calls the 'figured' city from the 'disfigured' city.

The figured city is the city 'composed as a series of carefully developed nodes generated from a set of design rules or patterns' (1995, 81). It is 'fragmented and hierarchized, like a grid of well-designed and self-enclosed places in which the interstitial spaces are abandoned or neglected' (ibid.). It is, in short, the highly valued archipelago of spaces and zones – financial and corporate districts, heritage zones, leisure, media and cultural areas, malls, festival market places, theme parks, affluent housing, hospital and university districts, research and development campuses, high-tech business parks, etc. – that are the focus of customised infrastructure development. As well as being favoured by unbundled infrastructure connections – airports, fast highways, telecommunications, energy and water – the figured city is 'imageable and remembered' to affluent and upper-income populations who live, work and play within it (1995, 82, original emphasis).

On the other hand, though, there is the disfigured city: the 'abandoned segments' that surround and interpenetrate the figured city (ibid.). Remaining 'unimageable and forgotten', the disfigured city is largely 'invisible and excluded'. Being detached from the well designed nodes and the prime infrastructure networks, the disfigured city actually has 'no form or easily discernible functions'. As 'the connecting, in-between spaces' with the weakest and most vulnerable network infrastructure the disfigured city remains easily forgotten by powerful groups inhabiting the figured city (ibid.). Within processes of infrastructural unbundling the disfigured city is likely to fair badly. Not only will it lose the cross-subsidies inherent within Keynesian and monopolistic models of network management, but it is likely to fail to attract new private entrants to transport, energy, communications and water markets.

infrastructural unbundling and urban fragmentation that we encompass within the umbrella term 'splintering urbanism'. In particular, we have identified the implications of LTS approaches, actor network and cyborgian perspectives, analyses of the changing spatial political economies of capitalist infrastructure, and relational urban theories, for our understanding of the interrelated transitions linking cities and infrastructure networks. It is worth briefly reviewing the usefulness of each in turn.

THE LARGE TECHNICAL SYSTEMS APPROACH

The LTS perspective develops powerful insights into contemporary shifts in the relations between cities and infrastructure. Whilst it has unfortunately largely ignored the inevitable spatiality of infrastructure networks, it has given rich insights into how groups of innovations become linked together to (sometimes) gain the systemic qualities of networked infrastructures. It has also helped explain the ways in which infrastructural technologies and services diffuse and become widely accepted 'black boxes' permeating and mediating large parts of the social and economic fabric (a process which leads us to take them for granted as banal, ubiquitous and permanent).

The LTS perspective, moreover, has begun to suggest that the current reconfiguration or unbundling of infrastructure may represent a reversal of long-term historical trends towards more territorially integrated and standardised infrastructures within and between cities. Thus we might read the unbundling of networks as a process through which infrastructures are taken out of their 'black boxes', to be socially and technically reconfigured. This is a 'messy' and complex process through which institutional efforts attempt heterogeneously to engineer social and technical artefacts in profoundly new ways.

ACTOR NETWORK THEORY AND CYBORGIAN PERSPECTIVES

Actor network theory and cyborgian perspectives add to such an understanding in three ways. First, they rightly suggest caution as to whether any meaningful distinction between 'technology' and 'society' or 'infrastructure' and 'cities' can now be made – so seamlessly are they interpenetrated. Just as the body blurs with technologies to become a cyborgian whole, so does the contemporary city. As everything within cities becomes seamlessly integrated into skeins of technical networks to distant elsewheres, we can no longer assume that 'cities', or, indeed, infrastructures, can be separated or compartmentalised, as in the pure modernist vision.

Second, such perspectives underline the tenuous fragility of infrastructure networks. They stress the immense, heterogeneous arrays of 'actors' that constitute them. And they emphasise the continuous effort needed to ensure that all the constituent elements operate to remake spaces and time in the desired fashion. If unbundled networks are constructing distant energy resources, water reserves, and transport and communications opportunities, to be targeted at particular parts of the world within and between cities, we should remember the sheer magnitude of social effort required by that task. Seeing the impact of the collapse of infrastructure networks, whether from war or natural disaster, forcefully hammers home their fragility and the constant efforts required to make them function.

Finally, actor network and cyborgian perspectives help us to understand how infrastructure networks weave the very constitution of geographical configuration through their subtle hybrid logic of offering provisional connections between certain points whilst offering nothing to the gaps, zones and interstices that fail to be enrolled on to them.

In so doing, actor network and cyborgian analyses help us to be sensitive to the subtle, capillary reach of networked infrastructures, especially as they unbundle into myriads of customised and fragmented time–space arrangements. We are left in no doubt about the profound difference between being spatially close to accessing a network – through a telephone, an electricity point, a motorway slip road, an Internet access point, an airport or whatever – and the reality of doing so. More and more, it seems, the logic of unbundling networks means that this infinite distance between spatial proximity and network access is closely woven into the fabric of urban life.

Thus many social and economic worlds within cities, particularly (but not exclusively) in the cities of the South, may be physically surrounded by water pipes, electricity grids, phone and cable lines, and rapid highways, but utterly excluded from the use of such networks. It really is a case, as the photograph of electricity pylons in Durban, South Africa shows (see p. 7), of being physically close but relationally severed.

THE SPATIAL POLITICAL ECONOMIES OF CAPITALIST URBAN INFRASTRUCTURE

Our third perspective, that of spatial political economy, takes such analyses of the power relationship surrounding infrastructural development further still. Analysts here offer a powerful, integrated view of how the production and reconfiguration of infrastructure networks is intimately bound up with the production and reconfiguration of cities and capitalist urban landscapes. Once again, this allows us to overcome overly separated and deterministic notions of technology and space. Instead, we see both as being intrinsically produced together within the dynamic political economies of contemporary capitalism.

'Technology' or 'infrastructure' thus become much more than materially impacting 'things'. They emerge, rather, as embedded instruments of power, dominance and (attempted) social control. They are intrinsically geopolitical and social phenomena that are 'sunk' into certain spaces and not others. Cities, moreover, emerge as contested terrains for capitalist political economy. They are dynamic spaces that are always in a state of uneasy tension with the infrastructure networks that fill and crosscut them. The traditional view of cities as bounded spatial 'containers' is thus in perpetual dynamic tension with the varying ability of infrastructure networks to crosscut and unevenly compress space and time barriers – as in the current period of 'time–space compression'. Urban spaces, and the geopolitical corridors between them, are thus full of struggles and contests over the highly uneven ability to overcome space–time constraints.

Moreover, with the arguments of Castells, Brenner, Swyngedouw, Gillespie and others, we can clearly begin to see how the changing political economies of urban and infrastructural development is supporting development logics based on the parallel unbundling of infrastructure networks and the fragmentation of cities. Castells's 'variable geometry' and Brenner's 'glocal scalar fixes' are based, at least in part, on restructuring the monopolistic, integrative infrastructures inherited from the modern ideal. Replacing them, using the power of new technologies to break down 'natural monopolies' in the process, are a widening array of customised, tailored, unbundled infrastructure networks targeted largely at highly valued economic and social spaces within and between cities.

Such valued spaces and users increasingly integrate into 'hub' and 'spoke' networks, through the 'tunnel effects' and 'global scalar fixes' of global airline, optic fibre, satellite, rail, road, sea and, increasingly, energy and water networks. All types of city demonstrate that the logic also supports the dislocation and distancing of non-valued spaces and users from networks. Here, again, we come across the relational perspective of time and space that demonstrates how physical proximity can be combined with relational severing, and how geographical distance can be combined with relational intensity (through the 'tunnel effects' of high capability infrastructures that operate on the principle of 'glocal bypass').

As a last point, spatial political economy perspectives usefully allow us to collapse dualist distinctions of local and global. They suggest that we are moving from a period when an infrastructure network was considered coterminously with a Cartesian space or territory to one where infrastructure hubs, spokes and tunnel effects will increasingly characterise the links between infrastructure and territory on all spatial scales, from global airline and optic fibre grid systems to local highway and utility grids. Certainly it is becoming increasingly difficult to assume that infrastructure networks are coterminous with any particular bounded definition of territory (be it neighbourhood, city, district, nation or whatever).

RELATIONAL URBAN THEORIES

Our final approach, relational urban theory, further illuminates how the ambivalent tensions surrounding the construction, use and reconfiguration of infrastructure networks are bound up within the social construction and development of cities both as places and as sociotechnical processes. Such theories serve further to debunk the notions that time and space are universally experienced wholes; that cities as places have any necessary coherence; that we can assume closely placed activities to be more related than far-off ones; or that we can simply and unproblematically characterise the 'wholeness' of cities.

Relational urban theories also do much to begin illuminating the ways in which the parallel worlds of intensely interconnected and high-value spaces interweave paradoxically with the worlds of exclusion and disconnection within the same urban fabric, as our discussion of Martin Pawley's 'terminal architectures' and Christine Boyer's 'figured' and 'disfigured' cities demonstrates.

CONCLUSIONS

To us the multidisciplinary theoretical journey that we embarked on in this chapter leads to four clear conclusions as to how we might understand the linkages between fragmenting cities and unbundling infrastructure networks. Crucially for this book, these four conclusions offer us clear foundations as we embark on the empirical chapters that comprise Part Two.

The first conclusion that all four perspectives have forcefully underlined is the intrinsic dynamism and seamless interdependence between the 'urban' and the 'infrastructural', or the 'social' and the 'technical'. 'Urban infrastructures' are heterogeneous assemblies of materials, technologies, social institutions, cultural values and geographical practices. Cities are held together by intimately linked social and technical assemblages that mutually construct one another in increasingly seamless ways. This mutual construction makes functionalist separations of 'city' and 'technology' as pointless as the separation of the 'body' and 'technology' becomes in the cyborg perspective. Cities, in fact, are the greatest 'sociotechnical hybrids' of all. Every aspect of urban life is utterly infused with, and dependent on, the heterogeneous filaments and capillaries of infrastructure networks, all working within subtle patterns of layered interconnection and mediation (Graham, 1998a).

Within this contemporary urban world, however, the modern infrastructural ideal founders. Its essentialist notions of Euclidean space and Newtonian time, of functional planning towards unitary urban order, of single networks mediating some 'coherent' city, are paralysed. It is largely incapable of dealing with the decentred, fragmented and discontinuous worlds of multiple space–times, of multiple connections and disconnection, of superimposed cyborgian filaments, within the contemporary urban world.

Second, urban infrastructure networks are only ever temporarily stabilised. Despite appearances, they are never stable or enduring, at least not without continuous effort to maintain the connective channels they support. Their complexity, and the difficulties involved in making them work to sustain control across time and space, often cause destabilisation, turmoil and even breakdown. Moreover, during periods of social and economic transformation, infrastructure networks are vulnerable to obsolescence, even though, as 'sunk capital', their physical legacies are often very enduring parts of the urban scene.

Third, the above discussion has added much to our understanding of the contemporary shift towards splintering urbanism, with its parallel and mutually constitutive processes of network unbundling and urban fragmentation. In the contemporary world, where distant infrastructural connections are becoming ever more highly valued in mediating exchanges and interactions, a central urban dynamic is emerging to assert the primacy of utterly reliable 'glocal' connection with places elsewhere. The central dynamic of urban fragmentation and infrastructural unbundling is thus a reduction of emphasis on standardised connective fabrics within cities. At the same time, emphasis on specialised, high-capacity connections between highly valued points within the city, and between those spaces and elsewhere, is massively inflated.

This transition manifests itself through innumerable constructions and reconstructions of actor networks and sociotechnical assemblies, and through the destabilisation and 'unblackboxing' of many previously taken-for-granted large technical systems. It is manifest through the changing spatial political economies of infrastructure development. And it is becoming subtly woven into the connective fabric within and between many cities.

Correspondingly, as the networked infrastructure within and between cities becomes reformulated in this way, physical and network spaces often tend to secure themselves off from the remainder of the urban fabric, either subtly (through access control technologies and intensifying surveillance), or overtly (through walls, ramparts, carefully controlled access ways, gates and 'zero tolerance' policing). Highly valued spaces, zones and buildings often retreat from the space–times and circuits of the wider city, which are seen as threatening and potentially disruptive of their seamless integration into the exchanges and interconnections of international capitalism. For such spaces and users 'mobility becomes the primary activity of existence' (Thrift, 1996a, 286). Transformations in the orientation and structuring of urban infrastructure must simultaneously involve reshaping the social organisation of power relations in both time and space and the heterogeneity of the networks.

We are thus able to understand why, currently, we see great attention to connecting parts of cities and certain urban spaces to distant elsewheres whilst, at the same time, there is a widespread parallel collapse of the notion of the integrated, coherent city served by single, harmoniously developed, infrastructure networks. As indirect relations mediated by infrastructure become increasingly important, drawing individual cities and urban sites into the planetary urban universe, so borders are increasingly mutually supported by the ramparts, walls and barriers of urban physical space and the relational disconnection of networked, infrastructural space. It has, in short, 'become increasingly difficult to imagine cities as bounded space–times with definite surroundings, wheres and elsewheres' (Thrift, 1996a, 290).

But connections elsewhere remain astonishingly uneven. They fold and warp the urban fabric into a multiplex of time–space circuits through which social agents try to wield power. In this scenario, to be disconnected from circuits or exchange and interaction threatens to be an ever more absolute form of exclusion as network-mediated forms of exchange become utterly normalised.

For those residents of Boyer's 'disfigured city', for example, who are not the privileged focus of customised infrastructures, of bypass optic fibre networks, of broadband telecom services, of modern water connections, of electronic financial and economic power or of rapid automobile or airline connectivity the greater porosity of their city space to flow and exchange may mean little more than buying foreign fast food and watching cheap, imported television (Massey, 1993). The rump of public network monopolies, be they for street space, energy, telecommunications, transport or water, threaten to offer little by way of salvation as they, too, re-engineer themselves into corporate or quasi-corporate bodies addressing the infrastructural needs of those within the circuits of the 'figured' city.

The final conclusion of this chapter draws from the emphasis of all four perspectives on dynamically working through social relations in action, rather than the use of essentialised and ossified notions of scale, space, technology, the city, agency, structure or identity. It is clear that only by maintaining linked, relational conceptions of infrastructural networks, technological mobilities and urban spaces and places shall we ever approach a full under-standing of the changing relationships between them. This is especially so in the present period of apparent infrastructural unbundling and urban fragmentation when older notions of the natural relationship between territories, temporalities and networks become utterly anachronistic.

For Latour's 'skein of networks' (1993, 120) involves relational assemblies linking technological networks, spaces and places, and the space- and place-based users of such networks, in complex, folded geographies of exchange, connection and disconnection. The linkages between place and technologically mediating networks are so intimate and recom-binatory that defining space and place separately from technological networks soon becomes as impossible as defining technological networks separately from space and place. It is especially so when linked processes of infrastructural unbundling and urban fragmentation are under way. Thus unbundled configurations of infrastructure networks continuously and subtly recombine with the production of new configurations of urban space; understanding one without the other soon becomes impossible.

Part Two takes these perspectives forward and applies them to a range of detailed empirical analyses of the linked dynamics of unbundling of networks and fragmenting cities in a variety of contexts around the world. This starts in the next chapter with an exploration of the changing social landscapes of contemporary cities.

PART TWO

EXPLORING THE SPLINTERING METROPOLIS

Plate 9 The World Financial Center at Battery Park City on the south-western tip of Manhattan, New York. A classic 'premium network space', the Center is an enclave configured for companies in global financial services and their employees. The development is equipped with its own infrastructure: schools, services, shops, marina, park, freeway connections, internal private winter garden and a skywalk link across the highway to the World Trade Center. *Photograph*: Stephen Graham

Authority produces space through . . . cutting it up, differentiating between parcels of space, the use and abuse of borders and markers, the production of scales (from the body, through the region and the nation, to the globe), the control of movement within and across different kinds of boundaries, and so on.

(Steve Pile, 1997, 3)

The network society has exploded limits, between human and non-human, between nation states, between the cultural and the material. But, under the motive force of the global corporations in the new sectors, it has created a new border: between those included and those denied access.

(Scott Lash, 1999, 344)

[Infrastructure] networks have tended to enhance the power of the powerful while reducing the negotiating power of the weak. It is also extremely important to recognise the part such networks play in allowing the separation of consumption from production, rich from poor, and a consequent apathy towards what were once considered to be civic and social obligations for one another.

(Ellen Dunham-Jones, 1999, 11)

The modern promise that sanctioning the pursuit of self-interest brings freedom is mocked by the uncrossable boundaries surrounding every territory of modernity's own cipher, the metropolis.

(Alan Waterhouse, 1996, xxi)

6 SOCIAL LANDSCAPES OF SPLINTERING URBANISM

Plate 10 Gated condominium complex in central São Paulo, Brazil. *Photograph*: Stephen Graham

A deepening spatial segregation between rich and poor, both within countries and in the world as a whole, defines our era, and enhances central power just as it peripheralizes those left behind, creating new polarizations of wealth and poverty that have only increased in the past two decades.

(Bruce Cumings, 2000, 19)

The globe shrinks for those who own it; for the displaced or the dispossessed, the migrant or refugee, no distance is more awesome than the few feet across borders or frontiers.

(Homi Bhabha, 1992, 88)

SOCIAL DIMENSIONS OF INFRASTRUCTURAL UNBUNDLING AND SPLINTERING URBANISM

With this chapter we begin to address the task of Part Two, which is to develop two thematic empirical analyses of processes of splintering urbanism in practice in a variety of urban contexts. The challenge will be to explore how the revolution in urban infrastructure and technology is bound up with urban social change in cities across the world. We will look, in particular, at shifts in the social 'landscapes' of contemporary cities, drawing on examples from North America, Europe, Australia, Asia, Africa and Latin America.

The chapter addresses two questions. First, how are processes of infrastructural unbundling involved in wider processes of urban social change? Second, in what ways are the social landscapes of cities being reconfigured in parallel with the widespread unbundling of infrastructure networks?

In attempting to explore how urban splintering and infrastructural unbundling mutually support each other, this chapter has three parts. First, we set the scene by analysing three key social aspects of parallel processes towards infrastructural unbundling and splintering urbanism: wider trends towards social polarisation and the construction of secessionary network spaces; the withdrawal of network cross-subsidies; and the socially polarising influences of information and communications technologies (ICT).

Second, we review how urban 'spaces of seduction' and safety are being 'bundled' together with advanced and highly capable premium networked infrastructure (toll highways, broadband telecommunications, enclosed 'quasi-private' streets, malls, and skywalks, and customised energy and water services). Together, these linked complexes of networks and spaces provide secessionary 'network spaces' for elites and upper-income groups in the contemporary metropolis – shopping malls, entertainment and leisure developments, gated communities, 'smart' homes and the like.

Finally, we look at some of the spaces being left behind by infrastructural unbundling and urban splintering. We explore the 'network ghettoes' of the contemporary metropolis and analyse those parts of cities which are home to the people who are being marginalised by the reconfiguration of contemporary cities.

URBAN LANDSCAPES IN THE 'AGE OF EXTREMES': SECESSIONARY NETWORKED SPACES AND THE WIDER METROPOLIS

Many parts of the 'Third World' today show Europe its own future.

(Ulrich Beck, 1999, 3)

The 447 mainly American dollar billionaires listed by *Forbes* magazine in 1996 had a stock wealth in excess of the annual income of the poorest half of the world's 6 billion people.

(Coyle, 1997, 11)

To understand the social importance of parallel trends towards splintered urbanism and unbundled infrastructure we need to stress three supporting trends. The first is the broader shift towards social and geographical polarisation within decentralising and polycentric urban landscapes across the world. Robinson and Harris argue that roughly 30–40 per cent of the population in 'core' developed nations, and rather less in developed countries, are now effectively 'tenured' within the global economy, with jobs that offer livable incomes, some degree of security and opportunities to maintain or expand consumption (2000, 50). In a second 'tier' some 30 per cent in the core and 20–30 per cent in the global periphery form a growing army of 'casualised' workers facing chronic insecurity and the absence of social or health benefits. And in the third 'tier' – representing some 30 per cent in the core and 50 per cent on the periphery – people are structurally excluded from productive activity and 'completely unprotected from dismantling welfare and developmentalist states'. Robinson and Harris define these people as the 'superfluous' population of global capitalism (ibid.).

The United Nations reported in 1999 that, between 1995 and 1999, the world's 200 richest people doubled their wealth to more than US$1,000 billion. At the same time 1.3 billion people continued to live on less than a dollar a day. In 1983 the resource disparity between the world's richest fifth and the world's poorest fifth stood at 30 : 1; by 1990 this had shifted to 60 : 1; by 1999 it was 74 : 1 and the picture was continuing to worsen (Denny and Brittain, 1999).

The new socioeconomies of all cities thus seem to be characterised by increasing rewards for socioeconomic elites and affluent professional classes but increasing impoverishment for social and geographical groups unable to qualify as the so-called 'symbolic analysts' of changing urban socioeconomies (Reich, 1992). The inevitable diagnosis, according to David Massey (1996), is that we live in an urban 'age of extremes'. The withdrawal of wholesale social redistribution, especially in Western nations, is combining with polarising urban labour markets, 'ushering in a new era in which the privileges of the rich and the disadvantages of the poor are compounded increasingly through geographic means'. Such trends are not at all surprising when they are placed against the backcloth of urban economic restructuring and the emergence of new, intensified patterns of urban poverty and social polarisation (see O'Loughlin and Friedrichs, 1996; Castells, 1998; Sassen, 2000b).

Across the cities of the developed world, for example, Enzo Mingione notes a 'growing conflict between new urban poverty and the system of citizenship and social inclusion' (1995, 196). Whilst there remains considerable variety of experience between nations and cities, dual labour markets have, in many cases, combined with welfare restructuring to undermine the fragile webs of more inclusionary urban development built up during the postwar boom and the elaboration of welfare states and public housing (Musterd and Ostendorf, 1998). At the same time, real incomes have often dropped for the poorest communities reliant on poor-quality, part-time service jobs and public or social housing (Sassen, 2000b).

Not surprisingly, this 'age of extremes' is being etched into social landscapes, both between and within nations and cities, especially as urban populations grow across the world (UNDP, 1999, 36). The result, in cities in virtually all areas of the world (developed, developing, newly industrialising and post-communist), seems to be an increasingly 'acute sense of relative deprivation among the poor and heightened fears among the rich' (Massey, 1996, 395).

Doel and Clarke (1998) call this the pervasive 'ambient fear' of the postmodern city, a feature related also to the international migration and mixing of wide ranges of ethnic and cultural groups in the city.

Such fears and practices threaten to support the separation of the socioeconomic circuits of the rich and the poor in the metropolis. 'In the social ecology now being created around the globe,' predicts Massey, 'affluent people increasingly will live and interact with other affluent people, while the poor increasingly will live and interact with other poor people. The social worlds of the rich and poor will diverge' (1996, 409). Shifts towards liberalised housing markets seem likely to encourage further such polarisation by pricing lower-income groups out of higher demand and higher-valued spaces whilst, at the same time, large-scale redistributive and social housing programmes are undermined or withdrawn in many countries (O'Loughlin and Friedrichs, 1996).

MUTUALLY SUPPORTIVE PROCESSES: THE 'UNBUNDLING' OF INFRASTRUCTURE AND THE 'REBUNDLING' OF CITIES

We create colonies of cohesion.
(Jon Jerde, mall and theme park architect, 1999, 152)

Ironically, however, in many cities, geographical distances between rich and poor may actually be shrinking as richer groups colonise and gentrify selected pockets of previously poor areas (see Caldeira, 1994; Smith, 1996). In such cases, richer and more powerful groups, and the real estate, retail, housing and infrastructure industries that target their markets, are attempting to use other strategies, as well as simple geographical distancing, to withdraw from what they perceive as threatening contact with poorer groups.

The result is the attempted piecing together of what we might call hermetically sealed 'secessionary networked spaces'. These intimately combine built spaces and networked infrastructures. They encompass malls and business parks as well as the highways and cars that carry the affluent to the heart of such spaces; gated communities as well as broadband optic fibre and highway connections elsewhere; downtown skywalk cities as well as customised energy, water and security services; airports, theme parks and city-size museum complexes as well as dedicated transit systems and railways.

The production of such secessionary networked spaces enrols security, urban design, financial, infrastructural and state practices in combination, to try and separate the social and economic lives of the rich from those of the poor. As secessionary enclaves become more grandiose and massive – encompassing housing, work spaces, resort and theme park activities, leisure, entertainment and cultural attractions – Dick and Rimmer (1998) argue, such complexes represent, in effect, a 'rebundling' of cities.

'SUBSTITUTING THE CITY BY THE BUILDING': URBAN MEGAPROJECTS

Jon Jerde – architect of some of the most massive and influential 'rebundled' complexes, such as LA City Walk, California, Canal City Hakarta in Fukuoka, Japan, and the nineteen-storey, 2 million ft² Core Pacific City complex in Taipei, Taiwan – captures the supporting argument perfectly. To him 'the "art" of citymaking disappeared with the segregation-by-use theories of contemporary real estate' practices and functional and modern urban planning. 'Instead of city,' he continues, 'we now have office park, cultural center, government district, etc.'. The task of his largest projects, as he sees it, is nothing less than to 'reassemble the city from the current disarray' (*sic.*; quoted in Wieners, 1999, 307; see Jerde Partnership, 1999).

So, just as infrastructure networks become 'unbundled', the built spaces of many cities are tending to become 'rebundled'. Both processes mutually support each other and the attempted secession of new, elite, sociotechnical configurations from the wider metropolis. The result in many contemporary cities – both Northern and Southern – is a mosaic of 'packaged developments' (Knox, 1993b): shopping and entertainment malls, affluent housing complexes, hotels and convention centres, business parks, theme parks, airport complexes, refurbished heritage and cultural zones, resort complexes, affluent housing enclaves, administrative districts, etc. Each space tends to be separated off by highways, design strategies and security practices from the poorer zones which often geographically surround or adjoin them. This new urban landscape, writes Rowan Moore:

is manifest in shopping malls, airports, new residential enclaves, and in hybrids like the themed shopping mall or the airport retail area. Each element creates a self-sufficient, artificial, all-embracing experience that is both controlled and controlling. The space between them is seen as background, as something you see through a a car window when travelling from one such space to another.

(1999, 10)

Such urban 'megaproject' developments represent new, specialised urban development products, which are rapidly diffusing across the globe through the operation of international-ising real estate capital and planning and design consultancies (Logan, 1993; Peacock, 2000). By way of example, Table 6.1 lists the urban 'megaprojects' that were under way or planned in the cities around the Pacific Rim in 1995. The efforts of city authorities to secure build-ings from 'star' architects such as Norman Foster, Frank Gehry or Rem Koolhaas serve to encourage further the construction of high-status 'edifice buildings' as symbolic capital to demonstrate the dynamism of the host city as a node on global networks of flow (Dunham-Jones, 2000).

In all such developments the goal is to maximise profitability by carefully packaging enclave-style 'total environments' (Crilley, 1993, 127) encompassing a range of uses, differentiated according to logics of geodemographic marketing (Knox, 1993b). In this rebundling of cities, casinos, theme parks, multiplex cinemas, virtual reality centres and hotels and transport facilities are built together with offices and affluent housing within integrated 'macrobuildings' (Cerver, 1998, 29) (see Figure 6.1). Whilst urban areas have spread out and dispersed virtually 'to the point where territory and metropolis are synonymous', writes Francisco Cerver, 'at

Table 6.1 Urban megaprojects under development or planned in Pacific Rim cities, 1995

Project	Estimated size (ha)	Developer	Estimated cost (US$ billion)	Approx. development period (years)	Approx. year of completion
Vancouver					
Pacific Place	80	Private	2.3	10–15	2005
Coal Harbor	41	Private	0.760	10–15	2005
San Francisco					
Mission Bay	127	Public/private	2.0	25	2018
Sydney					
Darling Harbour	54	Public/private	1.3	10	1993
Melbourne					
Bayside	31	Private	0.670	5–10	In dispute
Melbourne Docklands	150–2200	Public/private	0.860 minimum	20	Proposal
Adelaide					
Multifunction Polls	3500	Public/private	3.3–6.6	30	Uncertain
Jakarta					
Sudirman central business district	44	Private	3.2	10–15	1995–2007 (phased)
Singapore					
Suntec City	12	Private	1.5	10	1995–97 (phased)
Johor Baru					
Waterfront City	Unknown	Private	1.6	15–20	2013
Manila					
Asia World City	173	Private	22.0	25	Uncertain
Kuala Lumpur					
City centre	40	Public/private	1.2	15–20	2013

Table 6.1 continued

Project	Estimated size (ha)	Developer	Estimated cost (US$ billion)	Approx. development period (years)	Approx. year of completion
Bangkok Muang Thong Thani	750	Private	1.5	5	1995
Shanghai Lujiazui	170 (core); 2800 (district)	Public/private	Unknown	7–30	Established by 1997; complete by 2030
Tokyo Tokyo Bay waterfront projects (including Tokyo Teleport Town)	448	Public/private	64	5–20	2004
Yokohama Minato Mirai	186	Public/private	20	17	2000
Osaka Osaka Technoport Kansai Airport	775 511	Public/private Public/private	Unknown 13.5	20 25	2010 1994 (phase 1)
Rinku Town	320	Public/private	5.3	25	1994–2000

Source: Olds (1995), 1721.

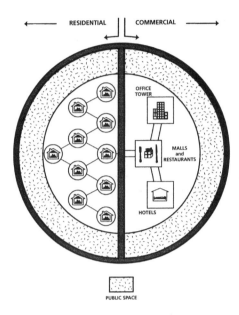

Figure 6.1 The rebundling of cities: how large, packaged developments under single ownership combine residences, workplaces, malls, restaurants and hotels in a single complex. *Source*: Dick and Rimmer (1998), 2313

the same time, within the city, new constructions have appeared which are more "city-like" than the city itself; that is, they are a distillation and intensification of the concentration that the city symbolises' (1998, 29). To him, this process is the 'substitution of the city by the building', a tendency that architects label, following Rem Koolhaas, the shift to 'bigness' (ibid., 30; Koolhaas and Mau, 1994, 510).

Thus the planned, corporate environments of business parks are developed with golf courses, lush landscaping, day care centres, fitness centres and integrated retail and entertainment spaces (Knox, 1993b). Airports and railway stations are starting to devote more space to leisure and consumption than to passenger processing. Resorts, malls and entertainment and theme park complexes mushroom to the scale of mini metropolitan areas in their own right, with dedicated hotel complexes, transport systems, utility grids, even airports. And gated communities 'are packaged with security systems, concierge services, exercise facilities, bike trails, etc.' (Knox, 1993b, 9).

AIR-CONDITIONING AND URBAN SPLINTERING

In tropical and warm climates, meanwhile, these increasingly integrated transport networks and urban spaces are now air-conditioned. This, again, tends to exaggerate the social and technical distance between the cool 'inside' – secured by security guards and often accessible only by air-conditioned car – and the stifling heat of the 'outer space' of the poorer districts and the traditional city (Dick and Rimmer, 1999). In South East Asian cities, for example (at least before the Asian financial crisis):

Privatised, high-rise urban space is the core of the market economy. This space commands the highest land price and so must attract custom. It therefore has to be a comfort zone. . . . The easiest way to achieve this is to air-condition the space. The aim is to create a microclimate which people will pay a premium to enjoy. . . . Air-conditioning and the built environment that goes with it has actually widened the gap between the 'man-in-the-street' [*sic*] and denizens in the urban comfort zone. . . . The exhalation of hot air [from the vents of air-conditioned buildings] becomes yet a further burden on the external environment.

(Dick and Rimmer, 1999, 322)

The result, in many emerging cases, seems to be what Badcock calls 'spatially partitioned and compartmentalized cities' (1997, 256). Whilst the United States tends to offer the emblematic examples of such trends, there are signs that broadly similar processes of social and spatial change can be observed in India (Madon, 1998; Masselos, 1995), China (Koolhaas, 1998b), South East Asia (Dick and Rimmer, 1998; Connell, 1999), Australia (Badcock, 1997), the Middle East (Aksoy and Robins, 1997) and Eastern and post-communist Europe (Castells, 1998; Herrschel, 1998). Such trends are also manifest, to a somewhat lesser extent, in Western Europe, where welfare and distributive policies, and the legacies of democratically organised public street systems, retain most power (Keil and Ronnenberg, 1994).

INTERNATIONAL REAL ESTATE CAPITAL AND ENTREPRENEURIAL PLANNING

Clear production-side forces are shaping the packaging of urban landscapes and the rebundling of cities. In all the above cases, local and international real estate interests seem to be intent on packaging together larger and larger luxury spaces of seduction or secession for the more affluent groups whilst, at the same time, they work harder to secure such spaces from incursion, or, perhaps more important, the perceived threat of incursion, from the new urban poor (Logan, 1993; Crilley, 1993).

The collapse of the comprehensive ideal in urban planning also serves to support the emergence of incoherent enclaves and clusters across the urban fabric. Urban planning and development agencies, keen to compete entrepreneurially for international tourists, conventions, sports events and favourable media exposure, are everywhere constructing spectacular flagship projects, set-piece developments aimed at revitalising downtowns or launching peripheral and 'postsuburban' spaces towards economic success (Knox 1993b, 11). Heavy public subsidies, infrastructural contributions and seductive grants are mobilised by public and public–private development agencies alike, to lure in the international real estate capital that has the muscle to make such projects work. The predilection of postmodern architecture for grandiose, inward-looking set pieces further strengthens such trends (Knox, 1993b, 14).

In general, as Boyer suggests, many contemporary planning practices serve to further the sense of fragmentation within contemporary urbanism. This is particularly so in the United States, where, as she writes, 'the city no longer plans for its physical development; it simply manipulates zoning bonuses and tax incentives that facilitate the building of huge real estate developments in ad hoc locations all over town' (1994, 113). Disjointed, decentralised urban landscapes thus tend to result, studded with grandiose real estate developments and their customised infrastructure links. To Michael Dear and Steven Flusty the North American 'city', at least, emerges as a 'partitioned gaming board subject to perverse laws and peculiarly discrete, disjointed urban outcomes' (1998, 66). In many cases, they continue, 'urbanization' can now be seen to be:

occurring on a quasi-random field of opportunities. Capital touches down as if by chance on a parcel of land, ignoring the opportunities of intervening lots. . . . [T]he relationship between the development of one parcel and the non-development of another is a disjointed, seemingly unrelated affair. . . . The traditional, center-driven agglomeration economies that have guided urban development in the past no longer apply. Conventional city form, Chicago style, is sacrificed in favor of a noncontiguous collage of parcelized, consumption-oriented landscapes devoid of conventional centers yet wired into electronic propinquity.

(Ibid.)

What Pierce Lewis (1983, 2) famously termed a 'galactic metropolis' thus emerges. Here elements 'seem to float in space: seen together, they resemble a galaxy of stars and planets, held together by mutual gravitational attraction, but with large empty areas between clusters'.

INFRASTRUCTURE NETWORKS AND ENCLAVE FORMATION

The establishment of large urban enclaves, whose evident purpose is protecting a specific territorial circle, reveals the rise of the new paradigm of the occupation and control of space in the 'network society'.

(Ezechieli, 1998, 7)

Secessionary enclaves appeal because they are grounded in 'the presence of an outside, unbounded, and opposing world against which [they] define the terms of [their] exclusion' (Pope, 1996, 96). But enclaves can exist only when they are connected to the networked infrastructures that allow them to sustain their necessary or desired socioeconomic connections with spaces and people in more or less distant elsewheres. Here the logic of unbundled infrastructure networks helps such spaces to connect very closely with the highly capable, yet socially highly exclusive, infrastructure networks necessary to support and sustain their operation – freeways, telecom networks, water and energy services, and direct links with international airports. These, as we have seen, are the instruments of local bypass, glocal bypass or virtual network competition. Thus networked infrastructure becomes directly embroiled in the secessionary process, supporting the material construction of partitioned urban environments. This is most visible with urban highways, with the production of 'substantial interurban spatial barriers [which] aggressively separate and exclude urban development from the greater urban continuity' (Pope, 1996, 181). In Box 6.1 we examine the particular ways in which contemporary car and highway systems contribute to such urban partitioning in US cities.

BOX 6.1 INFRASTRUCTURE NETWORKS AND ENCLAVE FORMATION: 'GRID EROSION', 'SPATIAL INUNDATION' AND THE COMPUTERISING AUTOMOBILE SYSTEM IN US CITIES

The results of the extreme dependence on the automobile that characterises many US cities are stark and clear: 'Vast parking lots, continuous or sporadic zones of urban decay, undeveloped or razed parcels, huge public parks, corporate plazas, high speed roads and urban expressways, the now requisite *cordon sanitaire* surrounding office parks, industrial parks, theme parks, malls and subdivisions' (Pope, 1996, 5).

GRID EROSION

As dedicated freeway drops start to define the linkage between the rebundled elements of the polynuclear urban landscape, so, as Albert Pope writes, each road termination or residential cul-de-sac increasingly becomes an exclusive destination on the 'roadplace' (1996, 189). This process Pope labels 'grid erosion' (see Figure 6.2). In car-dependent US cities like Phoenix and Los Angeles, the relatively even and publicly developed infrastructure 'grids' of the modern ideal are splintering and being remodelled into what Pope terms infrastructure 'ladders' where the terminal point of each link is geared towards exclusively servicing a single secessionary space. Pope argues that the contemporary closed urban systems in many US cities simulate the characteristics of open ones whilst at the same time carefully limiting and controlling unforeseen connections. In 'practically any street layout' in contemporary US cities, he writes:

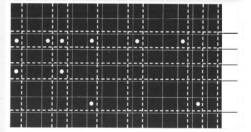

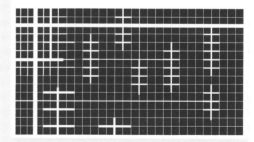

Figure 6.2 The process of grid erosion, according to Albert Pope (1996, 60, 94), through which open gridded streets (*upper*) give way to laddered streets (*lower*) where every destination has an exclusive highway entrance/exit

whether the cul-de-sac housing tract, a peripheral slab city, a gutted, skywalked CBD, or an upscale suburban office park, the traditional open urban grid . . . exists in fragmentary form as the remnant of a recognizable urban order. These reordered fragments establish, not traditional urban associations, but oblique suggestions, subtle omissions, minor exclusions, and incompletion.

(1996, 58)

SPATIAL INUNDATION

Thus a dialectical and mutually reinforcing process is established through which both constructed networks and constructed spaces turn their backs on the wider metropolis. As Pope further argues, in this process, which he terms *spatial inundation*, 'in the absence of open cities, closed developments no longer function as countersites which are both a reflection of, and a retreat from, the greater urban world. Rather they are now themselves obliged to *be* the greater urban world that was heretofore represented by the city and the metropolis.' To Pope all this means that, whilst the urban fabric has been 'opened up by space', ironically it has been punctured by countless closed and exclusive urban developments (ibid., 17). Intervening spaces and traditional streets, in the meantime, which fail to benefit from customised networked connections often decay and are downgraded as 'interstitial voids' (Pope, 1996, 109).

The result of these processes is that urban freeway links now funnel the mobile 'containers' of cars directly into the built hearts of rebundled or packaged developments. 'The soft interior of the car itself,' suggests Rowan Moore, 'is itself a kind of kindred space to the nerveless insides of the shopping malls and airports, so that it becomes possible to lead life as if in a continuous, carpeted, air-conditioned tube' (1999, 10). Cars are, in turn, now advertised as an elaborate form of mobile body-armour, secure cocoons against the (perceived) risks of the wider urban environment. Many are now replete with alarms and high-security locks (Ellin, 1997, 38). In cities like Johannesburg and São Paulo many middle- and upper-income drivers now insist on bulletproofing their vehicles as a response to the rising incidence of car-jackings and hostage taking.

RECONSTRUCTING THE CAR AS A PARALLEL VEHICLE OF PHYSICAL AND ELECTRONIC SPACE

At the same time as cars become more defensive, ever more elaborate navigational, communications and entertainment devices are being integrated into them, to while away the hours spent in worsening traffic. Luxury cars, at least, are gradually being turned into mobile 'cyborg' digital appliances that are intimately linked into complex computer communications systems to support communication, transactions, road pricing and information retrieval on the move. These days more money is spent on equipping such cars with computerised equipment than on the engine. 'The 500 million hours a week that [US] Americans spend in their cars represents a huge audience of consumers,' suggests

Figure 6.3 Advertisement for the Visteon voice-operated Internet system for US luxury cars, 1999. *Source: New York Times*, 14 February 2000

Lappin (1999, 127). This 'explains why an emerging constellation of auto makers, electronics manufacturers, and telecom providers convert transportation platforms into communications platforms that connect the driver, and the vehicle itself, to the rest of the datasphere' (ibid.).

Figure 6.3 shows a US advertisement for one such system – Visteon – which was launched in 1999. Effectively transforming the luxury car into a vehicle in both physical space and electronic space, Visteon supports wireless Internet access, allowing continuous streams of electronic navigation, communication, transaction and information retrieval while on the move. Thus the increasingly common experience of gridlock is rendered much less troublesome: a standstill in physical space allows users to extend their influence in electronic space via voice control. The worlds of electronic finance, navigation, e-mail and transactions all, in the words of the advert, become 'superintegrated into your [car's] interior', heralding the further isolation of the car's occupant (s) from the surrounding city space (Visteon advertisement, see Figure 6.3). At the same time, cars become ever more central means for drivers to express and construct their identities as advertising feeds into the dominant notion that cars represent 'freedom, autonomy, success, potency (sexual and otherwise) and mastery' (Lupton, 1999, 61).

PUBLIC SPACE, SOCIAL CONTROL AND SPLINTERING URBANISM

Finally for our first discussion, we also need to stress the degree to which splintering urbanism and unbundled infrastructures are matched by proliferating attempts at the explicit social control of 'public' spaces within the contemporary metropolis – that is, those streets and pedestrian spaces that were ostensibly configured for more or less free and open access within the modern infrastructural ideal.

Cities, of course, have always been contested spaces within which dominant power holders try to stipulate 'normative ecologies' of who 'belongs' (and who does not) where and when within the urban fabric (Norris and Armstrong, 1999). Urban 'public space' has never been truly 'public'. But the urban splintering process is making these normative ecologies more tightly defined and more self-reinforcing (Graham and Aurigi, 1997). The 'public' streets and spaces of many cities in North America, Australia, the United Kingdom and continental Europe, and increasingly of Asian, Latin American and African cities too, are giving way to instrumental quasi-public spaces geared overwhelmingly to consumption and paid recreation by those who can afford it and who are deemed to warrant unfettered access.

At the same time extraordinary efforts are made to control the incursions and behaviour of groups not seen to 'belong' in such spaces: young men in British towns and cities experience the scrutinising gaze of CCTV; poor groups are directly excluded from São Paulo shopping enclaves; thirty 'dangerous zones' have been designated in Berlin, giving police extensive new powers of eviction and search (Grell *et al.*, 1998, 211); homeless people are routinely expelled from the renovated and privately managed downtown districts of US city centres (Mitchell, 1997).

In many cases 'public space' is now under the direct or indirect control of corporate, real estate or retailer groups which carefully work with private and public police and security forces to manage and design out any groups or behaviour seen as threatening to the tightly 'normalised use'. This generally amounts to the recreation, consumption and spectacle of middle-class shoppers, office workers and tourists. Within secessionary street spaces those not seen to belong are actively pursued with the latest CCTV and surveillance technologies as attempts are made to 'sift' the quasi-public spaces in search of people transgressing the normative ecologies of the splintered metropolis. 'The proliferation of CCTV cameras surveilling city streets do not represent a panoptic gaze,' write Clarke and Bradford. They are, instead, 'a selective gaze, focused on those deemed to be "out of place" and "out of time"' (1998, 689).

Some now even doubt whether 'public space' still exists in certain Western cities (for example, Los Angeles; see Loukaitou-Sideris, 1993), as city centres become packaged and commodified for consumption, as enclosed malls start to dominate suburbs, and as middle classes retreat to cocooned houses and cars, linked via new communications infrastructures. 'Have we reached, then, the "end of public space"?' asks Don Mitchell. 'Have we created a society that expects and desires only private interactions, private communications, and private politics, that reserves public spaces solely for commodified recreation and spectacle?' (1995, 110).

However, we need to be careful not to assume the inevitable emergence of some universal urban dystopia. Many cities in Western Europe, for example, where the fabric of traditional streets and the operation of welfare states retain most resilience, are managing to maintain relatively open and socially democratic public spaces, even though commercialisation and privatisation may be encroaching. 'Spaces in London,' for example, 'do retain some semblance of "public" irrespective of whether they are enveloped in a privatised commercial environment' (Merrifield, 1996, 67).

THE WITHDRAWAL OF NETWORK CROSS-SUBSIDIES AND THE CONSTRUCTION OF INFRASTRUCTURAL CONSUMERISM

The second broad trend underpinning the social dimensions of splintering urbanism is the gradual withdrawal of the practices of social and geographical cross-subsidy that tended to underpin the extension of networked infrastructures under the modern infrastructural ideal (Schiller, 1999a). 'The present trend toward privatization is likely to end all such subsidies,' writes Kalbermatten. 'In the absence of effective regulation (a real risk in most developing countries), privatization is likely to result in efficiency gains and better service for those who already have service or who can afford to get connected to the existing system. The urban poor will again be overlooked' (1999, 15). An example comes from the South African town of Stutterheim. Here, post-apartheid privatisation of water 'was carried out in such a way that a large foreign firm "cherry picked" the lucrative white and colored areas which receive dependable water supplies at present, but which left much of the official Stutterheim township unserved' (Bond, 1998, 162).

As infrastructure services are developed as commodities to be offered at a price within markets, the ideals of universal tariffs and service are remodelled, just as the notion of the territorial monopoly unravels. Cross-subsidies between profitable spaces and routes are withdrawn and a project-by-project logic tends to replace the analysis of whole urban infrastructure grids and services. Affluent consumers often attempt to exercise their buying power by purchasing new, customised, private services that precisely meet their expanding needs to extend their powers in time and space.

Infrastructure industries, offering what were previously assumed in many cases to be public or quasi-public goods – city streets, urban highways, public transport networks, electricity or gas services, telecommunication links, and the like – thus start to 'mimic the market segmentation strategies of their private sector counterparts' (Christopherson, 1992, 274). Susan Christopherson, writing about retailing and banking services in the United States, captures this shift well when she says that:

the imperatives of the US investment system are associated with firm strategies that target low-risk, high-profit markets and discriminate among clients and customers, gearing product and service costs to profit and risk potential. Whilst these strategies are not new, the ability to target markets has been significantly enhanced by technological innovations and sophisticated market research.

(1992, 274)

As in other service industries, shifts are thus occurring towards differentiated ranges of highly symbolic infrastructural services, offering wider and wider choices of tailored infrastructure services, often within internationalising niche markets. Infrastructure services become less and less a basic means to sustain and socially reproduce modern life and more and more a means to support and construct diverse cultural identities and politics. Clearly, along with the automobile, it is broadcasting and telecommunications, as the dominant media of cultural production, that are leading this shift towards the splintering of mass markets under forces of global capitalism and privatisation (satellite television, pay-per digital television, the Internet, global telephone competition, mobile communications, interactive television, etc.). But street, transport, energy and water services are being remodelled in many cases, too.

FROM NETWORK CROSS-SUBSIDIES TO 'FISCAL EQUIVALENCE'

In some contexts like the United States, affluent social groups are tending to grow more resentful of traditional forms of redistributive local taxation. In particular, they may start to demand practices of 'fiscal equivalence' – 'where people and businesses get what they pay for' (Mallett, 1993b, 407) – rather than socially or geographically redistributive notions of universal service development. Thus, as cross-subsidies are withdrawn, increasingly 'if you can't afford services you do not get them' (Mallett, 1993b, 405).

Within liberalised infrastructure regimes, above all, infrastructures and services start to be developed on the basis of the price of delivery and the desire for maximum profits: explicit social redistribution tends to be withdrawn from the equation. In fact, it could be argued that cross-subsidies are often *reversed*; they now often go from poorer communities paying sales, income and municipal taxes, and towards the construction of secessionary network spaces, with their tax breaks, public grants and subsidies, and intensive private and public investment.

The dynamics of contemporary urban growth and change, and in particular the high levels of wealth of the new growth spaces of many cities, are leading to increasing infrastructure demands which cannot be met within the constraints of the modern public infrastructural ideal. In most, 'local government has not been able to respond quickly enough to satisfy the needs of developers' (Mallett, 1993b, 405). Instead, in many cases, private firms and developers are working together, often with financial and political support from official local governments, to construct networks and spaces that are customised specifically to the needs of the upper-income social and economic groups who are the target users. Mallett, speaking from the point of view of the growth of privatised, special infrastructure bodies in US cities, argues that this process of private governance formation amounts to the construction of a 'parallel local state'. He points out that:

some residents and businesses have been able to afford to buy services for themselves and, in the process, have been able to provide such services quickly and in a manner that gives them a high degree of control. . . . Thus, residents of postsuburbia are demanding and paying for exclusive services and regulations, and businesses are demanding and paying for enhanced mobility for workers and customers on the fringe, and better services in the core.

(1993b, 405)

Thus, as Glancey (1997) suggests, in many Western cities 'those of us with money and a degree of health and security are offered an ever increasing choice, not only of things but of ideas and ways of ordering our lives'. But people and spaces who tend to benefit from the modern ideals of universal connectivity and social access at standard tariffs tend to lose out as cross-subsidies are withdrawn – a process known as the 'rebalancing of tariffs'. Costs for them tend to rise and new investment focuses elsewhere on servicing the emerging secessionary enclaves of richer socioeconomic groups.

In short, infrastructural policy and regulation seem to be losing their social content as regulators in many cities, countries and regions construct markets for niche services developed unevenly across space. In many cases, neoliberal notions of social welfare have triumphed; competition is the means to maximise benefits for all in previous territorial monopolies. The United Kingdom's water services regulator, Ofwat, for example, has argued that:

it would be unfair to other water customers if general tariff policy were to reflect social objectives. These should be health and social service policy. Any costs from providing support to customers with particular needs should be met by the appropriate agency, and not by the water customers generally.

(Ofwat, 1990)

THE SOCIAL CONSTRUCTION OF INFRASTRUCTURAL CONSUMERISM

When power is moving between different bits of the value chain, you need to own the whole value chain.

(President of Time Warner, 1998; source: John Langdale, personal communication)

There has been a notable shift from treating the user population as a largely homogeneous group of citizens, with notional or formal rights, to a heterogeneous group of consumers, carefully differentiated according to how lucrative they are to serve. Essential resources like energy, street space, highway space, electronic communications and water become commodified to be distributed through markets. Not surprisingly, virtually all entrepreneurial activity and innovation attempts to meet the needs and desires of profitable groups, and the geographical spaces they inhabit, through 'the introduction of new tariffs, products and styles of service which vary across space, time and customer classification' (Marvin and Guy, 1997).

In telecommunications, for example, it has been shown recently that the newly liberalised communications markets in US city regions are marked by mosaics of extreme unevenness. Providers are seeking to 'cherry-pick only the most lucrative business and professional customers' from across the urban landscape (cited in Schiller, 1999a, 52). Upper-income spaces and buildings are targets of vast investment in broadband infrastructure and services (dedicated 'T-1' Internet trunks, cable Internet, digital subscriber lines, and the rapidly decreasing cost of international communications); poorer parts of cities are prone to underinvestment, deteriorating service quality and being disproportionately affected by the rising relative costs of local communications (Schiller, 1999a, 55).

Such dualisation is underpinned by the widespread shift from the standardised marketing of services to all to sophisticated marketing precisely targeted at socioeconomically affluent and highly profitable groups and areas. AT&T, for example, reorganised their approach after realising that they made 80 per cent of their $6 billion annual profits from 20 per cent of their customers (Schiller, 1999a, 54). The shift is also supported by the attempts of cross-media alliances – for example the merged AOL–Time Warner group – to take advantage of technological and financial 'synergies' in offering high-value customers whole baskets of services on a single 'one-stop shop' contract basis. In the United States, for example, John Donaghue, the CEO of MCI (part of WorldCom), stated that 'we're going to change our focus from being omnipresent to the entire market to talking to the top third of the consumer market that represents opportunities in cellular, Internet and entertainment' (quoted in Schiller, 1999a, 54). Dan Schiller calls these consumers the 'power users'. He defines them as:

high value residential customers who spend lavishly on a basket of telecommunications and information services, typically including (on an annualized basis) $650 on cellular; $500 on local wireline phone service; $400 on long distance telephony; $375 on cable, pay-per-view and video-on-demand; $250 on paging; as well as hundreds of additional dollars on online access, newspapers, magazines, and fiction.

(1999a, 54)

Clearly, there are likely to be mutually reinforcing connections between the construction of secessionary enclaves in cities and uneven investment and innovation within liberalised infrastructure markets. 'Premium' infrastructure services like electronically tolled highways, broadband and mobile telecommunications, and sophisticated water and energy services, are likely to be configured largely to the needs of the secessionary spaces and high-income groups of the changing metropolis. Infrastructure firms, now keen to construct or 'reposition' successful brands, are eager to adopt the advertising and public relations techniques of retailers, taking the homogeneous infrastructure product and customising its identity to specified socioeconomic groups, using the language of geodemographics. Users become carefully segregated and categorised according to profiles of what suppliers see as their needs, desires and profitability. Service provision becomes a means of tailoring packages of services to the most profitable target markets. All sorts of services are becoming 'unbundled' which previously offered a single set of prices and conditions to all customers. United Kingdom supermarkets have even suggested charging more for groceries at 'peak' times of the day to deter 'cash-poor but time-rich' customers from getting in the way of 'cash-rich but time-poor' professional people (Garner, 1997).

Customer loyalty cards, electronic 'smart' cards, bulk discount schemes, discounts for frequent users, geodemographic profiling, multiple tariffs and packages, concentrating on adding value for profitable users – the litany of marketing-speak quickly starts to define the uneven spatial and temporal access to previously 'bundled' and 'public' networks and services. Strange new alliances emerge here as infrastructure firms seek to survive the turmoil of market competition by connecting with other sectors like retailing and financial services (see Table 6.2).

In the process, individual infrastructure services are likely to become individually packaged with other services to tempt the loyalty of the most profitable users. In the UK utility industry, for example, 'a well-known brand need not be an existing utility; it could be Virgin, Tesco [supermarket], the Halifax [financial services] or the AA [automobile services]. Single diversified service providers could soon offer electricity, gas, water, telephony, cable TV, home security, home shopping and home banking' (Brooke and Nanetti, 1998, 56).

In short, multiservice 'rebundling' for selected affluent groups may replace the territorial and socially equalising 'bundling' of social infrastructure monopolies during the modern ideal. Infrastructure connections to profitable households here become a platform for more advanced and profitable incursions into value-added services: 'many see electricity and gas as little more than a low profit, low excitement product whose main purpose is to serve as a "foot in the door" for selling households a whole spectrum of more interesting and more lucrative commodities' (Brooke and Nanetti, 1998, 56).

In the case of energy services, for example, new technologies of virtual network competition are likely to allow services to be much more precisely tailored to the individual needs of affluent households. As Small argues, the key to maximising profitability for new market entrants will be using the new control capabilities of information technology to deliver carefully tailored energy, communications, financial, entertainment and security services, geared precisely to the needs of the most affluent segments of the market:

by making it possible to monitor and control individual appliances, telecommunications will allow energy tariffs to be tailored to an individual application, instead of being determined by the time of day. Encouraging new energy demand means catering for households with special applications – be it orchid houses, swimming pools, or applications yet unknown. Wireless technology is well suited to reaching a small number of households with special needs.

(1996, 20)

Table 6.2 The alliances formed in the UK's newly liberalised electricity market as previous territorial monopolies form alliances with other sectors – newspapers, retailers, municipalities, charities, financial services companies – and attempt to make inroads into each other's markets, 1998

Company	The situation so far	Alliances/Offers
British Gas	British Gas advertising dominates the opening of the electricity market. More than 400,000 customers have signed for electricity from British Gas, and, according to the company, a further 1.5 million have registered interest	British Gas has launched Goldfish, 'the fastest growing new credit card in Britain'. It has deals with Sainsbury's and Mencap
East Midlands (Sterling Gas)	Sterling Gas has been relatively successful in local areas, with around 150,000 customers, but East Midlands has not begun marketing electricity. Campaigns to date have been in local areas only	No deals
Eastern Electricity	Eastern claims 200,000 electricity customers. It was the most successful entrant in gas, with about 1 million customers. It is marketing nationally as areas open up, although the doorstepping it carried out in gas has been scaled down. It is carrying out more telesales for electricity	Eastern has a green tariff – Ecopower. It is offering Lionheart electricity in partnership with RJB Mining. Alliance with Barclaycard
London Electricity	London is about to launch a marketing campaign, and announce tariffs for outside its region. Its campaign will be in and around the Greater London area. London Electricity Gas has just under 250,000 customers	London has a deal with Alliance & Leicester – the free energy mortgage
Manweb	See ScottishPower	
Midlands Electricity	Midlands' offering is based around its 'Save your energy brand', a loyalty programme devised by MEB. Widespread advertising remains local, but it is targeting groups of customers through alliances with, for example, the Historic House Association, retailer PowerHouse and CalorGas	Midlands' Save Your Energy offers discounts on Orange telecoms, and energy-efficient equipment, for example
Northern Electric	Northern marketed its dual fuel deal aggressively and has signed up 350,000 customers both inside and outside its area. It is marketing nationally as areas open up	Northern has alliances with Saga, the *Telegraph*, local councils, Vaux and Granada Home Technology
Norweb (Energi)	Energi has marketed around the north-east, plus national areas as they open up	Norweb/Energi has a deal with Tesco. It will announce another at the end of the month
Scottish Hydro-electric	Hydro does not run a domestic gas business. It is beginning to extend its campaign out of the Grampian region into the rest of Scotland but has no plans to market widely south of the border	Scottish Hydro-electric offers Air Miles

Table 6.2 continued

Company	The situation so far	Alliances/Offers
Scottish Power	ScottishPower will not divulge figures. It is marketing nationally as areas open up	ScottishPower has alliances with Union Energy and Eaga. It has its own credit card
Seeboard (Beacon Gas)	Seeboard is starting its campaign on 28 October. This will include television, local radio and stands in shopping malls, stations, etc. It will concentrate on south-east areas, London, eastern and southern	Seeboard has an offer of money off First Choice holidays
Southern Electric	Southern will not divulge figures	
Swalec	Swalec won around 400,000 gas customers, around 150,000 of whom were outside its area. Many of them have signed up for both gas and electricity. For those outside, gas was sold as part of a dual-fuel package, so the company has a foothold outside its area	Southern offers Argos Premier Points. Its products will be promoted in Argos stores and catalogues
Sweb	Sweb has been concentrating on advertising to retain customers, although its tariffs, according to Offer's league table, are very competitive	No deals
Yorkshire Electricity	Yorkshire has signed up about 100,000 customers 'from Luton to the borders'	Yorkshire has a deal with Tandy

Source: Moore (1998), 18, from *Utility Week*, 16 October 1998.

THE SUPPORTIVE ROLES OF GEODEMOGRAPHICS

Beyond the automobile, media and telecommunications, in many urban areas even more prosaic infrastructural services such as energy and water suppliers are becoming open to a new set of cultural aesthetics where consumers may choose from a range of 'branded' competitors. Such firms now seek to differentiate themselves through the allure of lifestyle packages and precise geodemographic targeting to defined users within and between cities. As part of the shift from natural monopolies at urban, regional or national scales towards multiple, superimposed, competing infrastructures, geodemographic techniques adapted from retailing and financial services are providing the instruments to 'unbundle' user populations. Through the construction of complex user profiles, within geographical information systems (GISs), it becomes possible to disaggregate the social make-up of places (Goss, 1995; Pickles, 1995). For infrastructure service providers the imperative, as Golding observes, is now to:

decide how to divide up the market. Segment, then organise your business physically or virtually around that segmentation. Apply the principle right through the chain of activities surrounding a particular customer so that each channel (let us say an energy service), segment (the elderly, dual income families), or sector (utilities) receives an apparently seamless service.

(1998, 19)

Services can be tailored to particular consumer profiles and offered directly to the households they inhabit: Acorn demographic classifications range from some council estate residents to prosperous professionals in metropolitan areas.

ACORN CLASSIFICATIONS

A1 Wealthy achievers, suburban areas

A2 Affluent greys, rural communities

A3 Prosperous pensioners, retirement areas

B4 Affluent executives, family areas

B5 Well-off workers, family areas

C6 Affluent urbanites, town and city areas

C7 Prosperous professionals, metropolitan areas

C8 Better-off executives, inner city areas

D9 Comfortable middle-agers, mature home-owning areas

D10 Skilled workers, home-owning areas

E11 New home owners, mature communities

E12 White collar workers, better-off multi-ethnic areas

F13 Older people, less prosperous areas

F14 Council estate residents, better-off homes

F16 Council estate residents, high unemployment

F17 People in multi-ethnic, low income areas

Figure 6.4 Geodemographic targeting: the Acorn system used by UK utilities.
Source: adapted from Winter (1995), 15

An example, the Acorn classification which is now commonly used by UK utilities, uses seventeen categories to cover an entire city or region, from 'wealthy achievers, suburban areas', through 'skilled workers, home-owning areas' to 'people in multi-ethnic' areas and simply 'low income' (see Figure 6.4). The ways in which the metropolitan area of Washington

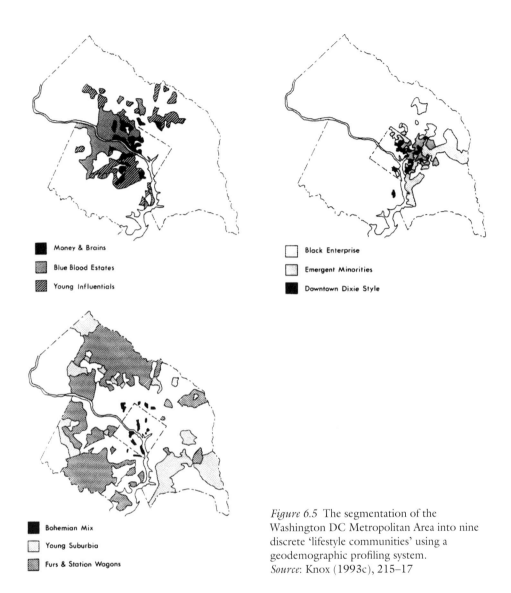

Figure 6.5 The segmentation of the Washington DC Metropolitan Area into nine discrete 'lifestyle communities' using a geodemographic profiling system. *Source*: Knox (1993c), 215–17

DC is segmented into nine distinct 'lifestyle communities' by a similar geodemographic profiling package is outlined in Figure 6.5.

With such packages, census data can be combined with real information of infrastructure consumption habits to support the direct classification of people and places according to the rigid categories of geodemographic profiling software. As more and more consumption of infrastructure is recorded using microelectronic devices, such 'transactional generated information' (TGI) can be used to develop real profiles of individual behaviour, further supporting the shift towards intensely personalised service and marketing (Crawford, 1996). The integration of customer databases with call centres and Internet portals allows the creation of automated mechanisms for subtly, and invisibly, treating consumers differently. A new system at the South West Water utility in the United Kingdom, for example, means that:

when a customer rings, just the giving of their name and postcode to the member of staff [a practice often now automated through call-line identification] allows all account details, including records of past telephone calls, billing dates and payments, even scanned images of letters, to be displayed. This amount of information enables staff to deal with different customers in different ways. A customer who repeatedly defaults with payment will be treated completely differently from one who has only defaulted once.

(*Utility Week*, 1995)

Infrastructure planning and the targeting of marketing and service delivery can then be tailored as directly as possible to meeting the more lucrative needs of profitable market niches, whilst supporting the reduction of cross-subsidies to unprofitable ones (Graham, 1997). The overall rationale tends to be to 'pinpoint concentrations of potentially high-spending customers' so that the costs of building or operating profitable infrastructure are minimised whilst the return is maximised (Winter, 1995, 14).

Thus we see a 'segmentation' of consumer identities based on intense scrutiny and surveillance, even though the categorisations used tend to be outdated, inaccurate and to reduce people's identity to a simple equation of anticipated consumption habits (Goss, 1995). Such a prospect is especially attractive to newly privatised firms or operators in newly liberalised markets, as the imperative of cost reduction and profit maximisation comes into play. Thus new roads, telecom lines, and water and energy infrastructures can be built only where direct, short-term profits are likely to ensue within the fragmented, decentralised terrains of the polynucleated metropolis.

As Lawrence (1996) suggests, from the point of view of utilities competition, geographical information and other expert systems offer powerful support for liberalised infrastructure competition. They allow advanced warning when consumers fall into debt, more efficient fixed asset management, more careful cross-selling of products and services to profitable market segments, the building up of lucrative databases to sell in the information 'market place' to other firms, and the tracking of incursions by competitors into the 'home' territory and of the excursions of the firm into others' 'home' territories. 'When utilities were monopolies,' writes Lawrence (1996, 20–1), 'there were few business reasons to invest heavily in building up intimate knowledge of the various types of customer. The main task to concentrate on was getting the utility service to the delivery point. Now, if you want to compete in utility markets across the country you must be in a position to understand what kind of organisations and users your customers are.'

Such geodemographic strategies support 'snuggling up' to the spaces and uses that are gaining through urban socioeconomic change and progressive distancing from those who are losing out. They are, in short, 'increasingly responsible for the spatial ordering of lifestyles'; they 'limit . . . individuality and offer . . . a limited number of spatially aggregated models of identity' within the city (Goss, 1995, 191). Such practices, in short, tend to 'petrify social inequalities' (Kruger, 1997, 20).

CYBERSPACE DIVIDES: INFORMATION AND COMMUNICATIONS TECHNOLOGIES AS SUPPORTS FOR URBAN SPLINTERING

This is an informational world where increasingly our self is linked to the world (or divided from it) through the screen – the glass pane of a car windscreen, the computer terminal or the television set.

(Crang and Thrift, 2000, 9)

Which brings us to our final supportive trend: the 'informatisation' of the city and urban life. It now seems very clear that new information and communications technologies like the Internet and mobile communications tend to support broader logics of urban splintering (see Castells, 1996, 1998). Three reasons can be identified why this is the case.

ACCESS CONTROL, SURVEILLANCE, AND THE COMMODIFICATION OF INFRASTRUCTURE

First, as we saw in Chapter 4, information and communications technologies provide the technological supports for the process of infrastructural unbundling itself. The ability to monitor and charge for the complex transactional paths of individual consumers within liberalised infrastructure markets is possible only because of linked computerised sensors, memory banks and telecommunications links. Information and communications technologies provide the technological means through which meaningful access to customised networks becomes possible in complex urban situations, whether it be to local bypass networks, global bypass networks or virtually competitive networks (Graham and Marvin, 1994).

The spread of electronic tracking and surveillance systems, in particular, enables access to networks to be monitored and controlled with unprecedented precision (Graham, 1998b). With electronic tagging already being applied to people, pets, babies, farm animals and cars, and with a range of biometric scanning devices emerging (smart cards, fingerprints, iris or retina scans, hand geometry scans, thumb scans, voice recognition, DNA testing, digitised facial recognition, wrist vein recognition, even odour recognition), the potential for city-wide, national, even international systems of intimate monitoring and control, classifying individuals into the spaces and times where they are deemed to 'belong', is clearly of great concern (Warf, 2000; Davies, 1995, 62).

In such circumstances Rose (2000, 22) worries about the emergence of a 'new biology of control' which threatens that the worth of individuals will become closely linked with technical judgements of their biological, genetic or reproductive fitness, or risk of aggression or deviance, by essentially eugenically constituted science and practice. There is a particular risk that generalised life and health insurance services will themselves become 'unbundled' in a similar way to networked infrastructures, with premiums and cover based on individual genetic diagnoses of future risk. After all, just as with infrastructure, 'cherry-picking' new entrants will be able to undercut incumbent insurers by targeting the most profitable and low-risk segments of the population, undermining cross-subsidy and mutuality in the process (Martin, 2000).

But individual profiling is only one element within a burgeoning field of personal surveillance and control. Biometric signatures, linked with large-scale personal databases, are already widely embedded in computer systems which are attempts automatically to control physical access and movement through spaces and infrastructure networks. Ann Davis (1997) reported that iris-scanning cash machines (ATMs) had been in operation in Japan since 1997. Inmate retina scanning was in operation in Cook County, Illinois, to control prisoner movements. The states of Connecticut and Pennsylvania were practising digital finger scanning to reduce welfare fraud. Frequent travellers between Canada and Montana were using automated voice recognition for speedier throughput. Hand geometry scans were made of immigrants entering San Francisco to check for illegal immigration. And Israel used hand-print biometrics to regulate the flow of workers to and from the Gaza strip (Lyon, 2000). 'The prospect of interoperable, even networked databases raises a frightening spectre,' writes Ann Davis. 'Our body parts and prints could soon be bought and sold like Social Security numbers by direct marketers, government clerks, or medical providers. One "harmless" little retina scan, some ophthalmologists warn, could indicate that a person has AIDS or abuses drugs' (1997, 174; see Warf, 2000).

AFFLUENT SOCIOECONOMIC GROUPS, ICTs AND THE CITY

> The consequence of cyberspace may be unfettered movement for some, but for everyone else the outcome is far less certain.
>
> (Fred Dewey, 1997, 272)

Second, personalised ICT services like the Internet, electronic cash, and computerised banking and shopping, tend still to remain the preserve of powerful, affluent minorities in nearly all cities (UNDP, 1999). Whilst 'new information and communications technologies are driving globalization,' writes the UN Development Programme, they are also 'polarizing the world into the connected and the isolated'.

THE INTERNET AS A 'GLOBAL GHETTO'

This process is creating, in a sense, a 'global ghetto' of affluent, largely metropolitan and technologically integrated users linked to the Internet and other technological systems (UNDP, 1999, 5). Whilst it is the fastest diffusing medium in history, the UNDP (1999, 63) still characterise the Internet as a 'global ghetto' encompassing only 2 per cent, or 250 million, of the most privileged and powerful of the the global population – over 80 per cent of whom live in OECD countries. This global 2 per cent – expected to rise to 700 million by 2001 – tends overwhelmingly to be relatively wealthy (90 per cent of users in Latin America come from upper-income brackets; 30 per cent in the United Kingdom have salaries over US$60,000). They are highly educated (globally 30 per cent have at least one university degree – in China the figure is 60 per cent, in Mexico 67 per cent, in Ireland almost 70 per cent). Male users dominate (62 per cent in the United States, 75 per cent in Brazil, 84 per cent in Russia, 93 per cent in China and 94 per cent in the Arab states). Users tend to be young (under thirty as an average age in the United Kingdom and China, thirty-six in the United States). Finally, dominant ethnic groups tend to dominate Internet use, as do English-speakers (in 1999 80 per cent of all global web sites were in English whilst only 10 per cent of the world's population spoke the language) (UNDP, 1999, 62).

Even in the United States, with one of the most 'mature' Internet diffusion patterns, only 40 per cent of all households had a computer in late 1999. However, only 8 per cent of those earning less than $10,000 had a computer and only 3 per cent of that group had Internet access (Lazarus and Marinucci, 2000, 3) (see Figure 6.6).

The overwhelmingly market-driven dynamics of the development of ICT networks and services, a pattern that is dominated by the construction of highly customised and personalised network solutions geared to the needs of mobile professionals, seems likely to exacerbate such unevenness. Friedmann (1995) argues that the emergence of such groups in Western, and increasingly in non-Western, cities needs to be seen as an integral element of a worldwide

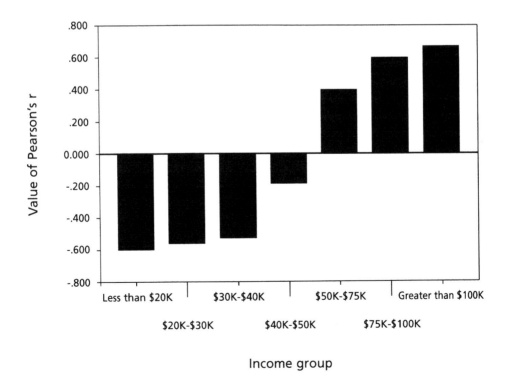

Figure 6.6 The correlation between high income and access to the Internet, 1998. Note positive correlation coefficients for incomes over US$50,000 and negative ones for incomes below US$50,000. *Source*: Moss and Mitra (1998), 29

shift towards the emergence of global spaces of capital accumulation, dominated by transnational corporations (TNCs), and their associated social elites (see Graham, 1994).

In effect, elite transnational groups are starting to experience interactive, empowering models of electronic democracy as they strive to be 'internally egalitarian and communitarian, and externally effective in exercising political and economic power' (Calabrese and Borchert, 1996, 250). In what is the closest contemporary approximation to the Habermasian notion of the bourgeois public sphere, Calabrese and Borchert believe, the:

cosmopolitanism of the new class will be enhanced by its activity on the superhighway. The high spatial mobility of its members will be mirrored by their high network mobility and activity in the formation and maintenance of political alliances and economic relations on a highly privatised, translocal and increasingly transnational basis.

(1996, 250)

PERSONALISED ICT SYSTEMS AS A 'PERSONAL BUBBLE'

Increasingly, affluent transnational groups are starting to benefit from the development of customised broadband ICT technologies through which tailored services are developed, geared intimately to their intense communication, service and mobility needs (see Kopomaa, 2000). What Eli Noam (1992) calls 'individually tailored network arrangements' are already developing, based on high-capability mobile and Internet services. These are geared to supporting intensified mobility and combined control and security at home, at work and on the move. Such systems:

will be packaged together to provide easy access to an individual's primary communications needs: friends and family; work colleagues; frequent business contacts, both domestic and foreign; data sources; transaction programs; frequently accessed video publishers; telemetry services, such as alarm companies; bulletin boards, and so on. Contact to and from these destinations would move about with the individuals, whether they are at home, at the office, or moving about.

(Noam, 1992, 408)

The Finnish telecommunication company Nokia is already planning a 'broadband' personal wireless communication system which acts as a:

'personal bubble'. The customized data environment will follow you everywhere: check in at a hotel and your Sony PlayStation™ games will be there, saved at the levels that you last played them. Your address book, correspondence, and favourite movies, and the contents of your company intranet, will be available anywhere there's a phone network.

(Silberman, 1999, 148)

PARADOXES OF CONNECTION WITH DISTANCIATION

For the groups that can benefit from such technology, access to telecommunication networks, fast transport networks and cocooned, affluent urban spaces and neighbourhoods becomes subtly combined. 'Physically separated by roads and cars, en route from gated community to enclosed shopping mall, telematics reinforce existing segregations by further reducing unplanned encounters' (Crang, 2000, 308). As Elwes puts it:

computer technology was designed to promote and speed up global communication. And yet the effect is somehow one of disconnection and distance. Individuals increasingly locked into the isolation of their own homes only make contact with the outside world through telecommunications and networked computer systems. Not so much distance learning as living at a distance.

(1993, 12, cited in Crang, 2000, 312)

Information and communications technology, enrolled into the urban splintering process, and backed up by the cocooning effects of cars, homes, offices and packaged leisure spaces, can thus provide an 'illusory escape into a private world. The tele-burbanite then is a villain and victim of telematics at the same time. An isolated individual, cut loose from the sociality of urban life, separated from the world by the pixilated screen' (Crang, 2000, 312). To Mark Dery (1999, 173) 'the digerati's clean-room fantasy of retreating into virtual worlds . . . parallels the theme-parking of urban space' because it, too, represents a search for risk-free and sanitised urban life and consumption – a retreat from the unpredictable encounters and exposures of the global metropolis.

As Richard Skeates (1997) suggests, we therefore need to maintain parallel perspectives on the collapsing sense of coherence in the city, the growing use of 'telemediated' exchange and the shift towards fragmented and privatised streets, highways and telecommunications and energy grids. To him:

the more the old ordered world of modernity is represented as having changed into a turbulent and dangerous postmodern place, the more the new world, represented as the virtual space of cyberspace, becomes an attractive option. However, this metaphorical domain of order, refuge and withdrawal manifests itself in the real world as an actual privatisation of space, a documentable removal of space from what was previously perceived as public, multi-functional and open to space which is private, monofunctional and closed. . . . Any withdrawal from the real streets of the real city will not be accomplished through its substitution by a virtual city, but through the mechanisms of privatisation, security and retreat into the mentality of the fenced and gated compound.

(1997, 15–16)

Christopher Lasch (1994) argues, from the point of view of the United States, that these trends need to be seen against the wider context of the changing nature of the urban public realm and labour markets discussed above. In the United States in particular, the mediation of urban life by information and communications technology has major implications because such processes may represent the disembedding of elite groups, not just from particular urban spaces, but also from whole systems of public service provision, public space and national consciousness:

at an alarming rate the privileged classes – by an expansive definition, the top 20 per cent – have made themselves independent not only of crumbling industrial cities but of public services in general. They send their children to private schools, insure themselves against medical emergencies by enrolling in company-supported health plans, and hire private security guards to protect themselves against mounting violence. It is not just that they see no point in paying for public services they can no longer use; many of them have ceased to think of themselves as Americans in any important sense, implicated in America's destiny for better or worse. Their ties to an international culture of work and leisure – business, entertainment, information, and 'information retrieval' – make many members of the elite deeply indifferent to the prospect of national decline.

(Lasch, 1994, 47)

INFORMATION AND COMMUNICATIONS TECHNOLOGIES AND ELECTRONIC SYSTEMS OF FINANCE AND CONSUMPTION

Finally, we need to address the shifts towards polarisation in access to electronic financial services. 'As money becomes information etched into computer memories', so favoured groups are rushing headlong into a widening universe of electronic transactions (credit and smart cards, electronic banking, telephone services, Internet retailing and share dealing, personalised media, e-cash, etc.) that are less and less dependent on location (Thrift, 2000, 282). Excluded groups, meanwhile, risk being marginalised to the physical cash-only and locally based economy (Kruger, 1997). These trends seem likely to support further the parallel logics of splintering urbanism and unbundling infrastructures.

In the burgeoning universe of electronically mediated consumption and transactions, driven by the credit card, the debit card, the smart card and the Internet, services are tailored to incorporate only those with bank accounts, credit cards, access to information and communications technologies, and the ability to extend their actions in time and space electronically (Solomon, 1998). More affluent consumers can thus use their access to 'virtual money' to extend their action spaces to achieve the best possible service and value for money in retail, financial and other services. Fuelled by liberalisation, Internet-based energy markets are also growing fast in some states of the United States and in the United Kingdom. As Nigel Thrift (1996c, 20) suggests, from the point of view of the withdrawal of bank branches from inner cities in the United Kingdom, a process of electronic 'super-inclusion' is occurring for favoured groups and spaces. At the same time, however, poorer people and communities face being further marginalised. He writes, for example, that for these groups:

information technology is not a panacea. Automatic Teller Machines (ATMs) can only be used to obtain cash. Telephone banking depends on access to a telephone, and, in any case, is specifically aimed at the relatively well off; indeed it is being used as a way of 'cherry picking' customers. The personal computer can only be used by the small (and affluent) percentage of the population with a PC and a modem.

In such circumstances banks, shorn of more and more of their physical branch assets, become, in a sense, 'nothing more than a conductor of transactions' (Gosling, 1996, 59). This further supports the interpenetration of finance capital and infrastructure and retail capital, within liberalising, internationalising contexts, as this 'facility might equally well be provided by telecoms corporations or on-line service providers' (ibid.).

SPACES OF SEDUCTION: THE PREMIUM NETWORKED SPACES OF THE SPLINTERING METROPOLIS

With our three broad supportive trends sketched out we are now in a position to look more closely at the premium networked spaces of the contemporary metropolis – at those sites and networked spaces that are splintering off from the wider urban landscapes whilst connecting intimately with international circuits of economic, social and cultural exchange.

In what follows, we explore a range of seven examples which illustrate the geographical and social diversity of such trends across a broad range of urban contexts. First, we look at the recent emergence of private, electronically tolled and 'premium' urban highways. Second, we analyse the ways in which the Internet itself is being splintered as it is reconfigured as a corporate entertainment, consumption and finance system. Third, we explore the proliferation of 'private public places' in atria and skywalk systems. Fourth, we follow this by looking at new ways of privately managing and regulating downtown street networks to maximise commercial consumption. Fifth, we look at the growth of theme parks, leisure enclaves and malls across a range of contemporary cities. Sixth, we examine the growth of gated residential communities in the cities of both the 'North' and the 'South'. And finally, we analyse the emergence of the home as a 'smart' terminal on customised infrastructure networks.

AUTOMOTIVE SECESSION: ELECTRONICALLY TOLLED 'SUPERHIGHWAYS' AND THE 'DIVERSIFICATION' OF URBAN HIGHWAY SYSTEMS

We have already established that the construction of public networks of urban highways tends to support splintering urbanism. Cities mediated by highways tend to reorient their urban space towards the dominant logic of freeway access, car parks and the entry of cocooned automobile users into urban spaces through strict hierarchies of 'laddered' flow. Albert Pope terms this the 'path to urban closure' 'which always terminates in an exclusive destination or end point' (1996, 189) – the mall, the corporate parking space, the suburban cul-de-sac, the fortified house garage.

To Pope, freeways have supported the extension and polynucleation of the metropolis, as new developments can 'leap over [the] edge' of the older urban order, to 'begin new autonomous nuclei of expansion' (1996, 111). The spaces between highways often then become urban interstices; they are residualised against the redeveloped nodes that the highways connect. 'For the inhabitant,' he writes, 'the city literally takes shape around the path. Through these spiralling vectors, the bewildering complexity of the sprawl is directly connected to the intimate center of countless private lives. Closure thus has a form, or at least a mechanism to achieve exclusion' (1996, 190).

But a new series of urban highways developing across the world in the 1990s take such exclusionary logics much further, based on the principle of electronic road pricing (ERP). In these, highway space has shifted from being 'dead', public and electromechanical; now it is 'smart', digitally controlled, privatised and sold as a priced commodity in a market for mobility that is increasingly diversified in both time and space. Highway networks and telecommunications and computer networks are, in short, converging as the latter are laid over the former. As a result, a vast process of industrial colonisation is under way as media, transport, defence and technology corporations seek to position themselves within the burgeoning global markets for 'road transport telematics' (Branscomb and Keller, 1996).

The capabilities of transport telematics technologies now mean that previously homogeneous highway networks can be splintered into diverse channels of customised lanes, from high occupancy vehicles (HOVs) to dedicated bus lanes and lanes only for transponder-equipped cars (Figure 6.7). Virtual electronic networks of automated sensors, CCTV, tracking and charging devices and computers are laid over such highways, controlling access and enforcing financial payment for each use.

This allows private firms, which are increasingly trying to gain political approval for the development of premium urban road spaces across the world's major metropolitan regions, to operate them profitably. Road networks, with all their complexity of flow and pattern, increasingly become computerised systems supporting new practices of commodification, control and exclusion. Such practices provide the basis for strategies which differentiate groups according to the power over space they are seen to warrant, within the new urban political economy. As Erik Swyngedouw puts it, 'road pricing, or other linear methods of controlling

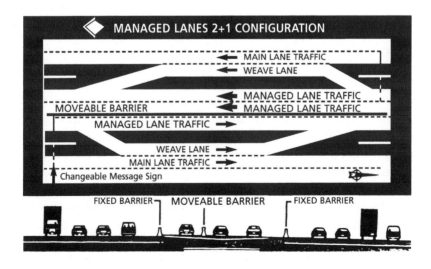

Figure 6.7 An unbundled highway, offering many types of lanes for specified classes of user. *Source: Toll Roads Newsletter*, March 1998, 1

or excluding particular social groups from getting control over space, equally limits the power of some while propelling others to the exclusive heights of controlling space, and thereby everything contained in it' (1993, 323).

Developed entirely by private firms to service only the most lucrative nodes within the urban fabric, these new premium urban highways are thus entirely commodified. Sometimes the price of the journey varies in real time with demand in the same way as the charges for telephone calls. This allows developers to guarantee the free flow of traffic, even in the most congested rush hours when surrounding public highways are gridlocked. Thus affluent commuters can, in effect, completely bypass the wider public street and highway network by using their purchasing power to enjoy premium networked connections within the metropolis (Solomon, 1996) (see Figure 6.8). In Box 6.2 we review a few leading examples of electronically tolled private highways in North America and Australia.

SPLINTERING THE INTERNET: TOWARDS A COMMERCIAL AND SOCIALLY SELECTING MEDIUM

While many still assume the Internet to be an intrinsically egalitarian and democratic medium, it is being remodelled as a corporately dominated communications medium, as part of a frenzy of network construction. Between 1990 and 2000, for example, the miles of optic fibre laid in the United States rose from 2.8 million to 17.4 million. As Dan Schiller argues, 'the web is rapidly being redeveloped as a consumer medium. This process is bound up with a triad of overarching trends within the media economy: multimedia diversification, transnationalization, and the extension of advertising and marketing' (1999b, 35). Increasingly these processes mean that the very physical and software architecture of the Internet is being redesigned so that it can subtly but powerfully discriminate between users, based on their perceived profitability as users of commercial entertainment and communications services (see Lessig, 1999). Emy Tseng (2000) argues that this is happening in three ways.

'CHERRY PICKING' LOCATIONS FOR BROADBAND INVESTMENT

First, as we shall see in further detail in Box 6.5, massive investment is creating a broadband Internet infrastructure that is (at least in the first instance) connected only to more affluent users and spaces. Because optic fibre access networks cost between US$6,000 and US$10,000 per home passed, access to new broadband Internet services is being carefully deployed only in more affluent neighbourhoods. New network technologies and protocols are emerging that allow faster speeds than ever for those connected. But, reflecting its commercialisation, these new network technologies tend to offer more bandwidth 'downstream' (for commercial entertainment and marketing) than 'upstream' (for interaction between citizens). This is known as 'the asymmetry of vendor and customer' (Tseng, 2000, 3).

Figure 6.8 'Wormholing' through the gridlock: the bypassing effect of electronically tolled private highways. *Source*: Solomon (1996), 42

BOX 6.2   ELECTRONICALLY TOLLED PRIVATE HIGHWAYS IN NORTH AMERICA AND AUSTRALIA

To understand the ways in which computerised and electronically tolled highways support urban splintering, it is necessary to explore the leading examples of such highways from cities around the world. Here we take a quick tour of some of these highways, stopping off *en route* at Los Angeles, Toronto, San Diego and Melbourne.

LOS ANGELES: THE RIVERSIDE SR 91 FREEWAY

Our first example of the private electronically tolled 'premium' highway is the Riverside SR 91 Freeway in Los Angeles, developed in 1995 by the company California Private Transportation. This offers an express lane accessible only to users with fitted electronic transponders, who receive itemised monthly bills for use of the highway. The route enjoys extra speed, reliability, maintenance and security over the public highways in the city. On it, in rush hour 'you're doing 65 miles an hour, passing everything on the freeway at 15' (Solomon, 1996, 42). Tariffs vary according to traffic conditions, from 25c to $2.50 a trip. A much wider network of private tolled highways is currently being planned in Los Angeles to 'bypass the congestion that is inevitable on untolled LA freeways' (*Toll Roads Newsletter*, August 1999, 1). Three major new toll roads are to be developed across main bottlenecks; a 220 km network of toll roads entirely for trucks was announced in January 1998.

TORONTO: ELECTRONIC TOLL ROAD 407

In Toronto an even more sophisticated highway is now in operation: the 70 km, C$800 million Electronic Toll Road 407 around the northern edge of the city. Built to ease congestion on one of the world's busiest public highways, to which it runs parallel, in-car transponders automatically charge all users around $1 per 11 km trip, without requiring them to stop. Tariffs vary automatically, to peak around main commute periods (at around 7c per km in 1997), so ensuring that the use of the highway never exceeds pre-defined limits, so allowing moving traffic to be guaranteed.

SAN DIEGO: THE 1–15 HIGHWAY

Our third example – the 13 km San Diego 1–15 highway – is the world's first to use fully automatic, real-time, dynamic congestion pricing. Charges for the highway actually vary between 50c and $8.40, according to traffic conditions. Charges change every six minutes and are notified on large electronic signs to drivers entering the highway. Drivers, in effect, are buying the right to drive 13 km with guaranteed flowing traffic, saving twenty to forty minutes per commute. This right is defended by the *Toll Roads Newsletter*, which argues against the 'dogmatic egalitarianism' of objectors who 'deny people the option to buy their way through congestion, and deny everyone the benefits of the kind of flexible pricing that we use to social advantage in producing our foods, our fuel, our housing, and indeed most of our goods and services' (4 April 98).

The social impacts of such networks on mobility patterns in the city are fairly easy to predict. Not surprisingly, electronically tolled highways are clearly exclusionary. They tend to be used much more frequently by higher-income than by lower-income groups, as the example of SR 91 in Los Angeles shows (see Figure 6.9). In that case, just over 10 per cent of the lowest income groups used the highway for more than 40 per cent of their trips along the urban corridor whilst over 50 per cent of the higher-income group did so.

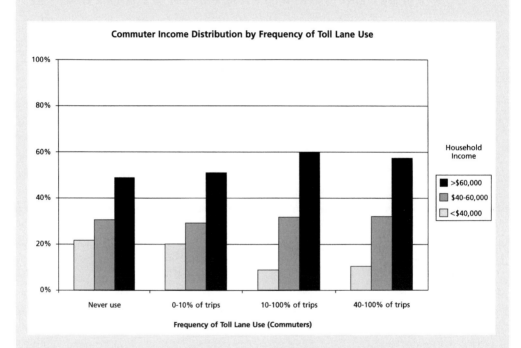

Figure 6.9 The social bias of the SR 91 highway, showing how it is used largely by wealthier socioeconomic groups. *Source*: SR 91 Web site at http://airship.ardfa.calpoly.edu/~jwhanson/sr91main.html

MELBOURNE: TRANSURBAN CITYLINK

Some of the wider linkages between electronically tolled 'superhighways' and splintering urbanism are illustrated in our final example of an e-highway: Melbourne's private Transurban CityLink network. This highway, which opened in 1999, offers a 22 km electronically tolled road system linking affluent suburbs and the city centre with a direct link to the city's international airport. David Holmes argues that the project 'normalises and legitimates new tolerances for what the broad base of the population may be prepared to see being commodified' (1999, 1). It also creates 'the conditions for even greater cycles of dispersal, concentrating shopping and consumption into fewer and fewer zones – the mega shopping malls' that the highway directly connects.

Holmes goes on to suggest that in 'by-passing the traditional agora-like markets and parking-based streetscapes' CityLink 'further privileges the superregulated private spaces of shopping complexes – another cocoon to which the freeway is the link'. To him this 'cocooning' of Melbourne's geography, now backed up with traffic restrictions on old streets parallel to CityLink, to force motorists on to the commodified highway, threatens to support an ever more fragmented and dispersed city where embodied association on the open street 'is replaced by the heightened private consumption of media of all kinds' (ibid.).

In effect, these examples suggest that we are gradually starting to see new technology and privatisation combining to force through the unbundling of integrated public highway systems. Extrapolating such trends forward, we can envisage the emergence of what *Toll Roads Newsletter* calls 'a diversified highway system' where 'operators can provide lanes to meet the needs of its customers . . . road operators will be able to give you weather reports, make or change motel reservations that suit your route, and give you spoken directions to the destination of your choice' (1998, 1).

MEDIA CONGLOMERATES AND THE CONSTRUCTION OF PRIVATE OVERLAY NETWORKS

Second, major media conglomerates are effectively building their own parallel Internet infrastructures so that they can offer their customers a portfolio of Web sites and services that they control and profit from. In order to overcome bottlenecks encountered on the Internet, companies like Akamai and iBeam are developing systems that allow large corporate customers effectively to bypass the public Internet infrastructure (Tseng, 2000). The new private networks deliver much faster services to selected users through local cacheing (that is, storing high-demand content on a range of servers distributed near metropolitan markets). They also allow media conglomerates and large content providers to offer integrated packages of services, products and e-commerce platforms to more lucrative consumers whilst subtly excluding access to the services and Web sites of competitors. As Emy Tseng argues, 'looking at the deployment maps of the Akamai content delivery network and InterNap's P-Naps [two major recent private Internet systems], it is apparent that these companies are deploying this private infrastructure at the same select group of highly connected US cities [that dominate Internet use]. Thus the private infrastructure serves to further the advantage these "global cities" have over other cities and regions' (2000, 6).

PRIORITISING 'PACKETS': USING NEW INTERNET PROTOCOLS AND ROUTERS TO DISCRIMINATE SOCIALLY

Finally, a new range of Internet protocols – the software codes and algorithms that route 'packets' of information around the system – are emerging which actively discriminate between different users' packets. This is especially so at times of congestion. Only recently all packets were treated equally on the Internet; now, at times of congestion, 'smart' routers can sift 'priority packets', allowing them passage, whilst automatically blocking those from non-premium users. Thus, in a striking parallel to the case of e-highways, high-quality services can be guaranteed to premium users irrespective of wider conditions. This further supports the unbundling of Internet services, as different qualities can be packaged and sold at different rates to different markets. As Emy Tseng suggests, 'the ability to discriminate and prioritize data traffic is now being built into the system. Therefore economics can shape the way packets flow through the networks and therefore whose content is more "important"' (2000, 12).

SPLINTERING STREETSCAPES (I)
SKYWALK CITIES, PLAZAS AND ATRIA

Broadly similar processes of infrastructural splintering are now apparent in the streetscapes of many metropolitan areas. Paralleling the emergence of specialised, premium freeways and bypass Internet systems, we are seeing the development of premium, privately managed street spaces that are increasingly secured off from the wider metropolitan fabric.

Urban streets, writes Trevor Boddy, 'are as old as civilisation'. They 'symbolize public life, with all its human conflict, contact and tolerance' (1992, 123). Grady Clay believes that the street is nothing less than 'the great carrier of information for a democratic society' (1987, 99, cited by Drucker and Gumpert, 1999). Across North America, however, 'downtown streets are now subject to attack, a slow, quiet, but nonetheless effective onslaught underground and overhead, by glittering glass walkways above the street, or tunnels beneath them' (Boddy, 1992, 123). Albert Pope argues that such trends are closely interconnected with the broader impact of urban freeways. 'As the freeway erodes the gridded urban fabric,' he writes, 'the discrete space of the conventional urban street begins to disappear' (1996, 125).

SKYWALK SYSTEMS: THE RECONSTRUCTION OF NORTH
AMERICAN DOWNTOWNS AS 'ANALOGOUS CITIES'

In some North American cities downtown areas are being effectively remodelled as 'analogous cities' (Boddy, 1992; Bednar, 1989). Corporate and consumption enclosures containing upscale retailing, theatres, convention centres and luxury housing are being directly interlinked with private, air-conditioned walkways, tunnels and 'skyway' bridges. Such networks are superimposed three-dimensionally below, above and within the traditional street system, whilst connecting with it only through limited numbers of highly surveilled and secured entrances. This is the strategy of 'building cities in the sky', a solution where 'the street is only for driving on, and parking under; where drivers and passengers move from house to office, shop or theatre without setting foot on the street' (Hamilton and Hoyle, 1999, 32). As Alan Waterhouse suggests, such trends amount to 'a suburbanisation of the centre' which, 'sanctioned by the state, brought an ambiguous landscape in its wake . . . as architecture turned upwards and inwards, retreating from the old civic thresholds' (1996, 300).

The best-known skywalk and atria complexes in the United States were the result of publicly sponsored downtown 'revitalisation' projects between the 1970s and 1990s. Such developments are secured from the wider urban fabric whilst being provided with customised connections elsewhere. They have dedicated slip roads and large car parks as well as 'fibre optic cables, private branch exchanges, satellite dishes, and other state-of-the-art transmission and receiving equipment' (Reich, 1992, 271). Steven Flusty, writing about the new Bunker Hill downtown complex in Los Angeles, tells the story of trying to reach the complex from the streets below by foot:

the Hill's designers are not too keen on pedestrians coming up from below (except janitors). . . . The entire Hill is . . . separated from the adjacent city by an obstacle course of open freeway trenches, a palisade of concrete parking garages and and a tangle of concrete bridges linking citadel to citadel high above the streets. Every path we try confronts us with the blank undersides of vehicular overpasses, towering walls studded with giant garage exhausts, and seating cleverly shaped like narrow sideways tubes so as to be entirely unusable. We could attain the summit from the south, but only by climbing a narrow, heavily patrolled stair 'plaza', studded with video cameras and clearly marked as private property.

(1997, 53)

Isolated corporate enclaves across metropolitan cores thus become interlinked into enclosed systems. These draw on modernist principles of separating pedestrian and highway traffic, but take such ideas to their extreme application. Tunnels also link directly with car parking garages and the nodes of public transit systems without recourse to the street. In some cases, for example Detroit, 'people-mover transit systems glide above the scuffling passions of street-bound cities', interconnecting plazas, corporate hotels and atria but completely bypassing the public street (Boddy, 1992, 124). In this case, however, the monorail runs virtually empty – a victim of the continued decline of Detroit's downtown (Vergara, 1997, 210).

Some of the most comprehensive examples of skywalk cities can be found in the downtowns of Minneapolis, Dallas, Montreal (see Plate 11) and Charlotte in North America (Bednar, 1989). Smaller parts of other downtowns, such as Battery Park City in Manhattan, have been rebuilt along similar lines but with the greater exclusivity of integrated winter gardens and upscale marinas (Grava, 1991, 11; see Plate 9). Similar developments have also been observed in some Asian cities, particularly Hong Kong (Cuthbert, 1995) and Tokyo (Tschumi, 1996). In Tokyo, Bernard Tschumi finds classic examples of urban 'bundling': 'multiple programs scattered through the floors of high-rise buildings; a department store, a museum, a health club, and a railway station, with putting greens on the roof. . . . Airports simultaneously integrate amusement arcades, athletic facilities, cinemas and so on' (1996, 42).

But it is probably in Houston, Texas, that this development logic has reached its extreme. Here 6.2 miles of mostly below-surface tunnels link 26 million ft² of city-centre office spaces with their car parks and hotel and retail malls (see Figure 6.10). Entry is policed by CCTV cameras and private security guards who work to suppress any 'inappropriate' entrants or 'threatening' behaviour – i.e. those which are seen to contradict the tight, normative ecology that the premium networked space is for middle-class consumption and formal work alone.

Plate 11 The interior of Montreal's Underground City, an interconnected complex of plazas, shopping centres, office complexes and civic buildings in the downtown core, served directly by the metro system. *Photograph*: Stephen Graham

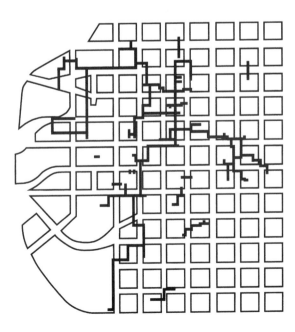

Figure 6.10 The 6.2 mile skywalk and tunnel system linking 26 million ft² of office space in downtown Houston, Texas. *Source*: Pope (1996), 114

Within such networks, amid the 'isolated empire within', suggests Boddy, 'we are inside, contained, separate, part of the system' (1992, 123). In this closure of the protected urban 'inside' from the wider 'outside' of the urban fabric the traditional, integrated street system – that legacy of the modern infrastructural ideal – becomes marginalised, a place of exclusion for those unable to enjoy access to the skywalked enclaves. Robert Reich's famed 'symbolic analyst' is thus able to 'shop, work, and attend the theatre without risking direct contact with the outside world – in particular, the other city' (Reich 1992, 271).

THE RESIDUALISATION OF TRADITIONAL STREET NETWORKS

Such secessionary networked spaces are, in a way, a self-fulfilling prophecy. Those inside the networked, secured enclaves increasingly fear the traditional street as a place of crime, disorder, poverty, insanity and danger. This, in turn, justifies their further social and technical distancing from the traditional street in their work and leisure lives. Retailing, and other spending within the downtown, becomes progressively internalised within the secessionary system. Not surprisingly, those high-income residents who make their homes in the condominium developments that are increasingly connected to enclave complexes – developments that are often now provided with customised and secured schools, health care, fitness suites, pools and commercial services – have been shown to 'have a lower level of civic concern than non-enclavers' (Bayne and Freeman, 1995, 419).

The traditional street system is thus often 'left to casual visitors, desultory shoppers, hangers-on, the young, the restless' (Boddy, 1992, 125). It, in turn, tends to fragment and be rendered marginal as sealed corporate enclaves erupt within it, eroding its connectivity and seamlessness. Remaining streets emerge as the 'fallen city' of the poor pedestrian unable to enter the corporate enclave via the normalised passage of automobile and freeway – the entry point to which buildings increasingly orientate (Pope, 1996, 1114). As Pope writes, the impact of the parallel skywalk and tunnel city is:

to drive the remaining pedestrian activity into an interior world of skybridges, atriums and tunnels. These elevated or subterranean pedestrian passages directly link freeways and garages to vast corporate interiors. . . . Against such a massive reorganisation of urban activity – the effect of turning the city outside in – the centrifugal fabric of the streets and sidewalks implodes. . . . An alternative city emerges as the closed, exclusive reorganisation of the urban sphere. No longer a presentation to or representation of the city as a whole, the centre of Houston is scarcely relevant to the surrounding city.

(1996, 114–15)

PLAZAS AND ATRIA

Within such parallel downtowns, massive new developments are emerging which bundle together offices, hotels, leisure activities, retailing and residences into giant, atriumed plazas under the control of single developers and private security. Many US cities now display such celebrated postmodern centres – the Peachtree Center, in Atlanta, Georgia (see Figure 6.11), Battery Park City in New York, the Renaissance Center in Detroit (complete with a monorail linking it with the rest of the downtown), the Bonaventure Hotel in Los Angeles. Each is directly linked to urban freeways with dedicated links. Addressing the Peachtree Center in Atlanta (Figure 6.11), Hamilton and Hoyle argue, the Center's major attraction is its:

isolation from the street: once you have entered the Center, it is possible to negotiate the whole complex without once having to set foot on a public street. This means that one could work in one of the many offices, eat in one of the many restaurants, shop in one of the many retail outlets for food, clothes, gifts or sport equipment, sleep in one of the several five-star hotels, and move between each, all in private space. And to make one feel extra secure, there are armed guards posted regularly throughout the Center.

(1999, 32)

Such spaces mimic the traditional iconography and semiotics of the urban street, even though their connections of tunnels and skywalks bypass the traditional street fabric (Bednar, 1989). 'Double-loaded corridors become "streets", interior partitions become "storefronts" which lead into a hotel lobby swollen to the proportions of a "plaza"' (Pope, 1996, 127). Above all, 'the "outside" is aggressively denied' (ibid., 127) – a 'hermetic *tour de force*' designed to

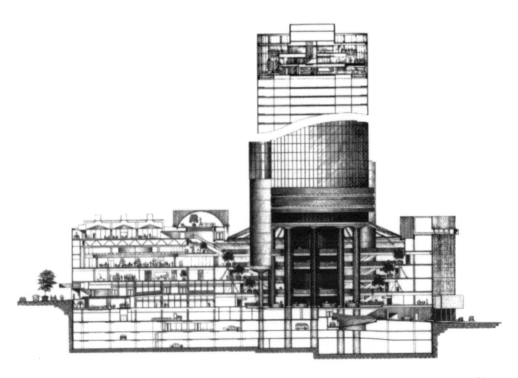

Figure 6.11 A cross-section of John Portman's Westin Peachtree Center in Atlanta, Georgia, opened in 1976. *Source*: Pope (1996), 130

maximise profits from tourists and corporate workers whilst excluding the poor. As Sze Tsung Leong suggests, such spaces are therefore 'entirely different from the intertwined and mutually dependent totality of the classical city' in that:

the success and proliferation of interiorized activity, and the fact that its nodes can be placed anywhere as islands whose connective tissue is a sea of formlessness and nothingness, has left the outside amputated, mostly inhabitable, and quite often a space of threat.

(1998, 196)

SPLINTERING STREETSCAPES (II) BUSINESS IMPROVEMENT DISTRICTS, TOWN CENTRE MANAGEMENT AND CCTV

It seems as if our best middle-class vision of the city today is that of an entertainment zone – a place to visit, a place to shop; it is no more than a live-in theme park. . . . This amounts to Urbanism Lite.

(Bender, 1996, cited in Soja, 2000, 247)

Even where traditional streetscapes remain economically successful or are gentrified, splintering tendencies are also in evidence as the regulators of such spaces try to compete with proliferating enclosed malls, plazas and atria for the custom of shoppers and tourists.

BUSINESS IMPROVEMENT DISTRICTS

One notable innovation which amounts to the splintering of a carefully selected system of traditional streets from the wider metropolitan fabric is the Business Improvement District (or BID). Originating in the United States, where there were over 1,200 in 1998, Business Improvement Districts are also diffusing widely around the world. By 1998 they were to be found in Europe, the Caribbean, Australia and South Africa (Hannigan, 1998a, 139). In the year 2000 they were starting operations in central London.

A tailor-made form of local government, Business Improvement Districts, essentially, involve the collaboration of local property capital to take control of a range of local municipal functions for their own private urban 'patch'. Such services encompass street cleaning, street lighting, public space management, garbage removal, public works, private policing, environmental improvements and marketing.

Business Improvement Districts have been characterised as 'cities in cities' or 'micropolises' (Vallone and Berman, 1995). Even though they are unelected bodies, BID boards are able to impose property taxes, which are enforced by law, and use them in a direct example of 'fiscal equivalence' – that is, all revenues are spent within the district. Free riders, and social or geographical cross-subsidies, are thus avoided. Richer Business Improvement Districts, as can be seen in Box 6.3, are increasingly active in controlling, managing and developing high-quality infrastructure networks and services, to tie in with the premium demands of those who control them.

BOX 6.3 'MALLS WITHOUT WALLS': BUSINESS IMPROVEMENT DISTRICTS AS SECESSIONARY STREETSCAPES

In a clear fragmentation of both urban streetscapes and regimes of public service provision and planning, richer Business Improvement Districts like some of those in midtown Manhattan are, not surprisingly, starting to undertake extensive public works and modify street systems, transport routes and utilities to the exact demands of BID members (Parenti, 1999, chapter 5). Central retailing districts, in particular, are using BID strategies to try and create 'the downtown as a mall' (Mallett, 1993a, 282). Such 'malls without walls' offer secure, privately managed spaces which strive to support the confidence of middle-class shoppers and tourists.

As with skywalk cities and atria, Business Improvement Districts are an explicit attempt to manage marginalised socioeconomic groups out of the urban scene. In the unevenly revitalised hearts of US cities they are a direct response to the fact that 'both the use of public space by the impoverished casualties of the post-industrial political economy and the crumbling, often dirty, public infrastructure threaten to frighten away corporate clients, gentrifiers, and suburbanite workers and shoppers' (Mallett, 1994, 277).

The Grand Central Business Improvement District in New York, for example, has closed streets to make outdoor eating spaces for rich commuters, and has employed private security companies and 'outreach' workers to ensure the immediate removal of homeless people or vagrants from the BID boundaries – part of the wider strategy of 'zero tolerance' in

the city (Zukin, 1995, 35). A report to the City of New York concluded that the Business Improvement District's security staff 'used excessive force against homeless individuals' (Vallone and Berman, 1995).

Crime and homelessness thus often move to the interstitial spaces between Business Improvement Districts – poorer areas which cannot develop their own improvement initiatives – or to the urban periphery (Smith, 1999; see Box 6.7). As a result, 'the rich BID's opportunity to exceed the constraints of the city's financial system confirms the fear that the prosperity of a few central spaces will stand in contrast to the impoverishment of the entire city' (Zukin, 1995, 36).

Many Business Improvement Districts are carefully themed, with uniform street furniture and streetscapes, signifying their secession from the wider city, whose 'inability to generalize improvement strategies' is widely cited by BID boards as a reason for seceding (Zukin, 1995, 36). Private police forces are often engaged by both commercial and mixed Business Improvement Districts as security becomes tailored to higher-income communities. 'The new community of like incomes,' writes Reich, 'with the power to tax and the power to enforce the law, is thus becoming a separate city within the city' (1992, 271).

In the longer run it is not difficult to envisage such Business Improvement Districts growing more powerful. It seems possible that they will eventually be able to take control of the wider development and management of both urban spaces and the infrastructure networks that connect and underpin them. They may even, in the long run, be able to secede completely from public taxation systems, 'replacing the city government' (Zukin, 1995, 36). As well as managing and customising their own private street and public transport systems it is likely that Business Improvement Districts will get involved in negotiating or constructing customised transport, energy, water and telecommunications services for users within their boundaries.

CONSTRUCTING DEVIANCE WITHIN CONSUMPTION SPACES: TOWN CENTRE MANAGEMENT AND CLOSED CIRCUIT TELEVISION IN THE UNITED KINGDOM

Even without the formal cession of tax-raising powers to Business Improvement Districts, other management strategies for the commercial and retailing cores of towns and cities in Europe are starting to replicate certain parts of BID strategies. In 1998, for example, the private managers of the Covent Garden market centre in London decided to 'exclude vagrants' from the piazza (Daly, 1999). Around the United Kingdom, private town centre management (TCM) strategies, similarly, try to harness corporate finances and aspirations to the cleansing of the 'inappropriate' behaviours and individuals from retailing streets (Reeve, 1996). Street theming, private policing, careful management and public CCTV are the usual tools here as competition intensifies between out-of-town malls and town and city centres for higher-income consumers. Users seen to be 'unaesthetic' or 'antisocial' are carefully managed out: 'junkies', 'down-and-outs' or others who, in the words of one town centre manager, 'make the town degraded' are not welcome (cited in Reeve, 1996, 70).

In practice, TCM schemes have been found to 'discriminate actively against [beggars and street people] in order to massage the social space of a town centre into something more socially conducive to consumers' (Reeve, 1996, 78). Moreover, in the United Kingdom, CCTV schemes, which back up such street management programmes, have been widely found to target people for 'no particular reason' other than 'belonging to a particular subcultural group' (Norris and Armstrong, 1998). Black people, in particular, 'were between one and a half and two and a half times more likely to be surveilled than one would expect from their presence in the population'.

In a detailed study of surveillance practices in CCTV control rooms, Norris *et al.* (1998) found that people targeted by CCTV were selected on the basis of the prejudices of CCTV operators, particularly their ideas as to which types of people 'belonged' in which spaces and times within the city. Through CCTV people and behaviours seen not to 'belong' in the increasingly commercialised, and privately managed, consumption spaces of British town and city centres tended to experience especially close scrutiny. Operators of current, non-digital CCTV 'selectively target those social groups they believe most likely to be deviant. This leads to an over-representation of men, particularly if they are young or black' (Norris *et al.*, 1998). CCTV control rooms were shown to be riddled with racism and sexism. Certain types of young men, targeted with socially constructed suspicion, were labelled as 'toe-rags' 'scumbags', 'yobs', 'scrapheads', '*Big Issue* scum' (after the magazine for the homeless),

'homeless low-life' and 'drug-dealing scrotes' and were scrutinised, followed and harassed. Malign intent was equated with appearance, youth, clothing and posture (Norris *et al.*, 1998; see Norris and Armstrong, 1999).

Thus CCTV operators are already attempting to impose a 'normative space–time ecology' on the watched parts of the city, stipulating who 'belongs' where and when, and treating everyone else as a suspicious 'other' to be disciplined, scrutinised, controlled (Graham *et al.*, 1996). 'For operators, the normal ecology of an area is also a "normative ecology" and thus people who don't belong are treated as "other" and subject to treatment as such' (Norris *et al.*, 1998, 43).

Taking this logic further, the latest digital CCTV system, the so-called 'Mandrake', in the east London district of Newham, even attempts to use computerised facial recognition techniques to target and track specified individuals through the main commercial centres. In sum, as Norris and Armstrong (1998) suggest:

the gaze of the cameras does not fall equally on all users of the street but on those who are stereotypically predefined as potentially deviant or, through appearance and demeanour, are singled out by operators as unrespectful. In this way youth, particularly those already socially and economically marginal, may be subject to ever greater levels of authoritative intervention and official stigmatisation.

SPACES OF SEDUCTION: THEME PARKS, SHOPPING MALLS AND URBAN RESORT COMPLEXES

The assemblage of car/credit card/shopper is the hypodermic of the shopping stream.

(Lerup, 2000, 5)

Economic and social inequalities remain as gross as ever, yet the global shopping mall renders them curiously invisible. Those without the passport of money are simply in absence. . . . Invisibility is a crucial feature of modern inequality.

(Wilson, 1995)

Such instrumental practices of attempting to 'sanitise' urban space, for the purposes of trouble-free consumption and paid leisure, are taken to further extremes by another range of fast emerging 'bundled' urban environments: the 'invented street' systems within shopping malls, theme parks and urban resorts (Banerjee *et al.*, 1996; Goss, 1995). Typically, such developments are strongly branded with tie-ins to leading sports, media and entertainment transnationals (Disney, Time-Warner, Sega, Sony, Nike, etc.). They make the most of merchandising spin-offs and 'synergize the sale of consumer products, services and land' (Zukin, 1995, 64).

Such developments cover larger and larger 'footprints' as the massive developers undertaking such projects attempt to bundle together the maximum number of 'synergistic' uses within single complexes (retailing, cinemas, IMAX screens, sports facilities, restaurants, hotels, entertainment facilities, casinos, simulated historical scenes, virtual reality complexes, museums, zoos, bowling alleys, artificial ski slopes, etc.). One of the largest malls in the world, the West Edmonton Mall in Alberta, Canada, added a live-rounds 'Family shooting centre' to complement its rooftop golf course, triple-loop roller coaster, artificial lagoon (complete with real dolphins), six night clubs, casinos, virtual reality centres, themed pirate area, forty restaurants and 700 shops.

Such major 'bundled' developments tend to be notably 'solipsistic; isolated from surrounding neighbourhoods physically, economically and culturally' (Hannigan, 1998a, 4). They use modular design approaches which 'mix . . . and match . . . an increasingly standard array of components in various configurations' (Hannigan, 1998a, 4). And they rely on major public subsidies for land acquisition and development costs.

As Hannigan (1998a) suggests, such 'entertainmentisation' developments are fast emerging as archetypal examples of splintered urbanism in North America, Europe, Asia, Australasia and Latin America. This 'new breed of entertainment centres [is] intended to anchor the "fantasy cities" of the future where tourism, entertainment and retail development are to be bundled together in a "themed" environment' (Hannigan, 1998b). Whilst US complexes have, once again, led the growth of these forms of urban development, they are rapidly diffusing across the globe, and some of the largest such developments are emerging in Australia (Sydney's Darling Harbour and Melbourne's Crown Entertainment Complex), Canada (with Toronto's $450 million Destination TechnoDrome complex), Europe (with the United Kingdom's MetroCentre and Blue Water), South East Asia (with Malaysia's new thirty-two-acre Star City and SunWay Lagoon complexes, and many other tourist mega-projects – see Cartier, 1998) and Japan, where a whole 'archipelago' of resort complexes have recently been developed across the country (see Rimmer, 1992).

RECONSTRUCTING CORPORATE LEISURE SPACES IN OLD URBAN CORES

Increasingly, however, transnational media conglomerates are also trying to take the attractions of theme parks and put them into cities. City core media complexes are a crucial part of the efforts of such corporations to stimulate higher returns and greater out-of-house consumption by the middle classes. This, in turn, increasingly ties themed city spaces seamlessly with in-home consumption in what is termed an 'inside/outside strategy' (see Davis, 1999). The result is a proliferation of reconstructed and privatised public spaces that are saturated by media content in old urban cores: 'entertainment-oriented retail' centres, mini theme parks, stores housing giant video walls, simulator rides, small water parks, IMAX cinemas, corporately constructed sports museums, converted traditional theatres, and movie and virtual reality centres (Davis, 1999).

Developers of such spaces face a dilemma. Whilst employing the latest surveillance techniques and careful entrance and exit control, to discourage 'undesirables' such as low-spending teenage boys or the homeless, such spaces must simultaneously encourage a feeling of urban vibrancy and (threat-free) social mixing. They must create a sense of destination by overcoming 'the sameness of the suburbs and the dispersal of the automotive city' (Davis, 1999, 46). As Susan Davis suggests:

at the heart of the location-based entertainment projects is this paradox: within the context of themed space, they aim to reproduce a sense of authentic space, and this means evoking the diversity and unpredictability of the older city using carefully calibrated recipes. The projects aim to reproduce a life and liveliness that look like the older commercial town centre. But in order to do this profitably, in order to turn out the right sort of crowd, they must control mixing and reduce real unpredictability.

(Ibid.)

CONTROLLING BOUNDARIES: MALLS AND LEISURE COMPLEXES AS 'LARGE URBAN CONTROL ZONES'

Whilst carefully controlling the relationship between their strictly controlled and carefully themed environments and the immediately adjacent cityscape, malls, theme parks and resort complexes increasingly also strive to shape their infrastructural connections with the wider urban region through customised highways, communications and water and energy infrastructures. Such developments thus provide a:

strongly bounded, purified social space that excludes a significant minority of the population and so protects patrons from the moral confusion that confrontation with social difference might provoke . . . and reassures preferred customers that the unseemly and seamy side of the real public world will be excluded.

(Goss, 1993, 26–7)

Like skywalk cities and Business Improvement Districts, then, the very appeal of these types of development is achieved 'by stripping troubled urbanity of its sting, of the presence of the poor, of crime, of dirt' (Sorkin, 1992). In fact, the very fact that such spaces are not truly 'public', and do not interconnect with public street systems, becomes an attraction to many middle-class consumers, because of the dangers now implicit to them in the term 'public street'.

Themed places are therefore what John Hannigan has labelled 'large-scale urban control zones' (Hannigan, 1998b, 36). They 'provide safe, secure environments where people can interact. [Such space] looks very much like public life, but in fact really isn't, because the environments are owned and controlled and heavily regulated by, generally, very large global corporations' (Dewey, 1994, quoted in Channel 4, 1994). Defensiveness, predictability and perceived safety are the watchwords here: 'what's missing is a sense of serendipity, diversity and the humanity of traditional street life' (Goldberger, 1996, 144, cited in Hannigan, 1998a, 6).

As an example, the largest mall in Europe, Bluewater, on the outskirts of London, announced that its 350 camera digital CCTV system would digitally record the face and car number plate of every visitor, heralding the prospect of computerised face and car recognition in the near future. As complex IT and surveillance practices are closely combined with the design and production of built space, Roger Burrows (1997, 41) even contends that 'buildings themselves become increasingly sentient as computerized systems of recognition include the new rich and exclude the new poor' (cited in Lyon, 2000, 81). Box 6.4 looks in detail at the mechanisms through which the operators of malls, theme parks and resorts attempt closely to manage their relationship with the surrounding cityscape.

INTEGRATING INFORMATION TECHNOLOGY INTO CONSUMPTION SPACES

Of course, within the new entertainment complexes of the city, huge efforts are also made to offer only the most sophisticated information infrastructures to the affluent and mobile target visitors and users. Hotel developers, in particular, now liaise closely with telecom and media firms to package the entertainment, security and communication applications that will give them a cutting edge in attracting high-tech clients. Leading-edge hotels in Los Angeles, for example, offer 'cybersuites' where computerised voice and face recognition systems mean that the 'entry system automatically recognises the guest and opens the door'. Environmental control systems 'can adjust lighting, temperature and draperies at the sound of a voice; with a phone call, it can even draw a bath' (Wolff, 1997). In the Clark Special Economic Zone north of Manila in the Philippines, meanwhile – a disused US military base – efforts to develop a massive consumption space have been bolstered by the distribution of 130,000 optical laser cards to selected middle-class shoppers which offer benefits to visitors who spend above defined limits.

FORTRESS SPACES: MASTER-PLANNED, GATED ENCLAVES ACROSS THE WORLD

Secessionary tendencies are taking over places of residence, too. To parallel the global diffusion of theme parks and malls, access-controlled, gated residential communities are now a feature of many large cities across the world. Such tendencies are vital in reshaping the configuration of cities because 'when the wealthy spatially segregate themselves within cities in exclusive suburbs and exurbs, they can replicate high-quality public services through private means' (Clarke and Gaile, 1998, 29). In the process, such transformations also tend to undermine the public service and infrastructure monopolies that were the legacy of the modern infrastructural ideal. Whilst such developments vary across the world in detail and context, they all tend to bring with them what McKenzie has called a 'privatopia' 'in which the dominant ideology is privatism; where contract law is the supreme authority; where property rights and property values are the focus of community life; and where homogeneity, exclusiveness, and exclusion are the foundation of social organization' (1994, 177).

BOX 6.4 STRATEGIES OF SECESSION: HOW MALLS, THEME PARKS AND URBAN RESORTS WORK TO WITHDRAW FROM THE WIDER URBAN FABRIC

A shopping mall is like a prison reversed: deviant behaviour is restrained outside.
(Mäenpää, personal communication)

A variety of common techniques of design, infrastructure development, disciplining and access control are widely used by developers and operators of malls, theme parks and urban resorts to withdraw from the wider urban fabric, whilst connecting as seamlessly as possible to wider middle-class markets. Six, in particular, can be identified.

PRICING ENTRY

First, when admission charges are in operation, as in theme parks, the price mechanism provides an effective instrument of social filtering.

PROACTIVE POLICING

Second, proactive and private policing, backed up by sophisticated surveillance technologies, is often geared to deterring users who are not seen to 'belong' within the spaces and times of the mall (Dovey, 1999). Target groups include groups of youths, or poorer consumers, or those with an appearance and demeanour that power holders judge transgress their own normative codes of 'acceptable' behaviour (Norris and Armstrong, 1998). In some cases these forms of disciplinary exclusion become stipulated as explicit rules which deny access to specified groups in space and time. For example, in 1996 one of the world's biggest malls, the Mall of America in Minneapolis, introduced a curfew preventing people under sixteen years old from entering the space after 6.00 p.m. unaccompanied by an adult (Drucker and Gumpert, 1999). Many other US malls have instituted similar curfews (Poindexter, 1997).

URBAN DESIGN PRACTICES

Third, practices of urban design, which place new, inward-looking developments within a cordon of car parks and highways, tend to exaggerate the sense of social and spatial separation from the wider urban fabric. To Mark Gottdeiner malls, in particular, rely on a form of 'fortress architecture': 'through the lack of signs they stimulate consumers to traverse the space of the parking lot quickly and enter within' (1997, 138). To him 'the problem with this kind of architecture is the way it ruptures the urban fabric by isolating buildings from both the surrounding landscape and the street' (ibid.).

GEOGRAPHICAL DISTANCING AND BIASED INFRASTRUCTURE DEVELOPMENT

The fourth practice is the traditional instrument of geographical distancing and the biased design and development of transport, street and highway connections. Very often developments 'are consciously situated beyond the geographic and financial reach of minorities and the poor in the exurban fringe' (Hannigan, 1998a, 190). Those built on the exurban fringe are often impossible to reach without a car, as developers try to connect such complexes as seamlessly as possible with major interurban highway networks in search of maximum middle-class spending power. By developers of such complexes easy access to major traffic arteries is considered absolutely vital.

Moreover the widespread absence of pedestrian sidewalks or pavements in suburban malls furthers the sense of atomisation and isolation. 'Without sidewalks, malls [are] accessible only by automobiles. In this environment of access roads, off ramps and parking spaces, traditional design elements such as scale, facades, and detailing become irrelevant' (Crawford, 1999a, 46). As at the Deira City Center complex in Dubai (see Plate 12), access for pedestrians without cars becomes all but impossible in many malls, not just because of the lack of public transport connections but because entry points are configured entirely for the car, the highway and the parking garage, without any pedestrian connections to walkable streets. In Asia's largest hybrid mall, the Ngee Ann City in Singapore, for example, 'nightclubs are entered through parking lots on the eighth floor' (Turnbull, 1997, 229, cited in Cartier, 1998, 172).

As with atria and skywalk complexes, then, malls and urban entertainment complexes often rupture traditional street patterns,

internalising the energy, circulation patterns and financial circuits that used to be tied closely

Plate 12 Deira City Center Mall, Dubai, United Arab Emirates, which is virtually impossible to access except by car or taxi. Pedestrian exits lead only to the car park and the complex is entirely surrounded by multilane highways. *Photograph:* Stephen Graham

to the fabric of such street patterns. Turnbull comments that 'as the traditional street pattern was violently displaced, folded, compressed, and replaced' by the Ngee Ann City complex, 'the heterogeneity of the street – its energy – was captured, contained and accelerated' (1997, 229, cited in Cartier, 1998, 172).

As we saw in the Prologue, the death in 1995 of Cynthia Wiggins as she clambered across a multilane freeway between a public city bus stop and Buffalo's Walden Galleria mall illustrates that peripheral malls often even work to exclude public transit buses from the dedicated highway infrastructures. In this case of 'bus route discrimination' the investigation into the death found that 'the plan for the immense suburban shopping complex intentionally avoided accommodating city buses, thereby making it difficult for residents of the inner city not only to shop there but also . . . to work there' (Gottdeiner, 1997, 134).

But even within the city centre new theme parks and entertainment complexes can be equally car-oriented, as they are often located in revitalised waterfront districts, far from public transport infrastructures. Poorer groups may even be forced away from parts of the city that they formerly utilised (Hannigan, 1998a) in the creation of what Muschamp (1995) calls a 'business class city' centre. John Hannigan (1998a) believes that this process may amount to 'an attempt to reinscribe secure, middle-class values within the urban center' as suburbanites start to expect controlled, predictable and socially sanitised consumption spaces right across the metropolis.

DEDICATED PREMIUM INFRASTRUCTURE LINKS

The fifth practice is the careful customisation of dedicated infrastructure links that allow entertainment complexes, malls, sports stadia, resorts and theme parks to withdraw from the immediate public city, whilst linking closely with target middle-class and upper-income markets. For example, the new stadium of the New England Patriots American football team, at Foxboro in the suburbs of Boston, will be equipped with a new access road for the exclusive use of corporate box holders. 'Ensuring easy access is a crucial element in the team's marketing strategy as it continues to hawk luxury suites to Boston's corporate community' (Boston Globe, 14 October 2000, A24). In Sydney the Darling harbour complex was equipped with its very own publicly financed monorail system linking it with the other key corporate and consumption nodes in central Sydney. The new Getty museum in Los Angeles has its own light rail system. And many other tourist and ski resorts are developing customised 'people movers' to enhance their economic prospects.

In North America, as elsewhere, multimillion dollar subsidies are received from local and central government to make 'road and transit additions and reconfigurations which are deemed necessary in order to make an entertainment complex accessible from the freeways which encircle and bisect many American cities' (Hannigan, 1998a, 135). For the Philadelphia Central Waterfront complex, for example, state and federal governments spent $58 million 'to build three major on/off ramps from the I-95 expressway to the newly emerging entertainment area along the Central Waterfront to Penn's Landing' (ibid., 134). Similar practices are common in Europe, Asia, Latin America and Australasia, as municipalities compete to redevelop their redundant spaces and new consumption megaprojects.

PRIVATISED GOVERNANCE REGIMES

Finally, at the most extreme level, theme parks and entertainment complexes can effectively *run their own private governments* – as Disney do in Orlando – to ensure that financial and infrastructural supports are tailor-made to their needs (Archer, 1995). In the Disney case a wholly private political structure, the Reedy Creek Improvement District, set up in 1967, 'has the power to levy property taxes, issue bonds to borrow money, organize and manage its own police and fire departments, develop and manage an airport, and provide utilities (including the ability to build a nuclear reactor if Disney sees fit)' (Archer, 1995, 326).

Developers of gated communities and condominium complexes in such diverse cases as North America (Zaner, 1997), Istanbul (Sandercock, 1998a), Mumbai (Bombay) and Delhi (Masselos, 1995; King, 1998), Jakarta (Dick and Rimmer, 1998), Johannesburg (Lipman and Harris, 1999), Manila (Connell, 1999), Shanghai (C. Smith, 1999), Tokyo (Waley, 2000) and São Paulo (Caldeira, 1996, 1999) are starting to take advantage of infrastructural consumerism and liberalisation by developing and 'bundling' their own customised utility, street, telecommunications and even transport services. In such spaces, once again, the promise of filtering local connectivity to the wider city (through walls, gating, closed circuit television and private security) is closely combined with enhanced local or 'glocal' connectivity (through direct highway connections to airports, premium water and power connections, and enhanced telecommunications).

Even nations that have not known the once widespread private gated streets for several generations are seeing their return. In the United Kingdom, for example, private streets for highly affluent groups have recently re-emerged in Buckinghamshire. In these examples, electronic gates recognise the transponders of residents' cars and burglar alarms. Video entry phones monitor all doors. And infrared motion detectors are directly wired to the local police station (Thomas, 1996). Redevelopment of inner city sites for secure, upscale, condominium-style housing in central London, meanwhile, is also increasingly common, fuelled by a rampant housing market and a cultural shift among urban professionals towards city-centre 'loft living' (Zukin, 1982). A recent analysis of this process found that it creates 'rich enclaves' which 'reinforce divisions between rich and poor' (Ballast Wiltshier, 1999).

To explore the scale and breadth of the global shift towards affluent, secessionary living spaces it is necessary to take a range of international examples. In what follows we analyse processes of secessionary residential development in condominium complexes and master-planned communities in North America; the development of new towns and planned 'communities' on the fringe of megacities in Indonesia, Turkey, the Philippines and post-apartheid South Africa; and the construction of fortified enclaves within the centres of São Paulo in Brazil and Mumbai (Bombay) in India.

MASTER-PLANNED, GATED COMMUNITIES IN NORTH AMERICA

Americans during the second half of the twentieth century focused their energies on preventing democracy in the built environment.

(Sennett, 1999, 278)

Although connectedness is the spirit of the city, and will probably remain so, the American version has always harbored a tendency to explode, to atomize and to spread itself as far as possible.

(Lerup, 2000, 80)

The US experience of suburban fortification is the best-known (see Blakeley and Snyder, 1997a, b). In the United States corporately constructed, gated and master-planned 'common interest developments' (CID) now account for up to 50 per cent of house construction around cities in the south and west (Pope, 1996, 178; see McKenzie, 1984). Figure 6.12 highlights for selected US cities the growing frequency of gating on new residential developments. Albert Pope calls this process the emergence of 'city-size corporate nuclei' (1996, 96). Such spaces are overwhelmingly geared to affluent and largely white groups seeking the ontological security that, for them, comes with living in 'communities' whose essential foundation rests on the regulating out of ethnic and social difference, diversity and chance encounters (see Dumm, 1993).

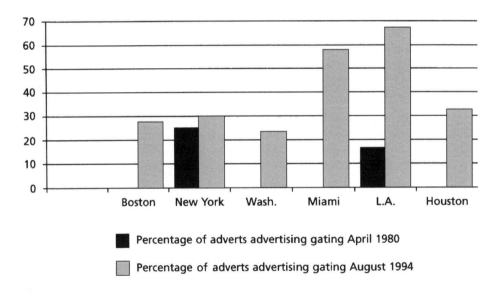

Note: The newspapers analysed were the Boston Globe, the New York Times, the Washington Post, the Miami Herald, the Los Angeles Times and the Houston Post

Figure 6.12 The gating of American residential developments: Luymes' (1997) analysis of the growing incidence of gating in advertised real estate developments in selected US cities. *Source*: adapted from Luymes (1997), 193

On a broader scale, the 'edge cities' within which most US gated residential developments are embedded operate as a form of 'shadow government able to raise taxes, legislate for, and police their communities' (Dear and Flusty, 1998, 55). In trying to meet the direct needs of the high-income residents for residences, recreation and retirement, common interest developments are taking over 'many municipal functions for those who can pay the price, offering a competing sector of pay-as-you-go utilities' (Lockwood, 1997, 62).

Their most obvious function is to restrict access tightly by maintaining and policing the strict, gated, cordon with the rest of the city's highway and street system. Internally, road networks branch in cul-de-sac, 'laddered' fashion, with each strand leading to a singular destination, to ensure the maximum closure of the social landscape (Pope, 1996). Common interest development responsibilities also extend to the construction and maintenance of internal roads, the enforcement of speeding and their own traffic regulations, landscaping, private security, CCTV and street lighting. For example, the gated community at Embassy Lakes in Cooper City, Florida, requires residents' cars to carry bar codes to allow entry. Visitors must show an identity photograph and wait for the entry guard to phone their friends or relatives inside before entering. Other spaces require a security key or code number to allow passage (Brunn *et al.*, 2000). Gumpert and Drucker note that, in one such development on Long Island, New York, the home owners' association has:

incorporated a rotating television monitoring system that watches and tapes activities on the street. The location of the twenty-six cameras is known only by the family on whose property it is mounted – not even the security guards employed by the home owners know the location. This way, if unknown visitors are not here for a legitimate reason, they are asked to leave by security guards. If they refuse, they are arrested, and the civic association will prosecute them.

(1998, 433)

As well as enforcing boundaries, common interest developments are starting to develop premium network spaces within those boundaries to tempt residents in and retain them. Even energy, water and telecommunications services are now being offered, in partnership with utility firms, as developers try to take advantage of utility liberalisation. In addition, direct highway connections to the gates of common interest developments are a common aspiration, often with favourable subsidies from local governments. The 10,000 population development at Weston, near Fort Lauderdale, for example, built by the Arvida corporation, has been furnished with dedicated expressway links straight to Miami and its airport (Lockwood, 1997). In 1988 the Florida state government passed the Community Development District Act to help developers of common interest developments pay for the high costs of dedicated transport and water networks by offering low interest rates and tax-exempt bonds issued by community development districts (Lockwood, 1997). Meanwhile, one common interest development on the edge of Phoenix, Arizona, has acquired its own fleet of electric vehicles for use within its boundaries; the vehicles cannot be legally used on the rest of the city's highway network (Kirby, 1998). But the most pronounced current intervention of the developers of gated communities and condominium complexes in infrastructure development is emerging in information technology and telecommunications. This we explore in Box 6.5.

BOX 6.5 PACKAGING ADVANCED INFORMATION TECHNOLOGY AND TELECOMMUNICATIONS INTO REAL ESTATE DEVELOPMENTS: 'SMART' CONDOMINIUM COMPLEXES AND 'TELECOMMUNITIES' IN NORTH AMERICA

'SMART' CONDOMINIUMS

Condominium developers in North America are seeking fresh income streams from diversifying into newly liberalised information, security, entertainment and utility markets. In this process 'alternative incomes may be[come] more important than rent' (Zaner 1997, 65). Zaner (1997, 63) argues that:

> enhanced telephone services, movies on demand – even Internet access – are taking their place alongside pools, fitness centers, and party rooms, as standard amenities in today's multifamily [and CID] communities. Thanks to technological innovations and the deregulation of telecommunications, these 'electronic amenities' are offering a new level of service to apartment residents as well as significant revenue generating opportunities for apartment developers, owners, and investors.

The key here is that developers house an affluent, relatively captive, market of people who are keen to benefit from the exploding possibilities of highly advanced infrastructure services. One developer, GE Rescom in Los Angeles, reported that 'our plan is to bundle products in such a way that virtually everything that someone would need would be provided "with lease"' (Zaner, 1997, 63). Services will extend to automatic credit checks, insurance, utilities, telecommunications, cable and broadband Internet, all designed and bundled to meet the needs and desires of a captive and affluent audience. 'One day,' Zaner continues, 'every building will have its own high speed local area network within the community, every building will have its own energy conservation equipment, and a great deal more money will be spent on sophisticated security equipment' (quoted in Zaner, 1997, 62). Bulk-buy deals from utilities and telecom firms allow developers to make considerable profits from such value-added services whilst offering enhanced IT services at below standard rates to their residents.

'BANDWIDTH IS LIKE A DRUG': CONFIGURING 'FAT' INTERNET PIPES FOR 'LOFT LIVING'

Apartment developers in New York, Washington, Seattle, Boston and San Francisco, meanwhile, are now offering trunk Internet connections to their highest-priced apartments, delivering speeds of between 100 and 150 times that experienced by Internet users over the traditional phone system. Having a direct connection in the apartment to a T-1 Internet trunk is proving a major draw for IT professionals for whom broadband Internet access is increasingly crucial to their multimedia work. 'Already, some of the bandwidth-obsessed are making decisions on where to buy their homes based on the availability of high-speed access' (Harmon, 1999, 2). Evans argues that 'bandwidth is like a drug; once you've experienced speed, you don't have the patience to download' (1999, 42). By February 2000 1.4 million consumers were estimated to have benefited from such high-speed Internet connections in the United States (*Boston Globe*, 2 February 2000, F4).

FROM PRIVATOPIAS TO 'TELECOMMUNITIES'

Meanwhile, on a larger scale, the developers of peripheral common interest developments of thousands of homes are also trying to develop themselves as 'telecommunities'. Like condominium developers, the developers of places like Playa Vista in Los Angeles, Desert Ridge in Phoenix and Montgomery Village in Toronto are attempting to attract highly communications-intensive professionals by carefully packaging dedicated broadband communications links into their communities (offering home local area networks and front-end services like ADSL and ISDN services).

Community intranets are also being developed within common interest developments. These offer everything from remote energy management and home security to liaison with local medical centres, receiving homework assignments from schools, booking slots on the CID golf course or teleshopping from local shops. Community bulletin board systems, meanwhile, advertise babysitting vacancies. As Pamela Blais (1998, 61) argues, 'IT infrastructure can support the growing trend for residential developments to offer a variety of services along with the bricks and mortar of houses, streets, parks and shops'.

DISNEY'S CELEBRATION: BROADBAND MEETS NEOTRADITIONALISM

At Celebration near Orlando, Florida, finally, the Disney Corporation is developing an idealised 1950s US townscape, replete with advanced telecommunications and a completely privatised governance structure (Dery, 1999). In the town 'a firehose bandwidth of fibre links every home to the Net, offering a quaintly familiar-sounding list of futuramas: home security, linking each resident to a central monitoring point, interactive banking, voting from home, virtual office, easy access to each other' (Hayman, 1996, quoted in Robins, 1999, 50). In fact, a whole suite of private communications, road, street, water, fire and security infrastructures are provided on a simply fiscal equivalence basis. The rationale of the Disney Corporation is that 'much of the unique infrastructure sought by Celebration was either unavailable through normal government channels, or simply too expensive for local government' (quoted in Drew, 1998, 181).

Of all secessionary network spaces, the risk of common interest developments prompting the collapse of the overarching municipal tax system is perhaps strongest. 'Those paying for and receiving the private services can be expected to resent paying for and receiving public services that they do not want' (Reich, 1992, 182). This is likely to create conflicts between private common interest developments and elected local governments. 'The result,' according to

Robert Reich, 'could be a gradual secession from the city that could leave it stripped of much of its population and resources' (1992, 183). Already many common interest developments are pushing for favourable tax rebates from local government property taxes, arguing that their residents do not use the full complement of local services (Luymes, 1997).

PERIPHERAL URBANISATION: 'BUNDLED' NEW TOWN COMPLEXES IN INDONESIA, TURKEY, THE PHILIPPINES AND POST-APARTHEID SOUTH AFRICA

Our second set of examples takes us to the exploding peripheries of Asian and South African 'megacities': specifically to Indonesia's capital, Jakarta, Turkey's 'global' metropolis, Istanbul, Manila in the Philippines, and South Africa's commercial capital Johannesburg. In all four cities, broadly parallel trends are under way towards the construction of private residential and leisure complexes for elite and upper-middle income groups on the urban periphery, which are carefully networked with the best available infrastructures whilst being secured off from surrounding urban spaces.

JAKARTA

In Jakarta, as in many South East Asian 'mega' cities, massive urbanisation and urban expansion are taking the form of giant speculative and oversupplied new town developments on the fringe of the city. Within them, elite, gated communities are being packaged together with 'bundled' retail, leisure and commercial complexes and strung together by public or private tolled highways geared entirely to the car-owning middle classes (Dick and Rimmer, 1998). A completely new urban landscape is being fashioned for the growing upper- and middle-income groups that allows them to secede in an absolute fashion from the central city with its poor infrastructure, pollution, perceived danger, contact with the poor and traditional street system.

There are clear fractures between this new city and the old Jakarta. Whilst 'most of these new towns are located close to toll roads, other links with the road network and with the public transport system remain tenuous' (Dick and Rimmer, 1998). Again, geographical distance is combined with the publicly supported construction of highly capable highway links that mean that spatial dispersal is no longer a barrier to distant integration. The elevated highways support vertical segregation, too, as:

driving through the elevated highways suggests an experience of flying over the top of the [old] city, escaping from its congested roads and leaving behind the lower classes who are routed through the crowded streets at ground level. From this suspended driveway, the details of the urban fabric are transformed into a series of blurred sketches, giving a sense of detachment from the 'worldly' streets below. [They allow] certain forms and spaces to be visualised and others concealed.

(Kusno, 1998, 163–4)

Thus urban life for middle-class car owners and new town residents in Jakarta becomes the negotiation of an interlinked complex of private, air-conditioned and secure spaces (home, office, leisure space) laced together by car travel (again air-conditioned) on private tolled highways. 'Gated residential communities, condominiums, air-conditioned cars, patrolled shopping malls and entertainment complexes, and multi-storeyed offices are the present and future world of the insecure middle class in south east Asia' (Dick and Rimmer, 1998).

To Dick and Rimmer (1998) this process represents a 'rebundling of urban elements' as ever more grandiose developments start to encompass previously discrete ranges of functions within single, secured, air-conditioned complexes: offices, housing, shopping, leisure activities, restaurants, hotels and, of course, multistorey car parks (see Figure 6.13). 'What has emerged,' they argue, 'is a pattern of new town developments integrated with industrial estates, toll roads, ports and airports,' in effect 'turning the city inside out. The innovation of the 1980s was the recognition by some of the richest South East Asian businessmen that enhanced profitability would flow from bundling as many as possible of these discrete facilities into integrated [air-conditioned] complexes' (1998).

In this process, which has been substantially impacted by the Asian financial crash of 1998, huge sums of risk capital were combined with massive speculative land acquisitions around the outskirts of Jakarta. The total area of speculative fringe development proposed or planned for self-contained private new towns, *kota mandiri* (90,000 ha) actually dwarfed the area of old Jakarta (66,000 ha) (see Figure 6.13). 'Driving around Jakarta's highways,' wrote Deyan Sudjic in 1995, 'clumps of high-rises crop up almost overnight at random' (35). Many, however, now lie empty or half built, as the financial collapse revealed a huge mismatch between the high-cost housing supplied and what the market could actually support (Firman, 1999).

ISTANBUL

Broadly similar processes of new town urban 'bundling' with dedicated infrastructures are emerging in the fast expanding periphery of Istanbul, Turkey's global 'megacity'. Asu Aksoy and Kevin Robins (1997) have provided a powerful critique of the splintering logics underpinning the Esenkent and Bogazköy new towns on the western edge of the city. These are separated off from the surrounding informal settlements 'by the *cordon sanitaire* of the main highway' (ibid., 21). To Aksoy and Robins these starkly dualised but geographically adjacent places demonstrate that the 'modernisation' of Istanbul's periphery is based on 'moving towards an ever greater segregation of the urban scene along class-based and identity-based lines' (ibid.).

On the edge of Istanbul techniques of modern urban planning, ostensibly designed to support order and coherence, are clearly implicated in fragmenting the metropolis into packaged, fortressed spaces for the growing middle class, on the one hand, and the surrounding landscapes of self-constructed squatter settlements on the other. 'For most Istanbulians,' they write, the latter are '*terra incognita*, a place too far (almost extraterrestrial)' (ibid., 23). Just across the highway, the purpose-built satellite towns of Esenkent and Bogazköy, on the other hand, are a 'new world of seemingly luxurious apartment blocks with

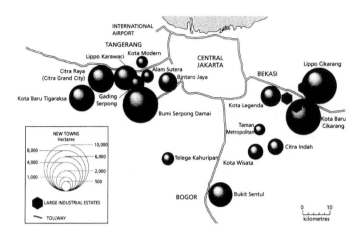

Project	Area (ha)	Launched (year)
North Jakarta		
Jakarta Waterfront City	2700	NA
West Jakarta (Tangerang)		
Teluk Naga	8000	NA
Bumi Serpong Damai	6000	1989
Kota Baru Tigaraksa	3000	1987
Citra Raya (Citra Grand City)	3000	1994
Lippo Karawaci	2630	1992
Bintaro Jaya	1700	1979
Gading Serpong	1000	1993
Kota Modern	770	1989
Alam Sutera	770	1994
East Jakarta (Bekasi)		
Lippo Cikarang	5500	1991
Kota Baru Cikarang	5400	NA
Kota Legenda	2000	1994
Bukit Indah City	1200	1996
South Jakarta (Bogor)		
Bukit Jonggol Asri	30 000	1996
Bukit Sentul	2000-2400	NA
Citra Indah	1200	1996-97
Kota Wisata	1000	1997
Telaga Kahuripan	750	NA
Taman Metropolitan	600	NA

Figure 6.13 Peripheral new towns, or rebundled cities, planned or under construction around the edge of Jakarta, 1998. *Source*: Dick and Rimmer (1998), 2313

familiar, pattern-book postmodern design features, and of spacious and comfortable villas with large gardens and swimming pools' (ibid.).

Huge investments in infrastructure have supported the emergence of these fortified spaces: a new highway straight to central Istanbul, some of Turkey's best 'intelligent' telecommunications links and dedicated water, sanitation and energy networks. 'These new developments,' writes Leonie Sandercock, 'are marketed as having all the amenities of "a small and modern American village", including private utilities and services, private buses into the city, private security and surveillance systems, electronic shopping facilities connected to the supermarket on site, sports, health and entertainment facilities, and schools' (1998a, 176).

These new infrastructures, especially the highly symbolic highway, were, according to the city authorities, designed to 'carry the lifeblood of commerce, communication and culture' (quoted in Aksoy and Robins, 1997, 27). The highway's slogan was 'the route to contemporary life' (ibid., 29); in practice it 'turned out to be an exit route for what came to be an increasingly disillusioned modernising vision' (ibid., 27). The urban plan was largely ignored and the new infrastructures were remodelled effectively to seal off the new towns from the wider squatter settlements, while linking them into wider circuits of exchange.

Middle-class Istanbulers arrived in the new towns and realised that they 'could accommodate a purified modern lifestyle, in retreat from everything that Istanbul had become as a consequence of its actual modernisation' (ibid., 31). Homogeneous, safe, orderly spaces offered a sense of social purity and retreat from the turmoil of metropolitan modernisation outside, with its perceived pollution, crime, danger, discomfort, poverty and ethnic mixing. Thus a new 'kind of self-contained, self-sufficient and self-regarding community' was created within which modern identities could be maintained in seclusion and isolation (ibid., 33). As in so many other cities, then, the modern infrastructural ideal was remodelled as a logic of splintering urbanism.

MANILA

In Manila in the Philippines, too, suburbanisation 'has produced new middle class consumer landscapes of exclusive suburbs – alongside tower blocks, offices, residential estates, shopping malls and golf courses – linked by freeways and flyovers' (Connell, 1999, 417). Gated suburban spaces, often constructed literally on the sites of forcibly demolished informal settlements, are 'designed and marketed as fragments of Europe in a global era' with 'enhanced security, exclusivity and isolation' (ibid.). These new developments have names like 'Little Italy', 'Harvard Avenue', 'Greenhills' and 'Britanny'. Huge efforts and investments are made to ensure that privatised, customised infrastructure networks – roads and kerbs, water supplies, drainage systems, highway links, power and telecommunications links – are available which befit the construction of such spaces as secure and exclusive 'islands of Europe or North America' within the peripheries of Manila (ibid., 426) (see Figure 6.14).

At the same time, tolled freeways and flyovers, such as the billion-dollar Metro Manila Skyway, are being completed between the new enclaves, malls and office complexes – often, once again, after the forced demolition of the squatter settlements in their path. Such freeway networks, developed by private–public partnerships, are routed and planned to 'tie together islands of affluence' (ibid., 435). In a classic example of the socially regressive splintering of circulation systems, such freeways are designed and regulated to exclude the vehicles of the poor – jeepneys, buses and tricycles – which cram on to the residualised, congested street system.

The cocooned, mobile isolation of the middle-class car driver is thus splintered off from the chaotic, risky openness of the street for the poor. 'When people with cars are stuck in traffic,

JESTRA VILLAS boasts of a fine location in the heart of one of Metro Manila's prime residential districts... in a quiet setting. Yet only 15 minutes away from the bustling business center of Makati.

It is easily accessible by car or public transportation via the South Expressway and Bicutan or Sucat Interchange. Exclusive schools, the Paranaque Medical Center, the Elorde Sports Center and the St. Anthony Parish Church are just some of the nearby establishments to cater your needs.

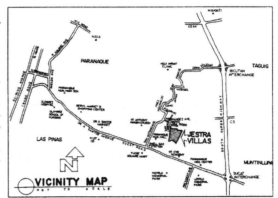

VICINITY MAP

(**HOUSING FEATURES**)

- Single attached, duplex and townhouse cluster type models
- Living room, dining room and kitchen
- 2 to 3 bedrooms with 2 to 3 bathrooms
- Storage and service areas
- Individual carport and patio
- Marble floor for living, dining and bath rooms
- First-class bathroom and kitchen fixtures
- Walk-in closets for Santino and TH-700 house models

(**DEVELOPMENT HIGHLIGHTS**)

- Landscaped entrance gate with guardhouse
- Wide concrete roads, curbs and gutters
- Paved sidewalks
- MWSS water supply with 2 elevated 20,000-gallon watertanks and deepwell back-up water system
- Underground drainage system
- Multi-purpose hall / chapel / function room
- Tennis court
- Swimming pool
- Children's playground
- Landscaped open spaces

Exclusive Marketing Companies

Owner-Developer

Fil-Estate Marketing Associates, Inc.

Fil-estate Sales, Inc.

Fil-Estate Network, Inc.

Fil-Estate Realty Corporation

JDM Corporation
Room 2E 2/F PDCP Bank Center
Herrera cor. Alfaro Sts.
Salcedo Village, Makati, Metro Manila

FOR MORE INFORMATION, PLEASE CONTACT **6329486 TO 88**

Figure 6.14 Advertisement for the Jestra Villas complex in Manila, the Philippines. *Source*: Connell (1999), 426

they can roll up the windows and turn on the air-conditioning. When the [relatively poor] get stuck in traffic, it is in an open-air jeepney with the sun beating down, fumes belching all around, a dozen passengers squeezed in, and the cacophony of horns and engines everywhere' (McGurn, 1997, 35–6, cited in Connell, 1999, 435). Thus, like Jakarta, Manila is being reconstructed as 'a decentralised spatial system resembling an archipelago whose islands are interconnected by [highway] bridges' but where the 'islands' are 'the exclusive, walled-in neighbourhoods where the upper strata are ensconced' (Tadiar, 1995, 298, quoted in Connell, 1999, 435).

JOHANNESBURG

Our penultimate example comes from the processes of urban restructuring that have followed the overthrow of the apartheid system in the major cities of South Africa. All South Africa's cities have since seen increasingly violent clashes between extremes of poverty and wealth (Lipman and Harris, 1999). Globally, South Africa is now ranked first in the number of murders, rapes, robberies and violent thefts *per capita* (*New York Times*, 15 May A10). Most startling is the case of the commercial capital, Johannesburg, where there has been a rapid growth in extremely violent militarised crime, especially in the areas of the city that previously were largely insulated from serious crime by the pass-law system: downtown and the middle-class white suburbs (Bremner, n.d.).

With the removal of the bureaucratic machinery of pass-law segregation, many poor blacks have moved from the appalling conditions in the townships to make their homes informally in the 'public' spaces of the city centre, in search of better prospects. At the same time, affluent whites, and increasingly the industries they work for and the services they use, are retreating to fortified spaces on the urban periphery to the north. Many planned 'communities' are being constructed there with bucolic-sounding names like High Meadow Grove and Brentwood Estate which emphasise security above all else (ibid., 62). In response to both real crime and the spiralling fear of crime, households are increasingly employing armed response teams, security firms and even armoured vehicles to back up the spontaneous or planned gating and walling of their living spaces. Often residents simply close off their own roads, destroy their systems of sidewalks, wall off their spaces, and bring in armed security companies – all without municipal permission. Almost overnight 'anything from a street to an entire suburb is excised from the public network' (Bremner, n.d., 58). To complement the closure and partitioning of previously relatively public networks:

households install ever more sophisticated security measures. They raise their low, picturesque garden walls by two, three or sometimes four metres, and top them with spikes or glass chips; they unfurl razor wire along the perimeters; they add electric fencing; they install automated driveways and intercom systems. Entry into houses and passages is barred by layer upon layer of metal security gates. Security has become a way of life.

(Bremner, n.d., 56)

A mushrooming range of speculative malls, consumption spaces and decentralised office parks – also located well way from the old city centre – are also emerging to cater for the perceived security needs of the white middle classes. To support such spaces, the state has given property owners broad legal powers to privatise the public realm through gates, CCTV and versions of the US Business Improvement District idea. Some new business spaces offer private schooling for employees' children. The South African stock market has also decided to abandon the old downtown core and move to a more secure developed enclave farther north.

Together, these trends amount to the effective collapse of spaces and networks in the new South African metropolis that offered at least limited hope of supporting the mixing of the nation's highly segregated ethnic and income groups. In contemporary Johannesburg, Bremner writes, the:

spaces between are simply movement channels along which the body must pass in moving from one insulated enclave to another. . . . A new security aesthetic dominates: walls, wire, barbs, locks, gates, intercoms, fortifications . . . fading into fantasy and pastiche. Combine this with signs of poverty, with squalor, irregularity, clutter, leaking sewer pipes, leaning corrugated walls, broken windows, and you have the image of the emerging South African city.

(Ibid., 63)

For our third and final set of examples of the construction of secessionary residential spaces, we turn to the global cities of São Paulo and Mumbai (Bombay), both of which demonstrate how urban bundling and the secession of the socioeconomically affluent can also go on within older urban centres of developing 'megacities'. These are discussed in Boxes 6.6–7.

BOX 6.6 CONSTRUCTING AFFLUENT ENCLAVES WITHIN THE CENTRE OF DEVELOPING WORLD 'MEGACITIES' (I) THE CASE OF SÃO PAULO

'FORTING UP' IN SÃO PAULO

In São Paulo fortified enclaves have grown up since the early 1980s within central *favela* (shanty town) districts, particularly around the Murumbi district to the south and west of the city core. Such complexes are designed to meet the demand from the growing middle class for perceived security, whilst supporting their withdrawal from public street and infrastructure systems (Caldeira, 1996, 1999). They 'appeal to those who are abandoning the traditional public sphere of the streets to the poor, the "marginal", and to the homeless' (Caldeira, 1996, 304). Displaying armed guards, high walls, electric fences, CCTV and automatic gates, the ostentatious wealth of such enclaves contrasts starkly with the extreme poverty that, in many cases, literally surrounds their walls. In few other cities do such extreme cheek-by-jowl contrasts of wealth and poverty exist.

These contrasts have intensified with the increasing income polarisation that has come with the 'opening up' of Brazil's economy to the forces of globalisation in recent years. Conspicuous consumption by the socio-economic elites that have most benefited has

created further tensions. 'The most visible social difference used to be between a Volkswagen Beetle and some locally made sedan,' recalls São Paulo's top official of security. 'Now it's between the same old Beetle and an imported Ferrari. Globalization has brought new models of consumption, relentlessly pushed on TV.' At the same time, crime levels are soaring: the city now has 8,500 murders per year, a rate ten times that of New York (Cohen, 2000).

São Paulo's condominium enclaves offer integrated spaces for residence, work and consumption to middle- and upper-income groups. They are 'private property for collective use; they are physically isolated, either by walls or empty spaces or other design devices; they are turned inwards and not to the street; and they are controlled by armed guards and security systems which enforce rules of isolation and exclusion' (Caldeira, 1996, 309). Their reliance on veritable armies of service personnel, on the time–space flexibility of the automobile, on dedicated energy and water connections, and on the most sophisticated telecommunications links available in Brazil, mean that they are locationally flexible within the metropolis. 'They possess all that is needed within a private, autonomous space and can be situated almost anywhere, independent of surroundings' (Caldeira, 1996, 309).

As with the US gated communities, the developers of condominium complexes are exploring the newly liberalising utilities and tele-communications markets to make the most of their highly favoured market position (Schiffer, 1997). With the collapse of public planning for energy, water and telecommunications infra-structure, and the concomitant withdrawal of cross-subsidies, there is a 'supply of sophis-ticated infrastructural services for top income groups in São Paulo' (ibid., 10).

This trend is focusing network development and service innovation on the new fortified enclaves – which have 'high concentrations of the infrastructure services – particularly telematics, optic fibre, cable TV and mobile telephone central stations' (ibid.) – whilst neglecting the wider city of the poor. Whilst the poorest districts suffer water and power shortages, these enclaves benefit from irrigated gardens, reliable power and highly capable telecommunications. As well as net-worked infrastructures, the common facilities in enclaves include drugstores, tanning rooms, psychologists, gym trainers, gardening, mas-sage, cooks, food preparation, car washing, transport and shopping (Caldeira, 1996, 211; see Silva, 2000).

As with the other examples in the United States and Indonesia, all these processes together mean that the public street is rendered problematic and is residualised. The relationship the enclaves establish with the wider city is

> one of avoidance . . . public streets become spaces for elite circulation by car and for poor people's circulation by foot and public transportation. To walk on the street is becoming a sign of class in many cities, an activity that the elite is abandoning. . . . Public space is increasingly abandoned to those who do not have a chance of living, working and shopping in the new, private, internalized, and fortified enclaves.
>
> (Caldeira, 1996, 314–19)

THE ULTIMATE BYPASS: THE HELICOPTER COMMUTE

The most visible and powerful symbol of the secession of the rich from the immediate space–time in São Paulo, however, renders the street and highway entirely avoidable. For the exploding use of personal helicopters enables the very wealthiest Paulistanos to fly between affluent enclaves and work spaces, in a bizarre parody of the predictions of Fritz Lang or Frank Lloyd Wright that mass individ-ual air transport would be realised within the twentieth century. Over 400 personal heli-copters existed in the city in late 1999; 100 were in the air at any one time; the market for them was the fastest growing and largest of any developing world city (Romero, 2000, 1).

'Like a fleet of airborne limousines, the helicopters are increasingly used by privileged Paulistanos to commute, attend meetings, even run errands and go to church. Helicopter landing pads are now standard features of many of São Paulo's guarded residential compounds and high-rise roofs' (ibid.). In effect, personal helicopters are the logical conclusion of all other secessionary processes in that they finally release users from depen-dence on sharing the city's highways and terrestrial surface with the rest of its 16 million inhabitants.

BOX 6.7 CONSTRUCTING AFFLUENT ENCLAVES WITHIN THE CENTRE OF DEVELOPING WORLD 'MEGACITIES' (II) THE CASE OF MUMBAI (BOMBAY)

Very similar processes of enclave formation are emerging in Mumbai (Bombay) – a cause of violent social disturbance in 1993 (Masselos, 1995; King, 1998). Faced with growing violence, population and the inability of urban planning and improvement programmes to provide basic services for the city's swelling population, socioeconomic elites have, as in São Paulo, benefited from the construction of modern, highly serviced and heavily fortified condominium complexes. These are equipped with pools, jogging tracks, tennis and badminton courts, croquet lawns, golf courses and multiservice retail and leisure malls (King, 1998, 28). Again, however, the new spaces, with their pseudo-aristocratic and Euro-American names, are surrounded by squatter settlements (or *zopadpatti*). 'Between them are walls and security guards but they inhabit the same locality, the one on the ground and the other in the air' (Masselos, 1995, 211).

Mumbai, long a symbol of India's hoped-for progress towards emancipation, has, in effect, become deeply dualised. All aspects of consensus between rich and poor, and their political groups, have broken down (Masselos, 1995, 206). Carefully networked high-rise structures provide a three-dimensional landscape of exclusion and polarisation. They create 'localities of the ultra-wealthy and the upper middle class' (Masselos, 1995, 209), groups who have benefited enormously from

the liberalisation of India's economy and Mumbai's key role in articulating that economy with the rest of the world.

Building work and public subsidy have gone overwhelmingly to serve the living, working, leisure and transport needs and desires of these groups. 'The new arrivals to the city have had to survive by living on the city's pavements or else shanties on whatever unbuilt land was available' (Masselos, 1995, 210). As we saw with the example of water networks in the Prologue, these processes are starkly woven into the material fabric of the city, with large water mains, threaded between high-rise enclaves, actually going through the shanty settlements. Whilst these give no access to the water that flows within to people living in the shanties, ironically, they provide walking routes in spaces where bitumenised roads are extremely rare.

Shanty residents, meanwhile, have to make do with highly inadequate stand pipes or, worse still, private water vendors, who charge exorbitant rates for water on a per litre or gallon basis (see Swyngedouw, 1995b). Above all, as in all the above examples, development processes in Mumbai mean that 'a sense of interconnectedness between the differing sections of the city's population as a whole' is being replaced 'by a sense of interconnectedness with certain parts only of the population, not all of it' (Masselos, 1995, 210).

'SMARTENING' THE HOME: DOMESTIC SPACES OF SECESSION AND CONTROL

Our final exploration of the premium networked spaces of the contemporary metropolis brings us to our smallest scale of analysis: the level of the domestic home. Here, too, affluent

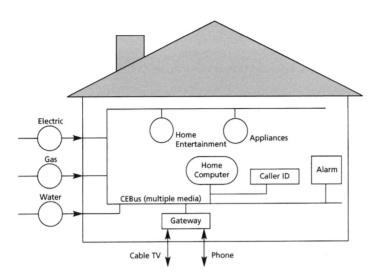

Figure 6.15 The 'smart home' concept in the late 1990s. *Source*: Schoechle (1995), 443

residences are being increasingly equipped with the highly capable networked infrastructures that allow them to secede from their immediate urban environments, particularly within the contexts of liberalising infrastructure market places (Lorente, 1997). In fact, such residences, and the developments within which they lie, are increasingly being integrated into the global spaces of exchange via telecommunications, transport, airport and, to a lesser extent, energy and water infrastructures (see Figure 6.15).

As their residents extend their communication spaces outwards to the international sphere they often simultaneously turn their homes away from public life through increasingly 'smart' security infrastructures. 'The more we detach from our immediate surroundings, the more we rely on surveillance of that environment,' suggest Gumpert and Drucker. As a result, 'homes in many urban areas around the world now exist to protect their inhabitants, not to integrate people with their communities' (1998, 429) (see Figure 6.16).

New media, transport and utility infrastructures support these parallel processes of local secession and securitisation and international interconnection, allowing the residents of detached homes to 'reach out through complex communications' to an 'emancipated disconnected world' (ibid.). The emphasis now falls on the ways in which living spaces are relationally connected with the 'right' infrastructures that link them into required circuits of flow and exchange across the wider urban region and the wider world. The house – Le Corbusier's 'machine for living in' in the modern infrastructural project – emerges as a terminal on myriads of carefully selected sociotechnical networks. Martin Pawley calls the contemporary house 'a single arterial service loop from which sewage, clean water, optical, data and telephone links can be accessed at any point required' (Pawley, 1997, 196).

Increasingly, then, the homes of more affluent socioeconomic groups are being transformed into secured sanctuaries and hubs for infrastructurally mediated exchange, communication, work and transaction. Purpose-built local area networks (LANs), home security systems,

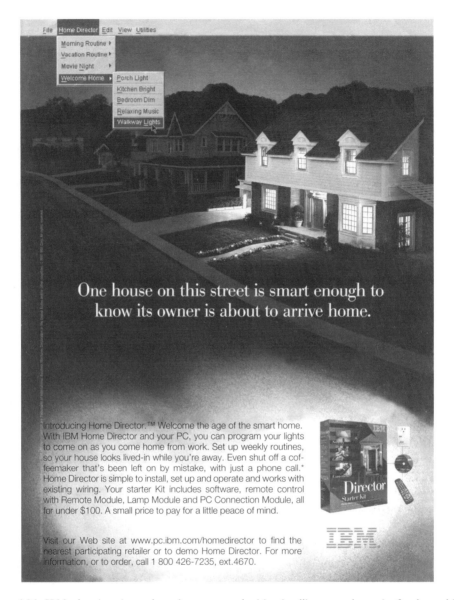

Figure 6.16 IBM advertise a 'smart home' system, emphasising intelligence and security for the mobile professional. *Source*: IBM Home Director

personal CCTV, motion detectors, automatic utility metering, energy management applications and domestic broadband Internet connections are some of the fastest growing industries across the advanced industrialised world. Some commentators even argue that 'every [high-income] new home built today without an internal [telecoms] network is obsolescent and represents a missed opportunity' (Gumpert, 1996, 41). And, as with the condominium and CID developments discussed above, newly liberalised utilities industries are increasingly keen to diversify into bundled services addressing many of these needs simultaneously.

Here we see utility providers 'snuggling up' to chosen households of the market, by offering loyalty discounts, extra services and 'bundled' packages, to 'cherry-pick' and maintain the most lucrative market segments (Marvin and Guy, 1999). As Jane Summerton suggests, 'a clear trend is to differentiate and individualise within previously anonymous consumer collectives, identifying specific categories of customers and giving them identities, needs and interests' (1995, 12).

For affluent consumers, in particular, 'utilities are realising that advanced services provide ready access to the home, its appliances and electronics. Data collected from these resources could lead to the development and provision of a host of new customer services' (Kirby, 1998, 28). In St Louis, for example, Union Electric has teamed up with Honeywell and CellNet data systems to test wireless home management systems. These allow 'customers to purchase off-peak electricity, authorise payments of utility bills, monitor energy use and, eventually, programme lighting and security systems' (Kirby, 1998, 28).

Elsewhere, utilities offer financial services, Internet and free local and long-distance calls in an attempt to keep affluent customers loyal. Other utilities in the United States, the United Kingdom and Sweden attempt to gain new customers by offering 'green electricity' guaranteed as generated from renewable sources, in effect 'matching individualised customers with individualised electrons' (Summerton, 1995; see Hirsch, 2000).

BEYOND SPACES OF SECESSION: MARGINALISED SPACES, NETWORK GHETTOES AND THE 'POVERTY OF CONNECTIONS'

> Whilst ghettos and slums are by no means new components of urban structure, [today's] landscapes of the excluded are unprecedented in the intensity of combined poverty, violence, despair and isolation.
>
> (Knox, 1993b, 231)

What, then, of the zones and spaces beyond and between the premium network spaces of the contemporary metropolis? How do people and places that are not so favoured experience the logics of splintering urbanism? What, more specifically, happens to those people and places that are being socially and economically marginalised within contemporary processes of urban development?

As we have seen already, most notably in the discussions of Bombay, Istanbul and São Paulo, these spaces tend to be bypassed by new trends in networked infrastructure development. They tend to be largely ignored in investment strategies and rendered marginal within practices of geodemographic targeting. And they are often relationally severed from secessionary network spaces by urban design and security practices.

Such 'disfigured' urban spaces thus tend to remain excluded and largely invisible within the contemporary metropolis, beyond the secured, well designed and carefully networked premium or 'figured' spaces of the premium-networked metropolis (Boyer, 1995, 82). For example, deindustrialised neighbourhoods of deepening social and economic marginalisation emerged in many large and medium-size Western cities in the later twentieth century (Massey 1996; Mingione, 1995; Law and Wolch, 1993; Doron, 2000). To Schwarzer such 'ghost wards', especially in US cities, are 'torpid remains of a market-oriented culture of speed, half-digested leftovers from a culture of plenty'. In such places 'history is heavy with departure. . . . Dreams have been replaced by nightmares and danger and violence are more common-place than elsewhere. Nonetheless in decay there is meaning' (1998a, 11–16). In the Developing World, meanwhile, shanty towns and *favela* districts, with improvised or rudimentary connections to paved street systems, water supplies, and power and communications networks, often encompass between a third and two-thirds of the entire urban population.

'ADVANCED MARGINALITY' IN NETWORK SOCIETIES

Loïc Wacquant (1996) terms the condition of such zones and spaces one of 'advanced marginality'. To him 'the resurgence of extreme poverty and destitution, ethnoracial divisions (linked to the colonial past) and public violence, and their accumulation in the same distressed urban areas, suggest that the metropolis is the site and fount of novel forms of exclusionary social closure' (1996, 121). Although Wacquant focuses on advanced Western cities, the words apply equally to many developing and post-communist cities (see Potter and Lloyd-Evans, 1998).

Within today's 'network societies' (Castells, 1996, 1997a, 1998a), then, we need to recognise that, in developing, developed and post-communist cities alike, marginalisation from the ability to use and configure networked infrastructures and technologies is as central to the experience of poverty as lack of food, money or formal employment. Not only are networked infrastructures configured in highly biased ways within and between the social architectures of places, but they often tend to work together in compound and mutually reinforcing ways. Service restructuring or technological innovations that enrol some people and places to premium status – like those discussed above – often simultaneously work system-atically to marginalise and exclude others from access to even basic services (Swyngedouw, 1993).

As the UK think-tank Demos put it, 'the poverty that matters is not so much material poverty, but rather a poverty of connections' (1997, 6). Such a 'poverty of connections' limits a person or group's ability to extend their influence in time and space, often condemning them to local, place-based ties and relationships. It prevents them from connecting socially and technologically with the premium networked spaces of the modern metropolis (which tend, as we have seen, to be actively distancing themselves from the wider urban realm). It adds major transactional and logistical burdens to the basic tasks of daily life like washing, cooking, securing food to eat, moving around the city, communicating and securing a decent place to live (Speak and Graham, 1999). And it works against people sustaining relations with the people and institutions that may help them to access services, markets, knowledge, skills, resources and employment opportunities.

Such a situation is especially problematic as many cities, or at least the premium network spaces of those cities, become remodelled largely as landscapes of consumption, entertainment, commodified pleasure and market-based housing, configured precisely through electronic financial systems, credit registers, geodemographics and transport, security, urban design and state practices for those who can pay. For, as Paul Knox suggests, 'If consumption has become the means by which people give existential meaning to their lives, it follows that low-income households do not exist! Unable to participate in the existentialism of the market place, they have become invisible to retailers, advertisers and, indeed, the rest of society' (1993a, 27). Marginalised spaces have thus become 'spatial gaps in our metropolitan narrations' as we fail to look beyond the 'well designed nodes of the [urban] matrix' analysed above, at the 'blank, in-between spaces of nobody's concern' (Boyer, 1992, 118). Such spaces often provide homes for countercultural movements of many sorts where lots and buildings can be appropriated relatively cheaply and beyond the intense disciplinary practices of premium network spaces (Doron, 2000).

THE RESTRUCTURING OF PUBLIC AND WELFARE SERVICES

To this context we must add the restructuring processes that are under way in systems of social welfare provision, particularly in Western cities, as the ideals of universal access are replaced in many cities by marketisation, cost cutting, minimum safety nets and active labour market policies (Pinch, 1989, 1997). One can add the tendency of many urban labour markets to polarise between the advanced service and knowledge-oriented professionals at the 'top' and the routine service employees or unemployed at the 'bottom'. Finally, we should stress the reorganisation or withdrawal of many social and public housing programmes or rent control systems under the impact of wider welfare restructuring and neoliberal attacks (Musterd and Ostendorf, 1998). Combined together, such processes of urban marginalisation become even more debilitating for the poor (Sassen, 1991, 2000b).

In their isolation in 'network ghettoes', marginalised groups often remain heavily dependent on whatever public, welfare and social services exist (see Thrift, 1995). But, under the influence of wider shifts away from universal, Keynesian welfare regimes to neoliberal, individualist ones, these, in turn, may be 'reconverted into instruments of surveillance and policing of a surplus population that is housed in the degraded enclaves where it has been relegated' (Boyer, 1996, 132). Such 'public institutions often thus tend to accentuate the isolation and stigmatization of their users, to the point where they effect a *de facto* secession of the ghetto from broader society' (ibid.). This is especially so for those social groups that 'appear to challenge the image of the clean citadel plazas' constructed in the revitalised urban cores and consumption districts: squatters, beggars, the homeless, 'squeegee merchants', mentally ill people and illegal street traders (Mayer, 1998, 68).

In the United States, for example, a widening lexicon of special districts provide a parallel institutional infrastructure of 'geographical patchworks' to the Business Improvement, common interest and Transport Improvement Districts that are, as we have seen, supporting the emergence of premium networked spaces (see Mitchell, 1997). Here land use regulations and criminal and civil controls merge in attempts to contain the social problems caused by urban spatial polarisation within formalised 'antigraffiti districts', 'antired light zones', 'homeless containment zones' and 'drug control zones' (Boyer, 1996).

THE FORTRESSING OF MARGINALISED SPACES

All these processes, of course, tend to work together and interact in complex ways. In extreme cases, as with the US ghettoes of the Bronx or south Chicago, marginalised spaces can effectively secede from the wider urban fabric, supported, as powerfully as any mall, theme park or skywalk city, by combined marginalisation trends in urban design, financial restructuring, technological innovation and institutional practices. In US cities, suggests Boyer, 'marginal [people] are left outside the protected zone of the shopping mall, the campus, the walled community' whilst also being excluded from 'the Internet and the credit card/ATM system. . . . It is these outsiders that haunt and invade the interior' (1997, 6).

In many cases these 'outsiders', like the fortified and secessionary enclaves of the rich and powerful, are also 'forting up' (Blakeley and Snyder, 1997a, b). They are adding their own secessionary forces to the splintering of the metropolis (Vergara, 1995, 1997). In US ghettoes, European 'inner cities' and the *barrios* of Southern megacities alike, road closures, fortification, CCTV and security practices are increasingly being mobilised as people, households and institutions try and insulate themselves from the drugs, crime and dangers that often inhabit their 'public spaces' (see McGrail, 1999).

FOUR CASES

Within the space constraints here, we cannot hope to document the diverse social experiences of network ghettoes, addressing the full range of infrastructures and the full range of cities across the world. Instead, we aim in what remains of this chapter to provide a series of illustrative vignettes of the processes at work. We focus, in particular, on four cases: transport in the splintered metropolis; neighbourhood exclusion from information technologies; water poverty in developing cities; and two very different examples of disconnection from essential power and energy supplies drawn from the United Kingdom and Russia.

Unmoved: transport and the splintering metropolis

First, let us turn to the multiple ways in which the reconfiguration of transport networks supports the splintering of the metropolis. For, as David Hodge writes, 'if the *raison d'être* of cities is to provide the opportunity for interaction', it follows that 'few aspects of urban infrastructure match the importance of the provision of urban transportation' (1990, 95).

We have already encountered in this book several examples of how urban transport systems, particularly urban highways, are being configured in ways that underpin the splintering of the metropolis. Rising car use, for example, helps turn the street from a place of personal meeting to a place of automobile traffic, rendering it denuded of people (and often therefore 'dangerous' from the point of view of drivers) (Hamilton and Hoyle, 1999). As automobiles have, in turn, become the *de facto* mode of urban transport across the world, the production of streets – and then highways – becomes biased towards car owners. As in the peripheral new towns of Istanbul and every out-of-centre mall ever built, highways are used as intentional or unintentional *cordons sanitaires* which tend to limit access to premium networked spaces for poorer and less mobile communities and people. Highways connect some places but work to sever the communities and places they pass through. Highways, in addition, tend to support the polynuclear decentralisation of cities. This makes public transport less viable and useful. Somehow, 'collective transportation networks, designed for the dense city, must be adapted to the diluted periphery' (Burgel and Burgel, 1996, 316).

What, then, of the mobility of those at the other end of the scale from those who inhabit the secessionary network spaces and mobility systems of the splintering metropolis? What of those people – often disproportionately the old, the female, the poor, the disabled – without the automobile in the motorised, splintered metropolis? What of those unable to access airports (with their fast-track immigration systems for the powerful), airlines (with their dedicated video screens, customised services, in-flight Internet access, satellite phones, masseuses, manicurists and, from 2000 onwards, first-class double-bed compartments) or fast trains, glitzy new toll roads or customised rail and 'people-mover' links? What of those places where the glossy ideologies of globalisation implying the universal 'death of distance' seem little more than science fiction?

The key point, of course, is that speed and mobility, as always in human history, are relative (Hamilton and Hoyle, 1998, 13). In stark contrast to the portrayals of power, mobility and the ubiquitous transcendence of time and space barriers that saturate our culture, media and dominant ideologies in these times of 'globalisation', for many low-income people the social experience of life in marginalised places is one of being tightly *confined* by time and space barriers rather than being liberated from them (Mitchell, 1997). In such places 'the space of flows comes to a full stop. Time–space compression means time to spare and the space to go nowhere at all' (Thrift, 1995, 31). As Doreen Massey puts it, 'these areas and groups tend simply to be on the *receiving end* of time–space compression' (1993, 62). In Box 6.8 we explore the particular experience of one such group: the relatively large population of homeless people in US cities.

Unwired: people and places beyond the information technology 'revolution'

Which takes us to our second example – the experiences of places and people beyond the so-called IT 'revolution'. In the marginalised spaces of all cities – North and South, newly industrialising and post-communist – there are disadvantaged groups living in poverty and structural un- or underemployment who remain, and seem likely to remain, altogether excluded from electronic communications networks. In such places, poverty and unemployment mean that access to *any* electronic network, from the phone upwards, is financially or technologically problematic. In Box 6.9 we start by looking at those excluded from access to the most basic and essential entry point into the 'information society': the humble telephone. Box 6.10 explores the urban 'digital divides' that surround the Internet.

BOX 6.8 'ZERO TOLERANCE' AND THE 'ANNIHILATION' OF THE SPACES OF THE HOMELESS IN US CITIES

As in all cities, homeless people in the United States tend to be largely immobile and vulnerable, with access to limited forms of transport. In the wake of the spread of 'zero tolerance' and 'quality of life policing' initiatives, aimed explicitly at their removal from US city streets, homeless people face challenges to the few basic rights they have left: using their foot power and bodies to try and occupy the street space necessary for their very survival (see Wodiczko, 1999; Smith, 1993). Don

Mitchell believes that the harsh new 'zero tolerance' legal regimes of many US cities, brought in as their authorities attempt to 'sanitise' their urban landscapes for the construction of premium networked spaces, are effectively outlawing 'just those behaviours that poor people, and the homeless in particular, must do in the public spaces of the city' ('loitering', urinating, camping, sleeping on the street, just being in the street) (1997, 305).

EXAMPLES OF ANTI-HOMELESS INITIATIVES

Examples of such policies abound. One of over forty US municipalities to have deployed legal statutes specifically to repel homeless people to other spaces, Santa Ana, in California, has developed a package of policies designed, in the words of the mayor, explicitly 'to move all vagrants and their paraphernalia out' (quoted in Smith, 1999, 102). Other wealthy communities such as San Marino, also in California, have simply imposed user fees on 'public'

parks as a useful expedient to exclude, creating what Davis and Moctezuma term 'corporately privatized recreation space' (1997, 37). And in 1999 Cleveland Mayor Michael White, following the lead of Mayor Guiliani in New York City, sparked a major protest by simply directing the city's police to move the growing number of the city's homeless people beyond city limits, arguing that it would improve 'safety for shoppers'.

PERIPHERALISING HOMELESSNESS

Such practices, in effect, are efforts to 'annihilate the only spaces the homeless have left' (Mitchell, 1997, 305). Above all, they are driven by a cosmetic 'out of sight, out of mind' approach to the 'sanitisation' of the premium

street spaces of postindustrial, consumption-oriented city cores. Such strategies do little or nothing to address the root causes of poverty and homelessness. With the reconstruction of downtown Manhattan through Business

Improvement Districts and privatised corporate plazas, for example, New York's still growing population of the homeless has increasingly been pushed to the social margins as to its physical margins. Neil Smith points out that:

As the 1990s progressed, homeless people – driven from the spaces of the central city, ever more desperate and no longer commanding the headlines – were pushed to places offering the last secluded shelter on the urban margins: coastal scrublands, boardwalks, highway onramps in the outer boroughs, the fenced-in desolation around airports, or the wooded bluffs of the Palisades in New Jersey. The political geography of eviction became an outer-borough phenomenon – out of Manhattan, out of the news.

(1999, 101)

BOX 6.9 URBAN POLARISATION AND THE 'UNPHONED'

Even in the age of the Internet, it has been estimated, 60 per cent of people in the world have never made a phone call (Graham, 2000b). As the phone becomes the crucial electronic medium of most information flows, communications and transactions in the modern metropolis, those without access to one will face increasing relative marginalisation (Milne, 1990, 365). In a world where meaningful connection to services increasingly means negotiating an extending web of telephone help lines, call centres, direct welfare services, electronic transactions, phone shopping, mobile telecommunications and the withdrawal of human-staffed offices scattered across the physical spaces of cities, to be without even a basic phone becomes ever more disabling (see Loader, 1998). This is especially so given that, in many Western cities in particular, home access to the phone is now mistakenly assumed to be a universal entry point into the 'information society'. In the United States, for example, whilst 94 per cent of all households had a phone in 1994, only 64 per cent of black households earning less than $5,000 had (Schement *et al.*, 1997). In the United Kingdom, meanwhile, over 40 per cent of those in the lowest income bracket (less than £50 per week) had no phone in 1992, compared with only 1 per cent in the highest income bracket (over £500 per week) (Oftel, 1994).

In addition, within liberalised markets, infrastructure providers are unlikely to target new investment, marketing and innovation on marginalised spaces, which also tend to face stark exclusion from formal financial services, insurance, retailing investment and other utilities (Speak and Graham, 1999). As Hallgren, a US community activist, has found, 'many communities and areas are not perceived as having a large enough customer base to attract a [telecommunications or Internet] company to offer a service, let alone multiple companies' (1999, 1). As marketing and infrastructure development strategies tend to reflect ever more the patchwork geographies of the splintered metropolis, the situation of Bell Atlantic in New Jersey, reported in the *Cybertimes* in 1997, is typical. By then the company, the report argued, had rolled out high-capacity optic fibre links with 'suburban business parks and large corporations' as well as 'setting a schedule for suburban neighbourhoods'. But it had 'not yet made specific plans for the thousands of poor people who live in the state's largest cities'. Worse still, it had 'let its network deteriorate in parts of Brooklyn and the Bronx, where corroded wires led to scratchy lines and service outages' (Schiller, 1999a, 56). Physical offices, used by many poor people without bank accounts to pay bills, have also been routinely closed by US telecommunications firms, whilst charges for directory assistance and local calls have been dramatically hiked to reflect 'cost-reflective pricing' and the withdrawal of social cross-subsidies (ibid., 57).

For those without home access to a telephone, public phones on the street become crucial. But here, again, investment is concentrating on equipping the premium networked spaces of the city, at the expense of network ghettoes and marginal spaces. Telecommunications liberalisation, in particular, allows entrants to the pay phone market to 'cherry-pick' 'commercially attractive sites in airports, hospitals, shopping malls, train stations and other busy city centre locations' with new public phone boxes (Warwick, 1999, 44). This often leads to the progressive deterioration of public phone services away from the highly connected nodes of the splintering metropolis, in the spaces where, ironically, access is most important.

Often, in UK and North American cities, being 'unphoned' tends also to mean being 'unbanked', as financial restructuring pulls away the financial infrastructure of poor communities (Dyer, 1995). In places like south central Los Angeles, for example, only a couple of miles from the highly networked premium space of the downtown, access to branches and ATMs is in decline. This is forcing inhabitants 'back on the informal system of cheque-cashing services, mortgage brokers, credit unions and cash' (Thrift, 1995, 31) – a process which further splinters this place off from dominant circuits of exchange and transaction.

BOX 6.10 BRIDGING 'DIGITAL DIVIDES'? MARGINALISED URBAN SPACES AND THE INTERNET

As the Internet starts to parallel the phone as an increasingly pervasive urban medium, the prospects of enrolling marginalised people and places into 'cyberspace' remain severely problematic. The key concern here is the polarising logic with which the global and disembedded medium of the Internet relates to the cultures, civil societies and landscapes of local spaces, cities and regions. This relationship tends to be one of extending the power of the powerful whilst further marginalising the less powerful within the same geographical spaces – a logic of intense polarisation. As the UNDP suggest:

the Internet is creating parallel communications systems: one for those with income, education and – literally – connections, giving plentiful information at low costs and high speed; the other for those without connections, blocked by high barriers of time, cost and uncertainty and dependent on outdated information. With people in these two systems living and competing side-by-side, the advantages of connection are overpowering. The voices and concerns of people already living in human poverty – lacking incomes, education and access to public institutions – are being increasingly marginalised.

(1999, 32)

For those without access to global electronic consumption and communications systems there are few opportunities to enter the world of customised infrastructure and services – even if you are lucky enough to live in a part of a city that benefits from the roll-out of these services (Pahl, 1999). 'The reliance on cash, or even cheques, helps to condemn individuals to life at the bottom end of the spending scale and can help keep them there,' suggests Kruger (1997, 21). Moreover 'the approach of electronic cash (or e-cash)', supporting renewable smart cards and Internet and telephone-based 'spending', 'presents a possible scenario in which only those judged to be appropriate by banks, credit companies or the like would be in a position to fully engage in consumer society'. Such people seem likely to be left with the poor, expensive services of incumbent monopolies, or beyond the reach of infrastructure services at all.

This is so at precisely the time when being on-line is becoming ever more critical to access key resources, information, public services and employment opportunities. Goslee (1998) estimated that, by the year 2000 in the United States, '60 percent of jobs will require skills with technology. Moreover, 75 percent of all transactions between individuals and the government – including such services as delivery of food stamps, Social Security benefits, and Medicaid information – will take place electronically. People without technology skills or access to electronic communications will be at a considerable disadvantage.'

THE CHALLENGES OF SECURING ON-LINE ACCESS FOR MARGINALISED COMMUNITIES

In marginalised spaces, efforts to get lower-income groups on to the Internet will continually face difficult issues. At the very least there are likely to be competing priorities, costly training needs, crime problems, relatively low levels of literacy, issues of technological intimidation, the rapid obsolescence of technologies, the high cost of continually upgrading software and hardware to meet the latest industry standards, and often, especially in developing nations, simply paying for electricity (Sparrow and Vedantham, 1996).

However, the relevance of Internet access can often be questioned for those facing the most severe social crises. 'Just giving someone time at a terminal with Internet capabilities – or, by extension, at a a kiosk in a public place – will not benefit anyone who feels confronted with a seemingly insurmountable problem, or who has no idea where to begin' (Rockoff, 1996, 59). Jones argues that 'connection to the Internet does not inherently make a community, nor does it lead to any necessary exchanges of information, meaning and sense-making at all' (Jones, 1995, 12).

Even when marginalised spaces of cities do gain electronic access, telecom and media firms are tending to offer much less capable electronic infrastructures and services than those being bundled into the packaged urban spaces, configured for affluent socioeconomic groups. Moreover, services for lower-income groups are more likely to be configured largely for the passive consumption of corporate entertainment and services. As Calabrese and Borchert argue, from the point of view of the United States, the worry is that, as a result, 'wage earners, the precariously employed and the unemployed will interact infrequently on the horizontal dimension, except primarily in commercial modes which are institutionally and hierarchically structured, and controlled for commercial purposes, such as games and shopping' (1996, 253).

UNWATERED: DUALISED SYSTEMS OF WATER CONSUMPTION IN DEVELOPING CITIES

Our third example addresses the highly dualised systems of water provision developing in the 'megacities' of the Developing World. We saw above how infrastructure services in cities like Jakarta, São Paulo and Bombay tend to be oriented to the new middle-class enclaves that are being constructed on their peripheries or within their shanty districts. The inevitable side effect of this process is a perilous infrastructural situation for those millions of people living in informal settlements beyond the reach of the formal street, highway, power, water and communications infrastructures of the city. Such people, very often, provide the economic dynamism that makes megacities grow. As Aldrich and Sandu suggest, 'the people living in slums and squatter settlements actually subsidise the formal economy (and perhaps the world economy) by not requiring large amounts of capital for housing and infrastructure services' (quoted in Baird and Heintz, 1997, 13).

Whilst processes of infrastructural dualisation in developing cities are far from new, the shifts towards privatisation, liberalisation and 'structural adjustment' often render the position of informal settlements even more perilous than under the regimes of developmentalist states and the (highly biased) infrastructural monopolies that characterised the Developing World's experience of the modern ideal. In most developing cities, water utilities, with inadequate financial, technological or political resources to extend their water grids to cover burgeoning urban peripheries, have therefore, in a sense, faced their own individual collapse of the modern infrastructural ideal (see Box 3.7). In the face of international constraints on borrowing and strict debt repayment conditions from the IMF, prospects for major extension seem bleak. Remaining resources and network infrastructure tend to concentrate overwhelmingly on the needs of social and economic elites, spaces for foreign direct investors and the emerging modern consumer spaces that are being packaged out of parts of the metropolis (Jaglin, 1997).

As Lee points out, the risk is that, far from moving piped water and sewerage services towards ubiquity, the 'extent of actual coverage [of infrastructure services] may actually shrink after privatisation' as loss-making parts of networks are withdrawn to cut costs and improve profitability (1999, 153). In a series of case studies of waste, water and transport privatisations in Asian cities he found that 'the target groups were invariably middle and high income neighbourhoods. The fact that private firms will only service richer communities may actually make the job of many municipal authorities more difficult' (ibid.).

In South Africa, of course, such extreme infrastructural inequalities were explicitly configured by the apartheid system. The issue now, in a post-apartheid context, is how universal, essential, networked infrastructures can be rolled out to the majority black population in the townships. This must happen at the same time as finances are secured to ensure that economic development areas also have the 'world class' networked infrastructures seen to be necessary to bring in foreign investment. However, Patrick Bond argues that, under the neoliberal pressures of the IMF, the post-apartheid South African state actually risks engineering a 'neo-apartheid' infrastructure policy in the country. This is because the target standards for networking the townships are excessively low. They include 'pit latrines (not water-borne sanitation), yard taps (not inside the house), 5–8 A electricity supply (not 20 A or 60 A, as in formerly white areas), untarred roads, no stormwater drainage etc.' (1999, 45).

For basic water needs informal settlements beyond the limits of piped water usually have to rely either on highly inadequate stand pipes or, worse still, on private vendors selling water as a commodity door to door by truck (Swyngedouw, 1995b). By the year 2000, it was estimated, 450 million urban dwellers would be deprived of potable running water across the world (UNCHS, 1991). In Jakarta, for example – the city of gleaming new toll roads and (often half built) packaged landscapes for the rich explored earlier – over half the population still obtain their water from vendors, at as much as thirteen times the cost of the piped water used to irrigate the golf courses and landscaped gardens of the extending new town complexes (Yeung, 1997, 99). In the city of Onitsha, Nigeria, meanwhile, private sector vendors distribute twice as much water as the public pipe system but make twenty-four times as much revenue (Kessides, 1993b, 37; see Lee *et al.*, 1999).

The result, commonly, is rapidly deteriorating sanitary conditions in many informal settlements. As Eric Swyngedouw found in his analysis of Goayaquil in Ecuador, 'the absence of water, and the exclusionary practices through which the urban water supply is organised, tell the story of the urban deprivation, disempowerment and repressive social mechanisms that turn slum life into the antithesis of modern urban life' (1995b, 388). In Goayaquil 400 water tankers serve over 600,000 people. The water retailers buy the water from the municipal water company at a highly subsidised price (Su 70/m^3 in 1993); they then sell it on at rate of Su 6,000/m^3 or Su 7,500/m^3. 'The price they charge is up to 400 times higher than that paid by low-volume consumers who receive their water from the public utility' (Swyngedouw 1995b, 389; see Box 3.7).

Unwarmed: marginal energy spaces

Our final two examples demonstrate how contemporary trends help to push people and places who are socioeconomically marginal further away from the fuel and power resources they need to live.

The first example of this phenomenon is the recent experience by low-income people of utility liberalisation in the United Kingdom. We saw earlier in this chapter how newly entrepreneurial utilities in the highly competitive UK market are striving to 'snuggle up' to socioeconomic elites, who can choose all manner of new services from the comfort of their homes. But a very different experience of liberalisation is emerging at the other end of UK utility markets, as we see in Box 6.11. Here, in a classic case of the socially biased configuration of technology, very similar IT technology is being organised very differently effectively to *distance* suppliers from lower-income, and unprofitable, users, who often now have to travel considerable distances physically even to maintain their consumption.

BOX 6.11 THE DUAL CONFIGURATION OF 'SMART' ENERGY CONSUMPTION SYSTEMS: THE URBAN SOCIAL EFFECTS OF 'PREPAYMENT' METERING IN UK UTILITIES

For more marginal consumers of energy services in the United Kingdom, utilities install electronic 'prepayment' meters. Through these, consumers electronically 'top up' smart cards, at post offices, in advance of utility use. Such smart cards must be inserted into the meter to allow the use of the service to continue (Drakeford, 1995). Over 3 million UK households used prepayment meters for electricity by 1994; the figure was expected to grow to 6 million by 1998 (Marvin and Graham, 1994). By 1994 330,000 British Gas customers had prepayment meters. And prepayment metering is also being increasingly pushed by some water firms on to marginal water consumers. Over 15,000 water users even had compulsory prepayment meters by 1996, a practice that has since been outlawed.

Prepayment meters provide utilities with a convenient and simple solution to the difficulties raised by poor and marginal consumers. They reduce the high transaction costs and bad publicity associated with utility disconnection (Speak and Graham, 1999). The technology uses much the same embedded microchip technologies as 'smart home' systems and 'smart meters', but in very different ways and for very different ends. As Figure

6.17 shows, whilst smart meters for affluent users use IT capabilities to help establish dense and reciprocal relations between utility operators and consumers, prepayment smart cards allow utilities to withdraw from having any direct contact with poorer users at all.

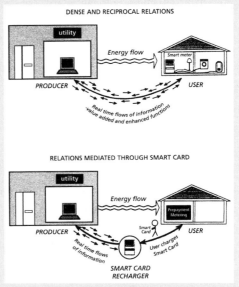

Figure 6.17 Contrasting configurations of more affluent utility users with smart meters (*upper*) and poorer users with prepayment meters (*lower*). *Source*: Marvin and Guy (1997), 126, 128

PROBLEMS OF PREPAYMENT METERING

Whilst prepayment meters may help low-income people to manage consumption, they also lead to three worrying problems.

First, users, often the most immobile people in society, must travel, often considerable distances, to 'top up' their tokens at recharge points.

Second, users of prepayment meters face higher charges. Users can be 'locked in' to expensive incumbent suppliers. They fail to benefit from the discounts accessible to those paying direct from bank accounts. And 'the expense of meters and other devices is passed directly on to the consumer in the electricity industry' through higher-than-normal tariffs (Drakeford, 1995, iv).

Third, prepayment meters effectively hide or disguise the issue of low levels of access to energy and water services, and the associated problems of water and fuel poverty, in marginalised urban communities. It is no surprise, for example, that the number of official gas disconnections fell dramatically from 0.4 per thousand consumers in 1985 to 0.1 per thousand in 1992. For, at precisely the same time, rate of prepayment meter use in the gas sector rose from 0.1 to 0.5 per thousand consumers (Ernst, 1994, 67). National water disconnections, meanwhile, fell from 10,047 in 1994/95 to 5,826 in 1995/96 at the same time as the number of water prepayment meters grew from a few hundred to 15,077.

'SELF-DISCONNECTION' AND THE PRIVATISATION OF SOCIAL COSTS

Two-thirds of households using prepayment water meters experienced 'self disconnection' during the first year but none of these 'self disconnections' was recorded in national disconnection statistics (which record only the physical disconnection of non-payers from networks). Despite the politically attractive decline in official disconnections, millions of households cannot afford to 'top up' their smart cards and access the fuel services they need to heat their homes and cook their food. In effect, prepayment metering has allowed a public, national policy issue to be quietly translated, through the use of consumerist rhetoric, into one based on the individual's or household's right to 'choose' which utilities to access when. As Mark Drakeford argues:

> the social costs of their decisions have been removed from the calculations of the commercialised utilities. These costs are then transferred elsewhere – either by 'privatising' them, too – by loading disadvantages on to those who, by accident of geography, ill-health or personal circumstances, incur additional commercial costs – or back to the public arena in terms of social services, housing and environmental problems.
>
> (Drakeford, 1995, 69)

Examples of the microsocial effects of prepayment metering are accumulating. In 1992, for example, two young girls died in a fire at a house in Essex because their single parent had 'self-disconnected' from the electricity network and was forced to use candles. Fire services in South Wales have reported being called out to fires caused by families who, unable to afford charging up their electricity prepayment meters, light a fire even when there is no fireplace. And, in the water sector, Harrison (1996, 3) reports that 'on some roads on estates in South Wales, and the south-west and the north-west [the areas with highest use of prepayment water metering], more than half the households have prepayment meters. People have to go round begging from neighbours for water for the kettle, the bath or the washing.'

continued

LONG-TERM IMPACT OF 'SOCIAL DUMPING' BY UTILITIES

By extrapolating current trends, it is no exaggeration to imagine whole urban zones which no longer fully benefit from social access to the distributive network grids painstakingly developed in the late nineteenth and twentieth centuries. Even the utility industries themselves have highlighted these likely eventualities. A utility executive has written that:

> When, in the new competitive market for electricity and gas, all the best consumers have been cherry picked, who supplies the won't-payers on Income Support on a

council estate in [the north-east industrial town of] Middlesbrough? Not a lot of utility managers are going to want to retain those customers. They won't want to buy added-value goods and services, and their addresses on mailing lists won't be worth much when you sell them on. Neither will they respond very well to direct mail. Everyone, unsurprisingly, is going to be after the 'ABC1s' [the professional and skilled social classes].

(Garrett, 1995, 16)

The collapse since 1989 of many of those paragons of the communist version of the modern infrastructural ideal – the rigidly standardised power and municipal heating systems developed in the cities of Russia and Eastern Europe – provides our second example of marginalised network spaces of energy and power (see Box 6.12).

BOX 6.12 THE 'CRASHING DOWN OF TECHNOLOGICAL SYSTEMS': POST-COMMUNIST RESTRUCTURING AND THE COLLAPSE OF ENERGY SYSTEMS IN RUSSIA

The withdrawal of maintenance and investment, growing corruption among state elites, and the incursions of criminal gangs and mafia operations have all led to the fragmentation and collapse of many power and energy systems in post-communist Russia (see Castells, 1998). The social, economic and ecological impacts of such collapse have been dramatic and immediate. In January 2000, for example, it was revealed that, since 1989, over 100,000 people have left the Russian city of Murmansk on the Arctic Circle, partly because the turmoil surrounding the collapse of the Soviet Union has led to the effective disintegration of the

centralised municipal heating systems that the citizens relied on (Bowcott, 2000, 3).

What is left of Russia's power and communications networks is also rapidly being undermined, as they are stripped by looters and sold as scrap on the black market by criminal gangs and people in desperate poverty. More than 15,000 miles of power lines were pulled down between 1998 and 2000 alone, yielding 2,000 tons of high-quality aluminium and even more copper. This haul, mostly exported to gain hard currency, was worth more than US$40 million. As a sign of the desperate poverty that afflicts so many

Russians, over 500 people died of electrocution in 1999 alone attempting such theft in what has been termed the 'Copper Rush' in some parts of Russia. Such theft is especially common in the coal-mining region of the Kuznetsk basin, where all the mines have recently shut down owing to World Bank-induced reorganisation of the traditional industries.

As a result of these collapses, large swathes of Russia have been plunged into power outages for weeks or months at a time in what the mayor of the town of Kiselevsk called the 'crashing down of the whole technological system' (quoted in Tyler, 2000, A10). In response, many of the newly affluent enclaves housing those who have prospered in Russia's predatory new economy, along with the growing number of foreign firms and nationals, have started to benefit from the construction of private and customised energy, telecommunications and water systems that are built for their spaces only (see, for example, Berlage, 1997). Thus a developing country model is emerging of protected, fortified and heavily networked enclaves, surrounded by spaces of infrastructural collapse and jarring poverty.

CONCLUSIONS
SPLINTERING URBANISM AND THE SPACES OF DEMOCRACY

In this wide-ranging chapter we have sought to trace the parallel social worlds of the splintering metropolis. Our broad coverage has allowed us to explore how the production of premium networked spaces is being combined with the configuration of infrastructure networks to support the urban splintering process.

For those enrolled into the premium networked spaces of the splintering metropolis, unbounded bounties seem to await. Seductive spaces of domesticity, leisure, consumption, travel and work jostle for attention. These are partitioned off from spaces of (perceived) danger, difference and poverty whilst being ever more seamlessly linked into the customised transport, energy, water and communications that allow users to extend the action spaces to distant elsewheres. Beyond the reach of these networks, however, in the places abandoned by the modern infrastructural ideal, there are worlds of intense localisation and largely invisible confinement and exclusion, where participation in the benefits of modern networked urbanism is ever more problematic.

This chapter has shown how geographical barriers, network configurations, software codes, sociotechnical assemblies of built spaces and built networks, and the new access control capabilities of electronic technologies, are increasingly configured to try and sever these two domains. The practices of designers, developers, operators and infrastructure firms seem intent on increasingly hermetic separation as the relatively open channels of flow and interconnection laid out under the modern infrastructural ideal close up and are patchily packaged into the emerging premium spaces of the metropolitan fabric.

Premium networked spaces anchor a city's formal participation in the globalising capitalist economy. They dictate the city's dominant 'image' and its representation in the media. They increasingly stipulate the experience of place for most visitors, travellers and those residents who are enrolled into the hermetic process. And they represent a wholehearted search by the rich and privileged for solutions to the anxiety, uncertainty and mistrust they feel when confronted by the exploding metropolitan fabric of many cities across the world.

Clearly, these new urban forms have major implications for the democratic possibilities of the city. They seem to signal the collapse of the coordinated public enterprise of interlinked infrastructural monopolies and comprehensive 'public' city planning. They mean the effective abandonment of the (always problematic) ideal of the cohesive, integrated and open city that can be characterised as having some organic unity. We are clearly losing the 'ideal of the city as a special place: the center of democratic exchange and a place where every person can do well and even expect certain basic social services' (Boyer, 1997, 82). Highly uneven commodified competition of the production of both networks and spaces becomes the single dominant ethos of the city; increasingly you are what the market dictates for you.

The urban forms of the splintering metropolis thus tend to undermine the principles of free openness and circulation 'which have been among the most significant organising values of modern cities' (Caldeira, 1994, 314). They utilise modernist design principles, along with new technologies, postmodern theming, high-tech security and customised transport, energy and communications, to articulate a new urban vision. This is based on sealing, closure, privatism and internalisation rather than on openness and free circulation. Arguably, in some cities, private enclaves and premium network spaces are becoming 'so pervasive that they effectively starve the life of public space in metropolitan regions' (Luymes, 1997, 201).

Above all, though, premium networked spaces and their customised arrays of technology and infrastructure threaten to feed off each other in a positive feedback spiral of interlinked secession (Graham, 1998b). The experience of urban life for the socioeconomically affluent increasingly becomes an interlinked, cosseted choreography where the networked interconnections of mobile phone, Internet, satellite television, electronic highway, air-conditioned car, parking garage, airport, airliner and glocal bypass rail link become ever more seamlessly fused into the rebundled plazas, atria, malls, resorts, gated communities and business parks that they increasingly orient towards. Overseeing all is an increasingly interlinked array of social and technological practices supporting surveillance, control, social purification, and allaying the ambient fear that pervades contemporary cities. As Mike Davis suggests, from the point of view of Los Angeles:

inevitably the workplace and shopping mall video camera will become linked with home security systems, personal 'panic buttons', car alarms, cellular phones and the like, in seamless continuity of surveillance over daily routine. Indeed yuppies' lifestyles soon may be defined by the ability to find *electronic guardian angels* to watch over them.

(1992, 5, original emphasis)

Finally, practices of urban splintering, and the urban landscapes which result, serve to undermine established notions of social and spatial redistribution across the metropolis. More and more effort goes into simply making the poor and marginalised people and spaces of the metropolis less and less visible (and threatening) to its interlinked constellation of premium networked spaces. This explains why the contemporary sense of the urban in many cities entails the experience of being geographically close to other social worlds whilst feeling technologically distanced from them. As Fred Dewey suggests, again, for the case of Los Angeles:

In Los Angeles, heavily secured and encased environments can be found a block away from intensely vulnerable areas; sophisticated communications technologies rise amid communities without telephones; trains for edge city 'information age' riders zip by overcrowded, filthy, and virtually unsubsidized buses for the laboring poor.

(1997, 270)

It follows that the key challenge is to imagine and construct an urban politics and a spatial imaginary for resisting the extremes of uneven development within the splintering metropolis. It is to this challenge that we turn in the last chapter, where we will explore the limits of splintering urbanism and begin to imagine how splintering cities might be rendered more democratic and equitable. In the next chapter, however, we complete Part Two of the book by turning our attention to the relationship between splintering urban economies and constructions of 'glocal' urban infrastructure.

7 'GLOCAL' INFRASTRUCTURE AND THE SPLINTERING OF URBAN ECONOMIES

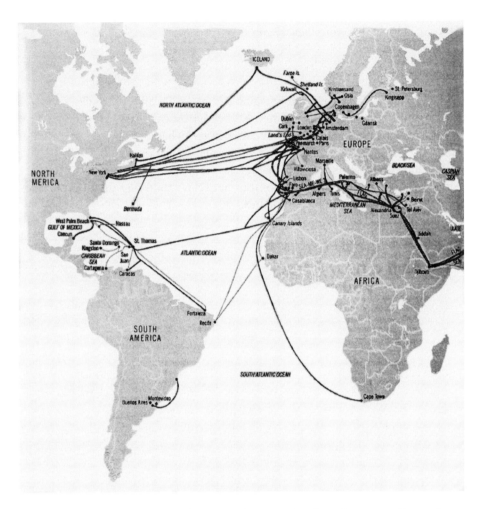

Plate 13 'Glocal' transoceanic optic fibre networks connecting the major metropolitan regions of Europe, the east coast of the Americas and Africa, 1997. *Source*: Vedel (1997), 34

National borders have ceased being continuous lines
on the earth's surface and [have] become nonrelated
sets of lines and points situated within each country.
(Andreu, 1997, 58)

CITIES AND THE 'ARCHIPELAGO ECONOMY'

It has been widely argued that, as urban economies integrate internationally, they are, in a sense, 'disintegrating' (Lovering, 1988, 150). In the old industrial cities of the North, for example, the tight, local interdependence between production units that characterised the earliest phase of industrialisation has, in many cases, largely unravelled. It has been replaced by an often largely disconnected series of economic and corporate spaces and spheres, many of which are increasingly oriented towards powerful connections elsewhere. Accelerated concentrations of growing industries in dynamic metropolitan zones contrast increasingly starkly with bypassed intervening spaces. As Pierre Veltz suggests, 'one increasingly has the impression of an "archipelago economy" in which horizontal, frequently transnational, relations increasingly outmatch traditional vertical relations with the [city's] hinterland' (2000, 33). Speaking about Northern industrial cities, particularly those in the United Kingdom, John Lovering uses a rather different metaphor. To him:

> If the local economy of the 'Old Model' was a skeleton in which each part was connected to all the others, under the new post-Fordist model it is more like a pile of bones. The bigger cities and towns are now centres of administration rather than production. The smaller ones are centres for a whole set of unrelated production activities . . . The 'local economy' is now a thing of fragments.
>
> (1988, 150)

Such fragmentation, according to Manuel Castells, is a tendency in virtually all contemporary city economies as they become enmeshed in what he calls the 'variable geometry' of the internationalising 'network society' (1996, 145–7). Within this logic – which tends to transcend traditional notions of scale and separation, 'core' and 'periphery', 'North' or 'South' – technological and economic integration is taking place in virtually all cities, but in extremely partial, uneven and diverse ways (see Sassen, 1991, 2000a, b). A logic of intense geographical differentiation is under way, within which people and places are enrolled in very different ways into the broadening circuits of economic and technological exchange (see Veltz, 1996).

The clear worry here, as the French communications scholar Armand Mattelart has written, is that 'the dynamic of the economic model of globalization now unfolding risks leading to a "ghettoized" world organized around a few megacities in the North, but occasionally in the South, called on to serve as the nerve centres of worldwide markets and flow' (1996, 304). Ricardo Petrella (1993), an ex-EU commissioner, is even more pessimistic. To him, current logics, based on the centralisation of wealth and power on key cities in the technological cores of the global economy, risk little less than a 'new Hanseatic phase in the world economy' riddled with a 'stark techno-apartheid'.

THE 'STICKY' PLACES OF GLOBAL CAPITALISM; GLOBAL AND SECOND-TIER CITIES

Very broadly, those global and second-tier cities, parts of cities, and the socioeconomic groups involved in producing high value-added goods, services and knowledge outputs, are tending to become intensively interconnected internationally (and sometimes even globally) (Sassen, 2000a, b; Markusen *et al.*, 1999). Using the capabilities of high quality information, transport, power and water infrastructures, zones of intense international articulation – business spaces, new industrial spaces, corporate zones, airports, new cultural or entertainment zones, logistics areas – are emerging in such cities, albeit to highly varying degrees.

Within some such 'sticky' spaces (Markusen, 1999; Markusen *et al.*, 1999) – global financial capitals like Manhattan or the City of London, 'high-tech' industrial districts like Silicon Valley, Cambridge, Seattle or Bangalore, government complexes like Washington DC, cultural production centres like Hollywood (Scott, 1997) or the emerging digital innovation clusters like New York's Silicon Alley – a tight degree of interaction on the 'industrial district' model may survive and prosper. In such places, flexible, continuous and high value-added innovation continues to require intense face-to-face learning and co-location in (the right) place, over extended periods of time (see Storper, 1997; Veltz, 1996; Markusen, 1999).

Ash Amin and Nigel Thrift (1992) have termed this the logic of 'neo-Marshallian nodes on global networks'. In their view, highly valued local production systems like the City of London or the 'Third Italy' manage to maintain their competitive advantage within the broader shift to pervasive, dominant, corporate networks. They argue, however, that it is extremely difficult, if not impossible, to generate 'artificially' such self-sustaining international economic nodes if the basic structures, production conditions and institutional cultures that make them grow organically are not already in place. To them it follows that 'the majority of localities may need to abandon the illusion of the possibility of self-sustaining growth and accept the constraints laid down by the process of increasingly globally integrated industrial development and growth' (ibid., 585).

ROUTINE PRODUCTION, SERVICE AND EXTRACTION CENTRES AND THE COMPETITIVE SCRAMBLE FOR INVESTMENT

A second layer of spaces in developed, developing or newly industrialising countries is also able to attain some sort of economic position within circuits of internationally integrated industrial and economic development. These spaces may be global nodes for the production of high-volume manufacturing goods and services; places that can deliver routine services on-line or via telephone links to the core city regions; or sources for the extraction and production of various types of raw materials. Here we see the familiar scramble of entrepreneurial and increasingly internationally oriented localities for foreign direct investment (FDI) in routine manufacturing, mobile services and resource extraction.

Because of their overwhelming external orientation, economic and technological connections elsewhere in many of these spaces – for example, the burgeoning clusters of call centres and back offices in North American, Caribbean and European cities – now tend to far outweigh connections with the local 'host' space.

The economic development of such urban spaces, and the splintering of infrastructure networks that reflects and supports their development, therefore bring new tensions between favoured parts of cities and their wider metropolitan areas. Customised spaces, linked into splintered infrastructure networks, increasingly tie global production chains and *filières* together, in telecommunications, transport and logistics, and even energy and water. New patterns of 'hubs', 'spokes' and 'tunnel effects' are emerging as infrastructure networks link up 'cherry-picked', favoured spaces across widening territories, whilst excluding and bypassing intervening spaces deemed to be less profitable. In fact, it could be argued that, in the context of regional trading blocs, global capital freedom and the growing dominance of transnational corporations (TNCs), such infrastructural chains, tying together corporate *filières*, made up of customised urban spaces, effectively *constitute* the dominant spaces and practices of the global economy.

SUBORDINATE AND BYPASSED TERRITORIES

Manuel Castells argues that the 'territories surrounding these nodes play an increasingly subordinate function' (1996, 380). Indeed, in some extreme cases, what he calls the 'redundant producers, reduced to devalued labour' (ibid., 147) that inhabit such spaces may become little more than 'irrelevant or even dysfunctional' as the labour or assets they possess are ignored or bypassed by the logics of the 'network society'. Neil Brenner has observed, for example, that 'world cities like London can become "delinked" from declining cities and regions' (1998a, 444; see Deas and Ward, 2000).

Complex patterns of relations emerge here. As the global financial networks linking London, Paris and New York, or the *train à grand vitesse* (TGV) rail networks connecting Paris and the French provincial capitals demonstrate, the infrastructure networks that support distant linkages, whilst always local and always embedded in space and place, may actually provide 'tunnel effects' which bring valued spaces and places closer 'together' whilst simultaneously pushing physically adjacent areas further 'apart' (Graham and Marvin, 1996). The global divisions of labour and telecommunications networks of transnational corporations provide another perfect example. For, as Paul Adam states, 'in this milieu of globalization, the buildings housing the various functions of a transnational corporation, although dispersed around the globe, are intimately connected, yet they may have little or no connection with offices or housing that are directly adjacent' (Adams, 1995, 277).

INTENSIFYING UNEVEN DEVELOPMENT:
WHY INTRAREGIONAL DIFFERENCES ARE STARTING
TO MATCH OR EXCEED INTERREGIONAL ONES

The key result of these trends is that all cities, whether they be 'global' cities like London and New York, 'mega' cities facing 'structural adjustment' policies in the Developing World, cities in post-communist Eastern Europe that are 'opening up' to foreign capital, or others, seem to be facing variations of the same broad logics of development. Everywhere, it seems, '*intra*-regional differentiations are often bigger than *inter*-regional ones' (Keil and Ronnenberg, 1994, 143).

Patterns of intensely developed and interconnected nodes are thus emerging which are increasingly attempting to secure themselves off from surrounding spaces of marginalisation and bypassed exclusion. This is not to deny, of course, the stark differences which exist between the situations faced by different cities. Each is embedded within a different economic, cultural, social and geopolitical context and history. The marginal spaces in cities like Tokyo tend to be much less extreme than those of, say, Johannesburg.

Of course, patterns of intense disconnection between internationally networked urban spaces and surrounding neighbourhoods are not new, either. As we saw in Chapter 2, they have, in particular, long been characteristic of developing and colonial cities, and the highly segregated cities of the United States. So it is very important to re-emphasise that we are not claiming complete convergence between contemporary cities. Nor do we claim some clean break in a binary transition model between more integrated and less integrated cities everywhere in the world. Rather, we suggest that the intersections between globalisation, liberalisation, new technologies and infrastructural practices have crucial implications for the development of urban economies in developed, developing, newly industrialising and post-communist cities alike. As a result, whilst major variations continue to differentiate individual cases, virtually all cities are starting to display intensifying unevenness based on the partial integration of their most valued elements towards global circuits of economic exchange, whilst their more peripheral and informal economic spheres face increased marginalisation at the very same time (see Hoogvelt, 1997).

THE AIMS OF THIS CHAPTER

In this chapter we seek to demonstrate how the construction of unbundled, 'glocal' infrastructures is intimately bound up with the splintering of urban economies in a wide variety of contexts. We analyse this question by undertaking an analytical journey across the most privileged and valued spaces and times of the splintering metropolitan economic landscapes of a wide range of cities.

The chapter has three parts. First, we explore the ways in which nation states, entrepreneurial urban agencies, infrastructural capital and corporate firms are all working to support the construction of 'glocal' urban infrastructures.

Second, we take a selective tour of those places that are emerging as highly valued and intensively connected 'glocally' within the splintering city. These are the spaces that are benefiting most from the erosion of the modern infrastructural ideal around the world, as they are being equipped with highly capable infrastructures on an increasingly private and self-contained basis.

We encounter seven such types of place on this journey: the enclaves in dominant 'global' financial service cities like London and New York; development enclaves in 'megacities' in the Developing World; emerging urban enclaves of innovation in multimedia; new industrial spaces for 'high-tech' innovation and production; spaces configured for inward investment in manufacturing; 'back office' enclaves for data processing and call centres; and, finally, spaces customised as logistics zones (airports, ports, export processing zones and e-commerce spaces). In each category we explore a range of current examples to analyse precisely how the production of dedicated spaces for these valued economic activities is bound up with the customisation of infrastructure networks that allow them to extend their influence internationally whilst carefully filtering the degree to which they connect with their host city.

We round off the chapter by looking once again beyond the favoured worlds of glocal infrastructure at those space–times of the urban economy which seem to be facing marginalisation from infrastructural connection and investment. Here we explore the economic fortunes of the urban 'peripheries' that are facing infrastructural and sociotechnical disconnection from the favoured 'glocal' spaces of the metropolis. In developed cities such spaces were the main beneficiaries of the cross-subsidies and universal service obligations that were inherent in the modern infrastructural ideal. In developing cities they are often the burgeoning unserviced, informally constructed economic spaces on the fringe of the metropolitan core. Here, too, we argue, networks are splintering, but for different reasons. In such spaces, micro-level entrepreneurship is emerging to try and address the failings of the infrastructural legacy of the modern ideal. In other words, rather than wait to be equipped with modern 'glocal' infrastructure, people and firms are trying to secure essential infrastructure themselves.

ECONOMIC PLAYERS IN THE CONSTRUCTION OF 'GLOCAL' URBAN INFRASTRUCTURE

A central argument of this book is that the broad logics of 'unbundling' urban infrastructure are working to provide the crucial material and sociotechnical underpinnings to these wider processes of urban splintering. Four key supports for this in the economic arena warrant further analysis here: the changing roles of nation states, urban municipalities, infrastructure capital and corporate capital.

NATION STATES

First, nation states in the developed, developing and post-communist worlds have largely abandoned the project of the modern infrastructural ideal with its ostensible goal of 'equalising life conditions on a national scale' (Brenner, 1998a, 445). Instead, they have tended to shift to 'the promotion of urban regions as the most essential level of policy implementation' (ibid.). Nation states have thus 'substantially rescaled their internal institutional hierarchies in order to play increasingly entrepreneurial roles in producing geographic infrastructures for a new round of capitalist accumulation' (Brenner, 1998, 476).

As we have seen earlier in the book, this process has meant a widespread shift to privatisation, liberalisation, opening up public infrastructure monopolies to private investment and allowing private capital the freedom to develop limited, customised infrastructures in specific spaces, without worrying about the need to cross-subsidise networks in less favoured zones.

URBAN DEVELOPMENT AND PLANNING AGENCIES

Second, and relatedly, entrepreneurial urban economic development planning is everywhere emerging as the key imperative of urban governance (Clarke and Gaile, 1998). City authorities are struggling to project their cities, or at least favoured parts of them, into internationalising circuits of exchange. The latest innovations in urban economic development strategies are concentrating on 'integrating local economies into global markets' through the provision of infrastructure, the development of customised spaces for global capital, place marketing and the assistance of training and human capital development (Clarke and Gaile, 1998, 181). Within the context of the collapse of meaningful notions of comprehensive urban planning in many contexts, ambitious real estate packages and project-oriented infrastructure improvements form essential elements within the wider packaging of sites and places to be enrolled into the uneven logics of the 'network society' (along with the customary place marketing and financial inducements). Neil Brenner suggests that:

Today municipal governments . . . are directly embracing this goal [of mobilising territory] through a wide range of supply-side strategies that entail the demarcation, construction and promotion of strategic urban places for industrial development – for example, office centres, industrial parks, telematic networks, transport and shipping terminals, and various types of retail, entertainment and cultural facilities.

(1998a, 446)

Thus urban agencies, too, are helping to support the practice of building 'glocal scalar fixes' by configuring infrastructural and urban spaces to the precise needs of valued spaces within the metropolis. This is what Shearer calls the apparently pervasive 'edifice complex' within contemporary urban politics, which tends to 'equate progress with the construction of high-rise office towers, sports stadiums, convention centres, and cultural megapalaces, but often ignores the basic needs of most residents' (1989, 289). Such 'glocal' urban economic strategies entail configuring spaces and infrastructures to connect seamlessly with dominant international circuits of exchange. Special-purpose private or quasi-private infrastructure development bodies are an increasingly popular policy option here, as they can be tasked with equipping strategic economic spaces with high-quality infrastructure without facing onerous political challenges or the imperatives of cross-subsidies and territorial equalisation (Foster, 1996; Mallett, 1993a, Nunn, 1996).

At the same time, however, 'there appears to be a paradoxical tendency towards the enforcement of local boundaries' (Ezechieli, 1998, 3). Fine-grained economic segregation within virtually all cities is increasing (Hack, 1997). Roger Keil asks if 'the only counterforce to the convergence of global capital interests [is] the tribalist fragmentation of diverging communities: guarded and fenced off from one another, crammed in between the barriers of high-speed traffic and humming to the deafening sound of electronic highways?' (1994, 132).

INFRASTRUCTURE AND REAL ESTATE CAPITAL

Third, in response to global moves towards liberalisation and/or privatisation, infrastructure and real estate capital is itself increasingly withdrawing from rolling out general networks across cities and regions to focus on 'glocal' infrastructural articulations for strategically favoured places and users, largely within metropolitan areas (Crilley, 1993; Logan, 1993). This is a reaction to the growing demand for, and profitability of, seamless glocal links that transcend national and municipal boundaries, to tie in with the wider development logics of interconnecting valued spaces at the expense of less valued ones. It also reflects a shift away from notions of universal service and towards a greater proactivity among utilities and infrastructure operators in ensuring the most profitable economic development of the spaces they serve (Graham and Marvin, 1995).

It is more and more common, then, for infrastructure operators to act as 'growth statesmen' for the valued spaces that they serve (Logan and Molotch, 1987, 74). They are increasingly eager to become involved in the growth politics and policies of their host cities or localities (Guy *et al.*, 1996). As Cox and Mair put it, 'public infrastructure networks are highly capital-intensive, and realizing the values locked up in fixed gas lines, power stations etc. requires the reproduction of a particular spatial pattern of customers who will provide the infrastructure's value inputs' (1988, 2).

At the same time, though, operators are developing international portfolios of mergers, alliances and strategic acquisitions, and aspiring to become 'global network firms' in transport, telecommunications, power, water or, increasingly, combinations thereof (Rimmer, 1998). Such firms are able to offer international corporate clients a 'one-stop shop' service, for example with global logistics solutions, 'flat rate' telecoms tariffs to anywhere on earth or CCTV surveillance or a single energy package for multiple sites. Figure 7.1 provides Peter Rimmer's analysis of the 'glocal' infrastructural configurations that such global network firms are aspiring to provide.

To infrastructure capital, serving a range of valued locations across a wider area helps to 'reconstitute local dependence at some broader geographical scale' (Cox and Mair, 1988, 3). But the most important point for our purpose here is that the geographically *embedded* nature of infrastructure networks is inevitable and impossible to avoid. Ways thus have to be found to minimise the risks it entails, either through careful targeting of the most profitable markets within a territory, extending it to serve an international portfolio of profitable spaces or diversifying into less vulnerable markets and sectors.

CORPORATE CAPITAL

Finally, it is clear that corporate capital is increasingly intervening directly to encourage the production of the infrastructural network spaces that most suit its internationalising and 'glocal' needs (Schiller, 1999a). Lobbying of states and providers perceived to be inadequate in opening up restrictions on the provision of customised corporate infrastructure, or of those who are deemed to offer inadequate infrastructural price, quality or reliability, is increasingly

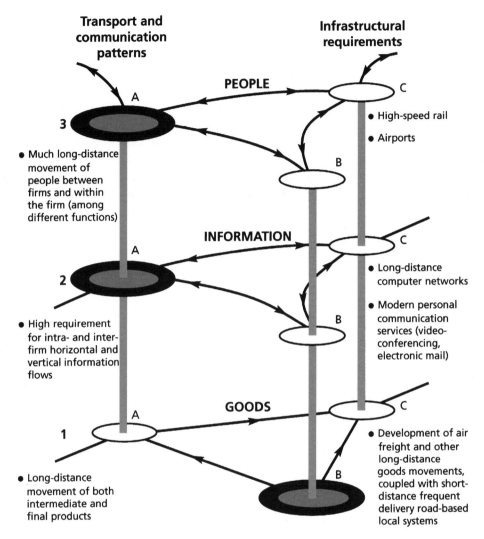

Transport and communication patterns

Infrastructural requirements

PEOPLE

A

C

3

• High-speed rail

• Airports

• Much long-distance movement of people between firms and within the firm (among different functions)

B

INFORMATION

A

C

2

• Long-distance computer networks

• High requirement for intra- and inter-firm horizontal and vertical information flows

B

• Modern personal communication services (video-conferencing, electronic mail)

GOODS

A

C

1

• Development of air freight and other long-distance goods movements, coupled with short-distance frequent delivery road-based local systems

• Long-distance movement of both intermediate and final products

B

Figure 7.1 Glocal infrastructure requirements of global network firms, focusing on transport and communications. *Source*: Rimmer (1998), 85

intense. Mobile corporations are also not slow to exploit the leverage they command to coerce entrepreneurial and ambitious municipalities and nation states to customise and configure infrastructural arrangements to their precise needs at little or no cost to them (Peck, 1996).

With this context in mind, we are in a position to explore the ways in which our seven chosen examples of premium infrastructural spaces are being 'globally' constructed in a range of cities across the world. The first example comes from the world's dominant 'global cities' (see Sassen, 1991, 2000b).

GLOBAL CONNECTIONS, LOCAL DISCONNECTIONS: CUSTOMISING INFRASTRUCTURE FOR GLOBAL FINANCIAL ENCLAVES

Any cursory examination of the dominant 'hub' positions of London, New York, Tokyo or Paris within their respective countries will quickly reveal that 'global' cities have long been central articulation points for all manner of networked infrastructures: rail, metro, water, power, airline, freight and telecommunications. But the combined processes of liberalisation, globalisation, technological change and the application of new urban design techniques are not only reinforcing the centrality of global city cores in global infrastructure networks. They are also, paradoxically, working carefully to secure the highly valued segments of global cities from their surrounding cityscapes. As Manuel Castells suggests:

the few nodal functions still located in central cities, around Central Business Districts (CBDs) and high quality urban spaces, can be bridged to national and global hinterlands via telecommunications, fast transportation and information systems, without needing to renovate their surrounding urban areas. Thus the central city's islands of prosperity and innovation can further isolate themselves from the city, whilst integrating into the space of flows and delinking themselves from their social and territorial environments.

(1999b, 31–2)

COMBINING GLOBAL CONNECTIONS AND LOCAL DISCONNECTIONS IN GLOBAL CITY CORES: THE CASE OF TELECOMMUNICATIONS

It is increasingly clear that the most highly valued spaces in global city cores are being provided with their own dedicated, high-quality infrastructural connections. These are configured to maximise the ease of connecting to other global city cores around the world. At the same time they are increasingly organised carefully to filter out unwanted connections with the surrounding metropolis – those that are judged to be 'threatening' or deemed to be irrelevant to the direct needs of the glocal enclave.

As we see in Box 7.1, the case of telecommunications presents perhaps the most potent illustration of how seamless connections can link powerful spaces and users 'glocally' with other powerful spaces and users, whilst helping them simultaneously to disconnect from the wider social and economic worlds of the surrounding metropolis. These processes, as Barney Warf suggests, show how telecommunications are being used to allow space to be 'stretched, deformed, or compressed according to changing economic and political imperatives' (1998, 225).

In global cities the most sophisticated, diverse and capable electronic infrastructures ever seen are being mobilised to compress space and time barriers in a veritable frenzy of network construction. Global city regions are heavily dominating investment in, and the use of, such

BOX 7.1 DEDICATED URBAN OPTIC FIBRE GRIDS AND THE COMPETITIVE STRUGGLE BETWEEN 'GLOBAL' CITIES

In the telecommunications field, the result of the combination of concentrated demand and customised infrastructure provision is the superimposition of many high-capacity optic fibre grids within the valued cores of global cities right across the world. The presence, or absence, of these networks, and the services which run on them, strongly defines the communications 'competitiveness' of global cities, an important consideration as they struggle to establish themselves as hubs of telecommunications traffic.

A survey by the Yankee Group, a US telecommunications consultancy, and *Communications Week International*, attempted to rank the competitiveness of telecommunications provision in early 1998 in twenty-five global cities encompassing 5 per cent of the world's population (see Finnie, 1998). Their scored rankings, shown in Table 7.1, were based on technical definitions of the pricing of services, the choice of physical infrastructure connections available, and the availability of the most advanced and sophisticated connections (for example, 'dark fibre', which is uncommitted to other users) and very broadband services.

Their results give a revealing portrait of the degree to which intense competition is focusing on the small number of global cities. Such cities concentrate particularly high demand, are located within the core geo-economic regions of the world, and are placed within nations that have enthusiastically embraced telecommunications liberalisation. The researchers concluded that 'cities large and small around the globe are integral to the fortunes of the world's economy, yet the [telecommunication] infrastructure in each can vary greatly. . . . Although the

gap between the best and worst of infrastructure is narrowing, particularly in the middle ground, it is still very wide' (Finnie, 1998, 20).

The five US cities included in the sample ranked highest and most competitive. New York led the way, with nine separate optic fibre infrastructures. London was the most 'competitive' city outside the United States, with six separate optic fibre grids. Cities that are experiencing a proliferation of urban fibre infrastructures, following liberalisation, came next (Stockholm, Paris, Sydney, Hong Kong, Frankfurt and Amsterdam – which has constructed its own municipally supported urban fibre ring called CityRing® in partnership with the Dutch PTT – see Figure 7.2).

Figure 7.2   The Amsterdam CityRing® initiative. *Source*: PTT Telecom Netherlands promotional brochure

Table 7.1 Ranked scores of global cities by the competitiveness of their telecommunications infrastructure, 1998

Rank city	Total score	Tariffs	Choice	Availability
1 New York	438	148	182	108
2 Chicago	428	154	166	108
3 Los Angeles	428	152	168	108
4 San Francisco/San Jose	425	149	168	108
5 Atlanta	409	141	160	108
6 London	391	131	161	99
7 Stockholm	386	129	149	108
8 Toronto	361	123	148	90
9 Paris	337	118	129	90
10 Sydney	331	123	118	90
11 Hong Kong	328	107	149	72
12 Frankfurt	321	78	135	108
13 Amsterdam	308	100	118	90
14 Tokyo	300	77	133	90
15 Brussels	294	97	107	90
16 Mexico City	283	93	118	72
17 Zurich	276	100	86	90
18 Milan	267	101	94	72
19 Kuala Lumpur	256	90	94	72
20 Tel Aviv	230	110	66	54
21 Singapore	206	108	44	54
22 Johannesburg	161	76	50	36
23 São Paulo	135	44	55	36
24 Moscow	134	26	72	36
25 Beijing	105	48	39	18
Maximum possible score	500	171	221	108

Source: Adapted from Finnie (1998), 21.

Note: Scores are based on a technical assessment of tariffs, choice of networks and availability of services.

The rest trailed further behind because of insufficient network competition, relatively high tariffs and lack of access to the most sophisticated services. Eleven of the twenty-five cities only had one optic fibre network, tying firms into sole, monopoly suppliers. Interestingly, though, the researchers believed that, such was the rate of the shift towards global archipelagoes of competitive global city optic fibre grids, all global cities would have 'at least five' optic fibre grids 'in the near future' (Fillion, 1996, 22).

Global cities in the 'Developing' World tended to be at the bottom of the table because of their nation states' general reluctance to privatise and/or liberalise their telecommunications regimes. The authors portrayed foreign-owned telecom infra-structures as the 'silver bullet' to such cities' lack of 'competitiveness', arguing that:

the 'poorer' cities in our survey – defined as such in terms of GDP *per capita* – trail far behind, victims by and large of local reluctance to allow competition. Of these five 'poorer' cities – Mexico City, Johannesburg, Beijing, São Paulo and Kuala Lumpur – only Mexico City makes a reasonable showing, mainly because it has been efficiently colonized by foreign-owned telecoms operators taking advantage of Mexico's liberal regulatory structure. The others still have a long way to go before they can join the global elite.

(Finnie, 1998, 22)

technologies (Graham and Marvin, 1996). A survey by the Yankee Group and *Communications Week International*, for example, found that around 55 per cent of all international private telecommunication circuits that terminate in the United Kingdom do so in London. About three-quarters of all advanced data traffic generated in France comes from within the Paris region (see Finnie, 1998).

But the 'wiring' of cities with the latest optic fibre networks is extremely uneven. It is characterised by a dynamic of stark dualisation. On the one hand, seamless and powerful global–local connections are being constructed by private communications operators within and between highly valued spaces of global cities – the downtown cores and newly constructed 'intelligent' corporate plazas and data processing areas (see Sassen, 2000b).

On the other hand, intervening spaces populated by poorer communities – even those which may geographically be cheek-by-jowl with the favoured zones within the same city – are often largely ignored by telecommunications investment plans. Such spaces threaten to emerge as 'network ghettoes' – places of low telecommunications access and social disadvantage. As with many contemporary urban trends, uneven global interconnection via advanced telecommunications becomes subtly combined with local disconnection in the production of urban space (see Amin and Graham, 1998). Moreover, such a situation seems likely to characterise developed countries (which are now fully liberalising telecommunications), developing and newly industrialised countries (which are increasingly liberalising telecommunications under structural adjustment pressures) and post-communist countries (where dedicated city networks are being built to bypass the obsolescent telecoms infrastructure left behind by communist regimes – see Berlage, 1997).

Last mile connectivity: the 'messy' material basis of the 'death of distance'

It is paradoxical, then, that an industry which endlessly proclaims the 'death of distance' actually remains driven by the old-fashioned geographical imperative of putting physical networks in trenches and conduits in the ground to promote market access. The greatest challenge of the multiplying telecommunications firms in global cities is what is termed the problem of the 'last mile': getting satellite installations, optic fibre 'drops' and whole networks through the expensive 'local loop'. In other words, the challenge is to thread fibre under the congested roads and pavements of the urban fabric, to the 'smart' buildings, dealer floors, headquarters, media complexes and stock exchanges that are the most lucrative target users.

Without the expensive laying of hardware in the financial and business districts of global city cores it is not possible to enter the market seriously and win lucrative contracts. Fully 80 per cent of the cost of a network is associated with this traditional, 'messy' business of getting it into the ground in congested, and contested, urban areas. There is a strong connection between the internal information infrastructures of the 'smart' buildings of global city cores – with their security, energy and communications management systems – and the global grids of fibre, satellite and transport infrastructure that link the buildings up across the planet.

Massive investment is planned to try and overcome the problem of the 'last mile' through the construction of fleets of 'flying base stations' which hover over major metropolitan cores twenty-four hours a day. Specially designed low-speed planes flying ten miles above the city, high-altitude airships and balloons, even dedicated geostationary satellites for major cities are all being planned to offer broadband connectivity over wireless links to the lucrative corporate markets in major city cores.

Connecting global city cores: integrating global archipelagoes of metropolitan fibre networks

Such is the pull of global city cores that they are strongly shaping the global geography of telecommunications investment. One of the world's fastest growing firms, for example, WorldCom (which incorporates MCI) is emerging as a global player by constructing dedicated fibre networks for 'global' city cores and few other places. This completely 'unbundled' solution avoids the costs of building networks to serve all but the most lucrative spaces. WorldCom have built over sixty fibre optic infrastructures in major city centres across the world, in carefully targeted, financially strong city centres (forty-five of them in the United States). A hundred and thirty WorldCom city grids are eventually planned – eighty-five in the United States, forty in Europe and the rest in Asia, Latin America and the Pacific. Each is carefully targeted on 'information-rich' global cities and parts of global cities which have a sufficient concentration of large corporate or government offices to ensure high levels of international revenue relative to miles of network constructed.

But WorldCom is also building the transoceanic and transcontinental fibre networks to tie the urban grids together into global archipelagoes – a global market which absorbed US$22 billion between 1988 and 1998 and which is expected to attract a further US$27 billion between 1998 and 2003 – largely on direct city-to-city global links (*Communications International*, July 1999, 47). As well as constructing a transatlantic fibre network known as Gemini between the centres of New York and London, WorldCom are building their own pan-European Ulysses network linking their city grids in Paris, London, Amsterdam, Brussels and major UK business cities beyond London. The strengthened importance of direct city-to-city connection is not lost on telecommunications commentators. As Finnie (1998) argues:

it should be no surprise . . . that when London-based Cable & Wireless PLC and WorldCom laid the Gemini transatlantic cable – which came into service in March 1998 – they ran the cable directly into London and New York, implicitly taking into account the fact that a high proportion of international traffic originates in cities. All previous cables terminated at the shoreline.

(Finnie, 1998, 20)

The increased 'filtering' of local connectivity: road pricing and 'rings of steel'

Thus the operations of global cities simultaneously 'reach out', extending their influence further across the globe via dedicated global fibre optic networks, whilst withdrawing into their ever larger, mixed-use corporate plazas. These 'electronic superbanks' are not skyscrapers but 'groundscrapers': 'huge nine-to-eleven-storey buildings with immense floor plates' to accommodate the remarkable IT needs of global financial institutions today (Pawley, 1997, 59).

Such processes are also supported by the growing shift towards filtering out 'unwanted' road traffic in the heart of global cities, either through police cordons and the electronic surveillance of car number plates (as in London – see Box 7.2), or electronic road pricing (as in Singapore). The Singapore scheme, which started in 1998, levies electronic tolls on car drivers commuting at peak periods into the core of the central business district (Soo, 1998; Seik, 2000). Obstensibly, this initiative is aimed at reducing traffic congestion. But the scheme, and others like it, also works as another form of local disconnection, as the toll mechanism filters out relatively 'cash-poor/time-rich' commuters, releasing space and improving the speed for wealthy 'cash-rich/time-poor' business commuters. Beneath the rhetoric that such road pricing is aimed at achieving environmental sustainability, the real objective is often therefore to create fast-flowing premium downtown road spaces as a boost to interurban competitiveness. Hong Kong, for example, is implementing a similar scheme to Singapore's, based on the fear that corporate head offices will select the uncluttered roads of central Singapore over Hong Kong's regular gridlock. High-profile cases of CEOs having to leave their air-conditioned limousines to walk the 'last mile' to meetings in searing heat and humidity are being explicitly used to justify the initiative (Khan, 2000).

BOX 7.2 GLOBAL CONNECTIONS AND LOCAL DISCONNECTIONS IN GLOBAL CITY CORES: THE CASE OF THE CITY OF LONDON

Few places exemplify how unparalleled global connectivity can be combined with highly selective local connectivity as well as the City of London. This space has the most powerful global telecommunications connectivity outside North America. Access to the world's airline networks is also exceptionally good, especially since the dedicated Heathrow Express rail link opened in 1998 connecting central London with Heathrow – the world's best connected international airport – non-stop in fifteen minutes. This link is due to be extended direct to the heart of the City early in the new century, further supporting the 'glocal' connectivity between the City and global airline networks.

GLOBAL CONNECTION: TELECOMMUNICATIONS

The overall telecommunications market for London was estimated in 1999 to be over £1,300 million, around the same as that for Paris and over four times that of Frankfurt (£253 million) (COLT communications, Web site, http://www.colttelecom.com/english/ corporate). As a result of global telecom firms scrambling to access this highly concentrated market, the City of London now has at least six overlaid fibre optic grids rolled out beneath the Square Mile and the rest of the main business areas of the City. They are operated by BT,

Mercury, City of London Telecommunications (COLT), WorldCom, Energis and Sohonet. Roads, canal pathways, old hydraulic power ducts, Underground railway tunnels, sewers and other utility pipes provide the conduits for this massive concentration of electronic infrastructure.

Increasingly, such urban networks link directly into transatlantic and international optic fibre grids, maximising the quality and reliability of transglobal connectivity. Detailed information on the urban geographies of these competing infrastructures is not easy to come by (Kellerman, 1993). But details are available of one of the networks – that operated by COLT (Figure 7.3). The geographies of the other five are unlikely to vary considerably. Figure 7.3 thus shows how dedicated fibre networks tend to be tightly focused, at least at first, on the central areas with the greatest concentration of communications-intensive activities. In the COLT network fibre is laid especially thickly in the City of London financial district. A broader grain of network coverage exists in the West End. An extension runs out to the new international business spaces in the Docklands.

Another of London's six optic fibre infrastructures has been developed since 1994 by WorldCom/MCI. This network has been particularly successful, providing a potent reminder of how powerful but geographically highly focused infrastructures in global cities can be at articulating large portions of the electronic flows of whole nations, even continents. With only 180 km of fibre constructed within the City, the London WorldCom network has already secured fully 20 per cent of the whole of the United Kingdom's international telecommunications traffic, which is, in turn, a good proportion of Europe's (Finnie 1998). WorldCom has been especially successful at building its own fibre networks across oceans and interurban corridors to link up its archipelago of global city networks. Direct and seamless glocal connections emerge which support the global interoperable operations of transnational finance and corporate capital whilst totally bypassing the old public phone systems laid out during the modern ideal. 'Bypassing incumbent carriers on both sides of the Atlantic, WorldCom's newly established transatlantic submarine cable facilities and

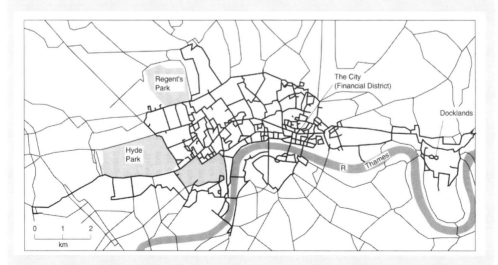

Figure 7.3 The optic fibre network in central London run by City of London Telecommunications.
Source: COLT Web site at http://www.colttelecom.com/english/corporate/mn_corp13.html

urban business networks will allow it to link directly some 4,000 business buildings in Europe with 27,000 such buildings in the United States' (Schiller, 1999a, 63).

LOCAL DISCONNECTION: THE 'RING OF STEEL'

At the same time as the City of London is being equipped to connect with ever greater power to (highly valued parts of) the world, however, it is also withdrawing from free-flowing and public local connections with the rest of London. This strategy 'guards the City [of London] so that it might continue to negotiate its path towards the increasingly cosmopolitan requirements of being a "global city"' (Jacobs, 1996). It is designed specifically so that the global financial core of the City of London can 'delineate its space and signal its exclusivity' as a centre of global, immaterial power (Power, 2000, 12).

This strategy applies especially to road connections and vehicular traffic. Since the two major IRA bombings in the early 1990s, the City Corporation has developed a strategic plan to protect what it calls 'the world's leading financial capital' by, effectively, erecting 'a modern version of the medieval Wall with security gates' (*The Times*, 27 April 1993, quoted in Jacobs, 1996). This is the so-called 'ring of steel' which carefully manages and scrutinises all incoming and outgoing vehicular traffic (Jacobs, 1996; Power, 2000). Electronic road blocks and armed guards now scrutinise every vehicle entering or leaving the City of London, as part of the corporation's efforts to 'make the City less vulnerable as an economic target' (Jacobs, 1996) (see Plate 14). As part of the process, entry points for vehicles have

Plate 14 The 'ring of steel' in the City of London. *Photograph*: Stephen Graham

been reduced from thirty to eight (Power, 2000). Car number plates are automatically recorded and a database has been created of all vehicles entering the area. Any vehicle not leaving the area after a specified time causes an alarm to ring, leading the suspect vehicle to be investigated.

ATTEMPTS TO DISCIPLINE URBAN BOUNDARIES ALGORITHMICALLY

More recently this computerised CCTV system has been upgraded so it can proactively search for any stolen vehicle reported in the United Kingdom. This takes four seconds between the car passing the CCTV camera and the computerised database being checked. In 1997–98 over 114,000 daily checks were made and 26 million checks were made against the national police computer for stolen vehicles (Power, 2000). Facial-recognition software has even been tested on the system. As Norris *et al.* argue, 'technology perfected during the Gulf War in 1991 has been utilised to track vehicles coming into the City of London and trigger an alarm when a car travels in the wrong direction on the one-way system' (1998, 8). By 1998 340 arrests and 359 stolen vehicles had been triggered by this proactive computerised scanning system.

In addition, an initiative called Camerawatch has been pursued, encouraging all private businesses in the City to install their own CCTV systems to monitor public areas of the City on a continuous basis. Over 90 per cent of the Square Mile is covered, involving 385 schemes and 1,280 cameras. A record of all the images captured by the cameras allows police to trace the movements of any suspected persons (Norris and Armstrong, 1999).

'BUNDLED' COMPLEXES AND SUPERBLOCK DEVELOPMENT

Finally, there is an architectural dimension to the selective local disconnection of global city cores from their immediate urban contexts. For, with the growing integration into enormous mixed-use urban redevelopment schemes like London's Broadgate and New York's Battery Park City (shown on p. 217), global cities are increasingly providing all the uses business executives need within single, bundled complexes or 'superblock' developments: state-of-the-art work space, upscale housing, retailing, schools, fitness centres, skating rinks, car parks, dedicated links to rail networks, etc. As Robert Reich observes in the US context:

Public funds have been applied in earnest to downtown 'revitalization' projects, entailing the construction of clusters of postmodern office buildings (replete with fibre optic cables, private branch exchanges, satellite dishes, and other state-of-the-art transmission and receiving equipment), multilevel parking garages, hotels with glass-enclosed atriums rising twenty storeys and higher, up-scale shopping plazas and gallerias, theaters, convention centers, and luxury condominiums. Ideally, these projects are entirely self-contained, with air-conditioned walkways linking residential, business, and recreation functions. The fortunate symbolic analyst is thankfully able to shop, work, and attend the theater without risking direct contact with the outside world – in particular, the other city.

(1992, 271)

COLONISING THE PERIPHERIES OF GLOBAL CITIES: CONFIGURING GLOBAL CONNECTIONS AND LOCAL DISCONNECTIONS FOR NEW FINANCIAL ENCLAVES

But the customisation of international links for highly valued parts of global financial capitals now extends far beyond the traditional central business district in the urban core. Increasingly, spaces are being redeveloped and configured for global financial services industries elsewhere in the metropolis (Crilley, 1993). In New York, for example, Longcore and Rees (1996) observe a 'doughnut' shape, with a restructured core remaining for headquarter functions and routine back offices and dealer floors moving to cheaper, more spacious locations further towards the urban periphery. 'As highly competitive major financial firms retreat to secretive, security-conscious structures and a building technology that stresses large horizontal over vertical spaces,' they write, 'the traditional tightly focused financial district and market has finally demonstrated geographical flexibility' (ibid., 368).

The development of new 'packaged' landscapes for decentralising financial services is particularly intense in the triumvirate of truly global financial centres: New York, London and Tokyo.

New York

In New York major new complexes have been constructed on the lower western tip of Manhattan (the World Financial Center at Battery Park City), and away from Manhattan, at Jersey City and Brooklyn, to accommodate the changing needs of financial services companies – especially for high-quality, lower-cost, relatively low-density space for headquarters and data processing functions. Each such 'smart building' is configured with new suites of infrastructure and high-security design and surveillance features, to secure them from the perceived risks of adjacent lower-income districts. Automated heating, cooling and humidity controls are tailored for the electronic equipment; back-up water tanks and air conditioning are provided. Three or four separate electricity grids are bundled together with emergency generators with at least three days of fuel. Building footplates are at least 40,000 m², to accommodate the needs of global financial institutions. And extremely generous conduits and spaces are provided for IT infrastructure – again with redundancy and several connections to the fibre networks of competing local providers (Longcore and Rees, 1996).

Across the Hudson river from Manhattan, in Jersey City, for example, public authorities have underwritten a major 6 million ft² complex of offices, elite condominiums, hotels, shops and a marina, to tempt major finance companies across the river from Wall Street. New rail, road, power and information infrastructures have been explicitly packaged to the needs of the complex. Merrill Lynch have moved a major back office facility there, as the site is only three and a half minutes from their Manhattan headquarters by commuter train (Longcore and Rees, 1996, 364). On the other side of Manhattan, in Brooklyn, meanwhile, at the new ten-block MetroTech development for corporate migrants from Manhattan, the utility Con Edison offer high-quality and individual utility connections to incoming companies.

On the one hand, all these developments exhibit a combination of highly regulated, policed and internalised 'public space' for corporate workers (with winter gardens, a marina and 'European' design features for the 30,000 people who work at the World Financial Center) (see p. 217). On the other, they are carefully removed from surrounding traditional streets. Instead, they articulate with integrated parking garages, skywalks linking them with other valued nodes, direct tunnels to transit systems, and malls (Crilley, 1993).

Tokyo

In Tokyo, a 1,100 acre artificial island known as Tokyo Teleport Town is the most obvious glocally connected reclaimed space (Obitsu and Nagase, 1998). The initiative is an attempt to construct an 'intelligent business centre', to 'prepare Tokyo to become a twenty-first century international metropolis for the future's advanced information oriented society' (Web site http://www.tokyo-teleport.co.jp/english/ttc/0-b.html). Centred around a massive twenty-four-storey dedicated satellite ground station complex, the site has its own highway network, light rail system, centrally controlled power and water infrastructures and, of course, a sophisticated suite of cable and telecommunications networks. 'The whole complex is a "smart building" with a fully integrated electronic facility management system relating energy supply, security systems and computer networks' (Riewoldt, 1997, 44). Over 70,000 workers are expected to be employed in the area; 'the land is gradually filling with exhibition centres, hotels, and office buildings for broadcasters and communications-intensive businesses' (World Teleport Association, 1999).

The urban nexus between state-of-the-art telecommunications and real estate speculation is increasingly forging similar 'teleport town' style urban enclaves, fuelled by a roving band of teleport consultants, real estate speculators and the World Teleport Association (WTA). Such spaces are designed to 'attract transnational corporations, international financing, trade and other international business activities' (Kim and Cha, 1996, 541), and are being developed in such diverse locations as Seoul, Korea (ibid.), Osaka (the 'technoport' project), and Rio, Brazil (Amborski and Keare, 1998).

Even more grandiose than the Teleport Town island are the 'artificial platform cities' that are planned in Tokyo by the Obayashi real estate firm. These are 1 km^2 platforms raised 31m above the existing cityscape, supporting all necessary modern infrastructures and super-high-rise mixed-use buildings (see Figure 7.4). Taking the logic of the 'packaged city' to its logical extreme, an 800 m tall 'millennium tower', a 1 million m^2 'building city' 'with enough space to accommodate the entire central area of a large city' (ibid., 328) has also been suggested on a reclaimed space in Tokyo Bay.

London

In London, finally, the development of new packaged landscapes for the global financial services industries has been just as dramatic. As the Thatcher government in the United Kingdom sought to establish a wholly new space for global finance capital in the London Docklands in the 1980s it adopted what Shane calls a 'free-market, deregulated, hyper-developmental enclave' model of urban development (1995, 63). This fuelled intense speculative development, supported by major public subsidies and tax breaks. Later, following the bankruptcy of the main developers, the government realised that only an immense amount of both private and state-backed infrastructure (to the tune of $2.7 billion) could make the project work (Crilley, 1993; Foster, 1999).

Docklands has now emerged with a carefully customised light rail system, a short take-off and landing (STOL) airport, two teleports, six competing fibre optic grids and dedicated

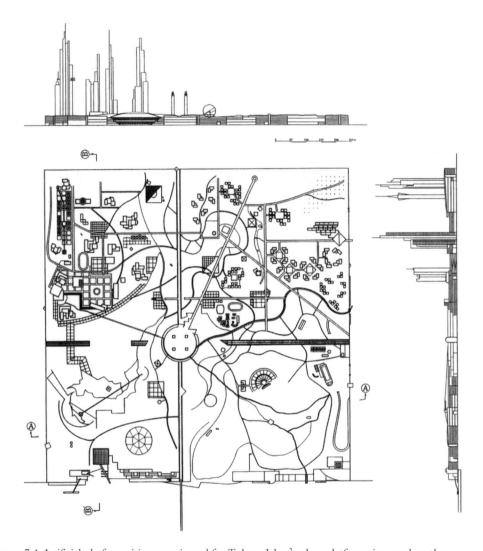

Figure 7.4 Artificial platform cities as envisaged for Tokyo: 1 km² urban platforms imposed on the cityscape to support new infrastructure and super-high-rise development. *Source*: Obitsu and Nagase (1998), 327

power, water, logistics and highway links. These allow high-income Docklands inhabitants and investors to connect with value spaces elsewhere whilst allowing them at the same time to secede relationally from the poor communities that geographically surround them (a strategy reinforced by the use of the old docks literally as moats (Avendano *et al.*, 1997)). (See Figure 7.5.)

Very notably, the Docklands light railway initially connected Docklands with the financial spaces of the City of London whilst avoiding most of the lower-income communities in the surrounding districts of Newham and Tower Hamlets. The United Kingdom's liberalised energy market allowed competing companies – for example, London Underground – to build

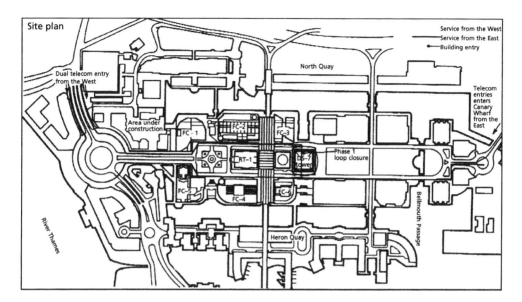

Figure 7.5 The carefully configured 'glocal' infrastructure connections of the London Docklands development. *Source*: adapted from Chevin (1991), 47

new electricity networks for Docklands, offering cheaper tariffs to the major companies located there. The highway link allowed commuter motorists to access Docklands seamlessly from professional housing spaces in the rest of the City. The newly built STOL airport provided direct connections with other European business capitals on the doorstep without having to access London's other airports. And a brand-new £1.3 billion Jubilee Tube line, completed in the year 2000, further improves the public transport to the West End of London and Westminster.

Particularly after the IRA bombing campaign of the early 1990s, access by road to the heart of Docklands, Canary Wharf, was carefully controlled by a so-called 'mini ring of steel' comprising CCTV cameras, a police cordon, a worker identity card scheme and a dedicated, patrolled tunnel road for approved goods deliveries (see Plate 15). All in all, Docklands was a paradigm example of how urban design approaches can be combined with security practices and highly selective infrastructural connections to configure a built space for certain users (global finance capital, allied industries and elite professional residents and workers) at the direct expense of others (adjacent multi-ethnic and low-income communities) (see Brownhill, 1990).

In Docklands customised infrastructural configurations are backed by intense electronic surveillance, 'fortress' architecture and private policing strategies in the new corporate enclaves. 'The rejection of a design framework for the area led to islands of development insulated from each other by security fences, stretches of open water, and the remnants of a derelict Docklands landscape' (Edwards, 1999, 23). Resulting commercial developments are 'inward looking and insular with "public" spaces on the inside. Externally they are forbidding' (ibid.). The emphasis is on securitisation and boundary control, to maintain and police the

Plate 15 London Docklands: a classic defensive glocal enclave with police cordons, digital CCTV surveillance, dedicated roads for goods access, defensive elite housing spaces and customised rail, air, energy, water and satellite connections. *Photographs*: Stephen Graham

stark social divides between wealthy and powerful and the marginalised and displaced. 'The landscapes of advantage and disadvantage are often only a security wall apart' (ibid.). The 'telehouse' development in Docklands, for example, boasts a 'sound-sensitive external fence which can detect a sparrow landing on it, infrared and videophone surveillance and cameras everywhere. Inside, customers [need] PIN numbers to head from chamber to chamber' (Quillinan, 1993, 14).

The 'mini ring of steel' around Docklands was, in fact, a direct echo of the strategy built up after the IRA's 1993 Bishopsgate bomb in the central financial core of London to partition carefully the core financial spaces of central London from the wider metropolis (Pawley, 1997, 153). In fact, as we show in Box 7.2, the global financial landscape of the City of London also represents something of a paradigm example of splintering urbanism. Whilst it is as electronically connected with far-off parts of the globe as any place on the planet, the City of London Corporation is simultaneously attempting to manage and remodel local connections by remodelling the 'public' streets inherited as part of the legacy of the modern infrastructural ideal.

INFRASTRUCTURE AND ECONOMIC ENCLAVES IN DEVELOPING CITIES

The linked construction of business and consumption enclaves and the networked infrastructures to sustain them are also a prevailing model of development in our second range of examples: aspiring 'global' cities in the Developing World. Many factors have combined to support this process: infrastructural liberalisation; the shift towards the construction of large, mixed-use 'superblock' enclaves in urban design; the shift towards extended, polycentric urban structures; a general process of social polarisation; and the predilection of local policy makers for large development projects to symbolise their modernising ambitions (so-called 'teleport' advanced telecommunications and satellite complexes, World Trade Centres, retail and commercial centres, new university precincts and the like).

In the largest 'megacity' in Latin America, São Paulo, for example, newly modernised and gentrified spaces of the city centre have been heavily supported by intense infrastructural investment by the state and private firms. As a result, there has been:

a remarkable increase in the gap between the areas where the advanced 'global' activities are located and the peripheral areas. Internally, the implementation of sophisticated systems of infrastructure [like optic fibre, cable television and mobile telephony] have been concentrated either on existing business districts or on new business developments, generating new centralities for the whole urban complex.

(Schiffer, 1997, 15)

In Bangkok, meanwhile, as we show in Box 7.3, this logic of interconnecting urban enclaves, at the expense of the wider city, is also taken to extremes.

BOX 7.3 INTERCONNECTING ENCLAVES OF NEW DEVELOPMENT IN AN EXTENDING MEGACITY: THE CASE OF BANGKOK

As the city of Bangkok explodes in a carpet of urbanisation stretching over 50 km from the original centre, all efforts to use infrastructure to integrate the city in a comprehensive manner have been abandoned. Instead, concessions are being offered for private developers to put in highways, metros and telecommunication lines connecting the places they most want to serve without real efforts to coordinate or integrate the resulting networks. Massive new private toll roads and expressways complement those operated by the state. These are oriented to the business enclaves and affluent residential spaces of growth corridors like those stretching out to the second Bangkok International and Don Muang airports at Chonburi. Until the Asian financial crash in 1998 separate, competing commuter and metro rail systems were being constructed, again by private firms seeking to cover the most lucrative spaces. Such networks will be 'uncoordinated in terms of fare structure and physical connection' (Kaothien *et al.*, 1997, 5).

Property companies are already taking 'advantage of high accessibility where the lines intersect to develop thematically oriented mixed-use "new towns in town"' comprising office, retail, leisure and housing spaces geared to the needs of affluent commuters – see Figure 7.6 (Kaothien *et al.*, 1997, 5). Over 50,000 low-income residents have been displaced over the past few years from shanty towns to clear the way for such 'mega-development' projects. Such people are expelled to the periphery, where they are poorly served by transit and infrastructure (Hack, 1997, 8). 'This process is being further fuelled by private redevelopment of inner city areas for high income residents and offices'

(ibid., 8). The partial liberalisation of tele-communications – formerly the city's greatest infrastructural deficiency – has allowed new entrants to meet the unsatisfied demand for wiring up and servicing the new middle-class spaces of the expanding city.

Outside the core area, the installation of fibre optics along the so-called 'intelligent corridor' round the major outer ring roads is reinforcing the linear expansion of the city into exurban areas (Hack, 1997, 11). A 'leapfrog' strategy is being encouraged, 'providing households and firms with fibre optic services

Note: 1. Government Subcentre
2. Transportation / Private Corporate Subcentre
3. Cultural / Information / International Interface Subcentre
4. New Bangkok International Airport

Figure 7.6 The four 'new towns in town' development: enclaves built on key infrastructure nodes in Bangkok. *Source*: Kaothien *et al.* (1997), 6

in high-income, educational, knowledge and high value industrial areas' at the expense of the wider city (Kaothien *et al.*, 1997, 14). Overall, the pattern of infrastructure in Bangkok, as in many other 'megacities' in the Developing World, shows a notable lack of horizontal coordination either within or between networks. The result, combined with the pattern of large-scale packaged development and market-oriented infrastructure providers, is 'oversupply in some areas and lack of services in others' (ibid., 14).

CONSTRUCTING 'HOME' DISTRICTS FOR CYBERSPACE: NEW MEDIA ENCLAVES IN GLOBAL CITIES

Our third type of emerging economic enclave, like the global financial cores, tends to be located in the 'global' cities of North America and Europe: the gentrifying 'cyber' district. Such spaces are now driving the production of Internet services, Web sites and the whole digitisation of design, architecture, gaming, CD-ROMs and music. The cities that are developing such enclaves tend to be those with very great strengths in the arts, cultural industries, fashion, publishing and computing: New York, San Francisco and London, to name but three (see Braczyk *et al.*, 1999; Zook, 2000).

THE 'INTERNETTING' OF MANHATTAN AND SAN FRANCISCO: CONSTRUCTING 'SILICON ALLEY' AND 'MULTIMEDIA GULCH'

Manhattan, for example, now provides one of the highest concentrations of Internet activity anywhere on earth, as the Internet and digital multimedia technologies weave in to support every aspect of the functioning of the city. According to Moss and Townsend (1997), Manhattan now has twice the 'domain density' (i.e. concentration of Internet hosts) of the next most 'Internet-rich' US city – San Francisco – and six times the US average.

In fact, the metropolitan dominance of the Internet in the United States is actually *growing* rather than declining, despite its association with rural 'electronic cottages' (Graham and Marvin, 1996). The top fifteen metropolitan core regions in the United States in Internet domains accounted for just 4.3 per cent of the national population in 1996. But they contained 12.6 per cent of the US total in April 1994; by 1996 the figure had risen to almost 20 per cent as the Internet was becoming a massly diffused and corporately rich system. As Moss and Townsend (1997) suggest, 'the highly disproportionate share of Internet growth in these cities demonstrates that Internet growth is not weakening the role of information-intensive cities. In fact, the activities of information-producing cities have been *driving the growth of the Internet* in the last three years' (emphasis added).

Manhattan is home to a booming set of interactive media industries. In particular, Manhattan's so-called 'Silicon Alley' – roughly the area south of Forty-first Street – is emerging as a dominant global provider of Internet and multimedia skills, design and high value-added content of all sorts. As in San Francisco's so-called 'Multimedia Gulch' district, several downtown urban neighbourhoods have been refurbished and gentrified to sustain the clustering demands of interlocking micro, small and medium-size firms in digital design, advertising, gaming, publishing, fashion, music, multimedia, computing and communications. In Manhattan, over 2,200 firms now provide over 56,000 jobs in these sectors, up 105 per cent between 1996 and 1998 (Rothstein, 1998).

Here, as with global financial service sectors, the need for on-going face-to-face contact to sustain continuous innovation is closely combined with exceptionally high use of advanced telecommunications to link relationally and continuously with the rest of the planet. Increasingly, too, certain downtown spaces are being constructed as the 'in' spaces of Internet innovation, places with a 'creative' urban ambience and 'milieu' that contrasts starkly with the sanitised campus landscapes of technopoles.

CYBERDISTRICTS, GENTRIFICATION AND URBAN SOCIAL CONFLICT

Such processes have set off spirals of gentrification, attracting considerable investment from restaurants, corporate retailers, property firms, 'loft' developers and infrastructure companies, and leading to the exclusion of lower-income groups from the newly 'high end' space (see Zukin, 1982). Rents have exploded and, somewhat ironically for an industry whose products can be sent on-line anywhere on earth, parking shortages have become critical.

In both New York and San Francisco major urban social and political conflicts have emerged as 'dot-commers', with their extraordinary wealth, along with real estate speculators and service providers, have colonised selected districts. This has, not surprisingly, dramatically driven up rents, leading to the eviction or exclusion of many poorer residents and to growing efforts at disciplining those who are not tapped into the high-tech, consumerist gentrification (in this case the poor and the black). As Dolgon (1999) suggests, the reconstruction of urban neighbourhoods as chic districts for young professional 'digerati' is often portrayed on the surface as the 'celebrating [of] a diverse and plural community' manifest in diverse ethnic restaurants, art spaces and shops. In reality, however, it tends to 'reinforce a class hierarchy that includes only those with access to new markets'. Furthermore the 'new landscapes of power' created in the process tend to 'further marginalize those whose downward mobility places them outside the marketplace of democracy, diversity, and identity except in their invocations as the hungry, the homeless, panhandlers, and the other "rude rabble"' (ibid.).

In San Francisco's 'Multimedia Gulch' district, centred on the SOMA area of the city, political coalitions such as the 'Yuppie Eradication Project' are already fighting back against the 'dot-com invasion' from Silicon Valley to the south (Solnit, 2000). Their campaign operates under the banner 'The Internet killed San Francisco' (see Figure 7.7). Paul Borsook (1999) outlines the symptoms of what he calls the 'Internetting' of the city: commercial real estate rates rose 42 per cent between 1997 and 1999; the median-price apartment was $410,000 by August 1999; the median rental for an apartment was over $2,000 per month;

Figure 7.7 Backlash against the colonisation of San Francisco neighbourhoods by affluent Internet and multimedia companies and their employees: lobbying by the *San Francisco Bay Guardian*

homelessness rates were rising fast. Landlords, backed by the relaxation of rent controls and tenant protection laws by the City Council in the 1990s, have instigated a huge rise in evictions. The rising stress levels which have resulted for older residents of gentrifying neighbourhoods have been linked with rapid rises in the death rates of elderly seniors (Nieves, 2000, 12). The result is a severe housing crisis, the expulsion of poorer people from the city (as many cannot afford to remain) and accentuating landscapes of social and geographical polarisation as pockets of the city are repackaged as places of work, leisure or living for Internet-based businesses and entrepreneurs.

'ULTIMATE GLOBAL CONNECTORS': CONSTRUCTING INTERNET-READY REAL ESTATE

Within such so-called 'digital districts', new types of work spaces, often with integrated living quarters, are also being configured. Within these, new infrastructural connections are closely combined with highly flexible and carefully configured office suites. Labelled 'Internet-ready' real estate by its inventors, a series of new complexes for interactive media firms is now emerging at the heart of the 'cyber districts'. The New York Information Technology Center, for example, a thirty-storey, 400,000 ft^2 building, sells itself as 'Manhattan's hottest wired building' and 'the ultimate global connector'. To its tenants of CD-ROM developers, Web companies, digital design consultancies and virtual reality artists it offers a dazzling suite of global telecommunications connections, from seven competing companies, direct from the desk, at bandwidths that few other buildings in the world can handle. Emergency power back-up, twenty-four-hour security and training, all-important meeting space, secretarial services and advanced fire suppression systems are also provided. The full suite of high-power electrical systems is especially important, as 'most buildings today are equipped with only 10 per cent of the necessary requirements' of an e-commerce or Web company (Bernet, 2000).

The city of New York has supported the emergence of the new media enclaves with tax holidays, grants, loan funds and financial support for the 'Plug 'n' go' programme to convert properties into Internet-ready real estate (see Figure 7.8). By 2000 millions of square feet of older commercial property across mid town Manhattan were being converted and customised for the new media industry (*New York Times*, 21 March 2000). To match the imperative of twenty-four-hour-a-day, year-round electric and electronic connections, these spaces are being equipped with 'massive quantities of electric power, advanced back-up power and security systems, and generator farms that allow tenants to install and manage their own generators' (*New York Times*, 21 March 2000, 6).

DIGITAL MEDIA CLUSTERS AND 'BYPASS' ELECTRONIC INFRASTRUCTURES: THE CASE OF LONDON

London, too, is extending its cutting-edge, customised telecommunications networks into its booming digital cultural and media industries. In Soho, for example, a tightly constructed media enclave is benefiting from dedicated infrastructure allowing it to extend to global markets in 'real time'. Called 'Sohonet', this system links the tight concentration of film and media companies, television broadcasters, publishers, Internet providers, graphic designers and recording studio headquarters in the West End directly with Hollywood film studios via seamless transatlantic fibre connections (see http://www.sohonet.co.uk).

Figure 7.8 Advertisements for Internet-ready real estate in Manhattan, New York: the 'Plug 'n' go' workspace programme and the New York Telecom Exchange. *Sources*: New York City Economic Development Corporation; New York Telecom Exchange

Sohonet allows on-line film transmission, 'virtual studios' and editing over intercontinental scales via highly capable, digital, broadband connections (see Plate 16). The network is seen as a critical boost to the broader global ambitions of the UK film and cultural industries. Other connections are planned with other global cities, leading to the possibility of a dedicated, global, interurban system for digital film and media production in the near future. Thus, once again, it is clear that patterns of tight geographical clustering, relying on intense, on-going, face-to-face innovation and contact, linked globally and locally through sophisticated telemediated networks, are a feature of many of the industries which concentrate in global cities (not just financial services and corporate services).

TECHNOPOLES AND THE CONSTRUCTION OF HIGH-TECH INNOVATION CLUSTERS

There is a great deal of interest in technopoles as economic growth engines, some interest in them as new forms of cultural representation, and practically no interest in their political governance, that is, addressing [them] as sites of political power, and their residents as citizens.

(Mosco, 1999a, 40)

Plate 16 The Sohonet under construction in the Soho district of London, a centre of media activity. *Source*: Sohonet Web page www.sohonet.co.uk

Our fourth example of carefully networked emerging urban enclaves encompasses the new spaces of 'high-tech' production and innovation that are emerging in new or renewed spaces of production in the North (such as southern California, Baden-Württemberg, and the Rhône Alps region of France); in the campus-like technopoles surrounding reconfigured global cities (such as London, Paris, Berlin); and in the newly constructed high-tech production and innovation spaces of the South (in places like Bangalore in India and the Multimedia Super Corridor south of Kuala Lumpur).

Such is the litany of imitators of Silicon Valley that virtually every region of the world now boasts a 'Silicon'-prefixed space or district, an alleged home to clusters of new high-tech firms and corporate research and development complexes, working (supposedly) in complex interdependence. The Siliconia Web site, which tracks the global diffusion of silicon or cyber prefixes to place marketing and urban boosterist strategies, listed fifty-one such sites in June 1999 worldwide, ranging from Silicon Prairie (Kansas City), Silicon Glacier (Kalispel,

Montana) and Silicon Glen (central Scotland), to Silicon Island (Taiwan), Silicon Plateau (Bangalore, India), Silicon Wasi (Tel Aviv, Israel), Silicon Plain (Kempele, Finland) and Silicon Beach (Santa Barbara, California) (available at http://www.tbtf.com/siliconia.html). However, it must be stressed that a much smaller number of spaces can be genuinely classified as 'new industrial spaces', in the sense that self-sustaining high-tech clusters of innovation are emerging there. Countless others are merely attempts symbolically to turn round the fortunes of ailing or peripheral spaces through decidedly optimistic place marketing.

In all these 'technopole' spaces, which Castells and Hall (1994) label the 'mines and foundries of the informational economy', highly customised and packaged 'edge city'-style landscapes are emerging. Within these, produced space is carefully combined with customised infrastructure whilst design practices, 'filtering' local infrastructures, surveillance and simple geographical distance are often used to connect selectively with only more prosperous parts of the host city or region. In fact the produced spaces, the customised infrastructures, the secure withdrawal and the supporting institutional and financial infrastructure of local agencies are seen to be central in supporting or nurturing the appropriate 'innovative milieux' or 'clusters' to create self-sustaining growth and development (Castells and Hall, 1994, 8).

These are the spaces, often distributed around the polycentric metropolis, where flexible production techniques flourish, where biotechnologies and information technologies are at the cutting edge, and where continuous research and development are necessary for non-stop innovation (Storper, 1997). We do not intend here to explore the technological dynamics of new industrial spaces (for reviews see Castells and Hall, 1994; Storper, 1997). Rather, we maintain our analysis of the central theme of this chapter, namely: how the new packaged landscapes underpinning technopoles and new industrial spaces are being produced in tight relationship with carefully configured, highly selective infrastructure networks. Whilst the binary distinction is a massive oversimplification, in what follows we divide our discussion of technopoles broadly between those in the Developed World and those in the Developing World.

INFRASTRUCTURE AND THE GROWTH OF HIGH-TECH CLUSTERS IN THE CITIES OF THE NORTH

The burgeoning new industrial spaces surrounding dominant Northern cities – the quintessential Silicon Valley, Route 128 to the west of Boston, Massachusetts, the Cambridge growth area north-east of London, Baden-Württemberg in Germany – are born out of a potent fusion of intense, on-going innovation, supportive finance capital, world-class labour market skills and universities, a little bit of serendipity and sophisticated, but highly partial, infrastructural links: state-of-the-art digital telecommunications, dedicated highway networks, excellent links with global hub airports, uninterruptible power supplies and, inevitably, generous water systems, both to fuel the water-hungry production processes and to irrigate the corporate lawns and atria.

In the United States, for example, real estate developers now routinely customise spaces with 'global connectivity' to try and lure in computing, multimedia, biotechnology and new materials companies, especially in the booming group of 'high-tech' cities like Boston, Austin, Silicon Valley, Seattle, Dallas, and Denver (Grogan, 1998). The 'Infomart' development in Dallas, for example, bundles unprecedented communications bandwidth into highly flexible office and production suites within a 1.6 million ft^2 complex catering to the needs of 120 small firms. Small, high-quality, flexible spaces with short leases are backed by many shared amenities, shared services and a high degree of infrastructural redundancy within such complexes. Such 'flex-tech' architecture is finding its expression in larger real estate strategies for whole innovation parks such as the Spectrum development in Irvine, California, which supports five major development clusters in computers, software, biomedical technology, medical devices and automotive engineering (ibid., 92). Custom-built high-tech office complexes have also emerged at major railway stations on the outskirts of major cities in Switzerland (Lehrer, 1994).

Strategies to generate new industrial spaces and clusters artificially on urban peripheries have long been supported in *dirigiste* countries such as France, Singapore and Japan, where vast new infrastructures have been combined with new urban complexes and universities in the 'technopole' and 'technopolis' programmes of national and regional governments. The Japanese technopolis concept, in particular, relies on a modular model encompassing a range of physical developments (R&D centres, higher education buildings, universities, etc.), tightly integrated with airport links, Bullet train connections, high bandwidth telecommunications, cable networks and dedicated water and power supplies (Rimmer, 1991; Markusen *et al.*, 1999).

Interiorised constructions and external delinking

Carefully configuring the infrastructure networks of new industrial spaces allows such places to extend their links to global markets and connections. But that also helps the innovation cluster itself to develop highly filtered links with its adjacent city. Whilst the 'clustering' of innovative firms encourages dense relations within new industrial spaces, they often have a semi-detached relationship with the wider urban landscape. The architecture of technopoles like Silicon Valley 'is shaped by land costs, parcel availability, road access, and business expansion and contraction rates' (Schwarzer, 1998b, 16). Resulting developments tend to be inward-looking. 'The real landscape of Silicon Valley,' writes Rebecca Solnitt, 'seems wholly interior, not only in the metaphor of the maze and the terrain of offices and suburbs, but in the much promoted ideal of the user never leaving the well-wired home or office and the goal of eliminating the world and reconstituting it as information' (1995, 231).

Typically, supporting infrastructures, services and labour are drawn in to such new industrial spaces, whilst connections with the poorer socioeconomic and sociotechnical spaces of the metropolis are neglected or undermined (often through the instrument of explosive rises in housing and living costs) (see Mosco, 1999a). 'Just across Highway 101' from the university–industrial complex of Silicon Valley, for example, 'is East Palo Alto, a ghetto in which chronic poverty and unemployment among its black residents seem beyond remedy.

. . . But for those in the white, self-actualizing utopia of Silicon Valley, the poor and black are of little concern' (Winner, 1992, 49). A whole network of hostels have emerged in the valley even to house the working poor – those who do crucial but relatively low-skill jobs but have no way of affording the rents or purchase prices for housing. (Median house costs in early 2000 were $410,000; median one-bed apartment rents were $1,700 per month.) Many caretakers and cleaners, unable to afford market rents, are squeezed into converted garages in overcrowded conditions. In addition, high-tech companies use the method of subcontracting to absolve themselves of responsibility for such workers' welfare. A spokesman for the IT firm KLA-Tencor, for example, when challenged to pay the company's janitors a living wage, stated that 'the janitors are not our employees, and we don't comment on other companies' employees' (Greenhouse, 2000, A12).

'ISLANDS WHERE EVERYTHING WORKS': CONSTRUCTING 'TECHNOPOLES' IN PERIPHERAL AND DEVELOPING WORLD CITIES

As microelectronics and software production plants are also gradually shifted 'offshore' to lower-cost locations in the newly industrialising and developing countries, technopoles and 'high-tech' spaces are increasingly a feature of cities in those countries – the products of increasingly elaborate development strategies by cities, regions and nations (see Van Grunsven and Van Egeraat, 1999).

Policy makers in Japan, for example, eager to secure the land, natural resources and cheap labour denied them at home, have even developed concepts of 'packaged' cities which are fully self-contained innovation and production spaces ready to be implanted in newly industrialising or developing nations (Rimmer, 1991). The Mitsubishi Electric Corporation developed a programme in the 1980s to export prepackaged technopolis cities to the main urban corridors of South East Asia (ibid., 253) (see Figure 7.9). In 1987 the Japanese Ministry of International Trade and Industry (MITI) also produced a grandiose vision of a 'multifunction *polis*', a 'high-tech' city of 100,000, with carefully customised infrastructure, which was proposed for a site to the north of Adelaide, South Australia.

BANGALORE: A PARADIGMATIC DEVELOPING WORLD TECHNOPOLE

In Developing World new industrial spaces, however, the precise configuration of infrastructure is even more important than in the North because the quality and reliability of the existing networks are often so poor. A good example is India's fifth largest city, Bangalore, an internationally important centre of software engineering and electronic commerce which sells itself to the world as 'India's Silicon Valley' (Wetzler, 2000). Here, extensive efforts have been made by real estate developers and local planning agencies to configure special software and technology campuses and enclaves to the needs of fast-growing inward investing

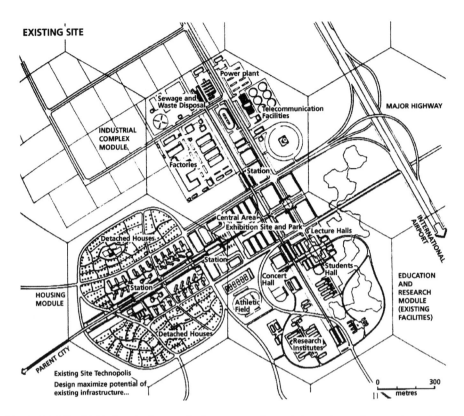

EXISTING SITE

Figure 7.9 Mitsubishi Electric Corporation's concept of a modular technopolis, complete with customised infrastructural connections. *Source*: Rimmer (1991), 259

and indigenous software and IT firms that deliver services and products to global markets.
The city's 300 high-tech companies employed over 40,000 people in early 2000 (ibid., 154).

The heightened wealth inequalities resulting from high-tech growth in Bangalore have
created an extremely fragmented and polarised urban structure. It is based on 'participation
in the information-intensive global economy by a core elite, and non-participation by the
masses' (Madon, 1998, 232). At the Electronics City complex, for example, three-quarters
of a mile from the centre, several hundred acres of 'offshore' technology campus have been
configured to house companies like Texas Instruments (undertaking circuit design), IBM,
3-M and Motorola. The Indian firm Wipro, another major presence, exploits advanced
communications to use India's cheap software programmers to service many of the world's
computers remotely. All these firms 'are insulated from the world outside by power generators,
by the leasing of special telephone lines, and by an international-style work environment'
(ibid., 234). With their on-site ATMs, soaring postmodern buildings and multiple redundant
infrastructures, such parks, in effect, are 'islands where everything works' within surrounding
spaces where modern facilities and networked connections are both very limited and extremely
unreliable (Dugger, 2000, 12).

Singaporean capital has also constructed an Information Technology Park on the
outskirts of Bangalore, equipping it with dedicated satellite ground stations, broadband

telecommunications, uninterrupted power supplies, back-up generators and international-standard private water, sanitation and waste disposal services (Wetzler, 2000). Because of the poor quality of the regional telecoms infrastructure, the park also serves a regional role as a hub linking global markets: 'companies within 30 km of the park can simply point their microwave antennae and connect by satellite link to clients anywhere in the world' (Rapaport, 1996, 105). Celia Dugger argues that most businesses within the new technology parks 'don't need decent roads: they can deliver their products via satellite links of fibre optic cables' (2000, 12). The Information Technology Park is also integrated with luxurious residential and leisure facilities, separating them even further from the prevailing poverty in the shanty towns which house the bulk of the city's in-migrant population (over 50 per cent of whom are illiterate). 'You won't see many Horatio Algers leaping from the shanty towns to workstations in Bangalore's infotech forms' (Wetzler, 2000, 166).

Indeed, whilst the bulk of public and infrastructural investment centres on linking the new parks globally and securing them locally, the local municipality has actively worked to bulldoze 'illegal' self-built housing areas in the name of a civic modernisation 'clean-up' programme. Thus it is clear that 'the recent internationalization of Bangalore has had a negative impact on the poor' (Madon, 1998, 236). The condition of shanty town areas is deteriorating and many have very poor access to mains water, communications, energy or metalled roads and motorised transport – a sharp contrast to the glocally configured modern landscapes of the new technopolis parks that they surround. In fact, a broader infrastructure crisis is emerging for the poorer majority of the city: shortages and interruptions of power are common, a water shortage is looming and the city authorities are desperately trying to attract private sector investment into the poorer districts.

The technology parks in Bangalore, however, are tiny compared with the most ambitious attempt to customise an entire metropolitan corridor to the needs of the international IT and multimedia industries: Malaysia's Multimedia Super Corridor (MSC). We explore the case of the MSC in more detail in Box 7.4.

CUSTOMISING INFRASTRUCTURE FOR FOREIGN DIRECT INVESTMENT IN MANUFACTURING

In our fifth range of examples, strikingly similar processes of customising infrastructure to the precise needs of export-oriented foreign direct investors are also widely established in an area where the race between cities and regions to lure in new investment is even more intense – the struggle for mobile routinised manufacturing (Dunning and Narula, 1996; Chan, 1995). Across the emerging urban and regional development strategies of North America, Europe, Asia, South Africa, the Middle East, Australasia and Latin America, there is one broadly consistent feature: intensive efforts to configure built space and infrastructure needs in parallel to the detailed desires and wants of manufacturing inward investors. This reflects the global mushrooming of flows of foreign direct investment from $77 billion in 1983 to $644 billion in 1998 (Robinson and Harris, 2000, 33).

BOX 7.4 CUSTOMISING A NEW URBAN CORRIDOR TO THE NEEDS OF GLOBAL INFORMATION CAPITAL: MALAYSIA'S MULTIMEDIA SUPER CORRIDOR

Giant among the emerging generation of urban planning initiatives that attempt to engineer new industrial and multimedia spaces is the US$20 billion Multimedia Super Corridor in Malaysia (see Figure 7.10). Here, in effect, at the heart of the ASEAN bloc in South East Asia, a whole national development strategy has effectively been condensed into a single, grandiose urban plan for a vast new urban corridor. The aim of the MSC is nothing less than to replace Malaysia's manufacturing-dominated economy with a constellation of services, IT, media and communications industries by turning a vast stretch of rain forest and rubber plantations into 'Asia's technology hub' by the year 2020 (see Bunnell, 2000). The MSC starts at the centre of the capital, Kuala Lumpur – itself a city symbolised by ambitious plans for a string of megastructure develop-ment projects, including 'GigaWorld' (a 2 km long, fourteen-level tube structure encom-passing office, retail and housing spaces and a 'themed water park') and the Petronas twin towers (momentarily at least, the world's tallest buildings) (Marshall, 1999). The MSC project ends thirty miles south at an immense new international airport strategically placed on the route to Singapore.

Driven by the dominating presence of the Prime Minister of nineteen years, Dr M. Mahathir, the MSC is one of the world's largest-ever planned urbanisation projects. Its scale and comprehensiveness are remark-able. Thirty miles long and ten miles wide, the corridor has an area of 750 km^2. In effect it is a greenfield site as big as the whole of Singapore island. Massive new highway grids, rail net-works, utilities and a very high capacity optic fibre web, all configured to the demands of

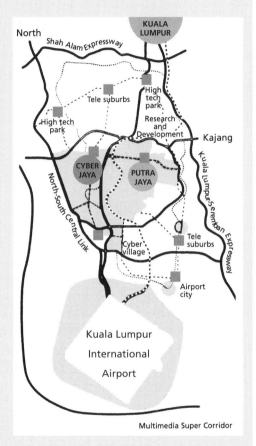

Multimedia Super Corridor

Figure 7.10 The Multimedia Super Corridor in Malaysia. *Source*: Allen (1999), 207

incoming multimedia transnationals, provide the infrastructural matrix for the plan, at a cost of US$10 billion to the state.

The key elements of the MSC plans include a network of 'smart schools', a new multimedia university, with surrounding research and development clusters, 'paperless' electronic government offices and campuses, on-line medicine and distance learning centres, advanced manufacturing centres, and teleservices and back office zones for inward investors. Each is associated with its own plans for new urban districts, carefully integrating the IT spaces with a strategic planning framework providing associated housing (inevitably labelled 'cybervillages'), parks, leisure, transport and retailing uses.

By May 1997 more than 900 companies, both foreign and Malaysian, had bid to participate in the MSC (Corey, 1998). As well as tax incentives, favourable cost structures and high-quality customised infrastructure for the space, Malaysia has even developed customised laws for the MSC. Incoming transnationals will have free in-migration for 'knowledge workers' from all over the world. And a special set of new 'cyber laws' surrounding intellectual property rights has been created to make sure that firms can recoup the investment costs of providing things on-line.

A specially built 'Intelligent capital' for Malaysia, twenty miles south of Kuala Lumpur – known as PutraJaya – is being built to spearhead the whole MSC dynamic. The city, with a planned population of 570,000, will be designed along 'garden city' lines and located on a newly created lake. Again, high-tech infrastructures and services – from building monitoring and control, traffic management, citywide information services and on-line government transactions and information – are being designed integrally with the physical, architectural and social aspects of the new capital. The movement of the entire administrative apparatus of the Malaysian government to the new city is a symbolic gesture of the importance of MSC to Malaysia's development.

PROBLEMATISING THE MULTIMEDIA SUPER CORRIDOR STRATEGY

The MSC will, according to a promotional brochure, be 'a global community living on the leading edge of the information society'. It will be a dazzlingly modern 'world of Smart Homes, Smart Cities, Smart Schools, Smart Cards and Smart Partnerships' (quoted in Allen, 1999, 209). But there are signs that the relationship between the MSC and the rest of Kuala Lumpur and Malaysia may become highly problematic.

First, as John Allen suggests 'the mere presence of state-of-the-art infrastructure guarantees nothing' (1999, 21). The advantages of existing global cities, with their subtle, embedded networks of ties, connections and ideas, are far more extensive than the simple accomplishment of infrastructure. Will transnationals genuinely transfer advanced R&D functions there rather than mere 'screwdriver' operations, keen to profit from government hand-outs and burgeoning ASEAN markets? Will Malaysia's work force really make the hoped-for 'quantum jump' in skill levels if corporations can bring in their own unlimited supplies of knowledge workers?

Second, there are dangers that the MSC will entrench a two-tier society, with Malaysian workers providing the low-value-added support for hermetically sealed corporate zones operating on global networks. In particular, there are major question marks over the fate of Malaysia's peripheral regions, and marginalised urban spaces, outside the MSC. Despite the implication in the prevailing discourse that the whole of the national space will benefit equally (Bunnell, 2000), the construction of the MSC is displacing plantation communities whilst configuring new spaces overwhelmingly for the elite corporate and IT professionals and their families who will be able to afford the new privately developed 'wired' homes (Bunnell, 2000). Low-skill, low-wage service staff – cleaners, security guards, gardeners – are being brought in from outside the corridor. Such concerns about the construction of a privatised, hyper-clean city for socioeconomic elites, gated off from the chaotic and polluted world of the rest of Kuala Lumpur, have emerged in criticism in the national press, as shown by the image in Figure 7.11.

Figure 7.11 'Urban utopia?' A critical reflection on the Multimedia Super Corridor from Kuala Lumpur's *Star* newspaper. *Source*: Tim Bunnell (personal communication)

Once again, the provision of new glocal connections that connect new plants into wider circuits of trade and exchange are dominating inward investment strategies here. Such a trend is part of a broader situation where 'localities compete to offer private investors ever speedier development approvals, ever larger tax concessions, ever weaker environmental regulations, and ever friendlier business environments' (Foster, 1996, 13). Indeed, in a world of liberalising trade flows and increasing locational freedom, supply-side inducements between competing local agencies often become a critical factor in shaping the specific patterns, and urban effects, of mobile manufacturing investments.

The logic is exemplified by South Africa's 'Spatial Development Initiatives'. These are a series of spaces across the country which are targeted for customised, glocal connections, 'aimed at setting up world class industries in regions of theoretical potential for high economic growth. These regions have been chosen because of their raw materials or underutilised infrastructure' (Hafajee, 1999, 52). Once designated, a Spatial Development Initiative offers fast-tracked development permissions, gives major financial incentives to investors, and offers state funds for major infrastructure investments. Efforts are being made to link Spatial Development Initiatives within 'development corridors' stretching into Namibia, Angola, Mozambique and Zimbabwe (ibid.).

In what follows we explore three examples of infrastructure-led strategies to tempt in manufacturing investment. First we look at the 'war' between federal states in Brazil to secure major automobile plants. Then we analyse more recent efforts to 'reindustrialise' old industrial cities in the north-east of England. Finally, we address the manufacturing-led 'hyper-urbanisation' now under way along the eastern seaboard of China and in the cross-border SiJoRi region around Singapore.

'CAR WARS': USING CUSTOMISED INFRASTRUCTURE TO ATTRACT MAJOR AUTOMOBILE PLANTS IN BRAZIL

Our first example takes us to the remarkable 'bidding war' which emerged in Brazil in the 1990s as federal states and development agencies representing cities and regions fought ruthlessly to lure major Western automobile manufacturers into Brazil's newly liberalised and, at least up to the crash of 1999, relatively stable economy (Rodríguez-Pose and Arbix, 1999).

Between 1980 and 1999 over US$25 billion flowed into Brazil from major global car makers, to build plants to serve the expanding Latin American market. Territorial competition to receive such plants is often portrayed as a panacea for major social and economic problems locally. Auto companies do everything in their power to encourage and exploit the already intense competition between cities and states, as politicians fight desperately against the worsening local unemployment brought on by liberalisation and structural adjustment. With urban and regional planning and coordination often largely abandoned, cities and states go to extraordinary and often financially crippling lengths to offer customised infrastructure and spaces that will tempt auto manufacturers to locate within their jurisdictions (ibid.).

Huge efforts are being made to combine financial inducements with extremely high-quality private infrastructure networks, configured along glocal lines, that often contrast starkly with the rudimentary networked infrastructures of the surrounding cityscape. In the state of Rio Grande do Sul, for example, the government succeeded in tempting General Motors to locate a major plant near the capital, Porto Alegre, in 1997. The $600 million plant, whilst an obvious local boost in terms of its 1,300 local jobs and possible multipliers, aimed to benefit from the free provision of land, a deferred $310 million loan at below-market rates, and fifteen-year tax breaks on fifteen local and thirty different state taxes. The plant was also to be equipped, at the cost of the local municipality and the federal state, with 'all the necessary infrastructure, including all utilities, sanitation, and links to the road system. Electricity, natural gas, telecommunications and sewerage disposal are to be subsidised (or as stated in the protocol "supplied at international costs")' (Rodríguez-Pose and Arbix, 1999, 20). Even more remarkably, however, 'it was agreed that the state was to build private port facilities for GM and to dig out an access canal to a minimum depth of twenty feet. . . . Finally, the protocol also includes a series of measures designed to reinforce security at the site and provide public transport to the factory' (ibid., 21). A dedicated railway link is also being provided.

Similar deals have been struck throughout Brazil's coastal strip (Mercedes-Benz in Minais Gerais, Renault and Chrysler in Paranà, etc.). What they effectively do is tie up scarce municipal and state resources in upgrading tiny portions of local space to meet the intense glocal infrastructural demands of rich and powerful transnational automobile firms. Secure 'islands' of powerful and gleaming infrastructural connection emerge, set apart by intense security and boundary enforcement from the surrounding context where even basic social access to sanitation, water, telephony or transport is often increasingly problematic (ibid.).

In fact, so unsustainable is the cost to local public sector institutions in configuring spaces in this way that basic social infrastructures are often withdrawn from the poorer sections of wider cities and regions to pay for servicing the transnational enclaves. Thus a logic of extreme uneven development emerges, with resources withdrawn from efforts to use infrastructure to equalise collective benefits across space, and piled into efforts to allow transnationals to connect glocally with extraordinary ease, cheapness, security, power and public subsidy from privileged sites. Worse still, the economic spin-offs of such investment are unlikely to be anything but 'a pure waste from a national perspective' as they allow transnational corporations to connect with external suppliers rather than sourcing from local ones (Rodríguez-Pose and Arbix, 1999, 25).

GLOCAL INFRASTRUCTURE PROVISION AND PARTIAL REINDUSTRIALISATION IN THE URBAN PERIPHERY OF THE NORTH: THE CASE OF THE NORTH-EAST OF ENGLAND

Our second example comes from the north-east of England, one of the world's oldest urban industrial regions, and one which is widely seen as a success story in terms of bringing in foreign direct investment in manufacturing as a spur to (partial) reindustrialisation after the collapse of the traditional industries of shipbuilding, coal mining and heavy engineering. Frank Peck (1996) has shown how here, reflecting broader practice in Britain and Europe, the customising of spaces with the tailored transport and utility infrastructure demanded by large transnational investors is increasingly sophisticated (see also Phelps *et al.*, 1998). With many locations offering similar grants, financial incentives, training packages and built space, the 'real internal competition' between regions is now, according to the chief executive of Invest in Britain Bureau, in the customisation and subsidisation of transport, energy, water and communications infrastructures (cited in Phelps *et al.*, 1998, 121). Thus the local elements of production are becoming heavily subsidised, as are the infrastructural chains necessary to tie global production sites into complex spatial divisions of labour across the globe.

THE ROLE OF UTILITY COMPANIES

In the north-east of of England, Peck further argues, 'the utilities are now significant actors in the promotion of regions and the attraction of foreign investment, and their autonomy enables them to respond flexibly to any specific demands made by inward investors' (1996, 330). Thus:

it is increasingly untenable to regard infrastructure as an independent variable influencing the regional distribution of mobile investment. Although the presence of certain basic infrastructure may be significant in attracting the initial interest of potential new investors, success in winning inward investment projects depends increasingly on the ability of pubic authorities to produce spaces which are customised to the changing needs of key firms.

(Ibid., 327)

The provision of infrastructure to inward investors is not new, of course. What is new is the way in which publicly funded infrastructure projects are configured to the needs of inward investors on an individual basis, within the context of increasingly fragmented physical development patterns of the region (Peck, 1996, 329). In particular, newly privatised utilities are now significant actors in configuring spaces to the specific needs of inward investors. Development agencies work on a site-by-site basis without worrying unduly about regional coordination or planning (as was the ethos in the 1960s). As Peck states:

large investors can exercise considerable control over the physical environment. . . . In some cases, public investment in infrastructure may create the 'collective' and 'integrative' basis of economic activity, but some forms of expenditure can become 'individualised' and 'exclusive' to a very narrow range of users. Inward investors may be interested not only in the general modernity of the infrastructure in a region, but also in the degree to which they can exercise control over its present and future development.
(Peck, 1996, 337)

During the production of a major site for Nissan in Washington, Co. Durham, for example, Sunderland City Council not only assembled the site, took it out of Green Belt planning restrictions and helped finance development. They also improved the local highway to handle 'just in time' logistics flows, removed the threat of the development of a major football stadium (which would have disrupted such traffic) and configured over 430 acres of extra land to be developed as a 'private industrial estate' by the company (ibid., 333). In addition, a major private port facility, with dedicated highway access to Nissan, was constructed on the nearby river Tyne to facilitate the export of motor cars to European markets.

WATER AS A FACTOR IN THE CUSTOMISATION OF GLOCAL NETWORK SPACES

Similar customisation of both built space and networked infrastructure is common in the strategies to secure further inward investment in the region. The abundance and high quality of the region's water supply, drawing from Europe's biggest human-constructed lake at Kielder, was used as a major selling point for bringing in the Fujitsu microchip manufacturing plant in the early 1990s (which has since closed). The region, especially the chemicals and petrochemicals concentration in the metropolitan area of Teesside, is directly targeting heavy water-using sectors as regions elsewhere in the United Kingdom start to experience water shortages. Here water services are being customised along with transport and built space under the banner 'Teesside: first choice for water in industry'.

But this picture of customising water services to the needs of inward investors is complicated by early moves towards liberalised competition for large users of water in the United Kingdom. Water companies are starting to compete in each other's areas, as with energy and telecoms. Large water users are also actively exploring ways of reducing costs and improving services for their sites by customising their own infrastructure. Larger hospitals, for example, are developing their own private water services, backed up by specialist firms like Enviro-Logic (Stedman, 1999). Many, including thirty to forty major hospitals in London, and some major pharmaceutical manufacturers, are exploring the possibility of sinking their own boreholes to bypass the urban and regional water infrastructure altogether.

CUSTOMISED INFRASTRUCTURE, MANUFACTURING INVESTMENT, AND 'HYPER URBANISATION' IN EAST AND SOUTH EAST ASIA

Our final example takes us to the unprecedented urbanisation under way along the eastern coast of China and in the so-called SiJoRi growth triangle centred on Singapore, much of it induced by an explosive in-migration of international manufacturing capital. Both regions demonstrate forcefully how manufacturing-oriented urbanisation is increasingly centred on the production of carefully customised glocal infrastructure packages. Our discussion of China follows directly here; the SiJoRi growth triangle is discussed in Box 7.5.

BOX 7.5 'GLOCAL' INFRASTRUCTURE INVESTMENT AND THE EXPLOSIVE URBANISATION OF THE SINGAPORE–JOHOR–RIAU GROWTH TRIANGLE

The 'SiJoRi' (Singapore/Johor/Riau) growth triangle exemplifies the role of uneven and carefully customised infrastructure packages in fuelling the growth of manufacturing and broader processes of urbanisation in South East and East Asia (Figure 7.12). Here the Singapore government is explicitly using targeted infrastructure investment to construct a manufacturing growth region around it in the adjacent parts of Malaysia and Indonesia, to create a transborder urban region with a sophisticated division of labour (Yeung, 1999; Parsonage, 1992).

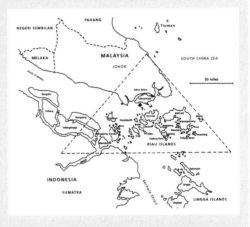

Figure 7.12 The SiJoRi growth triangle. *Source*: Parsonage (1992), 308

SIJORI AS A REGIONALISED DIVISION OF LABOUR FOR SINGAPORE

Through persuading the Malaysian and Indonesian states to liberalise investment restrictions, in line with the wider shift towards the ASEAN trade bloc, Singapore capital is now flooding in to exploit the relatively abundant and cheap land and labour in the surrounding region. New causeways, airport, rail and ferry links are being constructed to link Singapore with its new regional 'hinterland'. Water supply connections are being strengthened. Trade and customs restrictions are being removed. Special 'fast track' immigration systems are being put in place at airports and land borders to allow frequent business travellers to use 'smart cards' to move seamlessly across international borders. Singapore is also cooperating in training specialist technical workers.

Most important, however, investment is also moving in from Singapore and elsewhere to customised export-oriented 'flagship' manufacturing enclaves in Johor (Malaysia) and the Riau islands of Batam and Bintan (Indonesia), each of which is being equipped with the requisite 'glocal' packages of infrastructural connections. 'Each of these investment enclaves offers linkage to the Singaporean economy whilst minimising dependence on the . . . wider Indonesian environment,' write Grundy-Warr *et al.*; the parks are 'conceived as self-contained industrial townships' (1999, 310). For example, direct links are being made into Singapore's state-of-the-art telecoms infrastructure, allowing Indonesia's poor-quality telecommunications infrastructure to be completely bypassed. As a result, telephone calls from the enclaves across the national border to Singapore are classed as 'local'; those beyond the enclave walls to the rest of Indonesia, however, are classed as 'international' (ibid., 317).

Johor and Riau are both being rapidly urbanised in a polynuclear pattern characterised by the provision of 'glocal' infrastructure packages to valued nodes of manufacturing, leisure, tourist and luxury housing investment (Van Grunsven, 1998). The involvement of Singapore engineering firms in new business and technology estates in the wider region 'illustrates their expertise in establishing infrastructure for transnational manufacturing capital' (Parsonage, 1992, 312). The Batam park in Riau, for example, contains 500 ha of factories and dormitories for 8,000 workers, and is configured with high-quality water, energy, telecommunication and transport links with the 'hub' of Singapore which totally outclass the infrastructure of the surrounding regions (ibid., 312). The ability of Indonesia to use these developments to 'tap into Singapore's infrastructure, expertise and links with transnational corporations' is being heavily stressed, with an especially strong focus on telecommunication links (ibid., 313). 'The provision of telecommunications and other infrastructures has long been a problem for investors in Indonesia and the involvement of Singaporean [infrastructure capital] has been a main actor in attracting investment in the Riau islands' (Parsonage, 1992, 313).

BATAM ISLAND: SPACES OF SPLINTERED SEGREGATION

Transnational corporations like Philips, which are maintaining corporate headquarters and R&D centres in Singapore, are taking up space in Batam for routine manufacturing, functions for which land and labour are virtually impossible to find in Singapore itself. Such industrial parks are being combined with 'mega-resort' complexes, like the one at Bintan, which has twenty hotels, condominium complexes, a golf course and marinas, developed under the auspices of an Indonesian and Singaporean joint venture (ibid., 313).

Overall, the Batam and Bintan islands are emerging as places of 'sharp residential segregation between pockets of sparsely occupied, or even abandoned, executive housing and *ruli* [squatter] settlements' (Grundy-Warr *et al.*, 1999, 323) for the mass of in-migrants which are frequently cleared by the authorities, who are keen to provide the 'clean' modern spaces that are seen to be required to bring in Singaporean capital and tourists.

'SMALL ROADWAYS BRING SMALL RICHES, BIG ROADWAYS CREATE BIG RICHES, AND EXPRESSWAYS ENSURE FAST RICHES': CHINA'S PEARL RIVER DELTA

On the eastern Chinese seaboard, arguably the most awesome process of urbanisation ever seen on the planet is taking place. The Pearl River Delta, for example – the most spectacular example of all China's rapidly urbanising areas – will eventually reach a population of 34 million and is growing at a rate of 500 km² per year – the equivalent of Paris, doubled (Koolhaas, 1998b, 183) (see Figure 7.13).

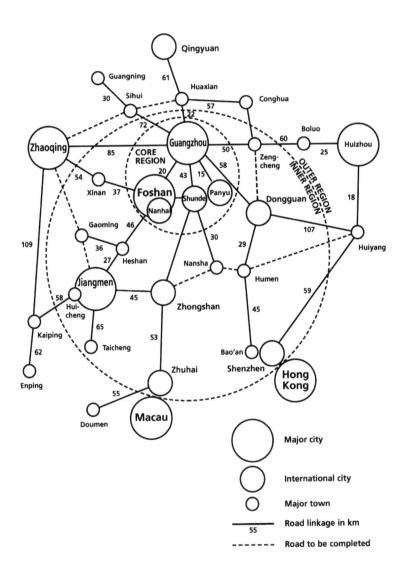

Figure 7.13 The Pearl River Delta megalopolis. *Source*: Woo (1994), 332

This 'hyperurbanisation' (Wu, 1998) is both fuelled by, and reflected in, a shift from the developmentalist tradition in community infrastructure planning to local entrepreneurialism led by newly powerful municipalities, working in partnership with international infrastructure and consultancy corporations and real estate developers (Zhang, 1996). Through joint ventures and ambitious efforts to enrol their spaces into the huge processes of export-oriented urbanisation and modernisation, municipalities are working to secure the high-quality energy, transport and communications infrastructures deemed necessary if they are to emerge as favoured zones within China's polycentric urban geography.

During the Mao years China's municipalities were 'reluctant to systematically invest in new infrastructure' (Zhang, 1996, 100). This was because the development of infrastructure was dictated by the central command economy. Now, however, under the influence of fierce intercity and intermunicipal competition, vast local investments are being made in showpiece infrastructure projects geared to the needs of local and international capital. This investment encompasses major airports (twenty-two in the Pearl River Delta alone) (Eng, 1996, 558), high-speed tolled highways, satellite ground stations, power infrastructures, ports, metro and light rail networks, and new dams and water management systems. Between 1981 and 1990, for example, 'the city government of Nanhai invested over 1.2 billion yuan (US$150 million) in infrastructure projects, resulting in the construction of about 300 km of roads, twenty new bridges, four power plants, and an imported telephone system of over 30,000 lines' (Eng, 1996, 558). The philosophy is exemplified by the words of a municipal official in Qingdoa city: 'To get rich, first build roadways. Small roadways bring small riches, big roadways create big riches, and expressways ensure fast riches' (quoted in Zhang, 1996, 100).

In China, loose-knit sprawling spaces like the Pearl River Delta megalopolis and the Yangtse corridor, up-river from Shanghai, are growing at the astonishing rate of 20 per cent per year. Widely scattered cities, business and technology parks, financial centres, resort complexes (newly owner-occupied), housing cocoons for the new-rich and leisure spaces are all laced together unevenly by gleaming new webs of infrastructure. 'Most developed are the facilities and amenities catering to foreigners' needs in investment and consumption, which to a great extent are also utilized by well-to-do locals' (Eng, 1996, 555).

The resulting urban landscape is one of disjointed and widely dispersed 'packaged' developments orientated more to infrastructural connections than to their immediate environment. 'Currently municipalities are putting a great deal of effort into site-clearing for packaged development' (Wu, 1996, 660). This reflects the widespread use of 'policy enclaves' and 'enclave estates' to shape development (special industrial districts, technology zones, tax-free havens, special development zones, affluent housing enclaves, and the like).

Such spaces are specifically built for foreign capital, foreign people and local new-rich. Eager to construct a new financial centre of global importance, Shanghai municipality, for example, has invested over US$3 billion in new infrastructure – bridges, tunnels, roads and subways, electricity and telecoms networks – specifically geared to the new Pudong development zone on the eastern banks of the Huangpu river. At US$80 billion Pudong is the largest construction project in the world (Wu, 1998, 154). With 150 skyscrapers sprouting 'out of the grey fields and smog-choked factories' of the old city, Pudong will include the US$460 million World Financial Centre – a mini city in itself containing a hotel, art gallery, fitness centre, shopping mall and restaurants, as well as offices and financial dealership spaces

(Tan and Low, 1998, 145). At the same time, the city is working hard to establish itself as the key connection point between global optic fibre grids and China's fast-developing telecommunications infrastructures (see Plates 8 and 21).

URBAN MEGALOPOLIS AND NON-CITY: BEYOND FUNCTIONING WHOLES

Overall, the result of these processes of development, as Keivani and Parsa (1999, 14) suggest, are extending regional patterns of polycentric growth corridors which are intricately, but highly unevenly, interconnected. This new urban form, which Piper Gaubatz labels the 'great international city', is modelled in Figure 7.14 (1999, 265). Within this type of city 'relational linkages are more horizontal than hierarchical' (ibid.). To Rem Koolhaas these startling new urban spaces mean that:

> we are confronted with a new urban system. It will never become a city in the recognizable sense of the word: each part is both competitive with and has a relationship to each other part. Now these parts are being stitched together by infrastructures, so that every part is connected, but not into a whole. . . . In this model, infrastructures which were originally reinforcing and totalizing are becoming more and more competitive and local. They no longer pretend to create functioning wholes, but now spin off functional entities. Instead of network or organism, the new infrastructure creates enclave, separation, and impasse.
>
> (1998b, 188)

Overcoming infrastructural bottlenecks and reliability problems for the mushrooming manufacturing belts, upper-income housing areas and the resort, leisure and shopping districts of the booming cities and Special Economic Zones is a central priority of China's urbanisation policy. Zhang outlines how city officials in Shandong Province sold municipally owned land to foreign developers in order to invest 'primarily on improving and constructing infrastructural facilities in the newly opened high-tech industrial park' in the city (a space that was forcibly evacuated of rural peasants beforehand) (1996, 102). Qingdoa, meanwhile, is 'building an economic and technological development zone, a tax-free zone (*boashui qu*), a high-tech industrial park with low tax rates, and a national-level zone for tourism and vacations' – all configured with state-of-the-art infrastructure built at municipal expense (Zhang, 1996, 107).

MARGINALISED GROUPS AND SPACES: 'FROM VULNERABLE TO TRULY VULNERABLE'

But these infrastructures are being starkly configured to meet the needs and spaces of the powerful; lower-income and poorer spaces within the emerging cityscapes remain very poorly served. Infrastructural tariffs are rising and new projects like highways (for example, the Hopewell highway between Hong Kong and Guangzhou) tend to be tolled to allow the

Great International City

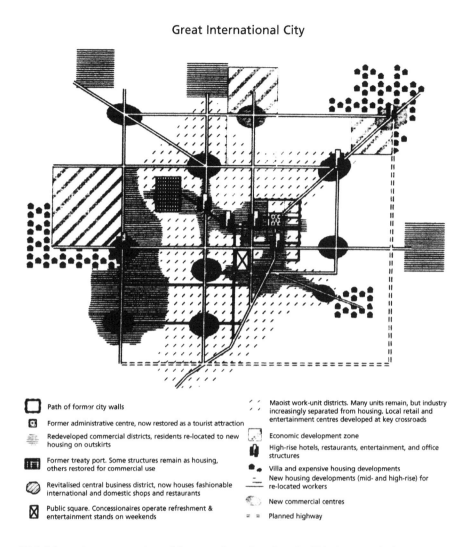

▢ Path of former city walls	⁄ ⁄ Maoist work-unit districts. Many units remain, but industry increasingly separated from housing. Local retail and entertainment centres developed at key crossroads
▣ Former administrative centre, now restored as a tourist attraction	
≋ Redeveloped commercial districts, residents re-located to new housing on outskirts	⌞⌝ Economic development zone
	▯ High-rise hotels, restaurants, entertainment, and office structures
▦ Former treaty port. Some structures remain as housing, others restored for commercial use	●ᵤ Villa and expensive housing developments
⊘ Revitalised central business district, now houses fashionable international and domestic shops and restaurants	⁻ New housing developments (mid- and high-rise) for re-located workers
	⌒ New commercial centres
⊠ Public square. Concessionaires operate refreshment & entertainment stands on weekends	= = Planned highway

Figure 7.14 Schematic representation of the emerging urban form in China's networked megacities on the eastern seaboard. *Source*: Gaubatz (1999), 265

debts incurred in construction to be repaid quickly, and profits to be shared between municipalities and foreign investors (Wu, 1998, 76). A whole grid of such toll roads is now planned to lace together the whole of the Pearl River Delta area.

Meanwhile, however, collectively provided basic improvements in road, water, electricity and communications infrastructures that are widely accessible as public goods lag far behind privately and municipally configured connections for elite cocoons and enclave spaces (Eng, 1996). In the new context 'the vulnerable have become *truly* vulnerable' (Wu, 1996, 661, original emphasis). But there are signs that social resistance to this strategy is growing: 'some question the wisdom of conceding the best business areas to "foreigners"' in the region (Zhang, 1996, 103).

'BACK OFFICE' SPACES AND DATA-PROCESSING ENCLAVES IN THE GLOBAL PERIPHERY

Away from the dominant cores of global cities, or the 'technopoles' of corporate innovation, our sixth range of examples demonstrate that other urban landscapes are being produced or reconfigured to take their allotted roles within the 'variable geometry' of Castells' (1996) 'network society'. These are the spaces of routine electronic transactions, communications and commerce, the blossoming landscapes of back offices, call centres and 'e-commerce' that are emerging in cheap-labour cities both on the periphery of rich countries of the North (Sunderland, Newcastle, Milwaukee) and in the newly constructed digital enclaves of the South (Jamaica, Mexico, the Philippines) (see Bristow *et al.*, 2000; Freeman, 2000).

Such spaces can be acutely disarticulated from the surrounding economic fabric of the wider host metropolis. After all, the only outputs of such spaces are often the streams of digital zeroes and ones that carry voice, data and transaction traffic on corporate computer networks to distant places and markets. A portal of optic fibre or a satellite link connecting the development with distant markets can, literally, articulate a back office or call centre's basic economic relationship with the rest of the world. Distant offices can thus be 'electronically integrated' to work seamlessly together as 'virtual single sites', delinking from the cities where they happen to be located in the process (Richardson and Marshall, 1996; Dabinett and Graham, 1994). Distance-independent phone tariffing for consumers, along with free phone services, is reinforcing such dynamics, as is the construction of advanced corporate networks integrating phone, data and database traffics. Back office spaces thus tend to exemplify very powerfully the glocal economic logics of splintering urbanism.

Very broadly, two types of back office space will be explored here: the back office spaces in old industrial cities of the North and the data-processing enclaves emerging in parts of the Developing World.

RECONSTRUCTING OLD INDUSTRIAL CITIES AS 'BACK OFFICE' SPACES

Following the collapse of the traditional heavy industries that sustained their economic fabric, urban development agencies in many old industrial cities in the North are working hard to recycle redundant industrial spaces as gleaming new business parks for call centres and back offices (Richardson, 1994; Richardson and Marshall, 1996). Such parks and spaces are being customised very precisely to meet the needs of data-processing, telemarketing and consumer service industries, which deliver services to far-off markets over highly capable telecommunications networks, linked into sophisticated customer data bases.

This so-called 'disintermediation' effectively means that consumers and providers of services no longer need to be located in the same place (OTA, 1995). Rather, electronic connections are customised between the parts of cities where call centres are directed and the regional, national and international markets that they serve. Office geographies distributing banks, insurance offices, travel agents and other consumer service outlets across cities and regions, roughly in line with population distributions, are thus being withdrawn, to be replaced by fewer but much larger concentrations of call centres and back offices, located in carefully customised zones of older industrial cities. The city of Milwaukee, for example, is developing a specialism in the third-party processing of on-line and retail transactions through its recruitment of major credit card and electronic commerce back offices.

BACK OFFICE SPACES IN THE PERIPHERAL REGIONS OF THE UNITED KINGDOM

Nowhere outside North America has seen call centres grow as rapidly as the United Kingdom. Over 200,000 people worked in call centres in the United Kingdom in 1997 (45 per cent of the European total); 5 per cent of the work force was expected to do so by 2005, as more and more consumer services start to be provided over the telephone and, increasingly, the Internet (Bristow *et al.*, 2000). Overwhelmingly these centres are located in declining industrial cities to the north and west, feted by the elaborate place promotion and attraction packages that these places offer, as well as low labour costs, low accommodation costs, low labour turnover and high-capability, customised electronic infrastructures.

Careful efforts are made by development agencies to customise back office spaces to the precise needs of inward investors. Specific training in call centre working is delivered in partnership with local colleges. Grants and loans are given to property developers to construct tailored call centre buildings (along with large floor footprints and the ability to take the latest technologies). Telecommunications operators are encouraged to provide state-of-the-art connections emphasising the glocal configurations required by the export orientation of the site. Unwanted incursions from the surrounding metropolis are managed out through the combined use of landscaping, urban design, CCTV and security practices. Road connections are configured as closely as possible to meeting the commuting patterns of workers. Two particularly good examples of how call centres contribute to the splintering of the urban landscape can be found in the major cities of the north-east of England (see Box 7.6).

TOWARDS TRANSGLOBAL OUTSOURCING

BACK OFFICE ENCLAVES IN THE DEVELOPING WORLD

The scramble to configure spaces to support the needs of back offices and call centres has now gone global (see Sussman and Lent, 1998). City and national governments throughout the developing, post-communist and newly industrialising worlds are competing directly with peripheral spaces in the North to attract highly mobile back office and call centre functions. A global offshore labour force is emerging in such cities, paid per keystroke (often at 10,000 keystrokes per hour minimum), marshalled by intense workplace discipline, and delivering services to distant markets and organisations instantaneously (Freeman, 2000; Wilson, 1998).

BOX 7.6 CUSTOMISING CALL CENTRE ENCLAVES IN THE NORTH-EAST OF ENGLAND: NEWCASTLE BUSINESS PARK AND DOXFORD INTERNATIONAL BUSINESS PARK

Development agencies in Newcastle and Sunderland – both old industrial cities facing the collapse of traditional shipbuilding and engin-eering – have worked hard to install call centre enclaves on the redundant riversides left by industrial collapse.

NEWCASTLE BUSINESS PARK

Newcastle's Business Park, adjacent to one of the city's poorest neighbourhoods, is particularly notable for the way it supports the connection of call centres with distant markets whilst virtually severing its relationships with surrounding neighbourhoods. The park has a work force of 5,000. Dedicated optic fibre grids from three companies support the major call centres from British Airways, AA Insurance and IBM, allowing each to link seamlessly into its respective electronic universes of planet-straddling 'virtual single offices', call management centres, customer assistance, telemarketing, travel reservations and electronic transactions.

At the same time, however, because the park is located across a main highway from one of the most deprived and stigmatised inner-city areas in the United Kingdom, great effort went into securing and filtering the space from its surrounding urban environment. When the idea of a large-scale development was first mooted, it was felt that big companies would be unlikely to move to one of the poorest parts of Newcastle to open up new offices. Even with the promise of grants and subsidised rents, the local crime rate presented a public relations problem. The solution was judged to be a 'fortress' approach, whereby the Business Park attracted clients on the basis of its high security. Defensive landscaping, a 3 m spiked fence, a state-of-the-art CCTV system, private security guards and highly restricted road and pedestrian access serve to withdraw the park from the surrounding neighbourhoods.

DOXFORD INTERNATIONAL BUSINESS PARK

Thirteen miles to the south-east an even larger call centre enclave is under development on the edge of the city of Sunderland: Doxford International Business Park (see Plate 17). Once again, the developers have attempted to configure the space for call-centre investors by offering intense glocal connectivity (in the form of state-of-the-art telecoms links) with intentionally filtered and compromised local connectivity (CCTV, landscaping, private security, fencing, urban design). In this case, however, the global connectivity has been bolstered by the development of a teleport: a major telecommunications hub explicitly designed to connect Doxford's call centres with distant markets (whilst not serving the rest of the city at all). Dedicated fibre connections from BT and Cable & Wireless connect all buildings with national and international networks as well as a proposed satellite ground station. £10 million of public money has supported telecoms infrastructure developments in the park (Southern, 2000). Further boosts to the site came when it was designated as a Enterprise Zone in 1990, allowing investors a ten-year tax holiday on all investments and enhanced flexibility in planning regulations. These attractions have led Doxford to be developed as the call centre location for London Electricity, Barclay's teleshopping, One-2-One mobile phone services, the National Lottery and Nike sportswear UK.

Plate 17 The Doxford International Business Park, configured for call centres and back offices, on the outskirts of Sunderland. *Source*: Twedco, 1999

In industries like airline reservations, insurance, data processing, animation, software development, remote diagnostics, language translation, proof reading, computer-aided design (CAD), telemarketing, logistics management, financial services, business information services and database management, seamlessly integrated global flows are created by linking back office enclaves around the world with dedicated, high-quality telecommunications (often islands of connectivity in a surrounding sea of very poor infrastructure and access). And, as Bannister writes, such dynamics are likely to grow significantly as telecommunications technologies improve. 'Once it becomes possible to transmit and switch video signals without overloading networks,' he writes, 'a whole new range of opportunities open up for developing countries. The World Bank, for example, has even trailed the idea of security cameras in American shopping malls being monitored in Africa' (1994, 1). The expected 'bandwidth glut' on major transnational routes now means that such video-based back offices are a realistic prospect.

One notable characteristic of the development of back office enclaves is the way in which time zone differences are used to create seamless working relationships, maximising the efficiency with which spatial divisions of labour are managed and exploited. Mark Wilson cites the example of an Ohio software firm which has maximised its use of a globally connected set of back offices:

the software code starts its day in the US Midwest, and at the end of the day it is electronically transmitted to a branch factory in Hawaii, where an additional six hours of processing by labor is performed. As the sun sets in Hawaii, the job is forwarded to yet another branch in Bangalore (India) for another day's work, before being returned by dawn to Ohio. Around the world in twenty-four hours, with seventeen hours' work performed.

(Wilson, 1998, 41)

The perceived development prospects of whole regions – most notably the Caribbean, India, Ireland and the Philippines – have been substantially boosted by the apparent opportunity to insert themselves and their work forces into such instantaneous chains of informational service production (Skinner, 1998). The imperative of forging electronic connectivity from the economic margins of the world to the core is clear: 'by staying out of telecommunications networks, nations face the real prospect of being left to fester on the edge of the world system' (Skinner, 1998, 64). The 'veneer of high-tech sophistication' also makes these services especially attractive to policy makers in developing cities and countries (ibid., 64).

DEVELOPING CITY BACK OFFICE ENCLAVES AS 'PORTS OF CALL FOR INFORMATION PROCESSING'

In this competitive rush between cities and nations, however, the collapse of spatial constraints at the international scale is closely bound up with the production of new types of spatial barriers at the local scale. On the one hand, telecommunications links are (unevenly) used to annihilate spatial distance; on the other, the precise infrastructural capabilities of the back office enclaves necessary to complete such labour become ever more important. This is especially the case as the prevailing telecommunications infrastructure in developing nations is very poor. 'The network does not reach all parts of the country; there is a tremendous disparity between urban and rural telecom availability, quality and variety' (Chowdray, 1998, 261). In addition, costs tend to be high, competition is limited, and reliability is usually poor (a critical problem when constructing back office networks).

In addressing these weaknesses, great efforts are therefore made by the developers of back office enclaves to construct what Skinner calls 'installations which are ports of call for information processing', with carefully customised infrastructural connections 'to the main lines of international telecommunications circuits in the service of [the] metropolitan economies' of the North (1998, 65). Advanced telecommunications links support the construction of local affiliate companies to control and operate back office spaces. As the case of one such customised enclave, the Jamaica Digiport International, shows (Box 7.7), whilst the greatest efforts here are concentrated on IT infrastructure, power, water and transport are also carefully improved.

BOX 7.7 BACK OFFICE ENCLAVES IN THE DEVELOPING WORLD: THE JAMAICA DIGIPORT INITIATIVE

An effort to diversify a fragile and tourist-dependent economy, the Jamaica Digiport is a classic example of a privately owned back office space constructed in a Developing World city to service markets in the North through the provision of customised, dedicated infrastructure and services (see Wilson, 1998, http://www.jadigiport.com/ main1.html).

The Digiport was developed after it was recognised that the expanding data-processing industries could 'make a major contribution to Jamaica's development if certain infrastructure requirements were met' (promotional Web site). It is part of an effort to bring in 10,000 back office jobs to the island (Skinner, 1998, 67).

Located in Montego Bay Free Trade Zone, the Digiport space has customised port facilities, freight forwarding services and high-quality schools, office space, trained personnel, medical services and water and electricity utilities. No taxes are due on profits, imports into the zone or exports to other countries (Skinner, 1998, 72). The work force is low-cost – US$0.34 per hour at 1991 rates – and English-speaking. Wage rates are calculated on a per-keystroke basis to stimulate productivity.

JAMAICA DIGIPORT AND 'ELECTRONIC LABOUR INTEGRATION'

But it is the telecommunications infrastructure of the Digiport which really distinguishes it from surrounding spaces. A dedicated satellite earth station, backed up by emergency power supplies, provides resident companies with high-speed communications and information processing links with over 200 countries that far exceed those available in the rest of the island. Citing the initiative as a powerful example of 'electronic labour integration', Edward Skinner notes that 'voice, text and data handled by JDI are beamed to a satellite off the African coast and then to clients in the United States, Canada and the UK' (1998, 71). The telecom facilities, developed jointly by AT&T of the United States and Cable & Wireless of the United Kingdom, also link the Digiport enclave directly with transglobal optic fibre networks. 'The geographic restrictions of the site have removed the need for JDI to invest in expensive technology platforms' (promotional Web site). A dedicated company has been formed to manage the telecommunications services that are specially designed to meet the data processing needs of resident export-oriented corporate clients. Tariffs are kept exceptionally low, with 'virtually every AT&T type service kept well below international rates. . . . Operators outside the zone must use the much more expensive Telecommunications of Jamaica services' (Skinner, 1998, 71–2). Network reliability is guaranteed at 99.95 per cent. Combined with low labour costs and low turnover, operating costs are roughly 40 per cent of those in the United States.

Data entry, telemarketing and information processing activities dominate the uses of the Digiport enclave. Data entry tasks for the US market can be turned round in less than twenty-four hours. Telemarketing can directly target US markets without major time zone problems. By 1992 there were 3,000 Jamaicans employed in Digiport; the target was 30,000 (ibid., 72).

LOGISTICS ENCLAVES, EXPORT PROCESSING ZONES, 'HUB' PORTS AND E-COMMERCE SPACES

Our final encounter with the glocal infrastructural spaces of emerging urban economic landscapes is with logistics enclaves. These are spaces within which the precise and rapid shipment of goods, freight and people across the planet are coordinated, managed and synchronised between various transport modes, along with supportive information and energy exchanges (Rodrigue, 1999). There is a proliferation of such spaces across the world, driven by the imperative of securing what we might term 'economies of conjunction' (Rondinelli, 2000). These are the efficiencies that arise when firms operate within premium network spaces which seamlessly interconnect virtual and physical systems of movement, allowing the precise and agile coordination of all forms of flow and transaction at the same time and space. Keller Easterling believes in fact that virtually all contemporary urban developments – transport interchanges, ports, airports, malls, economic franchises – can best be understood as dynamic sites for organising logistical processes. 'The primary means of making space' in contemporary America, she believes, they can be thought of as 'a special series of games for distributing spatial commodities' (1999b, 113). However, she also notes that 'the critical architectures of these spaces are not visible' but are woven into their extended technical and information systems and often hidden infrastructure networks. 'The real power of many urban organizations,' she continues, 'lies within their relationships between distributed sites that are disconnected materially, but which remotely affect each other – sites which are involved, not with fusion or holism, but with adjustment' (ibid.).

If the widening range of powerful 'glocal' infrastructures can be considered as amounting to a widening set of 'tunnel effects', bringing distant sites into close relational proximity, then these spaces are the points at which the 'tunnels' stop or interconnect: the global airports, major seaports, teleports, railway stations, e-commerce hubs and so on. The challenge for the developers and managers of such places is to make the transition from the 'tunnel' of the global airliner, freight transporter, telecoms link or fast rail network, either to the next 'tunnel' or to the selected, valued elements within the regional hinterland, as seamless an experience as possible.

In such a context it is no surprise that 'supply-chain management is moving to the top of the corporate agenda' (Bachelor, 1998, 1). Reflecting the shift towards extremely volatile, international markets and production techniques for high value-added and relatively low-weight logistics flows – microchips, scientific instruments, media products, technological equipment – emphasis now falls on combining highly flexible production strategies with the logistical ability to deliver goods quickly and accurately on a global basis from a series of logistic enclaves (OTA, 1995). As such, logistics enclaves obviously require privileged and high-quality infrastructural connectivity, especially for transport and telecommunications, with the emphasis on global connectivity to strategic centres and distribution hubs (Andersson and Batten, 1988). To the architect Paul Andreu the issue of 'how to enhance the value of such ruptures opens a huge field of reflection' for developers and planners. He continues:

With cities coming undone and losing their coherence, such points of interchange constitute lively sites full of energy and new possibilities. Projects abound in the field, whether they involve train stations, subways, bus or airport terminals, or any combinations thereof. For cities, such interchange points provide an occasion for reflecting upon and modifying themselves, for devising new models of organisation and new spaces.

(1998, 43)

Using the latest information technologies, combined with advanced logistics management techniques (notably 'just-in-time' and 'zero inventory' approaches), leading distribution hubs for road, rail, sea and air logistics (and the crucial connecting flows between these different modes) are emerging as mini cities in their own right. At the same time they have a tendency to delink from the immediate spaces around them (Pawley, 1997). Such spaces and buildings, in effect, are 'one-stop shops' which concentrate the service capabilities to help direct and organise the trade flows of the entire planet. They have 'ceased to be an investment and are simply expandable containers, linked to a chain of networks' (Bosma, 1998, 12). To succeed, they require intense concentrations of highly capable, interlinked infrastructure networks.

Here we touch on the four most salient types of logistical space emerging in the splintering metropolis: multimodal logistics enclaves dedicated to freight, export processing zones, passenger airports and fast rail stations, and teleport projects and e-commerce hubs.

MULTIMODAL LOGISTICS ENCLAVES: CONSTRUCTING 'FREIGHT EXCHANGE CITIES'

Teesside airport, a small-scale airport in the north-east of England, has long been used primarily by a small flow of charter passengers. By 2004, however, it is likely to be transformed into an international air freight hub covering 1,250 acres and costing £300 million. The airport will be equipped with improved links to international motorways and rail networks, as well as the tax breaks needed to make it a major 'offshore' location for international freight distribution (Nicholis, 1999).

Reflecting the fact that 50 per cent of global trade by value now goes by air, a range of specialist spaces are being constructed around the world to handle and organise the world's burgeoning aerial trade. Gethin (1998, 19) predicts that 'the next century will see just a few air cargo "superhubs" at strategic points across different continents'. Offering rapid freight services to connect seamlessly with just-in-time production methods, major logistics nodes offer direct links between air, land and sea transport. As Easterling suggests, 'new airport cities and superhubs that concentrate the intermodal transfer and storage of global or domestic goods, and that act as centers of distribution, have helped relocate a set of exurban switches for exchange between rail, highway, land and sea' (1999b, 120). Centred on the United States, for example, major international freight companies like Federal Express, DHL and UPS already run their own networks of massive multinodal logistics centres on a global 'hub and spoke' format, sometimes using freight-only airports. United Parcel Service employs over 13,000 at its Louisville base, for example, the central hub in a global operation network that delivers 3.1 billion items per year to over 200 countries and 600 airports via 500 aircraft (OTA, 1995, 155; Rodrigue, 1999). Such places are, in effect, 'freight exchange cities' and they are 'forming at critical junctures of the Interstate highway and global airline systems' (Easterling, 1999, 120).

THE 'DELINKING' OF MAJOR SEAPORTS

Major seaports, like Rotterdam, meanwhile, are also gradually 'delinking' from their surrounding hinterland in a struggle to emerge as global 'main port' hubs linking transcontinental systems of road, rail and air freight seamlessly on to the world's major sea lanes (Drewe and Janssen, 1996). Such IT-intensive container ports are now being 'transformed into largely silent and invisible operating environments' with ever more precarious relationships with their adjacent cities and immediate hinterlands (Taverne, 1998, 85). 'The hinterland from which the port draws its cargo or sends it is not secure. Ports on one side of a continent are in competition with ports on the other side as they both try to serve the shipping needs of inland areas' (McCalla, 1999, 248). Such tenuous links are likely to become more unstable with the latest developments 'that will use jet boats to guarantee on-time international cargo delivery in any weather, imposing just-in-time manufacturing from Houston to Kuala Lumpur' (Mau, 1999, 204).

Massive new port systems, laced with the latest electronic management infrastructures and meshed in close alliances with major shipping and inland transfer firms, are the way in which major ports are attempting to project their competitive power. The objective is to emerge as a node of seamless and ultra-cost-effective intermodal transfer between sea, land and, increasingly, air. Some cities have already developed logistics complexes, and the specialised infrastructures that underpin them, that far exceed the scale and sophistication of Teesside's plans. Seattle, for example, has built a dedicated sea–air logistics hub which now organises 23 per cent of the entire world's sea–air shipments (Kasarda and Rondinelli, 1998).

THE GLOBAL TRANSPARK PHENOMENON

But it is the municipalities and development agencies of North Carolina that have developed perhaps the most ambitious project of all – the North Carolina Global Transpark (Kasarda *et al.*, 1996). It integrates sea, road, rail and air within a single 15,000 acre complex (see Figure 7.15, http://www.ncgtp.com/). Its dual long-range airport runways offer twenty-four hour access to the world's freight aircraft fleets which can offload efficiently on to highway, rail and sea transporters. Customised information systems, telecommunications grids, water and energy services are designed, through service agreements with chosen operators, to meet the highest specifications for global business operators. Internal operations are supported by dedicated monorail, electronic point of sale, and IT and back-up power systems. Local tax breaks and reduced import duties apply.

The ultimate aim of the Transpark is to be 'connected to the world through state-of-the-art communications, utilities and transportation infrastructure' (promotional Web site). The end result, developers hope, will be a 'seamless environment' for manufacturing and international distribution (ibid.). Through imitation of the concept, it is hoped to develop a network of transparks throughout the economic 'hot spots' of the world. Already the developers have reached agreement to develop similar facilities in the Subic Bay area of the Philippines, in Thailand and in Germany (Kasarda *et al.*, 1996, 39).

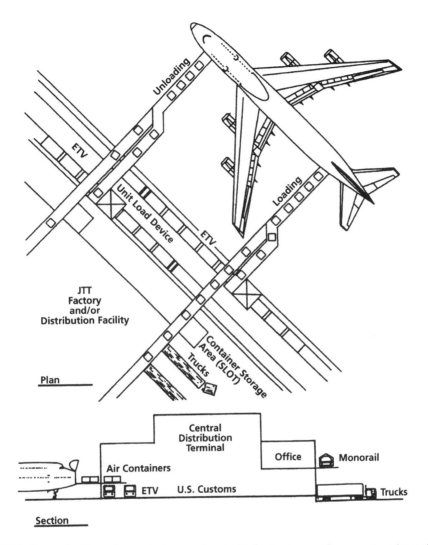

Figure 7.15 Customised interchanges between networked infrastructures and transport modes at the North Carolina Global Transpark. *Source*: Kasarda *et al.* (1996), 38

EXPORT PROCESSING ZONES AND THE INFRASTRUCTURAL IMPERATIVE

The Third World has always existed for the comfort of the first.

(Klein, 1999, xvi)

As Kessides suggests, 'the exigencies of modern logistical management in developed industrial countries pose similar requirements on developing countries wishing to compete in these markets' (1993, 13). The second emerging logistics enclave, long a mainstay of the development strategies for developing and newly industrialising cities, is a reflection of widening efforts to attempt just this: the export processing zone (EPZ) or free-trade zone (FTZ).

EXPORT PROCESSING ZONES AND FREE-TRADE ZONES

Essentially a low-tax and reduced-regulation haven for global trade and rudimentary production and processing, export processing zones – of which there were around 850 in 1999 – are attempts to equip cities, and parts of cities, with the high-quality infrastructural connections necessary to position them within global flows of trade and transaction (Dunham-Jones, 2000; Chen, 1995). Export processing zones and free-trade zones are also being adopted by cities in the Developed World, as the Transpark example above demonstrates (see Lee, 1999). Thirty of the US states now have export processing zones, for example (Chen, 1995, 587). In fact, there exists a whole typology of export-oriented trade spaces, from free ports to the advanced manufacturing spaces and technopoles covered elsewhere in this chapter (see Figure 7.16).

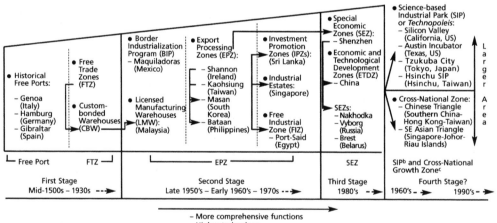

Source: Developed with numerous additions and modifications from the simple sketch in Wong and Chu (1984:14) and the textual classification in Gu, Wei and Wang (1993).

Notes:
a   The solid vertical lines differentiate the three major types and broad stages of FEZs, whereas the broken vertical lines separate the variants within each major type and broad stage of FEZs. The arrows point to directions of influence in form and function.
b   SIPs or *technopoleis* differ from FEZs in that they are distinctive urban areas linking technology and economic development, and are sometimes referred to as *technology centres* (IC² Institute, 1990). Therefore, SIPs are not formally incorporated into the classificatory scheme, but represent a potentially new stage of FEZ evolution.
c   Cross-national growth zones differ from intra-national FEZs because they stretch across border areas. Therefore, they are not formally incorporated into the classificatory scheme either. But cross-national growth zones herald a new phase and variant of the evolving FEZ. (See Figure 2 for the geographical location of several cross-national growth zones in the Asia-Pacific region).

Figure 7.16   A typology of export-oriented trade spaces. *Source*: Chen (1995), 599

Routinised production in export processing zones: 'work boiled down to a brutal essence'

As with customised spaces for foreign direct investment in manufacturing, export processing zones are generally constructed to combine low costs and minimised regulations and taxes, with an adequate supply of (cheap) labour and the best possible infrastructural connections, to allow the space to be integrated within global trade flows (Klein, 1999). It is in such spaces, far away from the corporate logos and postmodern style battles of consumption markets in the North, that many leading transnationals locate their subcontracted manufacturing work. Such 'distant grey sheds in the Far East and Central America offer work boiled down to a brutal essence' (Beckett, 2000, 17). In 1999 over 27 million people, usually young unmarried women, worked in such 'closed-off places across the world, where taxes and unions and regulations and the attentions of local politicians barely reach' (ibid.). In South East Asia 'hundreds of workers burn to death every year because their dormitories are located upstairs from firetrap sweatshops' (Klein, 1999, xvi).

Such spaces are therefore emblematic of the use of corporate networks to exploit spatial separation and geographical division. Talking to a seventeen-year-old woman in an IBM CD-ROM plant in Manila, for example, Naomi Klein told her she was impressed at the woman's skill in making such intricate machines. 'We make computers, but we don't know how to operate computers,' was the reply. Reflecting this, Klein urges us to debunk the corporate fetishism of 'globalization' that brings the endlessly repeated ideologies of the 'global village' and 'collapsing world'. 'Ours,' she writes, 'is not such a small world after all' (1999, xvii).

Whilst the poor infrastructural connections of many developing and newly industrialising cities often act as a deterrent to investment by global trade operators, the concessions and publicly supported infrastructure in export processing zones works to enrol them into the multinational locational decisions of shippers, transnational corporations and manufacturers. Export processing zones are therefore logistics enclaves whose high levels of infrastructural servicing, and connectivity elsewhere, tend to contrast sharply with their disconnection from their surrounding city or region and its poor infrastructure (Brenes *et al.*, 1997). This is especially the case in export-processing sub-Saharan economies. In Ghana in 1995, for example, international telecommunications infrastructure was limited to highly expensive, glocal connections for a few key players in international business enclaves in the capital city:

There were only ten 14,400 kbit/s leased lines linked to the UK costing about US $7500 per month each. One was used by the interbank clearing system, SWIFT, and another by the air traffic control network, SITA. By early 1996, a private network computer systems host in the capital city had 140 subscribers paying US $1300 each a year – the annual income of a Ghananian journalist.

(Mansell and Wehn, 1998, 106)

The planning of export processing zones invariably emphasises the modernity and quality of infrastructural connections, compared with the surrounding city (along with, of course, the low operating costs and 'flexible' on-site labour force). In the new Jinqiao Export Processing Zone, a 20 km^2 zone near Shanghai's port, for example, a whole suite of special advanced infrastructure networks has been put in place to tempt foreign capital, and foreign business people, to the enclave: a 500,000 line capacity telephone and optic fibre grid; a dedicated satellite earth station; eleven dedicated electricity substations with back-up; customised water, gas, road and public transport connections. Modern garden city-style residential and commercial spaces are backed by intense security as living spaces for foreign executives and business travellers.

As well as enhanced transport and infrastructure connections, export processing zones tend also to be provided with extra services befitting a modernised enclave such as 'twenty-four-hour security, garbage collection, maintenance systems, and, in case of emergency, water supplied from wells belonging to the EPZ administration' (Brenes *et al.*, 1997, 61).

PARTIAL ARTICULATIONS FOR THE 'KINETIC ELITE': AIRPORT CITIES, FAST RAIL STATIONS AND THE WIDER METROPOLIS

Our third type of logistics enclave has perhaps the most dramatic contrasts between intense global connectivity and the increasingly careful filtering of local connectivity: the international 'hub' airport or rail terminal. These are customised spaces *par excellence* for organising and housing global flows. In particular, such spaces are designed and regulated to meet the needs of affluent business and leisure travellers: the 600 million or so airline arrivals per year; the 300,000 people that are in the air at any one time above the United States (Urry, 2000b, 50). Borrowing from the German philosopher Peter Sloterdijk, Rem Koolhaas calls these people the 'kinetic elite' (see Wolf, 2000). In configuring themselves to meet these people's needs for seamless flow and painless interconnection with distant elsewheres, global network spaces like airports and fast rail stations start to show an ambivalent relationship with their host city.

Contemporary international airports concentrate a remarkable range of facilities and services within a spatial setting that is 'similar to that of an island connected with other distant regions only through very selective specialized systems of transportation' (Ezechieli, 1998, 18). Because it articulates more closely with the global flight-path network than with the surrounding city space, as Friedman suggests, 'today's airport is only partially connected to the environment around it. It is directly connected with other airports with which it is linked in an increasingly vital network' (1999, 14). Reflecting on this trend, Hans Ibelings believes that:

nowhere is the process of enclave formation stronger than in the world of airport architecture. All over the world, the major airports have grown into complex and multi-faceted mega-structures that not only offer space for more terminals, piers and hangars than ever before, but also accommodate a growing number of functions that have nothing whatever to do with aviation. In many cases, these other functions make a bigger contribution to airport turnover than activities directly related to air travel.

(1998, 78)

As Markus Hesse suggests, the tendency for the new logistics nodes and airport complexes to gravitate to the metropolitan periphery is helping to support a restructuring in broader urban forms:

New transportation concepts prove to be new locational concepts. And where transportation becomes the new bottleneck factor in regional development, those nodal points that promise frictionless movement of traffic are booming. [As a result] urban structures are being shifted: The center is wandering to the periphery, tertiary uses are moving magnetically to these locations.

(1992, 53)

In what follows we explore a few examples of how mega-airports are starting to emerge as exemplars of secessionary and premium network spaces.

Bypass immigration for premium passengers

In the first example, highly mobile and affluent business travellers can, increasingly, bypass normal arrangements for immigration and ticketing at major international airports. This allows them seamlessly, and speedily, to connect between the domains of ground and air, and through the complex architectural and technological systems designed to separate 'air' side and 'ground' side rigidly within major international airports (Virilio, 1991, 10). In fact, travel on an international airliner, 'with its portholes closed and movie screens on', can itself now be likened to a 'travelling segment of a tunnel' (Andreu, 1998, 59). The increasingly close integration of the 'air' and 'ground' experience means that this 'tunnel' can, in effect, extend right through the airport to highway or railway links with as little delay and as much comfort as possible – for selected, generally powerful, people.

After a pioneering agreement, for example, biometric hand geometry scans are now in operation for the most frequent business travellers at major airports linking the United States, the Netherlands, Canada and Germany and other OECD countries under the INSPASS (Immigration and Naturalisation Service Passenger Accelerated Service System). Selected 'premium' travellers are issued with a smart card that records their hand geometry. 'Each time the traveller passes through customs, they present the card and place their hand in a reader that verifies their identity and links into international databases', allowing them instant access to the plane (Banisar, 1999). In 1999 the scheme had 70,000 participants and the INS were already planning to extend the system globally. Such systems, of course, back up the extending infrastructure of luxurious airport lounges and facilities that are accessible only to elite passengers carrying special passes (see Figure 7.17).

Local bypass and seamless connection with dominant consumption and business spaces

Second, major international hub airports are increasingly connected up seamlessly with the major corporate and consumption spaces of the host city, many of which now tend to cluster around them. 'Thanks to all the offices, banks, hotels, restaurants, conference facilities, casinos and shopping centres in the immediate vicinity, the airport has developed into a significant economic centre that is sometimes so large that the airport starts to compete with the very city it was intended to serve' (Ibelings, 1998, 80). As Gary Hack suggests:

Figure 7.17 The Priority Pass programme: a fee-based system allowing access to luxury airport lounges at every international airport in the world

new business centers which are evolving in many cities have the tendency to spur the creation of 'elite corridors', with housing, entertainment uses and educational facilities oriented to high income groups. Virtually every city has such a new center, and often they are located near regional airports. In Santiago [Chile], such a cluster is emerging along the Amerigo Vespucci beltway. In the Randstad, the area around Schipol Airport is beginning to take this form. In Manila, the Makati area is becoming a new city center. . . . Often these new centers are the city's window to the international economy, and the preferred location for regional headquarters of multinational corporations.

(1997, 8–9)

As some key airports are developed on reclaimed islands (Osaka, Hong Kong) they demonstrate an even greater tendency to bundle uses and functions and emerge as global business cities in their own right. Rem Koolhaas's studio, the OMA, for example, has proposed the building of an entirely new airport city off the western coast of the Netherlands to replace the increasingly congested Schipol airport at Amsterdam. Geared specifically as a home base for high-frequency business travellers, the island would encompass the largest malls in Europe, beach resorts, a theme park, a technology park and elite housing. 'Walled off by the ocean, connected by a bridge, and governed by a charter, the airport–island [concept] is both futuristic and feudal' (Wolf, 2000, 310).

Developers and operators of key 'hub' airports are also finding ways of connecting more seamlessly with valued spaces farther afield across the metropolitan area. This is occurring in three ways. First, in many cases, local trains to airports, which stop at intervening stations, are being replaced by premium trains which connect airports directly with valued, downtown, locations, without stopping on the way. Such local bypass is occurring in Paris, where SNCF, the rail operator, has dramatically improved non-stop services between Roissy airport and the Gare du Nord in the city centre. The declared aim is to prevent 'train robberies' by criminals from intervening poor suburbs boarding trains packed with affluent tourists and business people. But the reality is a classic example of urban bypass through network redesign. The decrease in local trains has added to the isolation of the poor suburbs near the airport and has made it particularly hard for residents of those places to access the abundant jobs created around the airport space (Olivier Coutard, personal communication).

Second, we are seeing the construction of new traveller surveillance systems covering the key transport links between valued city cores and airports. In 1997, for example, British Airways tested a smart card system that locates passengers coming into Gatwick airport from London's Victoria railway station in real time. 'The idea is to try to improve the smoothness of flow through the rail transport system and the airport gate corridors,' so improving the seamlessness of the airport's connection with the West End (Lyon, 2000, 20).

Finally, some airport operators are going further and are constructing private, dedicated rail links that carry passengers much more directly between the premium city core and the airport. Unlike previous generations of underground and surface rail connections these are, almost literally, hermetically sealed from all intervening places, embedding a profound logic of glocal bypass into their design and operation. An excellent example of this logic is the privately developed Heathrow Express link, which opened in January 1998 (see Figure 7.18). Developed by the operator of Heathrow, British Airports Authority, at a cost of £450 million, this rail link offers an 'airline style ambience' with first-class space (complete with televisions and telephones). It connects all four of Heathrow's terminals direct with Paddington station in west London – a journey of fifteen minutes compared with the hour or more on the Underground – and offers a twenty-four-hours-a-day frequency of every fifteen minutes. No stops exist between the airport and the West End of London, totally bypassing intervening spaces. Automated baggage check-in at the stations allows seamless interconnection with the airport for air travellers (Spark 1998).

THE INTEGRATION OF AIRPORTS WITH FAST RAIL NETWORKS

In the longer run, BAA want to extend this logic of glocal bypass by connecting more distant business enclaves seamlessly with Heathrow. Their first target is to extend the link to the City of London, substantially adding to the glocal connectivity of the heart of London's finance district. Then they aspire to connect Heathrow with the growing European network of *trains à grand vitesse* rail links (a connection that has already happened at Frankfurt and Paris Charles de Gaulle airports). This, in turn, is all part of the wider project of selectively integrating Europe's 'glocal' infrastructures to support economic integration (see Johnson and Turner,

Figure 7.18 'Bringing Heathrow to the centre of London': the Heathrow Express link, an excellent example of glocal bypass, planned to link directly with the trans-European high-speed TGV rail network

1997). The growing interconnection of fast rail and air travel promises an increasingly seamless interchange for valued spaces and travellers between air and surface transport (with integrated ticketing, ownership, marketing and baggage forwarding under development).

In fact, the *trains à grand vitesse* and fast rail infrastructures of Europe and Japan provide our final example of how super-fast 'glocal' infrastructure networks only partially articulate with the landscapes and socioeconomies of the cities through which they pass. This is because, as John Whitelegg argues:

high speed rail developments pick out a few favoured parts of cities from a much larger number of possibilities and confer on them additional advantages in terms of accessibility and investment. Any resorting of the space economy which produces such a 'rush' of new investment inevitably leaves somewhere else high and dry. Building high speed rail links is of little relevance to the vast majority of women, the young, the elderly, the poor and the unemployed.

(1993, 10)

Such TGV networks tend to use massive public subsidies to benefit only small parts of a few cities. They overwhelmingly benefit largely the affluent, white, male users of the network. They underpin a polarisation of the space economy within and between cities because intervening spaces remain unconnected and can actually experience worsening accessibility (because local and regional trains are reduced to allow the fast ones to operate).

Not surprisingly, given the 'stretching' and 'warping' of space economies that accompanies TGV construction, highly customised spaces are tending to emerge at key connection nodes. The EuraLille complex in Lille, northern France, is perhaps the best example of the way in which Europe's growing fast rail network has provoked the customisation of adjacent business enclaves (Newman and Thornley, 1995; Peizarat, 1997). Set at the heart of the network, at 'the centre of gravity' between Paris, London and Brussels (ibid., 248), the 120 ha site has been equipped with 'all the necessary infrastructures and developments which are naturally implied by a communication interchange of this kind' (ibid., 243): 'intelligent' office spaces, a TGV station, luxury business and hotel accommodation, and dedicated telecoms, power and water infrastructure with back-up facilities and advanced technical control. The design of the space 'has taken into account all the essential elements for the smooth running of companies operating at a European scale' (ibid., 244).

E-COMMERCE SPACES AND THE SPLINTERING METROPOLIS

The final logistical zone we need to explore is that which inevitably burgeons with the explosive international growth of on-line retailing and e-commerce: the digitally connected Internet and electronic transaction facility. Such spaces are a reaction to the exponential growth of Internet traffic and electronic commerce, which is projected to double globally every year for the first decade of the new century. Three types of space are emerging here.

'Location, bandwidth, location': optic fibre lines and the customisation of urban 'telecom hotels'

First, and against the rhetoric that the Internet is somehow 'antispatial' (Mitchell, 1995), secure developments for the mushrooming telecommunications industry are proliferating, clustered around the invisible terminals to super-high-capacity interurban optic fibre trunk lines. These, in turn, tend to be laid alongside highways or railway tracks to minimise construction costs. The select high-bandwidth access points for the Internet trunk network run by the GTE company, which provide one set of high-capacity 'points of presence' used by many Internet server farms and e-commerce operators, is shown in Figure 7.19.

Higher Bandwidth
"Points of Presence"
Lower Bandwidth
Peering Point
Network Operations Department

Figure 7.19 The high-bandwidth access points of the US Internet trunk network run by the GTE company. *Source*: http://www.bbn.com

As development concentrates around such optic fibre 'points of presence' the edges of major global city cores are now being equipped with portfolios of anonymous, windowless buildings – massive, highly fortified spaces which house the computer and telecommunications equipment for the blossoming commercial Internet industry. Akamai, for example, one of the world's largest Web server management companies, operates 'the largest, most global network of servers in the industry, deployed across multiple carriers'. Its state-of-the-art server 'farms' are housed in highly secure building complexes located in the major global cities of the world. This offers the 'closest proximity to users possible', a factor of continuing importance in the location of heavily trafficked Web sites because of Internet congestion, bandwidth bottlenecks and the dominance of global telecoms capacity by major metropolitan regions (see www.akamai.com).

'What's critical to these companies is access to business centers, access to fiber routes, and access to physical transportation,' writes the *New York Times* (21 March 2000, 4). For example, as in other major US cities, many 'telecom hotel' projects – centres for the telecom switching and equipment of multiple competitors, housed in new or refurbished factory and office buildings – are being built in and around Boston, Massachusetts. They house the region's fast-expanding Internet operators, Web providers and telecom and multimedia firms; they cluster around the city's major optic fibre terminals, such as the Prudential Center over the Massachusetts turnpike. This reflects the new mantra of many real estate providers for IT-intensive users, a slight variation on the one for the industrial age: 'location, bandwidth, location' (Evans, 1999).

To occupying companies, the physical qualities of the chosen buildings (high ceilings, high-power and back-up electricity supplies) need to be combined with nodal positions on fibre networks. 'Whose fibre (and what type of fibre, for that matter) will be a major consideration in the site selection process. A perfectly built building in the wrong part of town will be a disaster' (Bernet, 2000, 17). In a frenzied process of competition to build or refurbish buildings in the right locations, a New York agent reported that 'if you're on top of a fiber line, the property is worth double what it might have been' (ibid.).

E-commerce distribution hubs

The wider explosion of e-commerce mediated by Internet or telephone transactions, and underpinned by advanced logistics systems distributing goods to customers, is leading to the proliferation of a second type of classic 'glocal' network spaces where connections elsewhere are far more important than links with the local urban landscape. Across the Western world, in fact, declining warehouse parks are being gradually reconstituted as 'virtual' warehouses – automated spaces, close to mail and highway hubs, that are linked seamlessly into the just-in-time logistics systems designed to serve national and even continental markets for Internet-sold goods.

In the United States alone it has been estimated that 60 million ft^2 to 100 million ft^2 of new warehouses, sited on major distribution hubs, will be needed between 2000 and 2003 to meet the exploding demand from business-to-consumer ('B2C') and business-to-business (B2B) e-commerce companies (*New York Times*, 14 June 2000, 8). Often such spaces are gravitating to 'piggyback' on existing UPS or Federal Express hubs. As with other digital economy complexes, multiple fibre loops, high-capability electrical infrastructure and back-up power are mandatory for e-commerce warehousing, both for major single-occupant centres and for multi-tenant developments. Such demand for customised network spaces for e-commerce seems likely to continue growing: by 2010, it has been estimated, one-third of the world's US$60 trillion B2B economy will operate on-line, mediated by Intra and Internets, and organised through e-commerce warehousing and logistical systems (Lohse, 2000, 21).

E-commerce distribution hubs are also growing rapidly in Europe. Slough, in the south-east of England, for example, is already emerging as a national and international e-commerce distribution hub. The town houses the UK version of Amazon.com – the major e-commerce bookseller. As e-commerce explodes with the mass diffusion of the Internet, and the growing sophistication of 'virtual malls' and on-line grocery shopping, the physical, hidden support, storage and transaction processing systems for virtually sold goods are likely to become an ever more important example of urban space.

DATA HAVENS: THE EMERGENCE OF ULTRA-SECURE E-COMMERCE ENCLOSURES

Finally, the imperative of security for data storage among many e-commerce and corporate firms is such that a wide range of peripheral, isolated and ultra-secure spaces are being configured for remotely housing the computer and data storage operations of major e-commerce operators (a process known as 'co-location'). There are several elements in this process. In the first element a variety of 'offshore' small island states – Anguilla and Bermuda, to name two – are packaging themselves as 'free Internet zones' – secure locations for Web server platforms which conveniently minimise corporate taxation vulnerability to Internet regulation, and operating costs.

In the second part of the process, old disused sea forts and oil rigs are being reconfigured by e-commerce entrepreneurs in an attempt to secede from the jurisdiction of nation states altogether. For example, the self-styled 'Principality of Sealand' – a disused World War II anti-aircraft fort six miles off the coast of England – is being touted as an ultra-secure space for corporate Web servers and e-commerce platform. Were the developers' plans to be realised, the platform would escape the intervention, taxation and regulatory powers of all national and supranational bodies and states. It would also be beyond interference from any company, pressure group or hacker whilst maintaining high-capacity 20 ms links with all the world's data capitals (Garfinkel, 2000; see www.havenco.com). The Oceania project, a much larger island city-state in the Caribbean, is also being mooted, aimed at creating an unregulated e-commerce space (see http://oceania.org/).

But perhaps even more bizarre is the third part of the process: the reconstruction of Cold War missile launching sites to offer the ultimate in security against the risk of both electronic and physical incursion (D'Antonio, 2000, 26). Developers of an old Titan facility at Moses Lake, in Washington state, for example, are exploiting the old ICBM launching and control bunkers to offer 166,000 ft^2 of the most dependable and secure data storage space on the planet. The buildings are 'tremor proof, fireproof and impervious to even the most powerful tornado. Their three-foot-thick concrete walls, reinforced with steel and lead, could withstand a truck bomb the size of the one that brought down the Murray building in Oklahoma City or a ten-megaton atomic explosion just one quarter-mile away' (ibid.). All infrastructure is backed up for guaranteed uninterrupted supplies. The space's computers are separated from the public Internet to deter hackers; the service and manufacturing firms that use the space are required to have private intranets that offer the best electronic firewalls available.

THE WORLDS BEYOND GLOCAL INFRASTRUCTURE

Our exploration of the seven ranges of premium networked space, characterised by glocal infrastructural connections, is now complete. But, as with our discussion of the social landscapes of splintering urbanism in the last chapter, the analysis would remain incomplete without investigating the economic prospects of the urban worlds that lie beyond the reach of these glittering new ranges of 'glocal' infrastructural connections. What economic fortunes face those lower-income people and spaces in cities that are clearly being bypassed as the new networks become tightly configured to the needs of powerful economic actors and the socioeconomically affluent?

In a sense these multiple global 'peripheries' face a broadly similar prospect, whether they be the sociotechnical and geographical peripheries of developed cities and spaces or the burgeoning peripheries of developing 'megacities' or post-communist urban regions. All lie beyond the increasingly isolated and secessionary enclaves that are connected like beads on a string to the extended archipelagoes of glocal infrastructure nodes and networks.

In general terms there are two choices for the new geographical or sociotechnical urban peripheries. The first is just to accept the (often deteriorating) legacy of the modern infrastructural ideal. In the developed context this generally means basic and robust connections with standard services (although often with rising costs of use, deteriorating maintenance and an increasing gap between the local infrastructure and that equipping the new glocal enclaves). In many of the poorest peripheral spaces of developing cities this may mean little or no formal connection with any networked infrastructures at all. Even when connections do exist, accepting the legacy of the modern ideal may mean accepting deteriorating services and higher charges for basic services because of the effects of liberalisation, privatisation and structural adjustment.

The second option is to take action to splinter oneself and one's space away from the modern, monopolistic infrastructure network, in the hope of achieving more suitable higher-quality or more reliable connections that way. Let us briefly look at some examples of how these choices are being made, from Developing and Developed World contexts.

OVERCOMING THE FAILURES OF THE MODERN IDEAL IN THE DEVELOPING WORLD: SELF-PROVISION OF INFRASTRUCTURE

In Developing World cities the infrastructural prospects of many economic spaces beyond the glocal enclaves where infrastructure investment is increasingly concentrating remain bleak. Networked services, when available, tend to be unreliable and of poor quality. Power and electricity shortages and outages are common. Water pressures, when water is available at all, are erratic. Road passability is unpredictable. And telephone calls and electronic connections are frequently interrupted. The results are production delays, damage to electronic equipment and the loss of perishable goods or outputs – problems which significantly constrain the prospects of economic growth and development in an internationalising, network-based economy (Kessides, 1993b, 10). One study of the economic impact of power shortages and interruptions in the Pakistani economy, for example, concluded that the total effect was a 4.2 per cent reduction in manufactured exports and a 1.8 per cent reduction in GDP (ibid., 10; see Lee *et al.*, 1999).

In consequence, it is now common practice for large and small firms, as well as wealthier households, effectively to splinter themselves off from the local legacy of the modern infrastructural ideal through the development of their own private infrastructures. But such options are very costly and inefficient. In Lima, Peru, for example, it was estimated in 1993 that private water storage facilities cost forty to eighty times more than services through the public utility (Kessides, 1993b, 11). In Nigeria in 1988 92 per cent of firms had their own electricity generators. More generally, 'private infrastructure provision (for generators, boreholes, vehicles for personnel and freight transport, and radio communications equipment) constituted 15 per cent of total machinery and equipment costs for large firms, but 25 per cent for small firms' in that country (Kessides, 1993b, 12).

The costs, inefficiencies and problems involved in self-provision significantly damage the prospects of peripheral economic spaces of the city 'engag[ing] in international trade, even of traditional export commodities' (Kessides, 1993b, 13). This is particularly so as this trade is increasingly mediated by complex information and logistics systems, which require high-quality uninterrupted infrastructure networks and electronic systems in order to function. Because of the poor infrastructure and utility systems of peripheral cities in Nigeria, for example, it has been found that they are unable to support the crucial 'seedbed' of small new firms. This is because such firms tend to locate near dominant urban spaces where infrastructure is more reliable and available (ibid.).

In São Paulo, meanwhile, basic infrastructure connections are gradually becoming much more widely available throughout the city. Mains water is now available to 88 per cent of users. But connections are scarce and unreliable in the rapidly expanding informal peripheries of the urban region. The core is thickly webbed with mains water arteries which make it immune from shortages and interruptions; peripheral areas, at best, 'are covered by single branches derived from the periphery of the redundant subsystems' (Silva, 2000, 11). Such networks are 'much more vulnerable to scarcity' (ibid.). The distribution of advanced telecommunications, meanwhile, is 'very unequal within the urban structure', with investment concentrated overwhelmingly on glocal enclaves for internationally oriented businesses and socioeconomic elites (ibid.).

BEYOND THE 'DIGITAL DIVIDE': INFRASTRUCTURE AND ECONOMIC DEVELOPMENT IN THE NON-FAVOURED SPACES OF THE NORTH

Beyond the glocal nodes and enclaves of developed cities in the North, infrastructural erosion and relative disconnections tend to be more subtle. But, in relative terms, they remain powerful, especially as the wider infrastructural logics of the city start to support 'bypass' configurations that connect valued zones and spaces with each other within and between cities, whilst at the same time pushing non-valued spaces relationally further away.

Thus intervening spaces between Heathrow and the West End and City of London have, in effect, been relationally pushed away by the Heathrow Express link, a classic local bypass or 'tunnel effect' which physically traverses west London whilst according no access. As we have seen, similar 'tunnel effect' logics characterise fast rail and *trains à grand vitesse*, teleports and glocal optic fibre grids within and between dominant metropolitan areas. The trend towards carefully managed road connections, intense electronic surveillance practices and defensive architecture and urban design serve further to distance and separate the valued 'citadels' of global cities from the interstitial, lower-income spaces that surround them (see Dear and Flusty, 1998).

ECONOMIC PROSPECTS IN THE DIGITAL ECONOMY BEYOND BROADBAND NETWORKED SPACES

Many telecommunications companies and service providers, eager to reflect the splintered geographies of contemporary cities in their investment plans, are cherry-picking only the most lucrative business and professional customers in their plans to invest in broadband infrastructures. In mid 1999 about 86 per cent of all Internet delivery capacity in the United States was concentrated in the prosperous suburbs and business areas of the twenty largest cities (Lieberman, 1999):

The private sector builds where the high volume and the money is. In most communities the fiber-optic rings circle the business district. If you're in a poor suburban neighbourhood or the inner city, you're at risk. What's more, providers that have spent years building their infrastructures don't come back and fill in the underserved neighbourhoods. That may be a shrewd financial strategy. But the social impact could be devastating.

(Lieberman, 1999)

In a context where bandwidth and connectivity are, quite literally, the lifeblood of 'glocally' organised electronic businesses, this works very directly to prevent new on-line and e-commerce-oriented small businesses from competing within less prosperous, peripheral towns and marginalised inner-city spaces. Quite simply, 'the phone line is too small' (Woodbury and Thompson, 1999).

Many small firms in all sectors, faced with the growing expectation by customers of offering the multimedia-based e-commerce services that only broadband connections can support, are finding themselves in neighbourhoods that are simply on the wrong side of the 'digital divide' (Lieberman, 1999). Phillip Burgess, president of the Center for the New West, a telecommunications advocacy group, argues that 'there is a growing digital divide' between the affluent enclaves, with their multiple, broadband glocal infrastructures, and other spaces, with their threadbare, obsolescent and expensive telephone systems. This, he says, will have 'dire implications for the social and economic fabric of many communities' (quoted in O'Malley, 1999, 22).

Figure 7.20 shows the extreme geographical unevenness of the 'dot-com' phenomenon in central Boston, Massachusetts, in 1999. Domain name registrations of small and large firms on the Internet are overwhelmingly concentrated in the central business district, in upscale and gentrified neighbourhoods like Back Bay and the South End, and in high-tech clusters around MIT and Harvard in Cambridge – all districts that are benefiting from several competing broadband Internet providers. Poorer, African-American neighbourhoods like Roxbury, meanwhile, remain almost totally absent from e-commerce and e-business and are outside the build-out plans of broadband Internet providers.

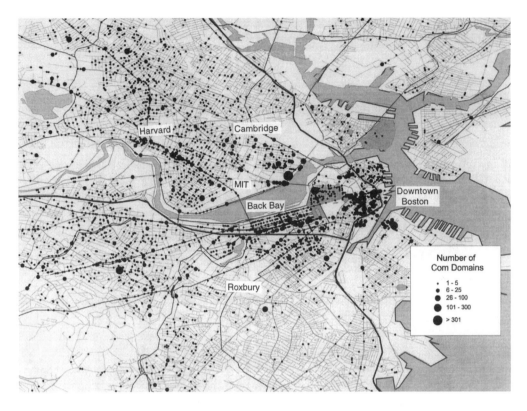

Figure 7.20 The uneven 'dot-com' geography of central Boston, Massachusetts, 1999.
Source: courtesy of Matthew Zook, University of California at Berkeley (personal communication);
see http://socrates.berkeley.edu/~zook/pubs/acsp1998.html

CONCLUSIONS

A broadly parallel logic seems to characterise the changing economic geographies of cities in the so-called developed, developing, newly industrialising and post-communist worlds. It is that of the increasingly defensive, self-contained and glocally oriented enclave, surrounded by social and economic spaces from which it seems increasingly disconnected. In many cases the divisions are increasingly salient and visible as walls, ramparts, fences, security cameras and public and private police forces attempt to secure and police the intervening boundaries, whilst at the same time maintaining the security and sanctity of the glocal infrastructural connections that traverse such boundaries.

In a sense the glocal enclaves in both 'global' cities and those in new urban peripheries are different results of the failure of the modern comprehensive infrastructural ideal. Global enclaves are increasingly securing their own high-quality, customised infrastructures to overcome the perceived inadequacies of the legacy of that ideal. The new urban peripheries, especially in the cities of the Developing World, often have to splinter themselves from the eroding and non-existent legacy of the modern ideal by paying heavily for smaller-scale, private infrastructural solutions such as generators, boreholes and the like. Certainly, our discussions in this chapter, which have engaged with an extremely wide range of cities, infrastructures and types of development project, underline that such logics of global partial interconnection, combined with growing local disconnection, seem to be a generic characteristic of contemporary urbanism.

Within increasingly pervasive neoliberal contexts, supporting the shift from the modern infrastructural ideal to project-based and 'pepperpotted' infrastructural improvements, geared to favoured and valued users, it is strikingly clear that networked infrastructures of all sorts are providing nothing less than the sociotechnical armoury which underpins and perpetuates these highly uneven urban development logics (see Pryke, 1999). It is the subtle, invisible but highly powerful configuration of the technologies and social practices of networked infrastructures that allows glocal enclaves to reach out to seamless interconnection with each other, across the polynuclear fields of the metropolis and the wider urbanising world. It is the 'bypass' configurations of water, road, rail, airport, power and telecommunications connections that allow the global city enclaves, the cyber districts, the manufacturing spaces, the 'technopoles', the back office zones and the logistics enclaves to remove themselves from the surrounding social world of the city with such apparent ease. And it is the erosion, withdrawal or neglect of the infrastructural fabric of peripheral cities and spaces that further undermines the economic prospects of such areas. This is the case even though such spaces may, geographically, be cheek-by-jowl with the gleaming glocal enclaves of the splintering city and the high-quality glocal infrastructures that interconnect such enclaves may actually pass above, through, within or underneath them. Adding the apparently pervasive shift towards private policing, the application of intense electronic surveillance and customised, closed access roads, and a heady cocktail of attempted economic secession and local disconnection emerges.

But the key word here is 'attempted'. For, as we shall see in Part Three, cities continue to be mixed economic and social spaces within which attempts at pure economic and technological secession are ambivalent, contradictory practices that are open to resistance and challenge. As we shall see then, such continued mixity and ambivalence offer a key hope to any attempt at working towards the economic democratisation of the twenty-first century splintering metropolis.

PART THREE

PLACING SPLINTERING URBANISM

Plate 18 Armed security personnel manning a gated checkpoint at the entrance to a South African suburb, 1996. *Source*: Bremner (n.d.)

8 Conclusion: the limits of splintering urbanism

   Postscript: a manifesto for a progressive networked urbanism

[Technical networks] are not objects as such, but projects, dreams, endeavours, or even entire societies.

(Latour, 1992, 2)

Cities are embodiments of desire, they are complex intersections of hopes and fears. Multi-layered and multi-faceted, they direct our attention to future and past. They are expressions of the geography of imagination; they are represented in imaginary geographies. They are tapestry-like fragments of past, present and future. It is here – in cities, in daily metropolitan life – that the unknown desires of anonymous strangers mingle.

(Lipman and Harris, 1999, 729–30)

Urbanization is more than a demographic, quantifiable phenomenon; it is civilization, and civilization determines our ways of living.

(Paquout, 2000, 86)

How is it possible to represent the networks and systems that comprise the city in the late twentieth century without a radical shift in the terms of representation itself?

(Allen, 1990, 35)

We are in the epoch of simultaneity . . . of the near and far, of the side-by-side, of the dispersed. We are at the moment when our experience of the world is less that of a long life developing through time than that of a network that connects points and intersects with its own skein.

(Foucault, 1986, 22)

8 CONCLUSION

The limits of splintering urbanism

Plate 19 Defensive office complex in Dubai, United Arab Emirates.
Photograph: Stephen Graham

This book has shown how we are starting to see a dramatic renewal in the physical, social, political and discursive salience of urban networked infrastructures and the diverse technological mobilities that they support and mediate. As a result, in many cities and parts of cities across the developed, developing, newly industrialising and post-communist worlds, networked infrastructures are, in a sense, being (re)problematised. The 'black boxes' surrounding them, built up through the uneven elaboration of the modern infrastructural ideal, are being 'reopened'. Certain powerful users are starting to look beyond the taken-for-granted point of consumption – the phone or Internet terminal, the electricity socket, the car ignition key, the water tap, the street – at the configuration of the whole technical and mobility system that supports their transport, street, communication, power and water needs.

In particular, in these times of 'globalisation', those users demanding intense local and global connectivity are starting, along with the internationalising infrastructure operators, real estate developers and urban development agencies that struggle to meet their every need, to pay considerable attention to how the whole of their networked urban infrastructures are configured, managed and developed. At the same time, in search of absolute security, privacy and control, local connections with the wider metropolis are being increasingly filtered through a widening array of walls, ramparts, security practices and access control technologies. In the process the relative infrastructural connections of less powerful users, and the spaces in which they live, seem to become more and more fragile and problematic.

To describe the dialectical and diverse sets of processes surrounding the parallel unbundling of infrastructure networks and the fragmentation of urban space we use the umbrella term 'splintering urbanism'. In response to the increasing scrutiny of local and translocal infrastructural connections, and to the parallel search for local closure, we have shown how standardised public or private infrastructure monopolies, (at least ostensibly) laid out to offer broadly similar services at relatively equal user charges over cities and regions, are receding as hegemonic forms of infrastructure management. In a parallel process, the diverse political and regulatory regimes that supported the roll-out of power, transport, communications, streets and water networks towards the rhetorical goal of standardised ubiquity are, in many cities and states, being 'unbundled' and 'splintered' as a result of a widespread movement towards privatisation and liberalisation.

These forces are combining with strategies to secede from free and open mixing with the wider metropolis to promote the widening construction of a wide range of premium and secessionary infrastructure networks – streets, transport systems, power systems, water networks and communications grids. These are customised to integrate and interconnect affluent and powerful spaces and users, and are increasingly being built and configured to bypass less valued intervening ones, where access to even basic networked services becomes undermined. Our question, in essence, has been: what becomes of cities *as a whole* in the context of the parallel dynamics of fragmentation and splintering that so often seem to accompany globalisation?

THE DANGERS OF SPLINTERING URBANISM

It is very clear that, unopposed, practices of splintering urbanism threaten to support a vicious cycle where attempts at sociotechnical secession lead to greater fear of mixing, so increasing pressure for further secession, and so on. The risk is that strategies of building 'defensible' networked spaces may:

exacerbate, rather than reverse, processes of spatial segregation and social polarisation. There is a clear redistribution of 'risks' involved in the entrenchment of 'suspect populations' in outcast, under-protected, disordered 'ghettos', and of 'innocent populations' in over-protected, ordered consumerist 'citadels' and residential 'enclaves'.

(McLaughlin and Muncie, 1999, 135)

The result, as Mike Davis (1992) has suggested for Los Angeles, may be the emergence of urban landscapes made up of layers of premium network spaces, constructed for socio-economically affluent and corporate users, which are increasingly separated and partitioned from surrounding spaces of intensifying marginality – spaces where even basic connections with elsewhere, and basic rights to access spaces and networks, are increasingly problematic. In this understanding of contemporary urban change, dominant practices of urban design, network configuration, electronic access control, and police, security and institutional enforcement, are increasingly seen to be working in parallel to support the sociotechnical partitioning of the metropolitan and, indeed, societal fabric.

To support and enforce such processes of change, electronic consumption and surveillance systems, with increasing degrees of automation, threaten to provide silent, invisible and pervasive networks which cybernetically police the boundaries between premium and marginalised network spaces, with unprecedented potential for exclusion. David Lyon asks the important question: 'is what faces us a world of electronic technologies that classify us clinically, include and exclude by consumerist criteria, and are backed up by police and welfare departments? . . . will the new "non-persons", segregated by surveillance systems, be failed consumers?' (1994, 211). As more and more network spaces in the city become commodified and electronically monitored and controlled, within broader contexts of neoliberal consumerism, Norris *et al.* (1998) warn, from a UK perspective, that:

those who cannot pay will be excluded from motorways; known troublemakers from football grounds; the unsightly casualties of [neighbourhood mental health programmes] removed from the decorous order of city streets and shopping malls; known shoplifters and fare dodgers excluded from shops and transport systems. . . . If the growing divide between those who have and have not and those who are included and excluded is intensified through the use of new technology, there is a real danger that our cities will come to resemble the dystopian vision so beloved by futuristic film makers.

Much of the evidence in this book certainly lends support to these depressing interpretations of the contemporary connection between infrastructure, technological mobilities and metropolitan and societal change. We are in no doubt that, whilst they are mediated through diverse cultural, historical and geopolitical contexts, dominant spatial practices across a very wide range of cities seem to be underpinning variations within the whole gamut of trends that we encompass by the term 'splintering urbanism'.

This is not to maintain some easy similarity across the wide range of contexts addressed in this book. Rather, we suggest that there are compelling reasons for attempting to understand the restructuring processes that we have explored across Western, developing and post-communist cities, in parallel. For example, in many Western and non-Western nations, emerging 'third way' politics, which assert the need to move beyond welfare and developmentalist states, seem to rest on a 'politics of exclusion' which brings shifts:

from the social welfare state to the social control (police) state, replete with the dramatic expansion of public and private security forces, the mass incarceration of excluded population (disproportionately minorities), new forms of social apartheid maintained through complex social control technologies, repressive anti-immigration legislation, and so on. It has also entailed, under the Third Way's deceptive discourse of 'local politics' and 'community empowerment', a shift in the responsibilities for social reproduction from the state and society as a whole to the most marginalized communities themselves.

(Robinson and Harris, 2000, 51)

But our story does not end here. For our many explorations of practices of splintering urbanism constitute dominant, but by no means all-encompassing and hegemonic, practices within the multiple social worlds of the contemporary urban condition. In this book we have necessarily centred largely on the dominant practices of the powerful, in their attempts to construct, manage and regulate secessionary and premium network and urban spaces. To avoid an overly narrow perspective, then, we now need to place a series of caveats against this book's dominant narratives.

Two tasks, in particular, remain for us in this concluding chapter. First, we need to reflect on practices of urban splintering by placing them within a wider, historical urban context. Second, we need to understand the ways in which the essential nature of contemporary urban life necessarily places limits on the degree to which practices of urban splintering can proceed, without reaching limits and running into strategies of resistance and the mobilisation of alternatives. It is to these two remaining challenges that we turn in this concluding chapter. They provide the basis of the Postscript which follows. In this we round off the book by exploring the broader implications of its analysis for urban theory, research, politics, and practice.

PLACING SPLINTERING URBANISM: THE WIDER CONTEXT

Our first challenge is to place our explorations of splintering urbanism within the wider urban and historical context. Four points must be stressed here.

AVOIDING SIMPLISTIC BINARY TRANSITION MODELS

First, it is necessary to reiterate that this book does not argue that all networked infrastructures in all places are somehow moving *en masse* from an era of standardised coherence to one of splintered fragmentation. We cannot reduce our analyses of a wide range of infrastructure networks and systems of technological mobility, within and between a broad diversity of cities, to some simple, phase-based, transition model. Such an approach would clearly risk the twin dangers of ethnocentricism and analytical reductionism. We would also be open to accusations that our analysis simply reified infrastructure networks as agents of urban transformation in their own right. And we would fall prey to simple functionalism by assuming that there will be some simple, bidirectional 'fit' between the 'needs' of contemporary global capitalism and the infrastructure networks that are now being constructed for it. In short, we are aware that, as Wacquant puts it, 'binary oppositions are prone to exaggerate differences, confound description and prescription, and set up overburdened dualisms that miss continuities, underplay contingency, and overstate the internal coherence of social forms' (1996, 124–5).

It is therefore necessary to be wary of the risk of overgeneralisation and oversimplification. As we stressed in Chapter 4, the many cases of splintering urbanism discussed in this book involve a wide variety of related but distinct processes: from the construction of completely new private infrastructures to the sociotechnical reconfiguration of old networks; from classic cases of wholesale privatisation to public 'reregulation' and the changing practices of continuing monopolies; from the combined secession of interlinked networks and built urban spaces to the 'virtual' construction of infrastructure markets over the single networks that were the legacy of the modern infrastructural ideal. The overarching notion of splintering urbanism is, we feel, a useful heuristic device to grapple with this diversity. But we do not intend it to be either a new urban 'metanarrative' or some reductionist analytical straitjacket.

INERTIAS AND CONTINUITIES WITH THE MODERN INFRASTRUCTURAL IDEAL

Second, the cases explored in this book tend to represent only the most visible and 'extreme' examples of the construction of premium and secessionary networked infrastructures, and the parallel marginalisation of less powerful users and spaces. The book has, for obvious reasons, not been able to address all cities everywhere. Our examples have been drawn primarily from UK, US, Asian, Australian, African and Latin American cities. We have said much less, for example, about cities of continental Western and Eastern Europe.

However, in all the cases discussed, we need to remember that the territorial monopolies, cross-subsidies and 'bundled' networks of the high modern period, along with the associated political practices and normalised 'black-boxed' consumption practices, have not been removed wholesale. In many cases they remain substantially intact; in a good many urban contexts the shift from emphasising the sociospatial homogenisation of networks to the 'unbundling' of different practices for different users of streets, power networks, communications, water and transport grids is in fact a good deal more subtle. Rather than the spectacular emergence of wholly new premium networked spaces, it is manifest in strengthened law-and-order practices on streets, in changed practices of marketing within continuing public or private monopolies, in the subtle exclusions of call centre queuing algorithms, or in the adoption of geodemographic targeting techniques (Graham, 1997; Offner, 2000). In all cities, therefore, the trends surrounding splintering urbanism amount not to some simple and wholesale urban 'revolution' but to:

a more complex patterning of old and new, of continuing trends and new forces. . . . New and old processes join together in complex ways and do so in particular places at particular points in the history of those places. . . . The spatial form of the city inhibits rapid and large-scale transformations. . . . In sum, the contemporary city hardly reflects postmodernism or post-Fordism in a one-to-one correspondence.

(Beauregard and Laila, 1997, 328–39)

HISTORICAL PRECEDENTS

Third, whilst they are no doubt in the ascendancy at present, it is also necessary to stress that customised and unbundled urban infrastructure networks, and secessionary socioeconomic enclaves geared to affluent groups, are not new. The dialectic of privacy and closure versus openness and mixing has been played out since the very beginning of urban life (see Waterhouse, 1996; Soja, 2000).

Users in social and economic enclaves for powerful groups have sought to construct their own 'closed' infrastructures throughout urban history. Processes of splintering urbanism seem to result in the kinds of fragmented and uneven infrastructure networks and partitioned urban spaces that were characteristic of most Western cities before the emergence of the modern infrastructural ideal (Graham and Marvin, 1995). Consider, for example, the closed and private Victorian streets that were common in the West End of London. Remember, too, the fragmented jurisdictions, privatised planning practices and secessionary tendencies of countless twentieth century US suburbs and the infrastructure networks that supported them, especially those which underpinned patterns of stark racial segregation (see Bayor, 1988).

Socioeconomic enclaves have also long been supported more subtly by uneven development practices *within* the discursive and technological construction of the standardised and 'homogeneous' infrastructural monopolies of the modern ideal (Cox and Mair, 1988). Thus, as we stressed in Chapter 2, the modern infrastructural ideal was never materially achieved in practice. It was always as much a symbolic and discursive construction as a technological reality. It was closely bound up with the wider legitimation of modern national and local states and urban professions, with the framing of particular cultures of urban territoriality, and with the production of dominant notions of scale. Unevenness and bias in the orderings of infrastructure networks – in terms of both quality and accessibility – remained in all cities within attempts to work discursively and materially towards standardisation, ubiquity and homogeneity.

The history of developing and colonial cities, in particular, as Balbo (1993) discusses, has also long been one of the use of constructed built form along with customised infrastructure to maintain the distant connections and local (dis)connections of socioeconomic enclaves. A classic example was the International Settlements of early twentieth century Shanghai, each of which had dedicated infrastructure networks serving dominant and elite spaces.

However, Noam Chomsky (1993, 8) has argued that intensifying globalisation and international neoliberalism are leading to such practices being exported from the Developing World to the Developed World, a perspective which supports this book's efforts to develop an integrated perspective on urban restructuring across such increasingly similar urban domains. The processes analysed above, in cases like global financial enclaves, foreign direct investment complexes and fortressed condominiums, represent a renewal and strengthening of old and established practices of distant infrastructural connection and attempts at filtering local connection, rather than something radically new. They simply do this with renewed degrees of intensity and global reach and, often, without the discursive pretence of eventually moving towards universal access or territorial ubiquity.

DANGERS OF ROMANTICISING THE MODERN INFRASTRUCTURAL IDEAL

Finally, we should be cautious not to fall into the trap of romanticising the modern infrastructural ideal, or the wider modern urban project within which it rose and fell (Bhabha, 1990). As we saw in Chapter 3, whilst in many cases it did succeed in overseeing the democratisation and roll-out of infrastructure networks, this tended to be achieved through masculinised, centralised and colonialist notions of 'order' and 'coherence' which were far from emancipatory of women, the disabled, indigenous and colonised populations or other marginalised social groups (King, 1996; Sandercock, 1998b). The normative concept of the networked city that was at the root of the modern ideal was totalising and centralised and driven by assumed and paternalistic notions of 'order' and (sometimes) redistribution. The modern ideal's great narratives of progress and emancipation for all tended to hide many practices of subjugation, repression and marginalisation. And often its decidedly paternalistic notions had little connection with the needs, desires or identities of the groups that were supposed to benefit most from its strategic, modernist vision.

CHALLENGING HEGEMONIES: RESISTANCE AND THE LIMITS OF SPLINTERING URBANISM

Cities are a mess by definition.

(Dave Hickey, quoted in Allen, 1994a, 39)

Nobody gets away with wrapping the breathtaking complexity of city building into a simple package.

(Waterhouse, 1996, xxii)

Whenever there is domination, there is resistance to domination. Whenever there is imposition of meaning, there are projects of constructing alternative meanings. . . . While the space of flows has been created by and around dominant activities and social groups, it can be penetrated by resistance, and diversified in its meaning.

(Castells, 1999a, 297)

A tunnel, any tunnel, directly threatens the integrity of that which rests above it. Such a passage erodes foundations from the bottom up, its presence rendering superstructures perilously heavy with contradiction.

(King, 2000, 269)

The second challenge in this concluding discussion is to explore and understand what we term the limits of splintering urbanism. For it is very important to realise that the sociotechnical secession of premium networked spaces from the wider metropolis will never be some simple, attainable, process. We can be certain that starkly polarised development logics are shaping the socioeconomic landscapes of cities as far afield and as different as Tyneside, São Paulo, New York, Istanbul, Guangzhou, Manila, Los Angeles, Mumbai, Jakarta or Bintan island. But it is important to realise that these are resulting from *strategies* of coalitions of interests within the contested and highly complex geopolitical, governance and socio-cultural contexts of their respective cities. These strategies do nothing to guarantee simple, total or easy secession. In fact, the dream of a totally purified, hermetically sealed world of splintered, premium networked spaces for affluent or powerful users in the contemporary metropolis is exactly that – a (modern or postmodern) urban dream which is just as unrealisable as the fantasies of clean, functional and perfectly geometric order promulgated by modernist visionaries in the early and mid twentieth century (see Castells, 1997a).

Take an example. In developing cities premium network spaces cannot entirely isolate themselves from the public health externalities associated with the poverty, poor sanitation and high rates of disease caused by poor or non-existent water and sewerage infrastructure in surrounding shanty districts. Even with the proliferation of private boreholes, generators, private medicine, supplies of bottled water and the usual techniques of geographic separation and fortification, 'the public health of one affects the other almost instantly' (Baird and Heintz, 1997, 13).

We therefore need to be wary of assuming the easy emergence of utterly separated premium network spaces; of completely integrated, all-seeing and perfectly effective electronic surveillance webs supporting the maintenance of urban social apartheid; and of completely segregated cities where different socioeconomic circuits cease to have any meaningful overlap or interconnection. Equally, we must be careful not to resort to cartoonish and deterministic readings of the technological 'impacts' of new infrastructural innovations (Graham and Marvin, 1996). The reality of technological innovation, the restructuring of urban landscapes, and social and cultural changes in cities, are a great deal more 'messy', difficult, contingent and open to contested interpretations and applications than such scenarios imply (Thrift, 1996b, c; Bingham, 1996).

There are at least six reasons why this is so, each of which deserves some discussion here.

PREMIUM NETWORK SPACES AND GLOCAL ENCLAVES AS CONTESTED URBAN ORDERINGS

First, the actual practice of attempting to configure infrastructure networks and urban spaces to support glocal enclave formation, and the construction of premium or secessionary network spaces, do not serve to insulate and separate such network spaces entirely from the social and economic dynamics of the wider metropolis. The differentiation of urban space, whilst often fuelled by efforts to use social power and technology together to create normative urban 'orders', is always uneven and porous. It is framed by wider efforts to manage and govern urban complexity and conflict and wider power-laden and contested representations of what constitute 'order', 'chaos', 'progress' and the 'good city' (Mooney *et al.*, 1999).

The fragmented and polynuclear geographies of many contemporary cities can mean that the social and technological worlds of the powerful and the marginalised are rarely far apart in terms of geographical distance. This is the case even in relatively new cities, where much of the urban landscape is actually made up of the tightly regulated architectures of secessionary network spaces. It is certainly the case in many older cities in Europe, Latin America, North America and Asia where the public street system, in particular, has maintained some overarching structuring power. In these cities the diverse social worlds of cities continue to intersect and cross-over each other, even if the relational connections between the rich and marginalised may weaken through the processes of splintering urbanism and privatisation.

We must also remember, of course, that premium network spaces and socioeconomic enclaves are not institutionally or politically unified as single actors. They, in turn, have their own complex and contested politics and spatialities. Supportive coalitions of developers, infrastructure operators, social and economic interest groups, politicians and their allies often meet internal as well as external resistance. Some Business Improvement Districts in New York, for example, have been plagued by conflict as store owners struggle against higher BID taxes and the perceived negative effects of gentrification and self-governance (Schulman, 1999). Many residents, particularly the young, are likely to resist and transgress the sort of strict restraints on behaviour now commonly imposed by the regulatory boards of gated communities (for example, curfews and the notification of visitors) or the operators of malls (Amin and Graham, 1998a; Poindexter, 1997). Dovey stresses the importance of this resistance among 'children who have not chosen such totalizing controls' and, to him, 'have a right to grow up in a public community' (1999, 153).

In fact, it is most fruitful to understand premium networked spaces and glocal socio-economic enclaves as particular attempts at ordering urban space–times within the context of what Mooney *et al.* call the 'jumbled orderings of the city' (1999, 346). Such attempts are (internally contested) efforts by powerful coalitions of interests to use their global power to try and construct purified and partial spaces. They are strategies to mobilise and enforce disciplinary practices of boundary control which impose what Norris and Armstrong (1998) call a 'normative space–time ecology' of who 'belongs' or can use the urban and network space where and at what time. They can therefore be understood as attempts to withdraw from the clashing social, economic and cultural differences that infuse contemporary urbanism.

AMBIVALENT CONNECTIONS:
THE NEED TO MAINTAIN LINKS TO THE REST
OF THE METROPOLIS

The new enclosures cannot endure because they cannot sustain themselves.

(Dumm, 1993, 192, cited in Soja, 2000, 321)

Second, premium network spaces must continue to maintain connectivity with wider, public, infrastructure networks and systems of technological mobility. Most require this in order to function, as in the cases of the City of London, FDI enclaves, gated communities and malls, the virtual utility services and e-highways. Despite the splintering of infrastructure networks, connectivity with wider, public networks, and the so-called 'network externalities' that accrue from such interconnection, remain absolutely central to supporting the overall functionality of infrastructural connection (Offner, 2000).

Glocal enclave spaces – the spaces of transnational economic globalisation – thus need to retain other connections with the wider city and beyond. They depend on labour inputs of various skill levels. They require a wide range of servicing (cleaning, catering, security guards, drivers, landscape staff). And they remain dependent on wider infrastructural inputs of power, water and financial and physical investment. In the case of malls, resorts and theme parks, premium networked spaces also need to draw on urban, regional, national and international visitor markets. Glocal enclave spaces are also supported and legitimised by a wide range of political actors and power bases within various scales of governance, from the whole edifice of financial and economic policy in the World Bank and IMF, through supportive nation states, to entrepreneurial and financial subsidies from municipalities and urban development bodies. Secessionary network spaces also achieve considerable support from the targeting strategies of aspiring international, as well as local, infrastructure companies themselves.

Residential secessionary spaces, too, often maintain a huge reliance on all aspects of state support and legitimation. Jennifer Light (1999), for example, has shown that US gated communities, despite their private security systems, local government associations and dedicated private police forces, continue to make disproportionate demands on public sector law enforcement and emergency services.

'SPILLOVERS' AND 'NEGATIVE EXTERNALITIES':
THE IMPOSSIBILITY OF 'PURE' URBAN BORDERS

To enter you pass through two security gates, each with its own intercom. The first lets you into a courtyard, the second into the house. Despite the vigorous security measures, on January 13th this year, the resident drove into his garage and, from his car window, found himself staring into the barrel of an AK47.

(R. Simmonds, 5 June 1998, describing a robbery in Johannesburg;

cited in Bremner, n.d., 60)

Third, the sheer diversity of identities, social worlds and political pressures in contemporary cities can quickly swamp crude efforts to impose simplistic notions of exclusion and purified urban order. Contemporary cities remain as sites of jumbled, superimposed and contested orderings and meanings; they are 'points of interconnection, not hermetically sealed objects' (Thrift, 1997a, 143). Multiple 'spillovers' can easily saturate and overwhelm simple attempts at establishing and maintaining 'hard' disciplinary boundaries. Virtually all boundaries remain to some extent porous. Perfect control strategies are never possible. 'Public' mixing can often still overcome strategies of separation and control.

The wider social worlds of the city therefore tend to find ways of overcoming the barriers erected around the 'dominant rhythms' of glocal enclaves (Pryke, 1999, 245). The mixity and diversity of the contemporary city tends to remain far too pervasive and powerful to be simply and starkly separated out through some simple construction of gleaming, sociotechnically 'pure' enclaves of dazzling modernity, customised connectivity and stark, secured boundaries. Very often 'the juxtapositions, combinations, and collisions of people, places, and activities' in the contemporary metropolis 'create a new condition of social fluidity that begins to break down the separate, specialized, and hierarchical structures' which dominant coalitions strive to build into city form (Crawford, 1999b, 34). Under some circumstances, such fluidity can help to encourage liberatory potential. As 'chance encounters multiply and proliferate, activities of everyday space may dissolve the predictable boundaries of race and class, revealing previously hidden social possibilities that suggest how the trivial and marginal might be transformed into a kind of micropolitics' (ibid., 34–5).

We must also remember that the cost of enforcing space–time boundaries around premium networked spaces can be hard to sustain. This was the case in 1995 when the policing cost of enforcing a ban on street traders in central Mexico City – put in place to 'sanitise' the space for international tourists – led, along with violent protests, to its removal. Instead, spaces were allocated that traders could rent at reasonable levels from the city authorities (Harrison and McVey, 1997, 323).

In newly industrialising, developing and some extremely dynamic Northern cities, glocal enclaves must also often contend with the overwhelming scale and chaotic logic of social and demographic shifts, with volatile economic cycles, and with spontaneous growth, all of which are part and parcel of 'hyperurbanisation'. Glocal enclaves, for example, often tend to become the focus of spontaneous mass migration from among the urban millions of unemployed or underemployed seeking work. They can simply be swamped within the wider processes of informal urbanisation that this triggers off. And they are vulnerable to sudden economic crises like those that afflicted many Asian cities in the late 1990s.

Grundy-Warr *et al.*, for example, describe how the attempts to develop sealed-off 'glocal' tourist and manufacturing enclaves in the Indonesian islands within the SiJoRi growth triangle, that we encountered in Box 7.5, have actually been overwhelmed by in-migration and the spontaneous construction of squatter settlements (1999, 324). As they suggest, all manner of attempts at boundary control in the new cityscapes have been effectively undermined. 'As well as the on-park housing and self-contained services [of the newly built tourist and manufacturing enclaves], tight border protection was expected to prevent an influx of job seekers and associated shanty town development. Residence was intended to be restricted to employed persons' (ibid.).

In this case, all manner of attempts at boundary control in the new cityscapes – erecting barriers, building walls, stipulating who has access and who does not, employing private security companies, customising premium infrastructure only to the needs of those inside the enclave – have been effectively undermined and rendered useless. Infrastructure networks have been 'illegally' accessed. 'The island's reputation as a booming economy has overwhelmed official controls. As a consequence, it has not been possible to entirely separate [the enclaves] from the surrounding development of the island. These processes draw attention to the difficulty of securing growth through protected enclaves' (ibid.).

It is also impossible to provide customised enclaves with high-quality environments that are disconnected from wider urban and global environmental issues. Do what they may, local enclaves cannot totally insulate themselves from all the scales of environmental problems and risks that they attempt to transcend. They are dependent on wider infrastructure services and, while these may be provided more 'greenly', and with less environmental impact, than conventional networked technologies, they have to operate within a wider urban, national and global environmental context. While the local enclave space may be high-quality, green and clean, such enclaves can never be totally insulated.

Secessionary network spaces are also open to wider environmental risks: climate change, sea level rise, chemical and biological pollutants, air pollution, noise, hazards, etc. Although they may be able to provide high amenity 'green' landscapes, and efficient or autonomous technologies, the wider urban environment of contemporary cities dominates and cannot be resisted through green enclaves.

STRATEGIES OF RESISTANCE AND THE POLITICS OF REPRESENTATION

No matter how terrifying a given system may be, there always remain the possibilities of resistance, disobedience, and oppositional groupings.

(Foucault, 1984, 245)

Fourth, and perhaps most important, the continuation of public mixing in cities means that spaces continue to exist within which strategies of resisting secession and urban splintering can be constructed. The life of major cities cannot be simply programmed like some computer by powerful socioeconomic or political interests, even within increasingly extreme and uneven capitalist contexts. Urban life is more diverse, varied and unpredictable than the common reliance on US-inspired urban dystopias suggests. This point deserves rather more elaboration.

ASSERTING THE CONTINUED 'PUBLICNESS' OF CONTEMPORARY URBANISM

As in so much of the literature on cultural hegemony, the social control of consciousness is alleged but never proven.

(Ley and Mills, 1993, 258)

Conflict means danger, often it means injustice, but always it means life and human yearning, human fascination.

(Merrifield, 2000a, 487)

Attacking the dystopian and *noir* portrayals of the likes of Mike Davis (1990) and Michael Sorkin (1992), Andy Merrifield lambasts their assumption that the diverse social groups of cities can simply be trammelled and controlled by all the accoutrements of postmodern urbanism, even within contexts of growing commercialism and social polarisation. 'It is patronising,' he writes, 'to believe that citizens of cities are easily hoodwinked by a city supposedly comprised of bits and bytes, texts and simulations, semiotics and CCTV systems' (1996, 67).

Merrifield argues that spaces do remain in cities, especially those which have more or less robust 'public' qualities and unified street networks, where difference and diversity can and do still come together under relatively free conditions. This is especially so in cities like many of those in Western Europe, which have not been subject to the massive construction of tightly policed and self-sufficient premium network spaces for affluent groups; where well financed (although rapidly restructuring) welfare states continue in some shape or form to help soften polarisation; and where traditional 'public' street systems still anchor and configure the overall urban experience, at least in the core (see Musterd and Ostendorf, 1998). To Merrifield:

great cities, by their very definition, have enormous diversity of ingredients and people, and they aren't mere passive pieces on a chessboard that big capital can move around or exclude at whim. Invariably, new forces of disintegration can be and are used as the medium for new forms of integration and affirmation. That is how and why people survive in cities and rebuild their lives out of so much rubble, injustice and disappointment.

(1996, 67)

THE INEFFECTIVENESS OF MANY DISCIPLINARY EFFORTS

Even in Western Europe, however, we must recognise that sharpened social and spatial inequalities, combined with rallying cries to 'urban competitiveness', growing racial tensions, the restructuring of welfare states, and growing experimentation with neoliberal management models, are leading to strengthened attempts to discipline public spaces. In Berlin, for example, as in so many other cities, the police and urban regulation and transport agencies have sought tightly to manage and control 'proper' behaviour in the city's increasingly high-profile and recently redeveloped 'prime spaces' – a reaction to the growing number of homeless and the efforts to relaunch Berlin as a 'global city' (Grell *et al.*, 1998). The first privatised public spaces have emerged in the city. As elsewhere in Germany, public CCTV is being mooted. Anti-homeless measures are now very strong. The police have designated thirty 'dangerous zones' within which they have intensified their powers of search and scrutiny. And the formal banning of people from public spaces and transport systems has skyrocketed (to 160,000 on the rail system alone in 1997) (ibid.).

However, as in most similar cases, the new regulatory packages are far from being 'coherent' and are of only limited effectiveness (Grell *et al.*, 1998, 211). They have failed in their desired aim of appropriating city-centre streets and spaces for their target 'clients': corporate office users, tourists and middle-class professional people. This is for two reasons. First, as in a wide range of cities, the initiatives have led to political resistance from such social movements as Inner!City!Action! which have successfully exposed processes of social and spatial polarisation in Berlin through highly publicised occupations, demonstrations and media interventions.

Second, it is clear that 'people are not willing, or able, to leave the central areas. People tend to come back and "reclaim" public spaces because they critically depend not only on the inner-city service structure but also on the inner-city spaces as places of income (panhandling, street trading, etc.) and communication' (Grell *et al.*, 1998).

DEMONSTRATIONS, SOCIAL AND CULTURAL MOVEMENTS, AND URBAN INVENTIVENESS

It is easy, in short, simply to forget 'the creativity, the sheer inventiveness of the inhabitants of cities' (Thrift, 1997a, 143). We must remember that 'human beings are not the unified subjects of some coherent regime of domination that produces persons in the form of which it dreams' (Thrift, 1997b, 136). The practices of urban life, and the organising power of social and cultural movements, offer channels through which logics of splintering urbanism can be resisted and transgressed (see Castells, 1983; Bell and Haddour, 2000; Mayer, 1999). Such practices tend to resist any normalising attempt by dominant power-holding organ-isations; resistance practices are 'integrated *within* the very relations and pathways produced by a pervasive and subtle disciplinarity' (Leong, 1998, 196, original emphasis). They can also subvert the dominant ideologies of splintering urbanism, so closely woven as they often are into ideological discourses of 'globalisation', 'competitiveness' and the supposed economic and technocratic imperatives of 'new technology' and the so-called 'Information Age'. And resistance practices can help shape the creation of places in the interstices between the logics of premium network spaces. These can often be fluid places which support counterhegemonic cultures and identities which tend at once to be 'spirited, threatening, dishevelled, dissenting, dynamic' and which help 'celebrate the force of social being' (Lipman and Harris, 1999, 733; Crawford, 1999b; Millar, 1999; Harrison and McEvey, 1997).

In short, social unrest, social movements and sometimes violent incursion, fuelled by the stark injustices as well as the complex identity politics of contemporary cities, constantly threaten premium and secessionary network spaces (see Pile and Keith, 1997; Body-Gendrot,

2000; Castells, 1997a). This has been demonstrated by the struggles of coalitions of homeless people and supporters against the enforced 'sanitisation' of public spaces and parks (as at Tompkins Square park in New York; Smith, 1993). The increasing power of international coordination in such protests was also shown by the powerful global 'Action against capitalism' demonstration in the City of London and thirty or so other 'global cities' on 18 June 1999 (see Slater, 1997). On that day financial districts were cordoned off and brought to a standstill, stock exchanges were barricaded, corporate headquarters were deluged with protest and parades and festivals were staged celebrating economic and ecological alternatives to global capitalism. Global anticorporate protests against the World Trade Organisation summit in Seattle in November 1999 and the IMF–World Bank meeting in Washington DC in April 2000 were even more intense and well coordinated (Cockburn and St Clair, 2000). Momentarily they succeeded in breaching what Beckett calls the 'wall of euphemism between the grey sheds' of transglobal corporate production chains and 'the grand corporate mission statements' of the transnationals themselves (2000, 17; see Klein, 1999).

We must be careful, at the same time, not to overromanticise or oversimplify the resistance efforts of social movements, or to assume the easy coherence of low-income neighbourhoods and spaces. As Erhard Berner suggests, 'the urban poor, in general, are not one collective actor, but a multitude of groups' who, in many cases, 'are largely indifferent to each other' (1997b, 195).

The limits of urban panopticism: resistance and the daily life of the city

There is a world of difference between minor transgressions of prescribed dress and behavior codes in [premium network street spaces] and those larger acts of spatial subversion in which the public had periodically reappropriated places of power for demonstrations of solidarity and oppositional practice.

(Crilley, 1993, 158)

Most practices of resistance are more prosaic and quotidian. Many are playful, representing a 'refusal to disappear beneath the imperatives of spatial regulation that favors select target markets' (Flusty, 2000, 156). Within malls, for example, groups of young people disrupt and resist the attempted panoptic control of CCTV and security regimes. People most targeted by CCTV in British town centre management strategies – for example, groups of young black men – have been shown to develop elaborate practices to exploit system 'blind spots' to undermine and transgress the attempts at social control that follow (Norris and Armstrong, 1998, 1999; Toon, 2000). Many young people 'develop strategies for maintaining their presence' against the disciplinary practices deployed against them, whether it be through asserting their rights as consumers, directly dealing with conflicts from security staff, developing their own covert spaces or coordinating through the Internet (Hill and Bessant, 1999, 46).

Similarly, Steven Flusty has shown how buskers, skateboarders and even poets in Los Angeles work to exploit the impossibilities of real urban panopticism. One busker, for example, says he 'knows where to find every security camera on Bunker Hill' (ibid., 152). All the people Flusty talks to exploit the fact that 'no matter how many "armed response" patrols roam the streets, and no matter how many video cameras keep watch over the plazas, there remain blind spots that await, and even invite, inhabitation by unforeseen and potent alternative practices. Even in a totally rebuilt and totalizing environment like Bunker Hill panopticism fails' (ibid., 157).

RESISTANCE AND THE POLITICS OF REPRESENTATION

> There exists no privileged vantage point from which to attain panopticity in representations of the city.
>
> (Flusty, 2000, 157)

The politics of representation are a key aspect of many acts of resistance to the reconstruction of parts of cities as premium network spaces (see Shields, 1995; Balshaw and Kennedy, 2000; Bhabha, 1994). Sophie Watson writes that 'cities are constituted in part by everyday interactions and conflicts as well as by struggles around collective consumption (1999, 2). But they 'are also constituted by struggles that may be less immediately observable . . . struggles around questions of representation, meaning and identity' (ibid.). Normative notions of the 'good city' become key here, as do representations and constructions of particular notions of 'order' and 'disorder' within the city, and representations of how particular infrastructural or technological configurations might be crucial in leading a city to such a future (Mooney, 1999; Bell and Haddour, 2000; Fainstein, 1999).

Many acts of resistance struggle not just against the actual material construction or maintenance of premium networked spaces or glocal enclaves (and the concomitant marginalisation of poorer or informal settlements). The wider ideological and political representations with which practices of splintering urbanism are associated are also a major focus. For example, residents of the *gecekondus* informal settlements that surround the ultramodern new towns on the edge of Istanbul have protested against the representation of their spaces by city authorities as 'backward', 'ignorant' and 'uncultured ganglands' which are 'run by the mafia' (see Chapter 6). They have struggled to resist the stereotyping by dominant power holders which places them and their spaces at odds with the ideological construction of 'the new Istanbul' that has been so instrumental in legitimising the construction of premium networked enclaves like Esenkent and Bogazköy (Mooney, 1999, 67; see Aksoy and Robins, 1997).

URBAN LIFE BENEATH THE 'SCAN OF GLOBALISM'

It follows that secessionary spaces and networks are not set in stone. Political and social responses to attempts at secession, including social movements, protests and resistance, can lead to the dismantling of premium network spaces and the instigation of more socially and spatially equalising regulatory, governance and technological regimes. Democratic resistance and social mobilisation can serve to balance the secessionary tendencies with more redistributive design, development, regulation and governance strategies.

Recent movements against CCTV installation in Bradford and Manchester have attempted just this (Lyon, 2000, 189). To some extent, it is also occurring in post-apartheid South Africa as infrastructure and transport regulators and urban social movements seek to establish norms for the basic networking of black townships (although see Bond, 1999, and Knosa, 1995, for critiques). São Paulo, too, has experienced a massive boom in social movements since the 1980s, processes through which 'excluded residents discover that they have rights to the city' (Caldeira, 1996, 324). Whilst largely failing to undermine the extension of guarded condominium and mall complexes for the rich, they have, in many cases, significantly improved the situation of many poorer neighbourhoods.

John Kaliski even believes that malls and privatised public spaces, in their own way, can be 'ultimately as conducive to urban life, ritual, and surprise, as dense urban neighborhoods' (1999, 94). Whilst this may be overstating the situation, he argues that processes of urban life, ritual and surprise are generally just harder to spot among the secessionary network spaces of contemporary cities, especially by urbanists brought up on Jane Jabobs and other idealised portrayals of traditional urban neighbourhoods. Even in his home town of Los Angeles, the favourite subject of urban *noir* dystopianists like Mike Davis, Michael Sorkin and Fred Dewey, Kaliski suggests:

curiously, many of the social transactions that are shaping the tenor of culture occur in the very places most subject to the scan of globalism. Shopping mall culture, gated enclaves (whether suburbs or rock houses), omnipresent recording, and surveillance of every aspect of daily life do not seem to limit ever new and evolving cultural expressions and mutations born of unexpected gatherings. The easy reduction of these places to unitary theories or definitions of globalized space overlooks the physical workings of their quotidian elements.

(1994, 7)

THE CONTINUED IMPORTANCE OF LOCAL AND NATIONAL REGULATION

The complex institutional fabric of urban governance, meanwhile, may work to resist the simple and easy secession of socioeconomically affluent groups, and their systems of infrastructure and technological mobility, from taxpaying and governance systems (McKenzie, 1984). In most cases, scope continues to exist at the level of local and national state and governance regimes to reassert and even strengthen leverage over the production and regulation of premium networked spaces. Local municipalities and planning agencies can renege on licence agreements and bring networks back into direct connection with public network operations. They can also undertake proactive initiatives to develop and maintain socialised and ubiquitous infrastructure and street networks for their cities (see McDowell, 2000; Southern, 2000; Offner, 2000, for examples).

Even Mayor Guiliani in New York tightened City Council control over the financing and regulation of Business Improvement Districts, as reports mounted of their financial abuses, unrepresentativeness and growing power to become 'micropolises' in their own right (Schulman, 1999, 6). Another good example comes from the startling Labor victory in the Federal state of Victoria, Australia, in 1999. It paved the way for the possible link-age of Melbourne's international airport to the city's public transit system, even though the developers of the CityLink e-highway had negotiated a contract with the previous Conservative government that promised them exclusive connectivity to that most lucrative of sites. With the right political backing, traditional policy intervention through the construction of public duct space, public investment, leeway rights and planning instru-ments can do much to (re)socialise benefits from premium networked infrastructure investments.

Efforts have also emerged to regulate mall spaces to improve the degree to which they support genuinely 'public' activities. For 'the more the mall deploys the signifiers of public space, the more it encourages democratic public use' (Dovey, 1999, 136). Legal challenges, admittedly tentative so far, have been mounted to mall owners in some US towns where malls have all but taken over from downtown street systems, to demand citizens' rights to undertake political activity in malls (Gottdeiner, 1997). The New Jersey Supreme Court, for example, has judged that citizens should have the right to distribute political leaflets in malls (Drucker and Gumpert, 1999). Whilst such social and legal challenges can be small in scale in relation to the overwhelming disciplinary power of the mall space, they are not to be ignored. Nor is the recent spread of municipal responses to untrammelled globalisation: blockages of corporate ownership of farms in the United States, resistance to 'big box' retailing, support for anti-sweatshop movements, and the emergence of a whole range of coordinated collaborative networks between cities and municipalities, especially in Europe (Graham, 1995).

RECOGNISING THE COMPLEXITY OF PREMIUM NETWORK SPACES

In sum, premium networked spaces like malls, Business Improvement Districts, skywalks, e-highways, 'fat' Internet pipes, international airports or CCTV-surveilled atrium complexes are far 'too complex to be reduced to a mechanistic function of the imperatives of the market' (Dovey, 1999, 135). Many practices of resistance, usually ignored by academic research in its portrayals of simple *Blade Runner*-style dystopias, can open such spaces to different uses and constructed meanings within and between cities, based on arguments and social movements about justice and rights as well as social practices which undermine their power to exclude socially (see Thrift, 1997a; Shields, 1989).

At the same time, it is important not to sentimentalise or romanticise such 'resistance' (Donald, 1999, 17). The 'implicit David versus Goliath romanticism' of such a view is, itself, a dramatic oversimplification (Thrift, 1997b, 124). It means that 'everything has to be forced into the dichotomy of resistance or submission and all of the paradoxical effects which cannot be understood in this way remain hidden' (ibid.).

Such are the extremes of contemporary disciplining and surveillance in many work and street spaces that we could ask, as Caroline Freeman does, whether 'mere survival for many people is not resistance' (2000, 206). Consider an example. Many desperately impoverished Nigerians are killed each year trying to extract petrol from overland pipelines around the Niger delta – either by the shoot-on-sight policy of security guards, or by the huge explosions and fires that often ensue. Such pipe vandalism – a practice known as 'scooping' – is clearly a form of 'resistance' to the huge inequalities and environmental injustices surrounding the operations of transnational oil companies in the delta. But, for the people involved, it is simply an attempt at mere survival rather than some romantic stand against the powers that be.

New monitoring technologies: exposure, shaming and critical research

As a last point in our discussion of resistance, we also want to stress that urban social movements and activists can themselves utilise the geographical information systems, CCTV and other surveillance technologies that are so central to the construction and (attempted) maintenance of secessionary network spaces. Ramasubramian (1996), for example, outlines how GIS techniques were used in Milwaukee to prove that an insurance firm was redlining African-American census tracts in the city. Kevin Robins has argued that, with the mass diffusion of consumer video, 'the city now constitutes a mosaic of micro-visions and micro-visibilities. With the camcording of the city we have the fragmentation and devolution of vision-as-control to the individual level' (1996, 139). Some have argued that citywide CCTV may, if properly configured and used, substantially reduce the abuse of power by law enforcement agencies (usually taking the Rodney King case in Los Angeles as an example of the democratising potential of widespread video technology).

Mobilising counterhegemonic channels of network provision and use

Penultimately, other challenges are emerging to the construction of premium network spaces based on water, power, transport, communications and urban street space networks. In Developing World and postcolonial cities, in particular, many poorer urban communities have developed sophisticated ways to resist the eviction and exclusion that so often come with the construction of premium network spaces (Berner, 1997a, 111). They are also actively tapping in to or accessing premium network spaces developed for affluent users only – streets, transport systems, power lines, water pipes and phone networks (Chougill, 1999). Holston (1998) terms these 'spaces of insurgent citizenship' which are constructed to resist the normalising and dominant practices of confining infrastructure access to the premium network spaces of the splintering metropolis. For example:

- Knosa (1995) shows how groups excluded from South Africa's apartheid-based public transport systems have undertaken a wide range of highly visible 'transport resistance' strategies against the high cost and appalling quality of services: bus boycotts, mobilising alternatives and political mobilisation.

- Soja (1999, 2000) shows how a union of 30,000 or so inner-city residents of Los Angeles joined forces within the Bus Riders' Union to sue the city's Metropolitan Transit Authority. The expensive and decrepit bus system upon which they relied was being ignored whilst the authority pumped $30 billion into a gleaming new light rail system for the suburban middle class. What is more, they won.
- Manandhar (1999) demonstrates how women's groups in Patan, Nepal, with support from non-governmental organisations, have constructed drainage and sanitation systems in their informal settlements which are directly designed to meet their needs rather than being imposed by distant male technocrats.
- Breslin (2000) demonstrates how a coalition of non-governmental organisations, backed by the Carvajal Foundation, have worked to arrest the vicious circles of poverty, violence and despair in informal settlements in Aguablanca, Colombia. The settlement of 400,000 has been incorporated. Low-cost provisions, building materials and services have been organised. Phones, metalled roads, drainage, sanitation, water and electricity have been delivered through low-cost schemes. And microfinance, education, business support and child care services, and parks, have been developed in partnership with existing small business owners.
- Grell *et al.* describe how strategies against the privatisation of street spaces in Berlin have sought to celebrate the notion that the management of urban street spaces should be based on the 'liberal idea of public space that the possibility of encounter is a positive aspect in terms of confronting prejudices and inducing communication and learning processes' (1998, 210).

In places such as the Orangi district in Karachi, Pakistan, meanwhile, the obvious failure of both the modern infrastructural ideal and the more recent elaboration of premium networked spaces for socioeconomic elites has led to the emergence of alternative strategies of community self-provision. Such community mobilising, backed up with finance and support from non-governmental organisations, has successfully equipped whole urban neighbourhoods with water and sewerage systems, which, in the case of Orangi, now have a 75 per cent connection rate (Chougill, 1999, 295). Community efforts have been shown to be much more sensitive to gender questions in the design and construction of network services than were the old technocratic, centralised regimes of the modern ideal (Makan, 1995).

MOBILISING THROUGH THE INTERNET

Finally, a wide range of the strategies of poor or excluded urban communities, social movements and unions are now directly mobilising the use of Internet applications in both developing and developed cities, to support socioeconomic participation and collective capacity building (Castells, 1999a; Schuler, 1996; Merrifield, 2000b). The hope is that widespread experimentation with counterhegemonic uses of the Internet may 'provide channels through which knowledge and information can be democratised, dispersed around the diversity of relational webs in urban regions. Could this . . . technology provide the basis for dispersing power out of current nodes, and empowering and articulating diverse democratic voices?' (Healey *et al.*, 1995, 277). There are certainly emerging positive signs. As Castells suggests, the widening diffusion of the Internet, with its:

creative cacophony and social diversity, with its plurality of values and interests, and given the linkage between places and information flows, transforms the logic of the space of flows, making it a contested space. And a plural and diversified space. . . . Through a blossoming of initiatives, people are taking on the Net without uprooting themselves from their places.

(1999a, 300)

This is especially so with the growth of informal and international alliances within the labour, environmental, anticorporate and progressive communications movements, where the Internet is allowing campaigns to coordinate around revealing the ways in which global capital is, itself, exploiting the power of information and communications technology to intensify extreme global divisions of labour, wealth and environmental quality (Kellner, 1999; see Figure 8.1). As Michael Moore, the US anticorporate campaigner, argues, 'corporate America has inadvertently given us an incredibly powerful tool to reach one another cheaply and quickly' (1996, 304, cited in Merrifield, 2000b, 33). In his book *Predatory Globalization* Richard Falk argues that 'potentially, such networks are visionary and dedicated to transformative political goals, offering the poor and vulnerable possibly their best hopes for the future' (1999, 6).

Efforts are also emerging to pluralise and democratise the content of electronic and digital media, a crucial step within broader efforts to rebalance international systems of biased cultural and economic power in this technologically mediated and increasingly commercialised and internationalised age. As the UNDP suggests, 'the information highway cannot be a one-way street. Websites need to be created locally, adding new voices to the global conversation and making content relevant to communities' (1999, 65).

Alongside broader international movements to urge the global regulation of unfettered global markets in information and communications technology and digital media, many local instances are emerging to support the incorporation of excluded voices into electronic domains. In Tamil Nadu, India, for example, keyboard standardisation and software in Tamil, the local language, are being promoted. In the United States the Seniornet project is providing applications, discussion space, services and support to the older population, usually so underrepresented in information and communications technology. Across the complex diasporas of the world, dedicated community networks such as Vietnet are emerging to allow cultures to survive in the context of extreme dispersal and cultural volatility.

Developing local content, and spaces for local expression, can also feed into wider efforts at incorporating marginalised voices into urban and local governance and the mainstream economy. This is important because the first efforts by most cities to use information and communications technology are in marketing and promotion to tourists and potential investors (Graham and Aurigi, 1997). In the United States the civic networking movement has now matured into a powerful set of spaces which support citizen mobilisation and action

Figure 8.1 The use of the Internet in global protest movements: Wasserman's view in the *Boston Globe*, 10 December 1999

(Schuler, 1996; Beamish, 1999). In Santa Monica, California, the Public Electronic Network (or PEN) system has allowed homeless and street people their own e-mail addresses, allowing them spaces of expression and contact points to support access to jobs, services and democratic rights (Doctor and Dutton, 1998).

More broadly, the Neighborhood Knowledge Los Angeles initiative is directly challenging the privatisation of local social, demographic and economic data in the city of Los Angeles. Integrating previously fragmented knowledge and information sources about the condition of at-risk neighbourhoods, NKLA offers bilingual services which help respond to social and economic deterioration, support community awareness of the problems neighbourhoods face, help challenge and expose 'redlining' by major firms and institutions, and democratise access to digital and geographical information.

Across Europe, finally, 'virtual' town halls, community intranets and community telecentres – from Amsterdam and Berlin to Athens, from Bologna and Barcelona to east London – are allowing marginal communities to assert their democratic rights within structures of governance, often for the first time (Tsagarousianou *et al.*, 1998). Perhaps the best-known example is *De digitale stadt* ('Digital City'), an Internet space in Amsterdam, which now has over 50,000 'residents' (both locally and globally). Using an explicitly urban metaphor of themed 'town squares' and 'cafés', *De digitale stadt* supports a vast range of specialist political, social, environmental and interest-based communities and discourses which, within the constraints of biased social access to the Internet, support a 'gigantic alternative and underground world' as well as 'an official city on the surface and in the open' (Lovink and Riemens, 1998, 183; see http://www.dds.nl). To its founders, tellingly, 'the city metaphor' used as the web interface for *De digitale stadt* 'stands for diversity. . . . What we have in mind are all those different "places" and localities that are possible in a real as well as a virtual city' (ibid., 185).

POSTSCRIPT

A manifesto for a progressive networked urbanism

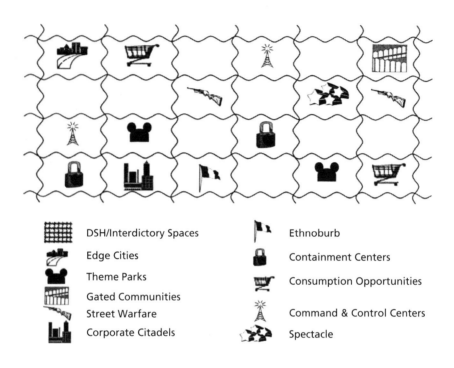

DSH/Interdictory Spaces	Ethnoburb
Edge Cities	Containment Centers
Theme Parks	Consumption Opportunities
Gated Communities	Command & Control Centers
Street Warfare	Spectacle
Corporate Citadels	

Plate 20 Michael Dear and Steven Flusty's conceptualisation of the fragmented landscape of postmodern urbanism, drawn from their work on US cities. *Source*: Dear and Flusty (1998), 61

Urban theory proliferates whenever the boundaries of the city are in flux.

(Waterhouse, 1996, 299)

We need to recapture the sense that a different global is possible.

(Smith, 1999, 105)

The tensions of heterogeneity cannot and should not be repressed. They must be liberated in socially exciting ways – even if it means more rather than less conflict, including contestation over the socially necessary socialization of market processes for collective needs. Diversity and difference, heterogeneity of values, lifestyle oppositions and chaotic migrations are not to be feared as sources of disorder. Cities that cannot accommodate the diversity, the migratory movements, the new lifestyles and the new economic, political, religious and value heterogeneity, will die either through ossification and stagnation or because they will fall apart in violent conflict.

(Harvey, 1996, 437–8)

The pattern of development of cities today is subject to control, it *is not* the result of uncontrollable forces, *is not* the result of iron economic laws whose effects states are powerless to influence.

(Marcuse and van Kempen, 2000, 272, original emphasis)

It is possible for the emergence of a politics of deterritorialization and reconnection, a politics in which arguments over space – its enclosures, exclusions, and internments – become subjects for debate and discussion, and, more important, for resistances and transgressions.

(Dumm, 1993, 192, cited in Soja, 2000, 321)

TOWARDS A SPATIAL IMAGINATION FOR THE SPLINTERED METROPOLIS

It is clear that strategies and practices of urban splintering must negotiate complex ambivalences: continued urban mixings, active resistance and the exploration of alternative and more democratic ways of constructing urban and network spaces. Such are their magnitude, pervasiveness and power, however, that it still seems likely that the processes of splintering urbanism outlined in this book will work to underpin more and more starkly polarised economic and social geographies of closely juxtaposed privilege and disconnection within many – perhaps most – contemporary cities.

We would therefore expect social and political tensions within many cities to increase. A central theme of urban politics and urban social movements in the first decades of the new millennium will therefore centre on the struggle between the 'glocal' forces of attempted, 'pure' boundary control and the customisation of premium, commodified network spaces, versus the imperatives of infrastructural, urban and technological democratisation and the need for more egalitarian and democratised practices and principles of development (see Sassen, 1996, 1998, 2000b). Kevin Robins asks the crucial question 'What kind of city can we imagine in this global context?' His answer is both stark and bleak:

This global city is the place where the newly mobilized and displaced populations gather in their millions. Each city contains within itself the dynamics of the new world disorder – its dramatic contrasts of rich and poor, its polarizations and segregations, and its encounters and concentrations. It also constitutes a new kind of city, as the coherent and ordered structure of the 'modern city' becomes overwhelmed and superseded by the sprawling, chaotic mega-city or megalopolis (the information and communications systems are ensnared and entangled in this urban anarchy).

(1999, 54)

The point leads us to the final task of the book: to establish some starting points for imagining frameworks of politics, planning, and what we might call a 'spatial imaginary', to support the challenges of addressing and researching the splintering metropolis. Rejecting the tendency of postmodern theory to withdraw from practical policy suggestions on issues of socialisation, urban planning and the desired nature of the state, the imperative here must be to explore how, in Holston's words, we 'can develop a different social imagination – one that is not modernist but that reinvents modernism's activist commitments to the invention of society and to the construction of the state' (1998, 39).

Our starting point is that the required spatial imagination and politics need 'to bridge the multiple heterogeneities, including most emphatically those of geography, without repressing difference' (Harvey, 1996, 438). It must be conscious of the complex ways in which networked infrastructure of all types, and the diverse technological mobilities they support, become bound up in the production of space, identity and meaning in urban life at various scales, within the context of globalisation and extending metropolitan regions (Graham and Marvin, 1999). It must directly engage with the complex superimpositions of ranges of sociotechnical connections and disconnections throughout the urban fabric that are such a characteristic symptom of contemporary urbanisation (Roberts *et al.*, 1999). Clearly, 'the emergence of new territories, which change the scale of understanding and intervention in urban projects, and multiplies their complexity, demands new planning styles and instruments and a new kind of architectural design' (Ezquiga, 1998, 7). Finally, the new spatial imagination must be fully founded on an appreciation of the diverse processes underpinning trends towards splintering urbanism in cities across the world: sociotechnical, geographical, political, legal and discursive (see Lefebvre, 1984).

THE CHALLENGES OF URBAN DEMOCRATISATION: ASSOCIATION, DIFFERENCE AND STATES

The very idea of struggle against inequalities, which at the world level have never stopped growing since the end of the nineteenth century, has now been called into question.

(Mattelart, 1996, 305)

If you want to change the city you have to control the streets.

(Protest poster, 'Reclaim the Streets!', London, 1997)

We believe that it is critical, above all, that any new urban spatial imaginary must actively seek to work towards urban democratisation in its fullest sense and in all urban contexts (Painter, 1999). To achieve this, there is a need, first, to struggle against the forms of (attempted) sociotechnical closure of urban network spaces of all kinds represented by strategies of splintering urbanism. New compromises will be required between perceptions and experiences of security and the perceived need for closure and the democratic ideals of openness and interconnection. For, as Albert Pope suggests, there are, in a sense, understandable reasons in contemporary cities 'why the people who can afford it chose the security of closed corporate development over the greater freedom and choices of the declining open city' (1996, 185). However, 'what ought to be disparaged', he believes, is 'the fact that one must make a choice between security and freedom at all'. To him:

The recent history of urban form tells us that such a compromise is not necessary, that cities have traditionally provided both. Despite the fact that we cannot return to historical urban forms, we must ultimately insist on cities that do not demand the surrendering of either security or freedom.

(Ibid.)

To some there is a sense in the contemporary metropolis of the uncontrollable in the apparently pervasive shift to extended, polynuclear metropolitan landscapes marked by deepening fragmentation, xenophobia and starkly uneven internal and external connectivity. Alan Waterhouse, for one, speaking especially about peripheral Toronto, believes that the price 'is a solitariness we long to escape, but [we] cannot avoid getting caught up in the centrifuge of modern urbanism as its spins other lives away from our own into inaccessible zones' (1996, ix).

ON THE MAINTENANCE OF MIXING: SUSTAINING EMBODIED AND SITUATED COPRESENCE

In this context it is clear that our new spatial imaginaries must stress the critical importance of the constitution – geographically, sociotechnically, politically, culturally and legally – of urban spheres of heterogeneous interaction and continued mixing – the very essence of the city. The dangers of the untrammelled commodification of networked spaces of mobility, of the on-going reconstruction of urban landscapes as closed, premium spaces which attempt to withdraw against the 'other' spaces of immobility and fear, must be recognised. The close connection between changing urban landscapes and global shifts towards neoliberalism and consumerism needs to be stressed. And all the associated financial and social costs, likely to stem from rising fear of crime, social unrest, polarisation and spiralling securitisation and fortification, must be underlined.

Rather, the new urban political imaginary must engage with, and actively support, the piecing together of the 'thousand tiny empowerments' offered through network spaces of resurgent citizenship (Sandercock, 1998a, 129) – from organised grass-roots mobilisations, through local exchange and trading systems (LETS) to a whole range of everyday practices that resist the simple secession of sociotechnical, commodified spaces from the wider metropolitan landscape (Cooper, 1998; see Fincher and Jacobs, 1998). Such democratisation, following Iris Marion Young (1990), must rest first on 'the possibility of expressing difference and "otherness" in unoppressive ways' (Swyngedouw, 1998, 120). To do this it will need to recognise and support the parallel construction of networked infrastructures and urban spaces that more sensitively support the many dimensions of the 'cultural politics of difference' thrown up by contemporary urbanisation, social and cultural change and migration: gender, age, ethnicity and 'race', (dis)ability, sexual preference and spirituality (Sandercock, 1998a, b).

Clearly, the growth of fragmented spaces of expression mediated by the Internet and other infrastructure networks will not suffice here. Successfully supporting an urban politics of difference will continue to require embodied and situated presence, proximity and contact – what (some) urban streets and spaces in certain cities have come to stand for and sustain. It will therefore need to work to overcome the widespread construction of premium network spaces – consumerised streetscapes, gated enclaves, electronically controlled highways, 'virtual' spaces – with their characteristic efforts at social, cultural or economic purification and controlled withdrawal from difference (Robins, 1999, 51). 'What is fundamental to urbanity,' writes Kevin Robins, is 'embodied presence and encounter. It is a question of . . . both the "individual" body and the "collective" body of the city' (1999, 52). Following some of the arguments of Richard Sennett (1970), Robins urges us to 'put a value on exposure [to difference] and its discomforts' (ibid., 54). Along the same lines, Jeremy Seabrook argues that 'security, if it arises anywhere, must arise from the tenderness and vigilance of people committed to the daily protection of one another' (1993, 12).

BEYOND THE PUBLIC REALM: EXPOSING AND REGULATING THE INEQUITIES OF GLOBALISATION

At the same time, however, such a politics of difference must be aware of the formidable obstacles provided by the common experience of the contemporary, capitalist urban condition in which 'difference is expressed and experienced as exclusion, domination, and repression' (Swyngedouw, 1998, 120). For this to happen, urban democratisation will need to look far beyond the traditional 'public realm' and, in particular, the challenge of reinstating and strengthening mixing in and on the street. It:

cannot be limited to protecting the right of individuals to participate equally in the public sphere, because systematic inequalities in the ability of individuals to gain access to, and have a voice in, the public sphere are produced in part outside the public sphere in civil society, the private economy and the private sphere.

(Painter, 1999, 42)

In particular, hidden mechanisms of negative redistribution and the exigencies of corporate power – practices like corporate tax breaks that are so closely bound up in the production of premium network spaces and processes of splintering urbanism – need to be exposed, including the many and diverse hidden public subsidies that go into the construction of premium and secessionary network spaces.

Without exposing these, and directing the reclaimed resources into strategies aimed at benefiting those places and people that are experiencing marginalised forms of network urbanism, what Erik Swyngedouw calls 'a humanized and just urbanization' based on 'a global reach that champions commitment to a more socially and ecologically inclusive urban life' will be but a pipe dream (1998, 120). In the age of the Internet, especially, the extreme difficulty of achieving real corporate regulation and transparency at the transnational scale is likely to make tax avoidance an ever more central element of corporate operations. For, increasingly, transnational corporations:

will install their web servers where taxes are lowest, disguise their trade in goods as a trade in services, and even launch their own virtual currencies. The tax burden, in other words, is shifting to those who are unable to move their assets offshore or out of the old economy into cyberspace. With little else to offer, poor countries [and spaces] end up giving everything away in a desperate attempt to attract 'investment'. If taxation is not to become wholly regressive, we will have to revolutionise the means by which the rich are charged.

(Monbiot, 2000, 13)

CHALLENGES OF URBAN DEMOCRATISATION

As Amin and Thrift (2000) argue, democratisation must therefore go far beyond the simple celebration of diversity and the recognition of the need for difference (associative democracy). It must also transcend Richard Sennett's (1970) and Kevin Robins's (1996, 1999) notion of the city as a creative clashing point of diverse social and cultural groups and identities (radical democracy). Above and beyond these, new forms of intervention by state and public institutions, at all geographical scales, are required directly to encourage democratic practices through (re)working towards equality of access to spaces, infrastructure networks, public services, opportunities of association, and systems for holding the wealthy and powerful accountable to public taxation.

To Amin and Thrift, it follows that the 'ideal city' must be considered as 'a place of equal access, mutuality, freedom and fulfilment of potential. . . . The city, more specifically its social and institutional set-up, must give us, *all of us*, the space and time to become something else, the right and opportunity to experiment, to enable lines of flight, to forge solidarities' (2000, 26, original emphasis; see Amin *et al.*, 2000). This involves major political issues such as reconstructing the balance between the state, the market and civil society, the need to nurture more democratic ways of economically organising, resisting the replacement of discourses of citizenship with those of consumerism, and connecting local and urban strategies with global practices and debates surrounding international economic governance, in an age of what Richard Falk (1999) calls 'predatory globalisation'.

Such challenges, of course, infuse all domains of contemporary governance and politics and are strictly beyond the immediate scope of this book (see Brecher and Costello, 1994; Mander and Goldsmith, 1996). But they raise critical questions about whether new, progressive and democratising relationships can be established in this 'post-Fordist' era between state social and economic regulations and internationalising regimes of economic, cultural and social governance, which are continuing to liberalise. In other words, can local activism and democratic and socially inclusive policy experimentation really work to counter significantly the broader shift towards a kind of global Social Darwinism that shapes the wider forces of contemporary urbanisation, with its premium network spaces and intensifying social and geographical partitions (Kipfer, 1998, 173)? How, moreover, can the multiple spatial scales of strategies that strive to resist global neoliberalism and social polarisation – from the body through to the street, the city, the region, the nation and the transnational – be brought to work in synchrony, in line with the dominant glocal scales articulated by premium network spaces themselves (Smith, 1993; Harvey, 2000)?

BUILDING NEW CONCEPTUAL UNDERSTANDINGS OF CONTEMPORARY URBAN LIFE

Globalization should not be fetishized and reified as a steam roller that rolls over 'local places'.
(Keil and Ronnenberg, 2000, 228)

Such an agenda is both massive and daunting. But, in order to work towards a creative imagination of practices of urban democratisation, it is also clear that a continued, thorough-going reconceptualisation of the contemporary nature of the 'urban' will also be necessary.

Obviously, such rethinking must resist the totalising concepts of order, progress and rationality inherent in modernist practices (and their legacies). It must simultaneously avoid tendencies to glamorise or reify the supposed 'global' imperatives of urban entrepreneurialism, imagineering and urban 'competitiveness' that were dominant obsessions among many planners and policy makers for the last two decades or so of the twentieth century. Far too often we still hear of 'the city' having to 'compete' globally as though it were some single reified actor on the planetary (neoliberal) economic stage. We are endlessly told of the urban and social 'impacts' of new technologies as though the Internet *et al.* were social meteors hitting our planet from outer space. Or we are told of the supposed 'dictates' that stipulate that socioeconomic elites must necessarily benefit from premium street, power, transport, water or communications spaces, and intensified mobility opportunities, and any associated real estate developments, public subsidies and discursive legitimations (Marcuse, 1997). The constant danger of such discourses, both practically and theoretically, is that 'we run the risk of treating the city [as well as technology and globalisation] unproblematically as a subject capable of action' (Jessop, 1998, 80).

In fact, the simplistic and parallel reification of cities, neoliberal 'market forces' and 'new technologies' often serves to obfuscate the broader power relations, political economies and practices bound up with the reconstruction of (parts of) cities as premium network spaces. Thus underlying ideas help to found and perpetuate policy, practice and the very imagination of the possible (INURA, 1998).

There has been much progress in reimagining the nature of the urban (see, for example, Ascher, 1995; Sandercock, 1998a; Healey, 1997; Douglass and Friedmann, 1998; Soja, 2000). But there remains a need to build further on the 'relational' conceptualisations of urban place which we reviewed back in Chapter 5. Instead of seeing the city 'as the stamp of great and unified forces which it is the task of the urban theorist to delineate and delimit', it must be viewed, rather, as 'a partially connected multiplicity which we can only ever know partially and from multiple places' (Thrift, 1997a, 143). 'Perspective views are multiple now,' suggests John Friedmann (1998, 29). 'The world cannot be lifted off its axis . . . the metropolis is incapable of being seen as a whole.'

BREAKING THE 'TYRANNY OF 'SPATIAL SCALE'

Relational perspectives, in particular, suggest a need to break 'the tyranny of spatial scale' in our imagination – the assumption that the urban scale must, *necessarily*, be the dominant scale of action and organisation. As much as anything else, this is the result of urban disciplines trying to bolster their own legitimacy in profoundly uncertain times. Instead, we must recognise that real democratisation must be pursued 'through the myriad connections in different networks within and beyond the city' (Painter, 1999, 43). Spatial scales, and geographical levels ('corporeal', 'local', 'urban', 'regional', 'national', 'international', 'global'), are in a sense being continuously 'telescoped' within the contemporary networked metropolis as premium networked spaces and new networked technologies are superimposed unevenly upon the more standardised and dispersed networks of the modern networked city (Offner, 2000).

This demands a dynamic, relational and multiscalar perspective on the city's multiple spaces and times. The 'city' that emerges is 'fluid, contingent and panoptically indescribable' (Flusty, 2000, 150). It:

is constantly changing (even to stay the same), [it is] a city that does not necessarily hold together, a city that is both little and large (since the idea of scale is replaced by the idea of partially connected networks), a city that is . . . a set of diverse, interacting, practical orders in which the interaction is more important than the order. This is the city in which the magic is still there.

(Thrift, 1997a, 143; see Sum, 1999)

Much planning practice and academic work by planners has, so far, failed to transform their conceptualisations of time–space and the nature of the urban, from the legacies of the modernist straitjacket of urban planning's postwar history. Too often, urban space and scale are themselves reified in these views; cities are still cast as 'differentiated, bounded (perhaps even organic) wholes' (Painter, 1999, 13). Joe Painter calls this 'the discursive construction of urban coherence' (ibid., 27). Such an objectified view of the urban leads, in turn, to the common implication that 'the city' can be 'impacted' as an objective 'thing' by external processes of economic, cultural or technological 'globalization'.

BEYOND CLASSICAL ANALYTICAL TOOLS AND SINGLE REPRESENTATIONS

A related problem is that architecture, planning and urbanism often still remain tied, however implicitly, to the classical analytical tools of the perspectival plan, or the formal geometric composition, even though awareness of the failings of such paradigms for representing contemporary urban life is increasingly widespread. As Leong suggests, 'the processes that constitute urban configurations can no longer be adequately represented by a Cartesian mapping system' (1998, 201). That 'the configuration of the contemporary city has transcended the realm of idealized geometry is by now an obvious fact – yet, surprisingly, one largely unheeded by an architectural and urban practice which still clings to the removed, visual techniques of formal composition' (Bell and Leong, 1998, 12).

The general result of this continuing reliance on single representations of urban coherence, as we have seen throughout this book, is to allow the space–time demands of articulate and powerful groups, who tend to have clear ideas about their (premium) space–time parameters and mobility requirements, and who dominate the minds of urban planners and real estate and infrastructure developers, to prevail utterly. Thus depictions of urban coherence tend to be 'always limited and partial'; they offer many 'systematic silences' (Painter, 1999, 28). Too often the relational time–spaces of powerful, affluent, corporate economic and social interests are presented as single, overriding imperatives for shaping the 'value-free' policy necessary for some organically coherent city to 'compete'.

The time–space demands of the powerful are therefore often used unproblematically to capture, represent and characterise a 'place'. The widespread enthusiasm for, and subsidy of, the new glocal premium network spaces of attempted purity and withdrawal – whether these be the 'multibundled' mini cities of the contemporary mall, the gentrified consumption or tourist spaces, the 'wired' or 'cyber' villages or the premium spaces of 'global mobility' – tend to be presented as naturalistic *faits accomplis* within urban politics and planning, as though there is no other choice.

But, as David Harvey (1996) reminds us, attempts to suggest that single time–space representations can somehow unproblematically capture the multiple space–time subjectivities of a city will inevitably have major distributive consequences. This will be especially so for those whose space–time parameters and relational orientations and aspirations are at odds with the modernising 'vision' or representation proffered by planners and infrastructure and real estate developers (see Shields, 1995).

CONSIDERING THE UNTHINKABLE? THE POSSIBILITY OF RESISTING NETWORKED CONNECTIONS

Given this, under some circumstances, it might even be more profitable for some local civil societies actively to *resist* incorporation into glocally configured premium network spaces, rather than devoting substantial public resources to fostering furthering local integration into such network spaces. Andrew Gillespie has argued that policy makers in peripheral or weak urban economies might consider *resisting* opening themselves up through grandiose and expensive 'glocal' infrastructure projects such as corporate telematics networks. He believes that more appropriate solutions might involve building more modest network applications up from the 'grass roots' within local economic and social spaces. He writes that we must:

face the possibility at least that the bias inherent in existing networks is unlikely, given the prevailing balance of power within society, to be deflected and that, in consequence, rather than embracing such networks with the intention of deflecting their deleterious impact upon local economies, such networks should instead be resisted.

(Gillespie, 1991, 255)

WAYS FORWARD FOR URBAN PLANNING AND PRACTICE: BEYOND URBAN AND TECHNOLOGICAL DETERMINISMS

If one were to find a place, and perhaps there are some, where liberty is effectively exercised, one would find that this is not owing to the order of objects, but to the practice of liberty.

(Foucault, 1984, 246)

A new 'historic compromise' between [infrastructure networks] and the regions must be found which preserves the policy-making capacity of the political powers of urban organization and regional development. To do this, greater attention must be paid to the occupation of public space and to the idea of service nodes, whilst at the same time encouraging the emergence of users as players in the regulation of [infrastructure networks].

(Offner, 2000, 167)

Penultimately, we would like to discuss the challenges thrown down by splintering urbanism for attempts at struggling towards progressive and democratic architecture, urban design and urban planning practices. For it is clear that urban design-based solutions often tend also to crystallise narrowly the time–space requirements of dominant interests within the built form. 'Given the ability of dominant interests to appropriate all architectural forms, there can be no such thing as an "emancipating design"; only the activity of design has any such potential' (King, 1996, 247, original emphasis). Rem Koolhaas (1995), lambasting the narrow environmental determinism of much of the 'new urbanism' movement in the United States, argues that any progress towards a genuinely 'new urbanism' must be centred on process rather than form, on openness rather than closure, on flexibility rather than order. Such a new urbanism, he writes:

will not be based on the twin fantasies of order and omnipotence; it will be the staging of uncertainty; it will no longer be concerned with the arrangement of more or less permanent objects but the irrigation of territories with potential; it will no longer aim for stable configurations but for the creation of enabling fields that accommodate processes that refuse to be crystallized into definitive form; it will no longer be about meticulous definition, the imposition of limits, but about expanding notions, denying boundaries; not about separating and identifying entities, but about discovering unnamable hybrids.

(Koolhaas, 1995, 969)

This leads us to three pointers for practices of urban design, policy and planning as they respond to and grapple with the challenges of splintering urbanism.

TOWARDS A FULLY NETWORKED URBANISM: BEYOND URBAN AND TECHNOLOGICAL DETERMINISMS

We should not repeat the mistake of CIAM and expect more from architecture than it can achieve.

(Ley and Mills, 1993, 267)

First, such practices must emphasise *relations and processes* rather than objects and forms. This is not to deny that changing urban form is a crucial product of processes of splintering urbanism. It is certainly true that the landscapes of the splintering metropolis throw down challenges to urban practice to 'assemble a landscape from the fragments of design created within individual developers' projects' (Fishman, 1990, 54). There is a desperate need, in particular, to imagine ways of weaving secessionary and glocal network spaces into the finer-grained fabric of the urban spaces and times that surround them. We must speculate as to how airports, malls, theme parks and the like 'may weave themselves into the local fabric to create social interaction and acceptance as opposed to continually reinforcing barriers' (Avedano *et al.*, 1997, 68).

But such attempts at 'reweaving' the urban fabric must be done in ways that recognise the full plurality of highly unequal voices in contemporary cities. They must, as a starting point, provide a range of supports for an 'open' set of network spaces to encourage heterogeneous mixing rather than monofunctional and socially purified secession. Strategies to overcome the disabling and exclusionary experiences of less powerful and lower-income groups, in accessing the essential network spaces of contemporary cities – streets, transport systems, water, communication and power networks – are also vitally required. Overcoming the interwoven, disconnecting logics of superimposed modernistic planning, glocal networked infrastructures, neoliberal infrastructure regimes and contemporary urban 'megastructures' will be a necessary, if daunting, agenda.

A starting point here is the imperative to develop a dynamic understanding of the ways in which infrastructure connections and nodes of all kinds can be creatively harnessed and integrated within the metropolitan fabric (see R.L. Cowan, 1997, for an early example). How can 'disarticulated' cities and city-regions be effectively integrated in democratic and relatively equitable ways (McGee, 1998, 45)? We would not deny that 'the conscious development of nodes in the infrastructure as new urban spaces with a strong public domain should become a key concern in our thinking' (Hajar, 1999, 33). But the challenge is wider than the imperatives of 'urban reconnection'; it is necessary to reimagine how the regulatory and legal treatment of all spaces of networked mobility (not just the traditional urbanist preoccupation with streets and the corporeal networks of transport) translates into the concrete spatiality of urban experience at the local scale (Offner, 2000).

In other words, a fully networked urbanism needs to be translated into ways of thinking about urban practice, teaching, policy and activism (Dupuy, 1991). Such practices must resist both technological and environmental determinisms. The complexities and contingencies of the relationship between form, infrastructure, mobility and the time–spaces of urban life suggest that the extent to which a proposed built or technological form – 'compact cities', 'urban villages' or 'new urbanism', 'wired city', 'community intranet' – will lead to particular social, economic and cultural outcomes needs to be demonstrated in terms of the relational dynamics of specific instances. They should not be assumed as universal generalisations.

Much of the vaunted 'new urbanism' movement in the United States fails, for example, because it adopts the naive assumption that the creation of idealised urban forms appropriated from history will inevitably lead to definitive social, cultural and institutional results. Whilst it 'capitalizes precisely on what many accept as being legible signs of community as they are iconographically inscribed into urban and architectural form', new urbanism says little about how urban processes and practices will address crises of difference and inequality (Leong, 1998, 207). In so doing, it often inadvertently works to support further the secessionary tendencies of socioeconomically affluent groups who can afford to access its neotraditional (but broadband connected) housing units.

URBAN SOCIOTECHNOLOGIES AS CONTROL RIGHTS: PROCESSES OF BUILDING AND MAINTAINING AUTHORISATIONS

Much care therefore needs to be taken to avoid investing any particular built form or network technology with some essential, reified capacity automatically to socially marginalise or socially democratise. In fact, we need to be extremely wary of such simple black/white predictions; of scenarios that assume totalised, geographic 'impacts' of new architectural, planning or technological techniques. Analyses of contingent practices and dynamic processes must finally triumph over deterministic and reified readings of static urban structures, forms and architectures. As Bernward Joerges argues:

Built spaces always represent control rights. They belong to someone and not to others, they can legitimately be used by someone and not by others. Variable control rights over built spaces constrain what can pass in and around these spaces. [But] only rarely can one show that such constraints are coupled to built form. . . . It is the *processes* by which authorizations are built, maintained, contested and changed which are at issue in any social study of built spaces and technology.

(1999b, 424, emphasis added)

Thus generalisations about desirable urban forms or the configuration of infrastructure networks and streets need to be replaced by polyvalent, plural and culturally sensitive appreciations of the dialectic and non-generalisable relations between social process and urban form. Once again, a similar challenge awaits in architecture, where the reification of urban form tends to be axiomatic. As Alex Wall (1994, 11) suggests:

As for architecture . . . , its future will rest less on Le Corbusier's dictum 'the magnificent, skilful art of pure volumes bathed in light' than on the organisation of programmes and process; in a city which depends equally on the programmatic activation of its voids as much as the maintenance of its built volume; a city represented by an architecture that is less and less material; an architecture that is primarily process and secondary fragments.

BEYOND SINGLE REPRESENTATIONS OF TIME AND SPACE

It follows, second, that planning and urban policy and practice must stress the *multiple meanings of space and time*. This requires careful attention to the representation of policies and projects in map form, and the expression of time periods. Where two-dimensional representation and fixed time periods (e.g. the 'master' or 'five-year' plan) are used, clarity is needed with respect to whose space and time this is and why it is helpful to use the particular form of expression. Project appraisal and policy development need to be informed by explicit recognition of the range of spatialities and temporalities in which they may be inserted, or which policy seeks to shape.

Within the complex terrains of superimposed connections and disconnections that emerge within the splintering metropolis, physical adjacency cannot be used as a proxy for identifying meaningful relationships and the impacts of a project or a policy. The time scale of a landowner is different from that of a small builder. People's spatial reach varies in daily, weekly, annual and life-span time. Some companies may be committed to a locale for a year or two. For others the locale is their permanent site of production, and they may be planning strategically over a time span of decades. Ecosystemic relations tie places into planetary relations over long time scales and into the micro-relations of species habitats. This implies the need for careful assessment of the many spatial and temporal experiences of a city, and how these flow across and into each other in shaping a place and filling it with value.

MOBILISING NEW CONCEPTS OF PLACES

Lastly, planning and urban policy practice need to represent places as *multiple layers of relational assets and resources*, which generate the distinct *power geometries* of places (Massey, 1993; Graham and Healey, 1999). This point emphasises the need to recognise that privileging one experience of space and time (e.g. TGV stations, optic fibre grids, super-highways, mega-airports, etc.) may necessarily undermine other, equally important but less powerful, network spaces. The multiple layering is thus neither neutral nor value-free.

The rich, multiple time–space fabrics of dynamic urban environments need to be carefully nurtured through fostering the development of relational exchanges across the layers, reducing the blockages and exclusionary practices which seal dominant groups into narrow relational networks and marginalise many others. This requires effort in imagining how planning processes can engage with many spaces and times – for example, providing world-class networked infrastructures and more basic community-level opportunities to engage and act at a distance.

CHALLENGES TO URBAN RESEARCH

Which brings us to the final task of this book: to suggest ways forward for urban research. This book has deliberately sought to bring together many debates, cases and examples into a single narrative. But it has not been able to explore in detail the complex and diverse processes of governance and politics that support, and resist, processes of splintering urbanism in the range of cities mentioned. As a consequence, whilst necessarily lengthy, our discussions in this book are still only a partial story, one where the general similarities of restructuring trends, in a variety of cities across the world, have been emphasised. This has been at the expense of an understanding of the detailed and contingent ways in which coalitions of actors are working, in different ways and in different contexts, to restructure the physical and sociotechnical fabric of cities.

This book suggests, then, that a central challenge for urban research is to undertake detailed and comparative empirical investigations into the ways in which physical and sociotechnical shifts towards splintering urbanism, and unbundled networked infrastructures, are being politically and socially constructed in profoundly different political, cultural, economic and historical contexts. Such research needs to encompass developed nations, newly industrialising nations, developing cities, and post-communist metropolitan areas embedded within different state, political, cultural and urban traditions. This is a huge research agenda which transcends many disciplines. We would like to round off the book by pointing to three particular questions which arise here.

THE CHANGING ECONOMIC TERRITORIALITIES OF STATES

First, there is the broad question of the changing economic territorialities of different types of supranational, national and local states and regulatory authorities (see Jonas and Wilson, 1999). For example:

- How are different types of national and local state working to construct 'glocal scalar fixes' in practice, in efforts to try and make their cities, or parts of their cities, more 'competitive' as nodes on global circuits of exchange?

- In what ways do territorial state interests connect with the rapidly internationalising interests of infrastructure, technology, real estate, and agents of transnational political and economic governance, in forging and developing premium network spaces?
- How do these processes vary between centralised unitary states, federal and decentralised states, the strong and less strong states in developing countries, and post-communist state systems?
- How are these processes changing the territorial and political nature of the states themselves as they become less concerned with managing national territories as a whole and more interested in the construction of multiply scaled premium network spaces, for connecting selected urban districts or city-regions to each other and to wider internationalising circuits of exchange?
- Will the political and territorial tensions surrounding such transitions undermine the legitimacy of both nation states and urban governance coalitions and regimes as premium network spaces threaten to 'delink' from surrounding and hinterland regions?
- Are some types of nation state managing to maintain strategies that preserve their territorial integrity better than others? If so, how and why?
- How can methodologies of urban analysis best capture and explore the reconstruction of social scales which parallels processes of splintering urbanism?
- Finally, what political strategies are open to those groups and spaces beyond the gleaming new glocal infrastructural fixes of contemporary capitalism – places that threaten to be literally bypassed by dominant technological and economic circuits of exchange? May such groups and spaces collaborate to lobby the growing institutions of global governance, bypassing their nation states in the process?

THE COMPARATIVE URBAN AND CULTURAL POLITICS OF SPLINTERING URBANISM

Second, and following on from these questions, we know very little about what we may term the comparative urban and cultural politics of splintering urbanism. For example:

- How do different state and intergovernmental systems, traditions of law, political cultures and civil societies support different experiences of urban and infrastructural change?
- How do the tensions between political and development coalitions supporting the privatisation and closure of urban streets and network spaces play out against social movements struggling to assert the social and territorial justice of the excluded urban poor, in different political and social systems and for different infrastructural networks?
- How do urban power holders' notions of urban 'order', the 'competitive urban space' or the 'good city' vary between different cultural and national contexts, in terms of their efforts to structure the configuration of urban spaces and networks normatively, or their attempts to control crime and disorder in urban spaces (see Body-Gendrot, 2000)?
- How are constructions of identity politics involved in supporting the splintering of infrastructure networks in different contexts?

- What roles do discourses and representation play in different cities, in terms of both highlighting the alleged virtues of constructing premium network spaces, and undermining the arguments of political and social oppositions?
- What, under conditions of apparently global urban polarisation, allows some dominant politicians and regulators into political compromises which serve to maintain or even enhance the democratic possibilities of urban mixing within urban network and social spaces?
- Can progressive urban and international coalitions successfully resist global tendencies towards the construction of secessionary network spaces for internationalising elites, tourists, and corporations? If so, how?
- Can social and governance traditions, which stress the importance of the equitable development of networks, survive within a wider international political economy stressing neoliberalism and underpinning social and economic polarisation? If so, how?
- What can groups and organisations that are resisting the construction of premium network spaces learn from each other?
- Lastly, how can practices and concepts for strengthening urban democracy be built, diffused and implemented, across spatial scales, cultural contexts, the disciplinary chasms separating those dealing with urban space from those dealing with urban infrastructure, and the infrastructure domains between the often sealed professional worlds of the water, energy, street, traffic and communications engineers?

SPLINTERING URBANISM, THE PRODUCTION OF URBAN SPACE, AND THE GLOBALISATION OF CAPITAL

Lastly, we need to understand better the relationship between the internationalisation of real estate, architectural, media and infrastructure capital, the growing influence of diaspora networks of social elites, and the widening application of techniques of urban splintering across the world. For example:

- What roles do the international diffusion of exemplar models of real estate development, infrastructure reregulation and liberalised network management play in the widening reach of premium network spaces?
- What roles do the international movements of urban designers (especially 'star architects') and economic models and cultural norms play?
- To what extent are the instruments, techniques, technologies and norms that go into constructing malls, theme parks, gated communities, liberalised infrastructures and premium network spaces diffusing from the North to the South, backed up by the growth of transnational infrastructure, real estate, financial and media corporations?
- How do such internationalising production forces interact, in the form of 'hybrid' cultures, with local traditions, in forging different 'glocal' styles of urban governance and the physical, social, economic and cultural transformation of cities?
- How do what Krasidy (1999, 471) calls the 'entangled articulations of global and local

discourses' become embroiled in the production and reconfiguration of urban space in different urban contexts?

• How do the variety of premium and secessionary network spaces analysed in this book fit into the 'cogredience' of contemporary metropolitan life, which, readers will recall from Chapter 5, David Harvey defined as 'the way in which multiple processes flow together to construct a single consistent, coherent, though multi-faceted time–space system' (1996, 260–1)?

END NOTE

There are clearly a myriad questions here which, despite its length, have only been touched on in this book. The hope, however, is that our broad exploration of processes of urban splintering in a variety of cities may help others to follow up some of these questions. Above all, we hope the book has revealed the merits of exploring the complex and dynamic interplay between networked infrastructures as a whole and contemporary urban space and societal change. We hope, too, that we have highlighted the merits of a perspective on these issues that manages to be simultaneously critical, transdisciplinary, international and sociotechnical. Such a perspective, we believe, helps forge powerful new views of the rapidly changing nature of urban life on our rapidly urbanising planet. In so doing we hope we have made a contribution to what we believe remains one of the salutary challenges for contemporary urban studies: constructing a critical urbanism of the networked city.

GLOSSARY

Plate 21 Defensive condominium complexes under construction beyond the traditional streetscape in the Pudong district of Shanghai, China, 1999.
Photograph: Stephen Graham

Actor network theory
A theory of the ways in which social and technological relations are mutually constructed and ordered through the elaboration of actor networks. Developed in the 1980s by Bruno Latour (1993), Michel Callon (1986) and John Law (Law and Bijker, 1992), among others.

ADSL (Asynchronous Digital Subscriber Line)
A new way of offering broadband telecommunications services over a normal telephone line.

Autonomous infrastructure
Refers to the development of stand-alone or decentralised technologies that allow users to lower their degree of reliance and increase their autonomy from centralised infrastructure networks. Examples include combined heat and power, renewable technologies, local water and waste treatment, and mobile communications technologies.

Back office
An office complex with no face-to-face contact with customers, which delivers services entirely via the telephone, Internet or private telecommunications connection (known also as a *call centre*).

Biometric scanning
The process of digitally scanning a representation of the human body (hand, fingerprint, iris, voice, face, etc.) in order to complete data bases which can be used in tracking, surveillance and access control.

Black box (or Technological black box)
A technological assembly including infrastructure networks whose inner workings are so completely unknown or hidden to its users that its successful functioning is totally taken for granted. See **Unblackboxing**.

Bundled infrastructure
A monopolistic infrastructure network geared to covering an entire geographical territory with broadly equal services offered at broadly equal cost (see **Natural monopoly** and **Modern infrastructural ideal**). See **Unbundling infrastructure**.

Business Improvement District
An urban policy instrument widely employed by the United States and other countries in recent years within which a local board of property owners is set up to raise specific local taxes to be spent on extra local services (sanitation, street cleaning, marketing, environmental improvements, new infrastructure, etc.).

Call centre
See **Back office**.

CCTV
Closed-circuit television.

CIAM
The Congrès Internationaux d'Architecture Moderne, an influential group of advocates of modernist architecture and planning during the the 1920s and 1930s.

Common pool goods

Goods that are **rivalrous** in consumption but have low **excludability** for other users. Examples include small rural roads and access to storm drainage.

Customised infrastructure

The process by which infrastructure networks are packaged together to meet the precise needs and demands of specific user groups, often inward investors and large commercial users.

Cyborg urbanisation

The notion that contemporary cities are seamlessly mediated by technological and cybernetic systems which completely mediate the city's relationship to society, culture and nature.

Delinking

The process through which one particular territory or social space (say a global city) becomes less and less related to its hinterland.

Denationalisation

The process through which the development, regulation and financing of infrastructure networks become less and less connected with the nation state.

Developmentalist state

A particular configuration of nation states in the Developing World, common between 1945 and the 1980s, characterised by the dominance of coordinated national programmes of infrastructural and economic development and planning.

Digital divide

Shorthand for the inequalities of socioeconomic and cultural access to, and use of, information and communications technologies (especially the Internet). Term coined in the United States in the early 1990s.

Disfigured city

A term coined by Christine Boyer (1995) to describe the spaces in the contemporary city that fall in between the highly designed, imageable and developed spaces of consumption, leisure and work. Usually encompasses spaces of dereliction and decay (see **Figured city**).

E-commerce

Umbrella term to describe the mediation of production, distribution and consumption by telecommunications (especially the Internet).

Electronic road pricing

The process of charging money to use and access previously public road space, usually through computerised systems of monitoring, tracking and transaction.

Electronic tagging

Attaching electronic transponders to a person (e.g. a baby, a low-level offender, an office worker), pet or object (e.g. car) in order to allow the tracking of its movements in time and space.

Embedded infrastructure
Urban infrastructure representing large-scale and highly capital-intensive investments that are sunk or embedded in particular locations. Networks that are rooted or entrenched in places so that they can easily or economically be moved to other locations. The level of embeddedness varies across networks: telecommunications networks are usually much more flexible and malleable than a transport system or trunk water and waste networks.

Euclidean plain
Following the Greek mathematician, Euclid, the concept of geographical space as a flat, geometric, gridded plain upon which events, activities and development occur.

Excludability
The feasibility of controlling access to an infrastructure. Usually, individual consumers can be excluded from transactions involving purely **private goods**. Such exclusion is usually not feasible, or very costly, in the case of **public goods**.

Export processing zone
A district or urban zone endowed with the financial and tax incentives, processing facilities, labour supplies and infrastructure networks to attract internationally oriented business, manufacturing and distribution and import–export activities.

Externalities
Unpriced effects which occur where the benefits and costs of producing or consuming a good affect persons other than those involved in the transaction. In the infrastructure sector *negative* externalities include air, noise, water and land pollution from motor vehicles and electricity production. *Positive* externalities include the public health benefits of access to water and sanitation infrastructure.

Fast track immigration
The use of special conduits in ports and airports, with or without biometric passports, to allow favoured individuals to bypass normal passport and immigration controls. Example of a **premium network space**.

Figured city
Christine Boyer's (1995) term for the highly imageable and highly designed and packaged spaces for work, consumption and leisure that dominate reconstructed and regenerated city centres (see **Disfigured city**).

Filières
An economic value-added chain linking networks of firms, across space and time, into interdependent systems.

Fiscal equivalence
The economic principle that 'you get what you pay for' in paying for goods, services and infrastructure – so avoiding the traditional systems of general taxation and social and geographical cross-subsidy associated with the **modern infrastructural ideal**. Applied in initiatives like **Business Improvement Districts** (see Mallett, 1993a).

Flying base station
An aircraft or airship that hovers permanently over a metropolitan region, delivering telecommunications services.

Fordism
The interconnected social, technological, cultural and political construction through which mass production, distribution and consumption societies were elaborated in Western countries between the 1920s and 1970s.

Foreign Direct Investment (FDI)
Direct investment in a city or nation by overseas capital.

Geodemographics
The range of techniques through which social, demographic and consumption data are mapped and processed through the use of **geographical information systems** to aid the development strategies of supplying firms.

GISs (Geographical Information Systems)
Computerised mapping systems which enable many geographically referenced data sets to be superimposed and analysed.

Glocal by-pass
The development of an infrastructure that is configured to support interaction between valued users and places across local and global circuits of infrastructure. Examples would include specialist international telecommunication services designed to service specialist ports and enclaves, customised transport services (most noticeably to international airports) to bypass public networks, and international power connections to support export processing zones.

Glocal infrastructures
Infrastructure networks configured to connect local geographical spaces seamlessly with international and global scales. See **Glocal scalar fix**.

Glocal scalar fix
Brenner's (1998b) concept of a political, institutional and territorial strategy which uses **glocal infrastructure** networks to help connect local spaces seamlessly with global circuits of economic flow.

Grid erosion
Pope's (1996) concept to describe the ways in which the traditional rectilinear grids of streets in many US cities have been eroded by development trends which tend to produce socioeconomic enclaves served by dedicated infrastructures and streets which terminate there.

Hubs, spoke and tunnel effect
A conceptual framework to capture the geographical effects of contemporary infrastructure networks. *Hubs* are the dominant nodes that articulate and connect the flows of major infrastructure systems. *Spokes* are infrastructure connections that radiate from hubs to serve subservient places. *Tunnel* effects are the 'wormholing' effects of fast and highly capable infrastructure networks that pass through, above or below a territory without allowing access.

ICTs (Information and Communications Technologies)
An umbrella term for all technologies of information or communication.

Infrastructural consumerism
A strategy for supporting competition, branding, diverse pricing and segmented marketing of a wide range of infrastructure products and services within markets which are more or less regulated.

Innovative milieu
A social, economic, cultural and institutional environment which fosters and sustains on-going economic and technological innovation. See **Neo-Marshallian node**.

Intranet
A communications system using Internet technologies and protocols which is closed off for the exclusive use of a firm or geographical or non-geographical community.

ISDN (Integrated Services Digital Network)
A first-generation attempt to offer digital and multimedia services over the conventional telephone line.

Keynesianism/Keynesian welfare state
A nationally driven political strategy, especially common in Western countries between the 1930s and 1970s, to stimulate economic growth through investing in coordinated infrastructure and economic development plans, at the same time as elaborating systems of universal public service provision in health, education and social services. Named after the British economist John Maynard Keynes.

LAN (Local Area Network)
A telecommunications network linking a local cluster of computers.

Local by-pass
The development of a parallel infrastructure, such as a cable network, mobile telephone or raised walkway, that selectively connects more valued users and places while at the same time bypassing less valued users and places.

LTS (Large Technical System)
A term coined by Thomas Hughes (1983) to describe an interconnected and coordinated infrastructure network encompassing technical artefacts, standards, practices and institutions working in harmony.

Megalopolis
A term coined by the geographer Jean Gottmann (see 1990) to describe the coalescence of cities and regions into large-scale polynuclear and extended metropolitan corridors.

Modern infrastructural ideal
The ideal of **rolling out** monopolistic, standardised and integrated infrastructure networks to cover a city, region or country that was associated particularly with the period 1850–1960. Closely associated with the idea of the **natural monopoly**, the theory of **public goods** and **Keynesian** policies.

Multiservice rebundling
The processes of diversification, alliance formation and mergers through which liberalised infrastructure companies make links with retailers and financial service companies to deliver more and more services on a one-stop-shop basis.

Natural monopoly
The Keynesian economic concept which drove the **modern infrastructural ideal** and the concept that infrastructure networks were **public goods**. A natural monopoly was seen to exist when the costs of rolling out a network were so huge that the rewards of a regulated monopoly were necessary to ensure the economic viability of the resulting infrastructure.

Neo-Marshallian node
Following the economist Alfred Marshall, an economic cluster of closely related firms within an **innovative milieu** who rely on untraded linkages to sustain global competitiveness.

Network ghetto
An urban space with relatively poor connections to infrastructure networks.

Normalisation
Sociological term for the process through which a social phenomenon becomes widely taken for granted and expected. Closely related to **black boxing**.

Pay-per revolution
Vincent Mosco's (1988) term for an aspect of infrastructural consumerism where new information technologies are used to transfer previously public services, charged through general tariffs, into ones that are charged on a pay-per-use basis. Examples include **electronic road pricing**.

Personal extensibility
Paul Adams's (1995) term to describe the use of infrastructure networks to extend one's influence in time and space.

Post-Fordism
Umbrella term for the social, economic and political regime that has existed since the collapse of **Fordism**.

Post-Keynesianism
Umbrella term for the social, economic and political regime that has existed since the collapse of **Keynesianism**.

Prepayment meter
A utility meter, especially common in the United Kingdom, requiring the user to charge up an electronic key in advance of use of the service.

Premium network space
A combination of urban and networked spaces that are configured precisely to the needs of socioeconomically wealthy groups and so at the same time are increasingly withdrawn from the wider citizenry and cityscape. (Also termed **secessionary network space**.)

Private goods

Goods or services, distributed within markets, that are usually consumed by one person at a time (e.g. food, consumer durables). Private goods have high **excludability** and are highly **rivalrous** in consumption.

Public goods

The economic notion that public authorities need to deliver certain goods and services which underpinned the concept of the **natural monopoly** and the practices of the **modern infrastructural ideal**. Public goods were seen to have three characteristics: *joint supply* (or non-rivalrousness), meaning that if goods were supplied to one person, they could also be supplied to all other persons at no extra cost; *non-excludability*, meaning that once a supply had been developed a user could not be prevented from consuming the service; and *non-rejectability*, meaning that once a service was supplied it must be equally consumed by all, even those not wishing to.

Real-time congestion pricing

See **Electronic road pricing**.

Rebundled city

Term coined by Dick and Rimmer (1999) to capture the movement away from functional single-use zoning and towards ever larger-scale buildings and complexes that encompass multiple uses and facilities under a single roof.

Regularisation/Haussmannisation

Urban planning doctrine to bring order and coherence to the perceived disorder of an unplanned cityscape through the imposition of standardised street systems and broader systems of planning and engineering control.

Relational urban theory

A broad range of contemporary urban theories which stress the importance of the dynamic and contingent worlds of social relations in structuring and making places, rather than the structuring effects of formal geometries or urban forms.

Rivalrous

The degree to which goods or services can be consumed by one person without being made available to others.

Rolling out

The process of extending an infrastructure network across space.

Secessionary network spaces

See **Premium network space**.

Segmenting infrastructure

There are three forms of segmentation. *Vertical* segmentation splits the core of networks from those parts connected with consumers (e.g. power generation from transmission and distribution). *Horizontal* segmentation involves the superimposition of new networks in parallel (e.g. the provision of radio and satellite phones separated from traditional telephones).

Virtual segmentation refers to the superimposition of new services on a monopolistic network through the use of new information technologies (e.g. driver information systems).

Skywalk
A dedicated passenger walkway linking office blocks, malls or leisure spaces, especially common in North American cities (see Boddy, 1992).

Smart home
A residential unit that has been equipped with the latest information technologies for managing security, the environment, lighting, entertainment, etc.

Sociotechnical hybrid
Sociological concept designed to overcome the modern dualism of society/technology. Stresses the subtle blending of the social and technical in contemporary societies (see **Actor network theory**).

System builder(s)
The key entrepreneur(s) driving forward the construction of a **Large Technical System**.

Technological determinism
The conceptual practice of suggesting that technologies and infrastructures directly shape social and spatial outcomes in a linear, cause-and-effect manner because of their abstract qualities.

Technological fetishism
The celebration of a technology and its aesthetic quality as an icon of transformation without paying attention to the social relations that produce and sustain it (see **Technological determinism**).

Technological sublime
The cultural aura of excitement, limitless power and emancipatory potential than can be constructed to surround new technologies.

Technopole
Planned space of high-technology innovation.

Terminal architecture
Pawley's (1987) term for the way in which contemporary buildings function primarily as terminals on infrastructure networks, rather than as disembodied monuments in urban space.

Territorial adaptor
Gabriel Dupuy's (1995) concept, designed to explain the automobile system, which describes the way infrastructure networks can be constructed to bring the extending landscapes of **megalopolis**-style urban regions into dynamic articulation.

Time–space compression
David Harvey's (1989) concept designed to capture the ways in which new infrastructure innovations like the Internet and air travel effectively reduce space and time barriers for their users.

Toll goods
Goods or services with high levels of **excludability** but a low level of **rivalrousness**. For instance, it is possible to control access to a piped sewage system, but consumption by one user does not usually lessen availability to others.

Town centre management
A UK movement, similar to that of **Business Improvement Districts**, to set up dedicated private and public management bodies for town and city centres for the purpose of upgrading their economic performance.

Unblackboxing
The process through which **blackboxed** technological systems become (re)problematised, (re)exposing their inner workings and performance to scrutiny. Infrastructure networks often become unblackboxed during periods of disruption through earthquakes, drought, war or the consequences of terrorist action.

Unbundling infrastructure
The process through which standardised and **bundled infrastructure** is broken apart or **segmented** technically, organisationally and institutionally into competitive and non-competitive elements to support **infrastructural consumerism**. Usually associated with privatisation and/or liberalisation.

Virtual network by-pass
Also known as *virtual competition*. The process through which information technologies are applied to single infrastructure networks to allow them to support infrastructural competition and consumerism.

BIBLIOGRAPHY

Plate 22 The fast lane on the Massachusetts turnpike in Boston: a classic example of the use of information technology to support local infrastructural bypass and the construction of a premium network space. Cars equipped with transponders enter the tolled highway without stopping. Tolls are electronically debited from users' bank accounts. *Photograph*: Stephen Graham

Abler, R. (1977), 'The telephone and the evolution of the American metropolitan systems'. In I. de Sola Pool (ed.), *The Social Impact of the Telephone*, Cambridge MA: MIT Press, 318–41.

Adams, P. (1995), 'A reconsideration of personal boundaries in space–time', *Annals of the Association of American Geographers*, 85 (2), 267–85.

Aibar, E. and Bijker, W. (1997), 'Constructing a city: the Cerdà plan for the extension of Barcelona', *Science, Technology and Human Values*, 22 (1), 3–30.

Akrich, M. (1992), 'The de-scription of technological objects'. In W. Bijker and J. Law (eds), *Shaping Technology/Building Society: Studies in Sociotechnical Change*, London: MIT Press, 205–23.

Aksoy, A. and Robins, K. (1997), 'Modernism and the millennium: trial by space in Istanbul', *City*, 8, 21–36.

Allen, J. (1999), 'Cities of power and influence: settled formations'. In J. Allen, D. Massey and M. Pryke (eds), *Unsettling Cities*, London: Routledge, 181–227.

Allen, J., Massey, D. and Pryke, M. (eds) (1999), *Unsettling Cities*, London: Routledge.

Allen, S. (1990), 'Mapping the postmodern city: two arguments for notation'. In M. Angeli (ed.), *On Architecture, the City and Technology*, Washington DC: Butterworth, 33–5.

Allen, S. (1994a), 'An interview with Dave Hickey'. In M. Avillez (ed.), *Sites and Stations: Architecture and Utopia in the Contemporary City*, New York: Lusitania, 31–42.

Allen, S. (1994b), 'From critique to construction'. In M. Avillez (ed.), *Sites and Stations: Architecture and Utopia in the Contemporary City*, New York: Lusitania, 6–13.

Amborski, D. and Keare, D. (1998), 'Large-scale development: a teleport proposal for Cordoba', *Land Lines*, September, 4–5.

Amin, A. (1994), *Post-Fordism: A Reader*, Oxford: Blackwell.

Amin, A. and Graham, S. (1998a), 'Cities of connection and disconnection'. In J. Allen, D. Massey and M. Pryke (eds), *Unsettling Cities: Movement/Settlement*, London: Routledge, 7–48.

Amin, A. and Graham, S. (1998b), 'The ordinary city', *Transactions of the Institute of British Geographers*, 22, 411–29.

Amin, A. and Thrift, N. (1992), 'Neo-Marshallian nodes on global networks', *International Journal of Urban and Regional Research*, 16 (4), 571–87.

Amin, A. and Thrift, N. (1999) 'Institutional issues for the European regions: from markets and plans to socioeconomics and powers of association'. In T. Barnes and M. Gertler (eds), *The New Industrial Geography*, London: Routledge, 292–312.

Amin, A. and Thrift, N. (2000), *The Democratic City*, London: Verso.

Amin, A., Massey, D. and Thrift, N. (2000), *Cities for the Many Not the Few*, Bristol: Polity.

Andersson, Å. and Batten, D. (1988), 'Creative nodes, logistics networks, and the future of the metropolis', *Transportation*, 14, 281–93.

Andreu, P. (1997), 'Borders and borderers', *Architecture of the Borderlands*, London: Wiley/Architectural Design, 57–61.

Andreu, P. (1998), 'Tunneling'. In C. Davidson (ed.), *Anyhow*, Cambridge MA: MIT Press, 58–63.

Angélil, M. and Klingmann, A. (1999), 'Hybrid morphologies: infrastructure, architecture, landscape', *Daidalos*, 73, 16–25.

Anton, D. J. (1993), *Thirsty Cities: Urban Environments and Water Supply in Latin America*, Ottawa ON: International Development Research Centre.

Archer, K. (1995), 'The limits of the imagineered city: sociospatial polarization in Orlando', *Economic Geography*, 71 (3), 321–33.

Ascher, F. (1995) *Métapolis, ou l'avenir des villes*, Paris: Jacob.

Atkins, P. (1993), 'How the West End was won: the struggle to remove street barriers in Victorian London', *Journal of Historical Geography*, 19 (3), 267–77.

Augé, M. (1995), *Non-places: Introduction to the Anthropology of Supermodernity*, London: Verso.

Aurigi, A. and Graham, S. (1997), 'Virtual cities, social polarization, and the crisis in urban public space', *Journal of Urban Technology*, 4 (1), 19–52.

Ausubel, J. and Herman, R. (1988), *Cities and Their Vital Systems: Infrastructure Past, Present and Future*, Washington DC: National Academy Press.

Avedano, S., Murphy, D. and Old, A. (1997), 'Between London city airport and Silvertown', in *Architecture of the Borderlands*, London: Wiley/Architectural Design, 67–9.

Bachelor, C. (1998), 'Moving up the corporate agenda', Supply Chain Logistics Survey, *Financial Times*, 1 December, 1.

Badcock, B. (1997), 'Restructuring and spatial polarization in cities', *Progress in Human Geography*, 21 (2), 251–62.

Bagnasco, A. and Le Galès, P. (eds) (2000), *Cities in Contemporary Europe*, Cambridge: Cambridge University Press.

Baird, D. and Heintz, R. (1997) 'Bi-national communities and the unregulated colonia', in *Architecture of the Borderlands*, London: Wiley/Architectural Design, 13–19.

Balbo, M. (1993), 'Urban planning and the fragmented city of developing countries', *Third World Planning Review*, 15 (3), 23–35.

Ballast Wiltshier (1999), *Landscapes of Change*, London: Ballast Wiltshier Construction Group.

Balshaw, M. and Kennedy, L. (2000), *Urban Space and Representation*, London: Pluto.

Banes, C., Kalbermatten, J. and Nankman, P. (1996), *Infrastructure Provision for the Urban Poor: Assessing the Needs and Identifying the Alternatives*, Washington DC: Transportation, Water and Urban Development Department, World Bank.

Banham, R. (1980), *Theory and Design in the First Machine Age*, Cambridge MA: MIT Press.

Banisar, D. (1999), 'Big Brother goes high-tech', *CAQ Magazine* (available at www.worldmedia. com/caq/articles/brother.html).

Bannerjee, T., Giuliano, G. and Hise, G. (1996), 'Invented and reinvented streets: designing the new shopping experience', *Lusk Real Estate Review*, summer, 18–30.

Bannister, N. (1994), 'Networks tap into low wages', *Guardian*, 15 October, 40.

Barakat, S. (1998), 'City war zones', *Urban Age*, spring, 11–15.

Barnett, J. (1986), *The Elusive City: Five Centuries of Design, Ambition and Miscalculation*, New York: Harper & Row.

Bateson, G. (1978), *Steps to an Ecology of Mind*, New York: Ballantine.

Bayne, J. and Freeman, D. (1995), 'The effect of residence in enclaves on civic concern: an initial exploration', *Social Sciences Journal*, 32 (4), 409–21.

Bayor, R. (1988), 'Roads to racial segregation', *Journal of Urban History*, 15 (1), 3–21.

Beamish, A. (1999), 'Approaches to community computing: bringing technology to low income groups'. In D. Schön, B. Sanyal and W. Mitchell (eds), *High Technology and Low Income Communities*, Cambridge MA: MIT Press, 349–70.

Beauregard, R. (1989), 'Between modernity and postmodernity: the ambiguous position of US planning', *Environment and Planning D: Society and Space*, 7, 381–95.

Beauregard, R. and Laila, A. (1997), 'The unavoidable incompleteness of the city', *American Behavioral Scientist*, 41 (3), 327–41.

Beaverstock, J., Smith, R. and Taylor, P. (2000) 'World-city network: a new metageography?', *Annals of the Association of American Geographers*, 90 (1), 123–34.

Beck, U. (1999), *World Risk Society*, Cambridge: Polity.

Beckett, A. (2000), 'Hate the firm – and buy its products', *Guardian Weekly*, 27 January–2 February, 17.

Beckouche, P. and Veltz, P. (1988), 'Nouvelle économie, nouveau térritoire', *Datar Letter*, June, supplement.

Bednar, M. (1989), *Interior Pedestrian Places*, New York: Whitney Library of Design.

Bell, D. and Haddour, A. (eds) (2000), *City Visions*, London: Longman.

Bell, M. and Leong, S. (1998), 'Introduction'. In M. Bell and S. Leong (eds), *Slow Space*, New York: Monacelli Press, 6–13.

Bender, T. (1996), 'Opinion', *Los Angeles Times*, 22 December.

Beniger, J. (1986), *The Control Revolution: Technological and Economic Origins of the Information Society*, Cambridge MA: Harvard University Press.

Benjamin, W. (1969), *Illuminations*, New York: Schocken.

Benjamin, W. (1979), *One Way Street*, London: New Left Books.

Benjamin, W. (1999), *The Arcades Project* Volumes I–II, Cambridge MA: Belknap/Harvard.

Berlage, M. (1997), 'The role of local access networks in regional economic integration in Eastern Europe'. In E. Roche and H. Bakis (eds), *Developments in Telecommunications*, Aldershot: Ashgate, 177–93.

Berman, M. (1983), *All that is Solid Melts into Air: The Experience of Modernity*, London: Verso.

Berner, E. (1997a), 'The metropolitan dilemma: global societies, localities and the struggle for urban land in Manila'. In A. Onçu and P. Weyland (eds), *Space, Culture and Power: New Identities in Globalizing Cities*, London: Zed Books, 98–116.

Berner, E. (1997b), *Defending a Place in the City: Localities and the Struggle for Urban Land in Metro Manila*, Quezon City: Ateneo de Manila University Press.

Bernet, B. (2000), 'Understanding the needs of telecommunications tenants', *Development Magazine*, spring, 16–18.

Bhabha, H. (1990), 'The third space: interview with Homi Bhabha'. In J. Rutherford (ed.), *Identity: Community, Culture, Difference*. London: Routledge, 208–24.

Bhabha, H. (1994), *The Location of Culture*, London: Routledge.

Bhabna, H. (1992), 'Double visions', *Artforum*, January, 82–90.

Bhatia, R. and Falkenmark, M. (1993), *Water Resource Policies and the Urban Poor: Innovative Approaches and Policy Imperatives*, Washington DC: Transport and Urban Development Department, World Bank – UNDP Water and Sanitation Program.

Bijker, W. (1993), 'Do not despair: there is life after constructivism', *Science, Technology and Human Values*, 18 (1), 113–38.

Bingham, N. (1996), 'Object-ions: from technological determinism towards geographies of relations', *Environment and Planning D: Society and Space*, 14, 635–57.

Black, M. (1995) *Mega-slums: the Coming Sanitary Crisis*, London: WaterAid.

Black, M. (1996) *Thirsty Cities: Water Sanitation and the Urban Poor*, London: WaterAid.

Blais, P. (1998), 'Getting wired', *Urban Land*, December, 60–3.

Blakeley, E. and Snyder, M. (1997a), *Fortress America: Gated Communities in the United States*, Washington DC: Brookings Institution Press.

Blakeley, E. and Snyder, M. (1997b), 'Divided we fall'. In N. Ellin (ed.), *Architecture of Fear*, New York: Princeton Architectural Press, 85–100.

Boddy, T. (1992), 'Underground and overhead: building the analogous city'. In M. Sorkin (ed.), *Variations on a Theme Park: The New American City and the End of Public Space*, New York: Hill & Wang, 123–53.

Body-Gendrot, S. (2000), *The Social Control of Cities? A Comparative Perspective*, Oxford: Blackwell.

Bond, P. (1998), 'Local economic development and the municipal services crisis in post-Apartheid South Africa', *Urban Forum*, 9 (2), 160–97.

Bond, P. (1999), 'Basic infrastructure for socio-economic development, environmental protection and geographical desegregation: South Africa's unmet challenge', *Geoforum*, 30, 43–59.

Borsook, P. (1999), 'How the Internet ruined San Francisco', *Salom.Com*, news feature (http://www.salon/com/nes/feature/1999/10/28/internet).

Bosma, K. (1998), 'Functional wrapping', *Archis*, December, 8–15.

Bosma, K. and Hellinga, H. (1998), 'Mastering the city II'. In K. Bosma and H. Hellinga (eds), *Mastering the City: North European City Planning 1900–2000*, Rotterdam: NAI Publishers, 8–17.

Bowcott, O. (2000), 'Russia's northern cities die as people leave in search of jobs and warmth', *Guardian Weekly*, 20–6 January, 3.

Boyer, C. (1987), *Dreaming the Rational City*, Cambridge MA: MIT Press.

Boyer, C. (1992), 'The imaginary real world of cybercities', *Assemblage*, 18, 115–27.

Boyer, C. (1993), 'The city of illusion: New York's public places'. In P. Knox (ed.), *The Restless Urban Landscape*, Englewood Cliffs NJ: Prentice Hall.

Boyer, C. (1994), *The City of Collective Memory*, Cambridge MA: MIT Press.

Boyer, C. (1995), 'The great frame-up: fantastic appearances in contemporary spatial politics'. In H. Liggett and D. Perry (eds), *Spatial Practices*, London: Sage, 81–109.

Boyer, C. (1996), *Cybercities: Visual Perception in the Age of Electronic Communication*, New York: Princeton Architectural Press.

Boyer, C. (1997), 'Cyber Cities and Regional Spaces'. Mimeo.

Boyer, C. (2000), 'Crossing cybercities: urban regions and the cyberspace matrix'. In R. Beauregard and S. Body-Gendrot (eds), *The Urban Moment: Cosmopolitan Essays on the late Twentieth Century City*, Thousand Oaks CA: Sage, 51–78.

Braczyk, H-J., Fuchs, G. and Wolf, H-G. (eds) (1999), *Multimedia and Regional Economic Restructuring*, London: Routledge.

Branscomb, L. and Keller, J. (1996), *Converging Infrastructures: Intelligent Transportation and the National Information Infrastructure*, Cambridge MA: MIT Press.

Braudel, F. (1967), *Civilisation matérelle et capitalisme*, Paris: Armand Colin.

Brecher, J. and Costello, T. (eds) (1994), *Global Village or Global Pillage? Economic Restructuring from the Bottom Up*, Cambridge MA: South End Press.

Bremner, L. (n.d.), 'Crime and the emerging landscape in post-apartheid Johannesburg'. In H. Judin and I. Vladislavic (eds), *Blank ——: Architecture, Apartheid and After*, Rotterdam: NAI Publishers, 49–63.

Brenes, E., Ruddy, V. and Castro, R. (1997), 'Free zones in El Salvador', *Journal of Business Research*, 38, 57–65.

Brenner, N. (1998a), 'Globalisation and reterritorialisation: the re-scaling of urban governance in the European Union', *Urban Studies*, 26 (3), 431–51.

Brenner, N. (1998b), 'Global cities, glocal states: global city formation and state territorial restructuring in contemporary Europe', *Review of International Political Economy*, 5 (1), 1–37.

Brenner, N. (1998c), 'Between fixity and motion: accumulation, territorial organization and the historical geography of spatial scales', *Environment and Planning D: Society and Space*, 16, 459–81.

Brenner, N. (2000), 'The urban question as a scale question: reflections on Henri Lefebvre, urban theory and the politics of scale in the late 1990s', *International Journal of Urban and Regional Research*, 24 (2), 361–78.

Breslin, P. (2000), 'From mud to markets', *Urban Age*, spring, 4–5.

Bristow, G., Munday, M. and Gripaios, P. (2000), 'Call centre growth and location: corporate strategy and the spatial division of labour', *Environment and Planning A*, 32, 519–38.

British Medical Association (1994), *Water: A Vital Resource*, London: British Medical Association.

Brooke, J. and Nanetti, P. (1998,) 'Success signals', *Utility Week*, 16 October, 56–8.

Brownhill, S. (1990), *Developing London's Docklands: Another Great Planning Disaster?*, London: Paul Chapman.

Browning, J. (1996), 'Who's what on the web', *Wired*, August, 33–5.

Brunn, S., Andersson, H. and Dahlman, C. (2000), 'Lanscaping for power and defense'. In J. Gold and G. Revill (eds), *Landscapes of Power and Defence*, London: Prentice Hall, 68–84.

Bunnell, T. (2000), 'Multimedia Utopia? A Geographical Critique of IT Discourse in Malaysia'. Mimeo.

Burgel, G. and Burgel, G. (1996), 'Global trends and city policies: friends or foes of uneven development?'. In M. Cohen, B. Ruble, J. Tulchin and A. Garaland (eds), *Preparing for the Urban Future: Global Pressures and Local Forces*, Baltimore MD: Johns Hopkins University Press, 301–36.

Burrows, R. (1997), 'Virtual culture, urban social polarization, and social science fiction'. In B. Loader (ed.), *The Governance of Cyberspace*, London: Routledge, 31–42.

Byatt, I. (1979), *The British Electrical Industry 1875–1914: The Economic Returns to a New Technology*, Oxford: Oxford University Press.

Calabrese, A. and Borchert, M. (1996), 'Prospects for electronic democracy in the United States: rethinking communication and social policy', *Media, Culture and Society*, 18, 249–68.

Caldeira, T. (1994), 'Building up walls: the new pattern of spatial segregation in São Paulo', *International Social Sciences Journal*, 147, 55–66.

Caldeira, T. (1996), 'Fortified enclaves: the new urban segregation', *Public Culture*, 8, 303–28.

Caldeira, T. (1999), *City of Walls: Crime, Segregation and Citizenship in São Paulo*, Berkeley CA: University of California Press.

Callon, M. (1986), 'Some elements of a sociology of translation: domestication of the scallops and the fisherman of St Brieuc bay'. In J. Law (ed.), *Power, Action and Belief: A New Sociology of Knowledge*, London: Routledge, 196–232.

Callon, M. (1991), 'Techno-economic networks and irreversibility'. In J. Law (ed.), *A Sociology of Monsters: Essays on Power, Technology and Domination*, London: Routledge, 196–233.

Calthorpe, P. (1993), 'The next American metropolis'. In J. Woodroffe, D. Papa and I. MacBurnie (eds), *The Periphery*, London: Wiley/Architectural Design, 19–24.

Calvino, I. (1984), *Invisible Cities*, London: Secker & Warburg.

Capello, R. and Gillespie, A. (1993), 'Transport, communication and spatial organisation: future trends and conceptual frameworks'. In G. Giannopoulos and A. Gillespie (1993), *Transport and Communications in the New Europe*, London: Belhaven, 24–58.

Cardoso, F. and Faletto, E. (1979), *Dependency and Development in Latin America*, Berkeley CA: University of California Press.

Cartier, C. (1998), 'Megadevelopment in Malaysia: from heritage landscapes to "leisurescapes" in Melaka's tourism sector', *Singapore Journal of Tropical Geography*, 19(2), 151–76.

Casey, E. (1998), *The Fate of Place: A Philosophical History*, Berkeley CA: University of California Press.

Cassidy, J. and Daly, M. (1998), 'Tourist sites plan to drive out the "violent vagrants"', *Big Issue*, 14–20 September, p. 5.

Castells, M. (1983), *The City and the Grassroots*, London: Edward Arnold.

Castells, M. (1989), *The Informational City*, Oxford: Blackwell.

Castells, M. (1996), *The Information Age: Economy, Society and Culture I, The Rise of the Network Society*, Oxford: Blackwell.

Castells, M. (1997a), *The Information Age: Economy, Society and Culture II, The Power of Identity*, Oxford: Blackwell.

Castells (1997b), 'Hauling in the future', *Guardian*, 13 December, 21.

Castells, M. (1998), *The Information Age: Economy, Society and Culture III, The End of the Millennium*, Oxford: Blackwell.

Castells, M. (1999a), 'Grassrooting the space of flows', *Urban Geography*, 20 (4), 294–302.

Castells, M. (1999b), 'The informational city is a dual city: can it be reversed?'. In D. Schön, B. Sanyal and W. Mitchell (eds), *High Technology and Low Income Communities*, Cambridge MA: MIT Press, 25–42.

Castells, M. (1999c), *The Culture of Cities in the Information Age*. Report to the US Library of Congress.

Castells, M. and Hall, P. (1994), *Technopoles of the World: The Making of Twenty-first Century Industrial Complexes*, London: Routledge.

Celik, Z., Favro, D. and Ingersoll, R. (eds) (1994), *Streets: Critical Perspectives on Public Space*, Berkeley CA: University of California Press.

Cerver, F. (1998), *The Contemporary City*, New York: Whitney Library of Design.

Chan, S. (ed.) (1995), *Foreign Direct Investment in a Changing Global Political Economy*, London: Macmillan.

Channel 4 (1994), *Once upon a Time in Cyberville* (programme transcript), London: Channel 4 Television.

Chant, C. and Goodman, D. (eds) (1998), *Pre-industrial Cities and Technology*, London: Routledge.

Chaoy, F. (1969), *The Modern City: Planning in the Nineteenth Century*, New York: Braziller.

Chatzis, K. (1992), 'A Conceptual Framework for Analysing the Long-term Evolution of Regulatory Control Practices within Large Technical Systems'. Mimeo.

Chatzis, K. (1999), 'Designing and operating storm water drain systems: empirical findings and conceptual developments'. In O. Coutard (ed.), *The Governance of Large Technical Systems*, London, Routledge, 73–90.

Chen, K. (1995), 'The evolution of free economic zones and the recent development of cross-national growth zones', *International Journal of Urban and Regional Research*, 19 (4), 593–621.

Chevin, D. (1991), 'All the right connections', *Building*, 19 July, 46–50.

Chomsky, N. (1993), 'The new global economy'. In D. Barsamian (ed.), *The Prosperous Few and the Restless Many*, Berkeley CA: Odonian, 6–16.

Chougill, C. (1996) 'Ten steps to sustainable infrastructure', *Habitat International* 20 (3), 389–404.

Chougill, C. (1999), 'Community infrastructure for low-income cities', *Habitat International*, 23 (2), 289–301.

Chougill, C., Franceys, R. and Cotton, A. (1993), 'Building community infrastructure in the 1990s', *Habitat International*, 17 (4), 1–12.

Chowdray, T. (1998), 'Telecom liberalization and competition in developing countries', *Telecommunications Policy*, 22 (5/5), 259–65.

Christopherson, S. (1992), 'Market rules and territorial outcomes: the case of the United States', *International Journal of Urban and Regional Research*, 17 (2), 274–88.

Clark, G. (1999), 'The retreat of the state and the rise of pension fund capitalism'. In R. Martin (ed.), *Money and the Space Economy*, London: Wiley, 241–60.

Clarke, D. and Bradford, M. (1998), 'Public and private consumption and the city', *Urban Studies*, 35 (5–6), 865–88.

Clarke, S. and Gaile, G. (1998), *The Work of Cities*, Minneapolis MN: University of Minnesota Press.

Clay, G. (1987), 'The street as teacher'. In A. Moudon (ed.), *Public Streets for Public Use*, New York: Van Nostrand, 95–109.

Cockburn, A. and St Clair, J. (2000), *Five Days that Shook the World: The Battle for Seattle and Beyond*, London: Verso.

Cohen, M. (1996), 'The hypothesis of urban convergence: are cities in the North and South becoming more alike in an age of globalization?'. In M. Cohen, B. Ruble, J. Tulchin and A. Garaland (eds), *Preparing for the Urban Future: Global Pressures and Local Forces*, Baltimore MD: Johns Hopkins University Press, 25–39.

Cohen, R. (2000), 'A Brazilian convict's path from poverty to a "very dark place"', *New York Times*, 29 April, A6.

Compton, K. (1952), 'Science on the march', *Popular Mechanics*, 97, January, 120.

Connell, J. (1999), 'Beyond Manila: walls, malls and private spaces', *Environment and Planning A*, 31, 417–39.

Cooper, D. (1998), 'Regard between strangers: diversity, equality and the reconstruction of public space', *Critical Social Policy*, 18 (4), 463–91.

Corey, K. (1998), 'Electronic Space: Creating and Controlling Cyber-communities in Southeast Asia and the United States'. Mimeo.

Corn, J. and Horrigan, B. (1984), *Yesterday's Tomorrows: Past Visions of the American Future*, Baltimore MD: Johns Hopkins University Press.

Coutard, O. (1996), 'Fifteen years of social and historical research on Large Technical Systems: An interview with Thomas Hughes', *Flux*, July–September, 40–7.

Coutard, O. (ed.) (1999), *The Governance of Large Technical Systems*, London: Routledge.

Cowan, R. (1983), *More Work for Mother: The Ironies of Household Technology from the Open Hearth to the Microwave*, New York: Basic.

Cowan, R. (1997) *A Social History of American Technology*, Oxford: Oxford University Press.

Cowan, R. L. (1997), *The Connected City: A New Approach to Making Cities Work*, London: Urban Initiatives.

Cox, K. (1993) 'The local and the global in the new urban politics: a critical view', *Environment and Planning D: Society and Space*, 11, 433–48.

Cox, K. and Jonas, A. (1993), 'Urban development, collective consumption and the politics of metropolitan fragmentation', *Political Geography*, 12 (1), 8–37.

Cox, K. and Mair, A. (1988), 'Locality and community in the politics of local economic development', *Annals of the Association of American Geographers*, 78 (2), 307–25.

Coyle, D. (1997), *The Weightless World: Strategies for Managing the Digital Economy*, Cambridge MA: MIT Press.

Crang, M. (2000), 'Public space, urban space and electronic space: would the real city please stand up?' *Urban Studies*, 37 (2), 301–17.

Crang, M. and Thrift, N. (eds) (2000), *Thinking Space*, London: Routledge.

Crawford, M. (1999a), 'The architect and the mall'. In John Jerde Partnership International (eds), *You are Here*, London: Phaidon, 44–54.

Crawford, M. (1999b), 'Blurring the boundaries: public space and private life'. In J. Chase, M. Crawford and J. Kaliski (eds), *Everyday Urbanism*, New York: Monacelli Press, 22–35.

Crawford, R. (1996), 'Computer-assisted crises'. In G. Gerbner, H. Mowlana and H. Schiller (eds), *Invisible Crises: What Conglomerate Control of the Media Means for America and the World*, Boulder CO: Westview Press, 47–81.

Crilley, D. (1993), 'Megastructures and urban change: aesthetics, ideology and design'. In P. Knox (ed.), *The Restless Urban Landscape*, Englewood Cliffs NJ: Prentice Hall, 127–64.

Cumings, S. (2000), 'The American ascendency: imposing a new world order', *The Nation*, 8 May, 13–16.

Curien, N. (1997) 'The economics of networks'. In D. Lorrain and G. Stoker (eds), *The Privatisation of Urban Services in Europe*, London: Pinter, 43–57.

Curry, M. (1998), *Digital Places: Living with Geographic Information Technologies*, London: Routledge.

Curwen, P. (1999), 'Survival of the fittest: formation and development of international alliances in telecommunications', *Info*, 1 (2), 141–60.

Cuthbert, A. (1995), 'The right to the city: surveillance, private interest and the public domain in Hong Kong', *Cities*, 12 (5), 293–310.

Dabinett, G. and Graham, S. (1994), 'Telematics and industrial change in Sheffield, UK', *Regional Studies*, 28 (6), 605–17.

D'Antonio, M. (2000), 'Bunker mentality', *New York Times Magazine*, 26 March, 26.

Daly, B. (1999), 'Covent Garden to "exclude vagrants"', *Big Issue*, March, 11.

Davies, J. (1972), *The Evangelistic Bureaucrat*, London: Tavistock.

Davies, S. (1995), *Big Brother: Britain's Web of Surveillance and the New Technological Order*, London: Pan.

Davis, A. (1997), 'The body as password', *Wired*, July, 132–40.

Davis, M. (1990) *City of Quartz*, London: Vintage.

Davis, M. (1992), 'Beyond *Blade Runner*: urban control, the ecology of fear', *Open Magazine*, Westfield: New Jersey.

Davis, M. and Moctezuma, A. (1997), 'Policing the third border', *Architecture of the Borderlands*, London: Wiley/Architectural Design, 34–7.

Davis, S. (1999), 'Space jam: media conglomerates build the entertainment city', *European Journal of Communication*, 14940, 435–59.

De Certeau, M. (1984), *The Practice of Everyday Life*, Berkeley CA: University of California Press.

Dear, M. (1999), *The Postmodern Urban Condition*, Oxford: Blackwell.

Dear, M. and Flusty, S. (1998), 'Postmodern urbanism', *Annals of the Association of American Geographers*, 88 (1), 50–72.

Deas, I. and Ward, K. (2000), 'From the "new localism" to the "new regionalism"? The implications of regional development agencies for city–region relations', *Political Geography*, 19, 273–92.

Deleuze, G. and Guattari, F. (1997), 'City/State', *Zone*, 1–2, 195–9.

Dematteis, G. (1988), 'The weak metropolis'. In L. Mazza (ed.), *World Cities and the Future of the Metropolis*, Milan: Electra.

Dematteis, G. (1994), 'Global networks, local cities' *Flux*, 15, 17–24.

Demos (1997), *The Wealth and Poverty of Networks*, Demos Collection 12, London: Demos.

Denny, C. and Brittain, C. (1999), 'UN attacks growing gulf between rich and poor', *Guardian*, 12 July, 14.

Dery, M. (1999), *The Pyrotechnic Insanitarium: American Culture on the Brink*, New York: Grove.

Dewey, F. (1997), 'Cyburbanism as a way of life'. In N. Ellin (ed.), *Architecture of Fear*, New York: Princeton Architectural Press, 260–80.

Dick, H. and Rimmer, P. (1998), 'Beyond the Third World city: the new urban geography of South East Asia', *Urban Studies*, 35 (12), 2303–21.

Dick, H. and Rimmer, P. (1999), 'Privatising climate: First World cities in South East Asia'. In J. Brotchie, P. Newton, P. Hall and J. Dickey (eds), *East–West Perspectives on Twenty-first Century Urban Development*, Aldershot: Ashgate, 305–23.

Doctor, S. and Dutton, W. (1998), 'The First Amendment online: Santa Monica's public electronic network'. In R. Tsagarousianou, D. Tambini and C. Bran (eds), *Cyberdemocracy: Technology, Cities and Civic Networks*, London: Routledge, 125–52.

Doel, M. and Clarke, D. (1998), 'Transpolitical Urbanism: Suburban Anomaly and Ambient Fear'. Mimeo.

Dolgon, C. (1999), 'Soulless cities: Ann Arbor, the cutting edge of discipline: post-Fordism, postmodernism and the new bourgeoisie', *Antipode*, 31 (3), 276–92.

Donald, J. (1997), 'Imagining the modern city'. In S. Westwood and J. Williams (eds), *Imagining Cities*, London: Routledge.

Donald, J. (1999), *Imagining the Modern City*, London: Athlone Press.

Doron, G. (2000), 'The dead zone and the architecture of transgression', *Archis*, 51 (4), 48–58.

Douglass, M. and Friedmann, J. (1998), *Cities for Citizens*, London: Wiley.

Dovey, K. (1999), *Framing Places: Mediating Power in Built Form*, London: Routledge.

Drakeford, M. (1995), *Token Gesture: A Report on the Use of Token Meters by the Gas, Electricity and Water Companies*, London: National Local Government Forum against Poverty.

Drew, B. (1998), *Crossing the Expendable Landscape*, St Paul MN: Graywolf Press.

Drewe, P. and Janssen, B. (1996), 'What Port for the Future? From "Mainports" to Ports as Nodes on Logistics Networks'. Mimeo.

Driver, F. (1984), 'Power, space and the body: a critical reassessment of Foucault's *Discipline and Punish*', *Society and Space*, 3, 425–46.

Drucker, S. and Gumpert, G. (1999), 'Public spaces and rights of association'. In *Free Speech Yearbook 1999*, Washington DC: National Communications Association.

Dugger, C. (2000), 'India's unwired villages mired in the distant past', *New York Times*, 19 March, 1–12.

Dumm, T. (1993), 'The new enclosures: racism in the normalized community'. In R. Gooding-Williams (ed.), *Reading Rodney King/Reading Urban Uprising*, London: Routledge, 178–95.

Duncan, S. (1979), 'Qualitative change in human geography: an introduction', *Geoforum*, 10, 1–4.

Dunham-Jones, E. (1999), Networking the Post-industrial Landscape'. Mimeo.

Dunham-Jones, E. (2000), 'Capital Transformations of the Post-industrial Landscape'. Mimeo.

Dunning, J. and Narula, R. (1996), *Foreign Direct Investment and Governments: Catalysts for Economic Restructuring*, London: Routledge.

Dupuy, G. (1988) (ed.), *Réseaux Territoriaux*, Caen: Paradigme.

Dupuy G. (1991), *L'Urbanisme des Réseaux: Théories et Méthodes*, Paris: Armand Colin.

Dupuy, G. (1995), 'The automobile system: a territorial adapter', *Flux*, July–September, 21–36.

Dyer, P. (1995), 'Who are the unphoned ?' *Telecommunications Policy*, January, 16–25.

Easterling, K. (1999a), *Organization Space*, Cambridge MA: MIT Press.

Easterling, K. (1999b), 'Interchange and container', *Perspecta*, 30, 112–21.

Edwards, B. (1999), 'Deconstructing the city: London docklands', *Urban Design Quarterly*, 69, 22–4.

Eisenstadt, S. (2000), 'Multiple modernities', *Daedalus*, winter, 1–30.

Ellin, N. (1996), *Postmodern Urbanism*, Oxford: Blackwell.

Ellin, N. (1997), 'Shelter from the storm of form follows fear and vice versa'. In N. Ellin (ed.), *Architecture of Fear*, New York: Princeton Architectural Press, 13–46.

Emberley, P. (1989), 'Places and stories: the challenge of technology', *Social Research*, 56, 741–85.

Eng, I. (1996), 'The rise of manufacturing towns: externally driven industrialization and urban development in the Pearl river delta of China', *International Journal of Urban and Regional Research*, (4), 554–68.

Entrikin, J. (1989), 'Place, region and modernity'. In J. Agnew and J. Duncan (eds), *The Power of Place*, London: Unwin Hyman, 30–43.

Enyedi, G. (1996), 'Urbanization under socialism'. In G. Andrusz, M. Harloe and I. Szelenyi (eds), *Cities After Socialism: Urban and Regional Change and Conflict in Post-socialist Societies*, Oxford: Blackwell, 100–18.

Ernst, J. (1994), *Whose Utility? Public Utility Privatization and Regulation in Britain*, Buckingham: Open University Press.

Evans, M. (1999), 'It's a wired world', *Journal of Property Management*, November–December, 42–7.

Everard, J. (2000), *Virtual States: The Internet and the Boundaries of the Nation State*, London: Routledge.

Ezechieli, C. (1998), 'Shifting Boundaries: Territories, Networks and Cities'. Mimeo.

Ezquiga, J. (1998), 'Cambio de estilo o cambio de paradigma', *Urban*, 2, 7–31.

Fainstein, S. (1999), 'Can we make the cities we want?'. In R. Beauregard and S. Body-Gendrot (eds), *The Urban Moment: Cosmopolitan Essays on the late Twentieth Century City*, Thousand Oaks CA: Sage, 209–41.

Falk, R. (1999), *Predatory Globalization: A Critique*, Cambridge: Polity.

Falkenmark, M. and Lundquist, J. (1995), 'Looming water crisis: new approaches are inevitable'. In L. Ohlsson (ed.), *Hydropolitics: Conflicts over Water as a Development Constraint*, Dhaka: University Press, 178–212.

Felbinger, N. (1996), 'Introduction'. In *Architecture in Cities: Present and Future*, Barcelona: Centre de Cultura Contemporània de Barcelona, 1–12.

Fillion, P. (1996), 'Metropolitan planning objectives and implementation constraints: planning in a post-Fordist and postmodern age', *Environment and Planning A*, 28, 1637–60.

Fincher, R. and Jacobs, J. (1998), *Cities of Difference*, New York: Guildford.

Finnie, G. (1998), 'Wired cities', *Communications Week International*, 18 May, 19–22.

Firman, T. (1999), 'From "global" city to "city of crisis": the Jakarta metropolitan region under economic turmoil', *Habitat International*, 23 (4), 447–66.

Fischer, C. (1992), *America Calling: A Social History of the Telephone*, Berkeley CA: University of California Press.

Fischler, R. (1998), 'The metropolitan dimension of early zoning: revisiting the 1916 New York City ordinances', *APA Journal*, spring, 170–86.

Fishman, R. (1982), *Urban Utopias in the Twentieth Century: Ebenezer Howard, Frank Lloyd Wright and Le Corbusier*, Cambridge MA: MIT Press.

Fishman, R. (1990), 'Metropolis unbound: the new city of the twentieth century', *Flux*, spring, 43–55.

Fishman, R. (1994), 'Space, time and sprawl'. In J. Woodroffe, D. Papa and I. MacBurnie (eds), *The Periphery*, London: Wiley/Architectural Design, 44–7.

Flusty, S. (1997), 'Building paranoia'. In N. Ellin (ed.), *Architecture of Fear*, New York: Princeton Architectural Press, 47–60.

Flusty, S. (2000), 'Thrashing downtown: play as resistance to the spatial and representational regulation of Los Angeles', *Cities*, 17 (2), 149–58.

Foot, J. (2000), 'The urban periphery, myth and reality: Milan, 1950–90', *City*, 4 (1), 7–26.

Foster, J. (1999), *Docklands: Cultures of Conflict, Worlds in Collision*, London: UCL Press.

Foster, K. (1996), 'Specialization in government: the uneven use of special metropolitan districts in metropolitan areas', *Urban Affairs Review*, 31 (3), 283–313.

Foucault, M. (1977), *Discipline and Punish: The Birth of the Prison*, New York: Vintage.

Foucault, M. (1980), 'Questions on geography'. In. C. Gordon (ed.), *Power/Knowledge: Selected Interviews and other Writings 1972–77*, New York: Pantheon.

Foucault, M. (1984), *Foucault: A Reader*, ed. P. Rabinow, New York: Pantheon.

Freeman, Caroline (2000), *High Tech and High Heels in the Global Economy*, Durham NC and London: Duke University Press.

Freeman, Chris (1990), 'Information technology and the new economic paradigm'. In H. Schutte (ed.), *Strategic Issues in Information Technology: International Implications for Decision Makers*, Maidenhead: Pergamon.

Friedman, K. (1999), 'Restructuring the City: Thoughts on Urban Patterns in the Information Society'. Mimeo.

Friedmann, J. (1966), *Regional Development Policy: A Case Study of Venezuela*, Cambridge MA: MIT Press.

Friedmann, J. (1995), 'Where we stand: a decade of world city research'. In P. Knox and P. Taylor (eds), *World Cities in a World System*, Cambridge: Cambridge University Press, 21–47.

Friedmann, J. (1998), 'The new political economy of planning: the rise of civil society'. In M. Douglass and J. Friedmann, *Cities for Citizens*, London: Wiley, 19–38.

Friedmann, J. (2000), 'The good city: in defense of utopian thinking', *International Journal of Urban and Regional Research*, 24 (2), 459–72.

Frobel, F., Heinrichs, J. and Kreye, O. (1980), *The New International Division of Labour*, Cambridge: Cambridge University Press.

Gandelsonas, M. (1999), *X-urbanism: Architecture and the American City*, New York: Princeton Architectural Press.

Gandy, M. (1998), 'Technological Modernism and the Urban Parkway in New York City'. Mimeo.

Garfinkel, S. (2000), 'Welcome to Sealand', *Wired*, July, 230–9.

Garner, C. (1997), 'Supermarkets prepare to charge more at prime time', *Independent*, 8 December, 4.

Garrett, P. (1995), 'Hanging on to your customers', *Utility Week*, 5 May, 14–17.

Garrison, W. (1990), 'Impacts of technological systems on cities', *Built Environment*, 6 (2), 120–30.

Gaubatz, P. (1999), 'Understanding Chinese urban form: contexts for interpreting continuity and change', *Built Environment*, 24 (4), 251–70.

Gavira, C. (1995), 'The image of water: lines, networks and graphs in the hydraulic cartography of Madrid', *Flux*, 19, 4–16.

Gershuny, J. (1983), *Social Innovation and the Division of Labour*, Oxford: Oxford University Press.

Gethin, S. (1998), 'Winning cargo business', *Jane's Airport Review*, March, 29–30.

Giannopoulos, G. and Gillespie, A. (eds) (1993), *Transport and Communications Innovation in Europe*, London: Belhaven.

Giddens, A. (1990), *The Consequences of Modernity*, Oxford: Polity.

Gilbert, A. (1992), 'Third World cities: housing, infrastructure and servicing', *Urban Studies*, 29 (3/4), 435–60.

Gilbert, A. (1994), *The Latin American City*, London: LAB.

Gillespie, A. (1991), 'Advanced communications networks, territorial integration and local development'. In R. Camagni (ed.), *Innovation Networks*, London: Belhaven, 214–29.

Glancey, J. (1997), 'Exit from the city of destruction', *Independent*, 26 May, 20.

Glewwe, P. and Hall, G. (1992), *Poverty and Inequality During Orthodox Adjustment: The Case of Peru 1985–90*, Working Paper No. 86, Washington DC: World Bank.

Gökalp, I. (1992), 'On the analysis of Large Technical Systems', *Science, Technology and Human Values*, 17 (1), 578–87.

Gold, J. (1997), *The Experience of Modernism: Modern Architects and the Future City 1928–53*, London: Spon.

Goldberger, P. (1996), 'The rise of the private city'. In J. Martin (ed.), *Breaking Away: The Future of Cities*, New York: Twentieth Century Fund.

Golding, G. (1998), 'Divide and conquer', *Utility Week*, 7 August, 18–19.

Golsing, P. (1996), *Financial Services in the Digital Age*, London: Bowerdean.

Goodchild, B. (1990), 'Planning and the modern–postmodern debate', *Town Planning Review*, 61 (2), 119–37.

Goodman, D. and Chant, C. (eds) (1999), *European Cities and Technology*, London: Routledge.

Goslee, S. (1998), *Losing Ground Bit by Bit: Low Income Communities in the Information Age*, Washington DC: Benton Foundation and National Urban League.

Goss, J. (1993), 'The magic of the mall: an analysis of form, function and meaning in contemporary retail environments', *Annals of the Association of American Geographers*, 83, 18–47.

Goss, J. (1995), '"We know who you are and we know where you live": the instrumental rationality of geodemographic systems', *Economic Geography*, 71 (2), 171–98.

Gottdeiner, M. (1997), *The Theming of America: Dreams, Visions and Commercial Spaces*, Boulder CO: Westview.

Gottmann, J. (1977), 'Megalopolis and antipolis: the telephone and the structure of the city'. In I. de Sola Pool (ed.), *The Social Impact of the Telephone*, Cambridge MA: MIT Press, 303–17.

Gottmann, J. (1990), *Since Megalopolis: The Urban Writings of Jean Gottmann*, Baltimore MD: Johns Hopkins University Press.

Goubert, J-P. (1989), *The Conquest of Water: The Advent of Health in the Industrial Age*, Cambridge: Polity Press.

Graham, S. (1994) 'Networking cities: telematics in urban policy – a critical review', *International Journal of Urban and Regional Research*, 18 (3), 416–32.

Graham. S. (1995), 'From urban competition to urban collaboration? The development of inter-urban telematics networks', *Environment and Planning C: Government and Policy*, 13, 503–24.

Graham, S. (1997), 'Liberalised utilities, new technologies and urban social polarization: the UK experience', *European Urban and Regional Studies*, 4 (2), 135–50.

Graham, S. (1998a), 'The end of geography or the explosion of place? Conceptualising space, time and information technology', *Progress in Human Geography*, 22 (2), 165–85.

Graham, S. (1998b), 'Spaces of surveillant-simulation: new technologies, digital representations, and material geographies', *Environment and Planning D: Society and Space*, 16, 483–504.

Graham, S. (1999), 'Global grids of glass: on telecommunications, global cities and planetary urban networks', *Urban Studies*, 36 (5–6), 929–49.

Graham, S. (2000a), 'Constructing premium networked spaces: reflections on infrastructure networks and contemporary urban development', *International Journal of Urban and Regional Research*, 24 (1), 183–200.

Graham, S. (2000b), *Bridging Urban Digital Divides? Urban Polarization and Information and Communications Technologies (ICTs): Current Trends and Policy Prospects*, background paper for the United Nations Centre for Human Settlements (UNCHS).

Graham, S. and Aurigi, A. (1997), 'Virtual cities, social polarisation and the crisis in urban public space', *Journal of Urban Technology*, 4 (1), 19–52.

Graham, S. and Healey, P. (1999), 'Relational concepts of space and place: implications for planning theory and practice', *European Planning Studies*, 7 (5), 623–46.

Graham, S. and Marvin, S. (1994), 'Telematics and the convergence of urban infrastructure: implications for contemporary cities', *Town Planning Review*, 65 (3), 227–42.

Graham, S. and Marvin, S. (1995), 'More than ducts and wires: post-Fordism, cities and utility networks'. In P. Healey, S. Cameron, S. Davoudi, S. Graham and A. Madanipour (eds), *Managing Cities: The New Urban Context*, London: Wiley, 169–90.

Graham, S. and Marvin, S. (1996), *Telecommunications and the City: Electronic Spaces, Urban Places*, London: Routledge.

Graham, S. and Marvin, S. (1999), 'Planning cyber-cities? Integrating telecommunications into urban planning', *Town Planning Review*, 70 (1), 89–114.

Graham, S., Brooks, J. and Heery, D. (1996), 'Towns on the television: closed circuit TV systems in British towns and cities', *Local Government Studies*, 22 (3), 3–27.

Granick, H. (1947), *Underneath New York*, New York: Fordham University Press.

Grava, S. (1991), *Battery Park City: Between Edge and Fabric*, unpublished dissertation, New York: Columbia University, School of Architecture, Planning and Preservation.

Greenberg, S. (1998), *Invisible New York: The Hidden Infrastructure of the City*, Baltimore MD: Johns Hopkins University Press.

Greenhouse, S. (2000), 'Janitors struggle at the edges of Silicon Valley's success', *New York Times*, 18 April, A12.

Grell, B., Sambale, J. and Veith, D. (1998), 'Inner!City!Action! Crowd control, interdictory space and the fight for socio-spatial justice'. In INURA (International Network for Urban Research and Action) (eds), *Possible Urban Worlds: Urban Strategies at the End of the Twentieth Century*, Berlin: Birkhauser-Verlag, 208–15.

Grimshaw, D. (1994), *Bringing Geographic Information Systems into Business*, London: Longman.

Grogan, B. (1998), 'Flextech', *Urban Land*, December, 56–60.

Grundy-Warr, C., Peachey, K. and Perry, M. (1999), 'Fragmented integration in the Singapore–Indonesian border zone: Southeast Asia's "growth triangle" against the global economy', *International Journal of Urban and Regional Research*, 23 (2), 304–28.

Guillerme, A. (1988), 'The genesis of water supply, distribution and sewerage systems in France 1800–50'. In J. Tarr and G. Dupuy (eds), *Technology and the Rise of the Networked City in Europe and North America*, Philadelphia: Temple University Press, 91–110.

Gumpert, G. (1996), 'Communications and our sense of community: a planning agenda', *Intermedia*, 24 (4), 41–4.

Gumpert, G. and Drucker, S. (1998), 'The mediated home in the global village', *Communications Research*, 25 (4), 422–38.

Guy, S., Graham, S. and Marvin, S. (1996), 'Privatized utilities and regional governance: the new regional managers?', *Regional Studies*, 30 (8), 733–9.

Guy, S., Graham, S. and Marvin, S. (1997), 'Splintering networks: cities and technical networks in 1990s Britain', *Urban Studies*, 34 (2), 191–216.

Hack, G. (1997), 'Infrastructure and Regional Form'. Mimeo.

Hafajee, F. (1999), 'South Africa invests', *Corporate Africa Business and Investment Guide 1998*, 52–3.

Hajar, M. (1999), 'Zero friction society', *Urban Design Quarterly*, 71, 29–34.

Hall, P. (1988), *Cities of Tomorrow*, Oxford: Blackwell.

Hall, P. (1998), *Cities and Civilization*, London: Weidenfeld and Nicolson.

Hall, P. and Preston, P. (1988) *The Carrier Wave: New Information Technology and the Geography of Innovation 1846–2003*, London: Unwin.

Hall, T. and Hubbard, P. (eds) (1998), *The Entrepreneurial City: Geographies of Politics, Regimes and Representation*, London: Wiley, 77–99.

Hallgren, M. (1999), personal communication, 17 March.

Hamilton, K. and Hoyle, S. (1998), 'Moving cities: transport connections'. In J. Allen, D. Massey and M. Prycke (eds), *Unsettling Cities*, London: Routledge, 49–94.

Hammack, D. (1982), *Power and Society: Greater New York at the Turn of the Century*, New York: Russell Sage Foundation.

Hannah, M. (1997), 'Imperfect panopticism: envisioning the construction of normal lives'. In G. Benko and U. Stohmayer (eds), *Space and Social Theory*, Oxford: Blackwell, 344–59.

Hannigan, J. (1998a), *Fantasy Cities: Pleasure and Profit in the Postmodern Metropolis*, London: Routledge.

Hannigan, J. (1998b), 'Fantasy cities', *New Internationalist*, December, 20–3.

Hanson, S. (1993), *The Geography of Urban Transportation*, New York: Guildford.

Haraway, D. (1991), 'A manifesto for cyborgs: science, technology, and socialist feminism in the late twentieth century'. In D. Haraway (ed.), *Simians, Cyborgs and Women: The Reinvention of Nature*. New York: Routledge, 149–81.

Hardoy, J., Mitlin, D. and Satterthwaite, D. (1992), *Environmental Problems in Third World Cities*, London: Earthscan.

Harmon, A. (1999), 'High-speed access begins to alter the role the Internet plays in the home', *New York Times*, 28 April.

Harris, N. and Fabricius, I. (1996), *Cities and Structural Adjustment*, London: UCL Press.

Harrison, D. (1996), 'Poor hit by "secret" water cut-offs', *Observer*, 25 August, 3.

Harrison, M. and McEvey, C. (1997), 'Conflicts in the city: street trading in Mexico City', *Third World Planning Review*, 19 (3), 313–24.

Harvey, D. (1985), *The Urbanization of Capital*, Oxford: Blackwell.

Harvey, D. (1989), *The Condition of Postmodernity*, Oxford: Blackwell.

Harvey, D. (1993), 'From space to place and back again: reflections on the condition of postmodernity'. In J. Bird, B. Curtis, T. Putnam, G. Robertson and L. Tickner (eds), *Mapping the Future: Local Cultures, Global Change*, London: Routledge, 3–29.

Harvey, D. (1996) *Justice, Nature and the Politics of Difference*, Oxford: Blackwell.

Harvey, D. (2000), *Spaces of Hope*, Berkeley CA: University of California Press.

Haug, W. (1986), *Critique of Commodity Aesthetics: Appearance, Sexuality and Advertising in Capitalist Society*, Cambridge: Polity.

Hayden, D. (1981), *The Grand Domestic Revolution: A History of Feminist Designs for American Homes, Neighborhoods and Cities*, Cambridge MA: MIT Press.

Hayman S. (1996), 'Two-dimensional living', *Independent on Sunday*, 30 June.

Healey, P. (1997), *Collaborative Planning: Shaping Places in Fragmented Societies*, London: Macmillan.

Healey, P., Cameron, S., Davoudi, S., Graham, S. and Madanipour, A. (eds) (1995) *Managing Cities: The New Urban Context*, London, Wiley.

Hepworth, M. and Ducatel, K. (1992), *Transport in the Information Age: Wheels and Wires*, London: Belhaven Press.

Herman, R. and Ausubel, J. (1988), *Cities and their Vital Systems: Infrastructure Past, Present and Future*, Washington DC: National Academy Press.

Herrschel, T. (1998), 'From socialism to post-Fordism: the local state and economic policies in eastern Germany'. In T. Hall and P. Hubbard (eds), *The Entrepreneurial City: Geographies of Politics, Regimes and Representation*, London: Wiley, 173–99.

Hesse, M. (1992), 'Logistik: Zauberwort der Raumpolitik', *Kommune*, 10 (3), 52–4.

Heynen, H. (1999), *Architecture and Modernity*, Cambridge MA: MIT Press.

Hill, R. and Bessant, J. (1999), 'Spaced-out? Young people's agency, resistance and the public sphere', *Urban Policy and Research*, 17 (1), 41–9.

Hinchcliffe, S. (1996), 'Technology, power and space – the means and ends of geographies of technology', *Environment and Planning D: Society and Space*, 14, 659–82.

Hirsch, J. (1991), 'Fordism and post-Fordism: the present social crisis and its consequences'. In W. Bonefeld and J. Holloway (eds), *Post-Fordism and Social Form*, London: Macmillan, 8–34.

Hirsch, R. (2000), *Power Loss: The Origins and Restructuring of the American Utility System*, Cambridge MA: MIT Press.

Hodge, D. (1990), 'Geography and the political economy of urban transportation', *Urban Geography*, 11 (1), 87–100.

Holmes, D. (1999), 'The Electronic Superhighway: Melbourne's CityLink Project'. Mimeo.

Holston, J. (1998), 'Spaces of insurgent citizenship'. In L. Sandercock (ed.), *Making the Invisible Visible: A Multicultural Planning History*, Berkeley CA: University of California Press, 37–56.

Hoogvelt, A. (1997), *Globalisation and the Postcolonial World: The New Political Economy of Development*, London: Macmillan.

Hooper, B. (1998), 'The poem of male desires: female bodies, modernity, and "Paris, capital of the nineteenth century"'. In L. Sandercock (ed.), *Making the Invisible Visible: A Multicultural Planning History*, Berkeley CA: University of California Press, 227–54.

Hopkins, A. G. (1973), *An Economic History of West Africa*, Harlow: Longman.

Hughes, T. (1983), *Networks of Power: Electrification of Western Society 1880–1930*, London and Baltimore MD: Johns Hopkins University Press.

Hughes, T. (1988), 'The seamless web: technology, science, et cetera, et cetera'. In *Technology and Social Process*, Edinburgh: Edinburgh University Press.

Ibelings, H. (1998), *Supermodernism: Architecture and Globalization*, Rotterdam: NAI Publishers.

Idelovitch, E. and Ringskog, K. (1995), *Private Sector Participation in Water Supply and Sanitation in Latin America: Directions in Development*, Washington DC: World Bank.

Ignatieff, M. (2000), *Virtual War*, New York: Metropolitan Books.

International Bank for Reconstruction and Development (1993), *Water Resources Management: A World Bank Policy Paper*, Washington DC: IBRD.

INURA (International Network for Urban Research and Action) (eds) (1998), *Possible Urban Worlds: Urban Strategies at the End of the Twentieth Century*, Berlin: Birkhauser-Verlag.

Israel, A. (1992), *Issues of Infrastructure Management in the 1990s*, World Bank Discussion Paper 171, Washington DC: World Bank.

Jackson, K. (1985), *The Crabgrass Frontier: The Suburbanization of the United States*, Oxford: Oxford University Press.

Jacobs, J. (1961), *The Death and Life of Great American Cities*, New York: Vintage.

Jacobs, J. M. (1996), *Edge of Empire: Postcolonialism and the City*, London: Routledge.

Jaglin, S. (1997), 'La commercialisation du service d'eau potable à Windhoek (Namibie)', *Flux*, 30, October–December, 16–29.

James, G. (2000), 'Cry 2K wolf', *Improper Bostonian*, 3–16 May, 8.

Jay, M. (1988), 'Scopic regimes of modernity'. In H. Foster (ed.), *Vision and Visuality*, New York: New Press, 3–28.

Jerde Partnership International (1999), *You are Here*, London: Phaidon.

Jessop, B. (1998), 'The narrative of enterprise and the enterprise of narrative: place marketing and the entrepreneurial city'. In T. Hall and P. Hubbard (eds), *The Entrepreneurial City: Geographies of Politics, Regimes and Representation*, London: Wiley, 77–99.

Jessop, R. (2000), 'The rise of the national spatio-temporal fix and the tendential ecological dominance of globalising capitalism', *International Journal of Urban and Regional Research* 24 (2), 333–60.

Joerges, B. (1999a), 'Do politics have artefacts?', *Social Studies of Science*, 29 (3), 411–31.

Joerges, B. (1999b), 'High variability discourse in the history and sociology of large technical systems'. In O. Coutard (ed.), *The Governance of Large Technical Systems*, London: Routledge, 258–90.

Johnson, D. and Turner, C. (1997), *Trans-European Networks: The Political Economy of Integrating Europe's Infrastructure*, London: Macmillan.

Johnson-McGrath, J. (1997), 'Who built the built environment? Artifacts, politics and urban technology', *Technology and Culture*, 38 (3), 690–7.

Jonas, A. and Wilson, D. (eds) (1999), *The Urban Growth Machine: Critical Perspectives Two Decades Later*, New York: State University of New York Press.

Jones, M. (1997), 'Spatial selectivity of the state? The regulationist enigma and local struggles over economic governance', *Environment and Planning A*, 29, 831–64.

Jones, S. (1995), 'Understanding community in the information age'. In S. Jones (ed.), *Cybersociety: Computer Mediated Communication and Community*, London: Sage, 10–35.

Joy, W. (2000), 'Why the future doesn't need us', *Wired*, April, 238–60.

Kaika, M. and Swyngedouw, E. (2000), 'Fetishising the modern city: the phantasmagoria of urban technological networks', *International Journal of Urban and Regional Research*, 24 (1), 122–48.

Kalbermatten, J. (1999), 'Should we pay for water, and, if so, how?', *Urban Age*, winter, 14–16.

Kaliski, J. (1994), 'Liberation and the naming of paranoid space.' Foreword in S. Flusty, *Building Paranoia: The Proliferation of Interdictory Space and the Erosion of Spatial Justice*, Los Angeles: Los Angeles Forum for Architecture and Urban Design.

Kaliski, J. (1999), 'The present city and the practice of city design'. In J. Chase, M. Crawford and J. Kaliski (eds), *Everyday Urbanism*, New York: Monacelli Press, 89–108.

Kaothien, U., Webster, D. and Lukens, J. (1997), 'Infrastructure Investment in the Bangkok Region'. Mimeo.

Kasarda, J. and Rondinelli, D. (1998), 'Innovative infrastructure for agile manufacturers', *Sloan Management Review*, winter, 73–83.

Kasarda, J., Rondinelli, D. and Ward, J. (1996), 'The global transpark network: creating an infrastructure support system for agile manufacturing', *National Productivity Review*, winter, 33–41.

Keil, R. (1994), 'Global sprawl: urban form after Fordism?', *Environment and Planning D: Society and Space*, 12, 131–6.

Keil, R. and Ronneberger, K. (1994), 'Going up the country: internationalization and urbanization on Frankfurt's northern fringe', *Environment and Planning D: Society and Space*, 12, 137–66.

Keil, R. and Ronneberger, K. (2000), 'The globalization of Frankfurt am Main: core, periphery and social conflict'. In P. Marcuse and R. van Kempen (eds), *Globalizing Cities: A New Spatial Order?*, Oxford: Blackwell, 228–48.

Keivani, R. and Parsa, A. (1999), 'Globalisation, Growth Corridors and the Development of New Urban Forms'. Mimeo.

Kellerman, A. (1993), *Telecommunications and Geography*, London: Belhaven.

Kellner, D. (1999), 'New technologies: technocities and the prospects for democratization'. In J. Downey and J. McGuigan (eds), *Technocities*, London: Sage, 186–204.

Kerf, M. and Smith, W. (1996) *Privatising Africa's Infrastructure: Promise and Challenge*, World Bank Technical Paper 337, Africa Region series, Washington DC: World Bank.

Kern, S. (1986), *The Culture of Time and Space*, Cambridge MA: Harvard University Press.

Kerr, C. (ed.) (1989), *Community Water Development*, London: Intermediate Technology Publications.

Kessides, C. (1993a), *Institutional Options for the Provision of Infrastructure*, World Bank Discussion Paper 212, Washington DC: World Bank.

Kessides, C. (1993b), *The Contributions of Infrastructure to Economic Development: A Review of Experience and Policy Implications*, World Bank Discussion Paper 213, Washington DC: World Bank.

Khan, A. (2000), 'Reducing traffic density: the experience of Hong Kong and Singapore', paper presented at the New York Academy of Sciences conference *Moving People, Goods and Information in the Twenty-first Century*, Brooklyn, 26–7 June.

Kim, Y. and Cha, M. (1996), 'Korea's spatial development strategies for an era of globalisation', *Habitat International*, 20 (4), 531–51.

King, A. (1990), *Urbanism, Colonialism and the World Economy*, London: Routledge.

King, A. (1996), *Re-presenting the City: Ethnicity, Capital and Culture in the Twenty-first Century Metropolis*, London: Macmillan.

King, A. (1998), 'Writing the transnational city: the distant spaces of the Indian city'. In H. Dandekar (ed.), *City, Space and Globalization: An International Perspective*, Ann Arbor MI: College of Architecture and Urban Planning, University of Michigan, 25–31.

King, M. (2000), 'Tunnels'. In S. Pile and N. Thrift (eds), *City A–Z*, London: Routledge, 268–9.

King, R. (1996), *Emancipating Space: Geography, Architecture and Urban Design*, London and New York: Guildford.

Kipfer, S. (1998), 'Urban politics in the 1990s: notes on Toronto'. In INURA (International Network for Urban Research and Action) (eds), *Possible Urban Worlds: Urban Strategies at the End of the Twentieth Century*, Berlin: Birkhauser-Verlag, 172–9.

Kirby, A. (1999), 'The new private city: recreating civil society at the millennium', paper presented at the Association of American Geographers' conference, Hawaii, March.

Kirby, J. (1998), 'Start the meter', *Utility Week*, 16 October, 28–9.

Kirkup, G. and Smith, K. (1992), *Inventing Women: Women in Science and Technology*, Cambridge: Polity.

Kirsch, S. (1995), 'The incredible shrinking world? Technology and the production of space', *Environment and Planning D: Society and Space*, 13, 529–55.

Klein, N. (1999), *NoLogo*, New York: Picador.

Knight, R. and Gappert, G. (eds) (1989) *Cities in a Global Society*, London: Sage.

Knosa, M. (1995), 'Transport and popular struggles in South Africa', *Antipode*, 27 (2), 167–88.

Knox, P. (ed.) (1993a), *The Restless Urban Landscape*, Englewood Cliffs NJ: Prentice Hall.

Knox, P. (1993b), 'Capital, material culture and socio-spatial differentiation'. In P. Knox (ed.), *The Restless Urban Landscape*, Englewood Cliffs NJ: Prentice Hall, 1–34.

Knox, P. (1993c), 'The postmodern urban matrix'. In P. Knox (ed.), *The Restless Urban Landscape*, Englewood Cliffs NJ: Prentice Hall, 207–35.

Konvitz, J., Rose, M. and Tarr, J. (1990), 'Technology and the city', *Technology and Culture*, 284–95.

Koolhaas, R. (1995), 'Whatever happened to urbanism?' In R. Koolhaas and B. Mau, *S, M, L, XL*, New York: Monacelli Press.

Koolhaas, R. (1998a), 'Discussion 2'. In C. Davidson (ed.), *Anyhow*, Cambridge: MIT Press, 94.

Koolhaas, R. (1998b), 'Pearl river delta'. In C. Davidson (ed.), *Anyhow*, Cambridge: MIT Press, 182–9.

Koolhaas, R. and Mau, B. (1994), *S, M, L, XL*, Rotterdam: 010 Publishers.

Kopomaa, T. (2000), *The City in your Pocket: Birth of the Mobile Information Society*, Helsinki: Gaudemus.

Kostof, S. (1992), *The City Assembled: The Elements of Urban Form Through History*, London: Thames & Hudson.

Kostof, S. (1994a), 'His majesty the pick: the aesthetics of demolition'. In Z. Celik, D. Favro and R. Ingersoll (eds) (1994), *Streets: Critical Perspectives on Public Space*, Berkeley CA: University of California Press, 9–22.

Kostof, S. (1994b), *The City Shaped: Urban Patterns and Meanings through History*, London: Thames & Hudson.

Kotamyi, A. and Vaneigem, R. (1961), 'Elementary program of the Bureau of Unitary Urbanism', *Internationale Situationiste*, 6, 16–19.

Krasidy, M. (1999), 'The global, the local, and the hybrid: a native ethnography of glocalization', *Critical Studies in Mass Communication*, 16, 456–76.

Kruger, D. (1997), 'Access denied'. In Demos (1997), *The Wealth and Poverty of Networks*, Demos Collection Issue 12, 20–1.

Kusno, A. (1998), 'Custodians of (transnationality): metropolitan Jakarta, middle-class prestige and the Chinese'. In H. Dandekar (ed.), *City, Space and Globalization: An International Perspective*, Ann Arbor MI: College of Architecture and Urban Planning, University of Michigan, 161–70.

Lahiji, N. and Friedman, D. (1997), 'Introduction'. In N. Lahiji and D. Friedman (eds), *Plumbing: Sounding Modern Architecture*, New York: Princeton Architectural Press, 7–14.

Laporte, D. (2000), *History of Shit*, Cambridge MA: MIT Press.

LaPorte, T. (ed.) (1991), *Social Responses to Large Technical Systems*, Dordrecht: Kluwer.

Lappin, T. (1999), 'The new road rage', *Wired*, July, 127–81.

Larkin, B. (1996), 'Home systems', *Urban Land*, March, 41–6.

Lasch, C. (1994), 'The revolt of the elites: have they canceled their allegiance to America?', *Harpers Magazine*, November, 39–49.

Lash, S. (1999), *Another Modernity, a Different Rationality*, Oxford: Blackwell.

Lash, S. and Urry, J. (1994), *Economies of Signs and Space*, London: Sage.

Latour, B. (1987), *Science in Action*, Cambridge MA: Harvard University Press.

Latour, B. (1992), 'Introduction'. In Collection du Centre de Sociologie de l'Innovation (CSI), *Ces Réseaux que la Raison Ignore*, Paris: Harmattan.

Latour, B. (1993), *We Have Never Been Modern*, London: Harvester Wheatsheaf.

Latour, B. (1997), 'On Actor Network Theory: a few Clarifications'. Mimeo.

Latour, B. and Hermand, E. (1998), *Paris: Ville Invisible*, Paris: La Découverte.

Law, J. and Bijker, W. (1992), 'Postscript: technology, stability and social theory'. In W. Bijker and J. Law, *Shaping Technology, Building Society: Studies in Sociotechnical Change*, London: MIT Press.

Law, J. and Hassard, J. (1999), *Actor Network Theory and After*, Oxford: Blackwell.

Law, L. (1998), 'Cebu and *ceboom*: the political place of globalisation in a Philippine city'. In P. Rimmer (ed.), *Pacific Rim Development*, London: Allen & Unwin, 240–64.

Law, R. and Wolch, J. (1993), 'Social reproduction in the city: restructuring in time and space'. In P. Knox (ed.), *The Restless Urban Landscape*, Englewood Cliffs NJ: Prentice Hall, 165–206.

Lawrence, S. (1996), 'Marketing to win', *Utility Week*, 4 April, 20–2.

Lawson, M. (2000), 'Mistaken attachment', *Guardian Weekly*, 11–17 May, 11.

Lazarus, D. and Marinucci, G. (2000), 'Clinton plan to wire the have-nots', *San Francisco Chronicle*, 21 January, 1–3.

Le Corbusier (1929), *The City of Tomorrow and its Planning*, New York: Dover Publications.

Lee, K., Anas, A. and Oh, G. (1999), 'Costs of infrastructure deficiencies for manufacturing in Nigerian, Indonesian and Thai cities', *Urban Studies*, 36 (12), 2135–49.

Lee, Y-S. (1999), 'The Masan free trade zone: conflict and attrition'. In A. Markusen, Y-S. Lee and S. DiGiovanne (eds), *Second Tier Cities: Rapid Growth Beyond the Metropolis*, Minneapolis MN: University of Minnesota Press, 183–98.

Lefebvre, H. (1979), 'Space: social product and use value'. In J. Frieberg (ed.), *Critical Sociology: European Perspectives*, New York: Irvington.

Lefebvre, H. (1984), *The Production of Space*, Oxford: Blackwell.

Lehrer, U. (1994), 'Images of the periphery: the architecture of FlexSpace in Switzerland', *Environment and Planning D: Society and Space*, 12, 187–205.

Leonard, P. (1997), *Postmodern Welfare: Reconstructing an Emancipatory Project*, London: Sage.

Leong, S. (1998) 'Readings of the attenuated landscape'. In M. Bell and S. Leong (eds), *Slow Space*, New York: Monacelli Press, 186–213.

Lerup, L. (2000), *After the City*, Cambridge MA: MIT Press.

Leslie, J. (1999), 'Powerless', *Wired*, April, 119–83.

Lessig, L. (1999), *Code – and other Laws of Cyberspace*, New York: Basic.

Lewis, P. (1983), 'The galactic metropolis'. In R. Platt and G. Macuriko (eds), *Beyond the Urban Fringe*, Minneapolis MN: University of Minnesota Press, 23–49.

Lewis, T. (1997), *Divided Highways: Building the Interstate Highways, Transforming American Life*, New York: Penguin.

Ley, D. and Mills, C. (1993), 'Can there be a postmodernism of resistance in the urban landscape?' In P. Knox (ed.), *The Restless Urban Landscape*, Englewood Cliffs NJ: Prentice Hall, 255–70.

Leyshon, A. (1994), 'Access to financial services and financial infrastructure withdrawal: problems and policies', *Area*, 26 (3), 268–75.

Lieberman, D. (1999), 'America's digital divide', *USA Today Tech Report*, 11 October.

Liegland, J. (1995), 'Public infrastructure and special purpose governments'. In D. Perry, *Building the Public City: The Politics, Governance and Finance of Public Infrastructure*, London: Sage, 138–68.

Light, J. (1999), 'Fortress America? Home Security Systems, Defensible Space, and Citizen–Police Relations 1970–1998', unpublished Ph.D. thesis, Cambridge MA: Harvard University.

Lipman, A. and Harris, A. (1999), 'Fortress Johannesburg', *Environment and Planning B: Planning and Design*, 26, 727–40.

Little, P. (1995), 'Changing utilities', *Utility Week*, 16 October, 9.

Loader, B. (1998), 'Welfare direct: informatics and the emergence of self-service welfare?' In J. Carter (ed.), *Postmodernity and the Fragmentation of Welfare*, London: Routledge, 220–36.

Lockwood, C. (1997), 'A jewel of a community', *Urban Land*, April, 55–62.

Logan, J. (1993), 'Cycles and trends in the globalization of real estate'. In P. Knox (ed.), *The Restless Urban Landscape*, Englewood Cliffs NJ: Prentice Hall, 33–54.

Logan, J. and Molotch, H. (1987), *Urban Fortunes: The Political Economy of Place*, London: University of California Press.

Lohse, G. (2000), 'The state of the web', *Wharton Real Estate Review*, spring, 19–24.

Longcore, T. and Rees, P. (1996), 'Information technology and downtown restructuring: the case of New York City's financial district', *Urban Geography*, 17 (4), 354–72.

Lorente, S. (1997), 'The global house', *Trends in Communication*, 3, 117–41.

Lorrain, D. (2000), 'The construction of urban service models'. In A. Bagnasco and P. LeGalès (eds), *Cities in Contemporary Europe*, Cambridge: Cambridge University Press, 153–77.

Lorrain, D. and Stoker, G. (1997), *The Privatization of Urban Services in Europe*, London: Pinter.

Loukaitou-Sideris, A. (1993), 'Privatisation of public open space: the Los Angeles experience', *Town Planning Review*, 64 (2), 139–67.

Lovering, J. (1988), 'The local economy and local economic strategies', *Policy and Politics*, 16 (3), 145–57.

Lovink, G. and Riemens, P. (1998), 'The monkey's tail: the Amsterdam digital city three and a half years later'. In INURA (International Network for Urban Research and Action) (eds), *Possible*

Urban Worlds: Urban Strategies at the End of the Twentieth Century, Berlin: Birkhauser-Verlag, 180–5.

Lovins, A. (1977), *Soft Energy Paths: Toward a Durable Peace*, New York: Penguin.

Lucan, J. (1992), *Eau et Gaz à Tous les Étages: Paris 100 ans de logement*, Paris: Picard.

Luke, T. (1994), 'Placing power/siting space: the politics of global and local in the New World Order', *Environment and Planning D: Society and Space*, 12, 613–28.

Luke, T. (1997), 'At the end of nature: cyborgs, "humachines", and environments in postmodernity', *Environment and Planning A*, 29, 1367–80.

Lupton, D. (1999), 'Monsters in metal cocoons: "road rage" and cyborg bodies', *Body and Society*, 5 (1), 57–72.

Lupton, E. and Miller, J. (1992), *The Bathroom and the Kitchen and the Aesthetics of Waste (A Process of Elimination)*, New York: Kiosk.

Luymes, D. (1997), 'The fortification of suburbia: investigating the rise of enclave communities', *Landscape and Urban Planning*, 39, 187–203.

Lyon, D. (1994), *The Electronic Eye: The Rise of Surveillance Society*, London: Polity.

Lyon, D. (2000), *Surveillance Society: Monitoring Everyday Life*, Buckingham: Open Unversity Press.

Madon, S. (1998), 'Information-based global economy and socioeconomic development: the case of Bangalore', *The Information Society*, 13 (3), 227–43.

Makan, A. (1995), 'Power for women and men: towards a gendered approach to domestic energy policy and planning in South Africa', *Third World Planning Review*, 17 (2), 183–98.

Mallett, W. (1993a), 'Managing the post-industrial city: Business Improvement Districts in the United States', *Area*, 26 (3), 276–87.

Mallett, W. (1993b), 'Private government formation in the DC metropolitan area', *Growth and Change*, 24, 385–415.

Manandhar, L. (1999), 'Women in planning and implementation of a drainage construction project', *Trialog*, 60, 24–9.

Mander, J. and Goldsmith, E. (eds) (1996), *The Case Against the Global Economy and for a Turn to the Local*, San Francisco: Sierra Club.

Mansell, R. and Wehn, U. (1998), *Knowledge Societies: Information Technology for Sustainable Development*, Oxford: Oxford University Press.

Marcuse, P. (1996), 'Privatization and its discontents: property rights in land and housing in the transition in Eastern Europe'. In G. Andrusz, M. Harloe and I. Szelenyi (eds), *Cities after Socialism: Urban and Regional Change and Conflict in Post-socialist Societies*, Oxford: Blackwell, 119–91.

Marcuse, P. (1997), 'The enclave, the citadel, and the ghetto: what has changed in the post-Fordist US city', *Urban Affairs Review*, 33 (2), 228–64.

Marcuse, P. and van Kempen, R. (2000), 'Conclusion: a changed spatial order?' In P. Marcuse and R. van Kempen (eds), *Globalizing Cities: A New Spatial Order?*, Oxford: Blackwell, 249–76.

Markusen, A. (1999), 'Sticky places in slippery space: a typology of industrial districts'. In T. Barnes and M. Gertler (eds), *The New Industrial Geography*, London: Routledge, 98–123.

Markusen, A., Lee, Y-S. and DiGiovanne, S. (eds) (1999), *Second Tier Cities: Rapid Growth Beyond the Metropolis*, Minneapolis MN: University of Minnesota Press.

Marshall, R. (1999), 'Kuala Lumpur: competition and the quest for world city status', *Built Environment*, 24 (4), 271–9.

Marshall, T. (1997), 'Futures, foresight and forward looks', *Town Planning Review*, 68 (1), 31–50.

Martin, P. (2000), 'Bio-technology and Surveillance', paper presented at a seminar at Hull University, 5 July.

Martin, R. (1999), 'Selling off the state: privatization, the equity market and the geographies of shareholder capitalism'. In R. Martin (ed.), *Money and the Space Economy*, London: Wiley, 261–83.

Martinotti, G. (1997), 'Dimenticare i navigli', *La Repubblica*, 17 October.

Marvin, C. (1988), *When Old Technologies Were New: Thinking About Electric Communication in the Late Nineteenth Century*, Oxford: Oxford University Press.

Marvin, S. J. and Graham, S. (1994), 'Privatization of utilities: the implications for cities in the United Kingdom', *Journal of Urban Technology*, 1, p. 16.

Marvin, S. J. and Guy, S. (1997), 'Smart metering technologies and privatized utilities', *Local Economy*, August, 119–29.

Marvin, S. J. and Guy, S. (1999), 'Towards a new logic of transport planning?', *Town Planning Review*, 70 (2), 139–58.

Marvin, S.J. and Laurie, N. (1999) 'An emerging logic of water management, Cochabamba, Bolivia', *Urban Studies*, 36 (2), 341–57.

Marx, K. (1976 edition), *Capital I*, New York, International Publishers.

Marx, K. (1978 edition), *Capital II*, New York, International Publishers.

Masselos, J. (1995), 'Postmodern Bombay: fractured discourses'. In S. Watson and K. Gibson (eds), *Postmodern Cities and Spaces*, Oxford: Blackwell, 199–215.

Massey, Doreen (1992), 'Politics and space/time', *New Left Review*, 196, 65–84.

Massey, Doreen (1993), 'Power-geometry and a progressive sense of place'. In J. Bird, B. Curtis, T. Putnam, G. Robertson and L. Tickner (eds), *Mapping the Future: Local Cultures, Global Change*, London: Routledge, 59–69.

Massey, Doreen (1995), *Space, Place and Gender*, London: Polity.

Massey, Douglass (1996), 'The age of extremes: concentrated affluence and poverty in the twenty-first century', *Demography*, 33 (4), 395–412.

Mattelart, A. (1994), *Mapping World Communication: War, Progress, Culture*, Minneapolis MN: University of Minnesota Press.

Mattelart, A. (1996), *The Invention of Communication*, Minneapolis MN: University of Minnesota Press.

Mau, B. (1999), 'Getting engaged'. In C. Davidson (ed.), *Anytime*, Cambridge MA: MIT Press, 202–7.

Mayer, M. (1998), 'The changing scope of action in urban politics: new opportunities for local initiatives and movements'. In INURA (International Network for Urban Research and Action) (eds), *Possible Urban Worlds: Urban Strategies at the End of the Twentieth Century*, Berlin: Birkhauser-Verlag, 66–75.

Mayer, M. (1999), 'Urban movements and urban theory in the late twentieth-century city'. In R. Beauregard and S. Body-Gendrot (eds) (1999), *The Urban Moment: Cosmopolitan Essays on the Late Twentieth Century City*, Thousand Oaks CA: Sage, 209–40.

Mayntz, R. (1995), 'Technological Progress, Societal Change and the Development of Large Technical Systems'. Mimeo.

Mayntz, R. and Hughes, T. (eds) (1988) *The Development of Large Technical Systems*, Frankfurt: Campus.

McCall, M. K. (1977), 'Political economy and rural transport: an appraisal of Western misconceptions', *Antipode* 9 (2), 98–110.

McCalla, R. (1999), 'Global change, local pain: intermodal seaport terminals and their service areas', *Journal of Transport Geography*, 7, 247–54.

McCarter, R. (ed.) (1987), *Building Machines*, New York: Princeton Architectural Press.

McDowell, S. (2000), 'Globalization, local governance, and the United States Telecommunications Act of 1996'. In J. Wheeler, Y. Aoyama and B. Warf (eds), *Cities in the Telecommunications Age: The Fracturing of Geographies*, London and New York: Routledge, 112–29.

McGee, T. (1998), 'Urbanisation in an era of volatile globalisation: policy problematiques for the twenty-first century'. In J. Brotchie, P. Newton, P. Hall and J. Dickey (eds), *East–West Perspectives on Twenty-first Century Urban Development*, Aldershot: Ashgate, 37–49.

McGowan, F. (1999), 'The internationalization of large technical systems: dynamics of change and challenges to regulation in electricity and telecommunications'. In O. Coutard (ed.) (1999), *The Governance of Large Technical Systems*, London: Routledge, 130–48.

McGrail, B. (1999), 'Communications technology, local knowledges, and urban networks: the case of economically and socially disadvantaged "peripheral" housing schemes in Edinburgh and Glasgow', *Urban Geography*, 20 (4), 303–33.

McGurn, W. (1997), 'City limits', *Far Eastern Economic Review*, 160 (6), 6 Febuary, 34–7.

McKenzie, E. (1984), *Privatopia: Homeowner Associations and the Rise of Residential Private Government*, New Haven CT: Yale University Press.

McLaughlin, E. and Muncie, J. (1999), 'Walled cities: surveillance, regulation and segregation'. In S. Pile, C. Brook and G. Mooney (eds), *Unruly Cities?*, London: Routledge, 103–48.

McShane, C. (1988), 'Urban pathways: the street and highway 1900–40'. In J. Tarr and G. Dupuy (eds), *Technology and the Rise of the Networked City in Europe and North America*, Philadelphia: Temple University Press, 67–89.

Melosi, M. (2000), *The Sanitary City: Urban Infrastructure in America from Colonial Times to the Present*, Baltimore MD: Johns Hopkins University Press.

Merrifield, A. (1996), 'Public space: integration and exclusion in urban life', *City*, 5–6, 57–72.

Merrifield, A. (2000a), 'The dialectics of dystopia: disorder and zero tolerance in the city', *International Journal of Urban and Regional Research*, 24 (2), 472–89.

Merrifield, A. (2000b), 'Phantoms and spectres: capital and labour at the millennium', *Environment and Planning D: Society and Space*, 18, 15–36.

Miles, M. (1997), 'Another hero? Public art and the gendered city', *Parallax*, 5, 125–35.

Millar, N. (1999), 'Street survival: the plight of the Los Angeles street vendors'. In J. Chase, M. Crawford and J. Kaliski (eds), *Everyday Urbanism*, New York: Monacelli Press, 137–50.

Miller, D. (1995), *Acknowledging Consumption*, London: Routledge.

Milne, C. (1990), 'Universal telephone service in the UK: an agenda for policy research and action', *Telecommunications Policy*, October, 365–71.

Mingione, E. (1995), *Urban Poverty and the Underclass: a Reader*, Oxford: Blackwell.

Minkley, G. (n.d.), '"Corpses behind screens": native space in the city'. In H. Judin and I. Vladislavic (eds), *Blank ——: Architecture, Apartheid and After*, Rotterdam: NAI Publishers, 203–19.

Mitchell, D. (1995), 'The end of public space? People's park, definitions of the public and democracy', *Annals of the Association of American Geographers*, 85 (1), 108–33.

Mitchell, D. (1997), 'The annihilation of space by law: the roots and implications of anti-homeless laws in the United States', *Antipode*, 29 (3), 303–35.

Mitchell, W. (1996), *City of Bits: Space, Place and the Infobahn*, Cambridge MA: MIT Press.

Mitchell, W. (1999), *E-topia: Urban life, Jim, But Not as we Know it*, Cambridge MA: MIT Press.

Monbiot, G. (2000), 'Devil take the hindmost', *Guardian Weekly*, 30 March–5 April, 13.

Montgomery, J. (1988), 'The informal service sector as an administrative resource'. In D. A. Rondinelli and G. S. Cheema (eds), *Urban Services in Developing Countries*, Hong Kong: Macmillan, 89–111.

Mooney, G. (1999), 'Urban "disorders"'. In S. Pile, C. Brook and G. Mooney (eds) (1999), *Unruly Cities?*, London: Routledge, 103–48.

Mooney, G., Pile, S. and Brook, C. (1999), 'On orderings and the city'. In S. Pile, C. Brook and G. Mooney (eds) (1999), *Unruly Cities?*, London: Routledge, 345–68.

Moore, K. (1998), 'Pick 'n' switch', *Utility Week*, 16 October, 16–18.

Moore, M. (1996), *Downsize This! Random Threats from an Unarmed American,* New York: Harpers.

Moore, R. (1999), 'Introduction'. In R. Moore (ed.), *Vertigo: The Strange New World of the Contemporary City*, London: Laurence King, 10–59.

Mosco, V. (1988), 'Introduction: information in the pay-per society'. In V. Mosco and J. Wasko (eds), *The Political Economy of Information*, Madison WI: University of Wisconsin Press, 3–26.

Mosco, V. (1999a), 'Citizenship and the technopoles'. In A. Calabrese and J-C. Burgelman (eds), *Communication, Citizenship and Social Policy*, New York: Rowman & Littlefield, 33–48.

Mosco, V. (1999b), 'Cyber-monopoly: a web of techno-myths', *Science as Culture*, 8 (1), 5–22.

Moss, M. and Mitra, S. (1998), 'Net equity: a report on income and Internet access', *Journal of Urban Technology*, 5 (3), 23–32.

Moss, M. and Townsend, A. (1997), 'Manhattan leads the Net nation', available at http://www.nyu.edu/urban/ny_affairs/telecom.html.

Moss, T. (2000), 'Unearthing water flows, uncovering social relations: introducing new waste water technologies in Berlin', *Journal of Urban Technology*, 7 (1), 63–84.

Moyer, J. (1977), 'Urban growth and the development of the telephone: some relationships at the turn of the century'. In I. de Sola Pool (ed.), *The Social Impact of the Telephone*, Cambridge MA: MIT Press, 318–41.

Mumford, L. (1934), *Technics and Civilisation*, London: Routledge.

Mumford, L. (1961), *The City in History*, New York: MJF Books.

Murdoch, J. (1995), 'Actor networks and the evolution of economic forms: combining description and explanation in theories of regulation, flexible specialization and networks', *Environment and Planning A*, 27, 731–57.

Muschamp, H. (1995), 'Remodeling New York for the bourgeoisie', *New York Times*, 24 September, 38.

Musterd, S. and Ostendorf, W. (eds) (1998), *Urban Segregation and the Welfare State: Inequality and Exclusion in Western Cities*, London: Routledge.

Narayan, D. (1993) *Participatory Evaluation: Tools for Managing Change in Water and Sanitation*, Washington DC: World Bank Technical Paper 207, Washington DC: World Bank.

Nelkins, D. (1979), *Controversy: Politics of Technical Decisions*, Beverly Hills CA: Sage.

Neuman, M. (1998), 'Does planning need the plan?', *APA Journal*, spring, 208–20.

Neumeyer, F. (1990), 'The second-hand city: modern technology and changing urban identity'. In M. Angeli (ed.), *On Architecture, the City and Technology*, Washington DC: Butterworth, 16–25.

New York Times (2000), *Commercial Real Estate: High Tech, High Touch*, special supplement, 21 March.

Newman, O. (1972), *Defensible Space*, London: Architectural Press.

Newman, P. and Thornley, A. (1995), 'EuraLille: "boosterism" at the heart of Europe', *European Urban and Regional Studies*, 2 (3), 237–46.

Newton, P. (1991), 'Telematic underpinnings of the information economy'. In J. Brotchie, M. Batty, P. Hall and P. Newton (eds), *Cities of the Twenty-first Century*, London: Longman, 95–126.

Nicholis, B. (1999), 'Airport's "offshore" scheme nearer', *The Journal*, 14 April, 32.

Nieves, E. (2000), 'In San Francisco, more live alone, and die alone, too', *New York Times*, 25 June.

Noam, E. (1992), *Telecommunications in Europe*, Oxford: Oxford University Press.

Noam, E. and Wolfson, A. (1997), *Globalism and Localism in Telecommunications*, Amsterdam: Elsevier.

Norris, C. and Armstrong, G. (1997), 'Categories of Control: The Social Construction of Suspicion and Intervention in CCTV Systems', unpublished report to the UK Economic and Social Research Council.

Norris, C. and Armstrong, G. (1998), *The Unforgiving Eye: CCTV Surveillance in Public Space*, Hull: Centre for Criminal Justice, Hull University.

Norris, C. and Armstrong, G. (1999), *The Maximum Surveillance Society: The Rise of CCTV*, Oxford: Berg.

Norris, C., Moran, J. and Armstrong, G. (1998), 'Algorithmic surveillance: the future of automated visual surveillance'. In C. Norris, J. Moran and G. Armstrong (eds), *Surveillance, Closed Circuit Television and Social Control*, Aldershot: Ashgate, 255–67.

North Eastern Electricity Board (1967), *A Powerful Plug for Progress*, publicity brochure, Newcastle upon Tyne: North East Electricity Board.

Nunn, S. (1996), 'Urban infrastructure policies and capital spending in city manager and strong mayor cities', *American Review of Public Administration*, 26 (1), 93–112.

Nunn, S. and Schoedel, C. (1997), 'Special districts, city governments, and infrastructure: spending in 105 metropolitan areas', *Journal of Urban Affairs*, 19 (1), 59–72.

Nye, D. (1994), *American Technological Sublime*, Cambridge MA: MIT Press.

Nye, D. (1997), *Narratives and Spaces: Technology and the Construction of American Culture*, Exeter: University of Exeter Press.

Obitsu, H. and Nagase, I. (1998), 'Japan's urban environment: the potential of technology in future city concepts'. In G. Gloany, K. Hanaki and O. Koide (eds), *Japanese Urban Environment*, Oxford: Pergamon, 324–36.

Office of Technology Assessment (1995), *The Technological Reshaping of Metropolitan America*, Washington DC: Congress of United States.

Offner, J-M. (1993), 'Le développement des réseaux techniques: un modèle générique', *Flux*, July–December, 11–18.

Offner, J-M. (1996), '"Réseaux" et "Large Technical System": concepts complémentaires ou concurrents?', *Flux*, 26, 17–30.

Offner, J-M. (1999), 'Are there such things as small networks?' In O. Coutard (ed.) (1999), *The Governance of Large Technical Systems*, London: Routledge, 217–38.

Offner, J-M. (2000), '"Territorial deregulation": local authorities at risk from technical networks', *International Journal of Urban and Regional Research*, 24 (1), 165–82.

Oftel (1994), *Households without a Telephone*, London: UK Office of Telecommunications.

Ofwat (1990), *Paying for Water: A Time for Decisions*, consultation paper on future charging policy for water services. London: Office of Water Services.

Ogborn, M. (1999), *Spaces of Modernity: London's Geographies 1680–1780*, New York and London: Guildford.

Ogle, M. (1999), 'Water supply, waste disposal, and the culture of privatism in the mid-nineteenth-century American city', *Journal of Urban History*, 25 (3), 321–47.

Olalquiaga, C. (1994), 'Paradise lost'. In M. Avillez (ed.), *Sites and Stations: Architecture and Utopia in the Contemporary City*, New York: Lusitania Press, 43–50.

Olds, K. (1995), 'Globalization and the production of new urban spaces: Pacific Rim megaprojects in the late twentieth century', *Environment and Planning A*, 27, 1713–43.

O'Loughlin, M. and Friedrichs, J. (1996), *Social Polarization in Post-industrial Metropolises*, Berlin: de Gruyter.

O'Malley, C. (1999), 'The digital divide', *Time Magazine*, 22 March, 22–5.

Pahl, J. (1999), *Invisible Money: Family Finances in the Electronic Economy*, Bristol: Policy Press.

Painter, J. (1999), 'The Aterritorial City: Diversity, Spatiality, Democratization'. Mimeo.

Paquot, T. (1999), 'The post-city challenge'. In R. Beauregard and S. Body-Gendrot (eds), *The Urban Moment: Cosmopolitan Essays on the Late Twentieth Century City*, Thousand Oaks CA: Sage, 79–98.

Parenti, C. (1999), *Lockdown America: Police and Prisons in an Age of Crisis*, London: Verso.

Parsonage, J. (1992), 'Southeast Asia's "growth triangle": a subregional response to global transformation', *International Journal of Urban and Regional Research*, 16 (2), 307–17.

Pawley, M. (1997), *Terminal Architecture*, London: Reaktion Books.

Peacock, C. (2000), 'The globalization of real estate companies', *Wharton Real Estate Review*, spring, 25–30.

Pearce, F. (1992), *The Dammed: Rivers, Dams and the Coming World Water Crisis*, London: Bodley Head.

Peck, F. (1996), 'Regional development and the production of space: the role of infrastructure in the attraction of new inward investment', *Regional Studies*, 28, 327–39.

Peizarat, C. (1997), 'The Meanings of Europe and Integration: A Discursive Interpretation of the Development of Business Sites', unpublished Ph.D. dissertation, Newcastle upon Tyne: Department of Town and Country Planning: University of Newcastle upon Tyne.

Perry, D. (1995), 'Introduction'. In D. Perry (1995), *Building the Public City: The Politics, Governance and Finance of Public Infrastructure*, London: Sage, 1–20.

Petersen, G. E. (1984), 'Financing the nation's infrastructure requirements'. In R. Hanson (ed.), *Perspectives on Urban Infrastructure*, Washington DC: National Academy Press, 110–42.

Petrei, A. H. (1989), *El Gasto Publico Social y sus Efectos Distributivos: Estudio Conjuntos de Integracão Económica da America Latina*, Rio de Janeiro.

Petrella, R. (1993), 'Vers un "techno-apartheid" global', *Le Monde Diplomatique*, 18, 4.

Phelps, N., Lovering, J. and Morgan, K. (1998), 'Tying the firm to the region or tying the region to the firm?', *European Urban and Regional Studies*, 5 (2), 119–35.

Pickles, J. (ed.) (1995), *Ground Truth: The Social Implication of Geographic Information Systems*, New York: Guildford.

Picon, A. (1998), *La Ville Territoire de Cyborgs*, Paris: L'Imprimeur.

Picon, A. and Robert, J-P. (1998), *Un Atlas Parisien: le Dessus des Cartes*, Paris: Picard.

Pile, S. (1997), 'Introduction'. In S. Pile and M. Keith (1997), *Geographies of Resistance*, London: Routledge, 1–32.

Pile, S. and Keith, M. (eds) (1997), *Geographies of Resistance*, London: Routledge.

Pile, S. and Thrift, N. (1996), 'Mapping the subject'. In S. Pile and N. Thrift (eds), *Mapping the Subject: Geographies of Cultural Transformation*, London: Routledge, 13–51.

Pile, S., Brook, C. and Mooney, G. (eds) (1999), *Unruly Cities?*, London: Routledge.

Pinch, S. (1985), *Cities and Services: The Geography of Collective Consumption*, London: Routledge.

Pinch, S. (1989), 'The restructuring thesis and the study of public services', *Environment and Planning A*, 21, 905–26.

Pinch, S. (1997), *Worlds of Welfare: Understanding the Changing Geographies of Social Welfare Provision*, London: Routledge.

Platt, C. (2000), 'Re-energizer', *Wired*, May, 114–30.

Platt, H. (1988), 'City lights: the electrification of the Chicago region'. In J. Tarr and G. Dupuy (eds), *Technology and the Rise of the Networked City in Europe and North America*, Philadelphia: Temple University Press, 246–75.

Poindexter, G. (1997), 'Saturday night at the mall', *Wharton Real Estate Review*, 1 (2), 59–66.

Polo, A. (1994), 'Order out of chaos'. In J. Woodroffe, D. Papa and I. MacBurnie (eds), *The Periphery*, London: Wiley/Architectural Design, 24–9.

Poole, C. (1998), 'Private toll roads', *Public Works Management and Policy*, July, 5–10.

Pope, A. (1996), *Ladders*, New York: Princeton Architectural Press.

Postel, S. (1992), *Lost Oasis: Facing Water Scarcity*, Worldwatch Environmental Alert series, New York: Norton.

Potter, R. and Lloyd-Evans, S. (1998), *The City in the Developing World*, London: Longman.

Power, M. (2001), 'Technology and structuring the financial district of London'. In S. Brunn and S. Leinbach (eds), *Worlds of Electronic Commerce*, Chichester: Wiley (forthcoming).

Prendergast, C. (1992), *Paris in the Nineteenth Century*, Oxford: Blackwell.

Preston, P. (1990), 'History Lesson 2: Some Themes in the History of Technology Systems and Networks'. Mimeo.

Pririe, G. H. (1982), 'The decivilising rails: railways and underdevelopment in Southern Africa', *Tijdschrift voor Economische en Sociale Geografie*, 73 (4), 22–34.

Pryke, M. (1999), 'City rhythms: neo-liberalism and the developing world'. In J. Allen, D. Massey and M. Pryke (eds), *Unsettling Cities*, London: Routledge, 229–69.

Quillinan, J. (1993), 'Curse of the money's tomb', *Telecom World*, spring, 13–15.

Rabinow, P. (1994), 'On the archaeology of late modernity'. In R. Friedland and D. Boden (eds), *NowHere: Space, Time and Modernity*, Berkeley CA: University of California Press, 402–18.

Ramasubramanian, L. (1996), 'Building communities: GIS and participatory decision making', *Journal of Urban Technology*, 3 (10), 67–79.

Ranmert, W. (1997), 'New rules of sociological method: rethinking technology studies', *British Journal of Sociology*, 48: 171–91.

Rapaport, R. (1996), 'Bangalore', *Wired*, February, 56–107.

Ravetz, A. (1980), *Remaking Cities: Contradictions of the Recent Urban Environment*, London: Croom Helm.

Reeve, A. (1996), 'The private realm of the managed town centre', *Urban Design International*, 1 (1), 61–80.

Reich, R. (1992), *The Work of Nations*, New York: Simon & Schuster.

Reid, A. and Allen, K. (1970), *Nationalised Industries*, London: Penguin.

Reiff, F., Roses, M., Venczel, L., Quick, R. and Witt, V. (1996) 'Low-cost safe water for the world: a practical interim solution', *Journal of Public Health Policy*, 17 (4), 389–408.

Reiser, J., Allen, S., Apfelbaum, P. and Umenoto, N. (1996), 'The water project: foggy geographies'. In P. Phillips (ed.), *City Speculations*, New York: Princeton Architectural Press, 73–5.

Richardson, R. (1994), 'Teleservice Cities? "Second Wave" Back Offices and Employment in European Cities'. Mimeo.

Richardson, R. and Marshall, J. N. (1996), 'The growth of telephone call centres in peripheral areas of Britain: evidence from Tyne and Wear', *Area*, 28 (3), 308–17.

Rider, G. (1999), 'Watershed', *Utility Week*, 30 April, 16.

Riewoldt, O. (1997), *Intelligent Spaces: Architecture for the Information Age*, London: Laurence King.

Rimmer, P. (1991), 'Exporting cities to the western Pacific rim: the art of the Japanese package'. In J. Brotchie, M. Batty, P. Hall and P. Newton (eds), *Cities of the Twenty-first Century*, London: Longman, 243–61.

Rimmer, P. (1992), 'Japan's "resort archipelago": creating regions of fun, pleasure, relaxation and recreation', *Environment and Planning A*, 24, 1599–625.

Rimmer, P. (1998), 'Global network firms in transport and communications: Japan's NYK, KDD and JAL?'. In P. Rimmer (ed.), *Pacific Rim Development*, London: Allen & Unwin, 83–113.

Robbins, E. (1998), 'The New Urbanism and the fallacy of singularity', *Urban Design International*, 3 (1), 33–42.

Roberts, G. and Steadman, P. (ed.) (1999), *American Cities and Technology*, London: Routledge.

Roberts, M., Lloyd-Jones, T., Erickson, B. and Nice, S. (1999), 'Place and space in the networked city: conceptualizing the integrated metropolis', *Journal of Urban Design*, 4 (1), 51–67.

Robins, K. (1996), *Into the Image: Culture and Politics in the Field of Vision*, London: Routledge.

Robins, K. (1999), 'Foreclosing on the city? The bad idea of virtual urbanism'. In J. Downey and J. McGuigan (eds), *Technocities*, London: Sage, 34–59.

Robins, K. and Webster, F. (1999), *Times of the Technoculture: From the Information Society to the Virtual Life*, London: Routledge.

Robinson, J. (1999), 'Divisive cities: power and segregation in cities'. In S. Pile, C. Brook and G. Mooney (eds), *Unruly Cities?* London: Routledge, 149–200.

Robinson, W. and Harris, J. (2000), 'Towards a global ruling class? Globalization and the transnational capitalist class', *Science and Society*, 64 (1), 11–54.

Rochlin, G. (1997), *Trapped in the Net: The Unanticipated Consequences of Computerization*, Princeton NJ: Princeton University Press.

Rockoff, M. (1996), 'Settlement houses and the urban information infrastructure', *Journal of Urban Technology*, 3 (1), 45–66.

Rodrigue, J-P. (1999), 'Globalization and the synchronization of transport terminals', *Journal of Transport Geography*, 7, 255–61.

Rodríguez-Pose, A. and Arbix, G. (1999), 'Globalization, Regional Development, and Territorial Competition: Bidding Wars in the Brazilian Automobile Sector'. Mimeo.

Rogerson, R., Findlay, A., Paddison, R. and Morris, A. (1996), 'Class, consumption and quality of life', *Progress in Planning*, 40, 1–66.

Romero, S. (2000), 'Rich Brazilians rise above rush-hour jams', *New York Times*, 15 February, A1–A4.

Rondinelli, D. (2000), 'Cities and Smart Infrastructures'. Mimeo.

Rose, M. (1988), 'Urban gas and electric systems and social change 1900–40'. In J. Tarr and G. Dupuy (eds), *Technology and the Rise of the Networked City in Europe and North America*, Philadelphia: Temple University Press, 229–42.

Rose, M. (1995), *Cities of Light and Heat: Domesticating Gas and Electricity in Urban America*, University Park PA: University of Pennsylvania Press.

Rose, N. (2000), 'The biology of culpability: pathological identity and crime control in a biological culture', *Theoretical Criminology*, 4 (1), 5–34.

Rosston, G. and Teece, D. (1997), 'Dynamics of the new local communications markets'. In E. Noam and A. Wolfson (1997), *Globalism and Localism in Telecommunications*, Amsterdam: Elsevier, 1–23.

Rostow, W. (1960), *The Stages of Economic Growth: A Non-Communist Manifesto*, Cambridge: Cambridge University Press.

Rothstein, M. (1998), 'Offices plugged in and ready to go', *New York Times*, 4 February.

Rutsky, R. (1999), *High Techné: Art and Technology from the Machine Aesthetic to the Posthuman*, Minneapolis MN: University of Minnesota.

Rybczynski, W. (1983), *Taming the Tiger: The Struggle to Control Technology*, New York: Penguin.

Sachs, W. (1992), *For the Love of the Automobile*, Berkeley CA: University of California Press.

Samarajiva, R. and Shields, P. (1990), 'Integration, telecommunications, and development: power of the paradigms', *Journal of Communications*, 40 (3), 84–105.

Sandercock, L. (1998a), *Towards Cosmopolis: Planning for Multicultural Cities*, London: Wiley.

Sandercock, L. (ed.) (1998b), *Making the Invisible Visible: A Multicultural Planning History*, Berkeley CA: University of California Press.

Sassen, S. (1991), *The Global City: New York, London, Tokyo*, Princeton NJ: Princeton University Press.

Sassen, S. (1996), *Losing Control? Sovereignty in an Age of Globalization*, New York: Columbia University Press.

Sassen, S. (1998), *Globalization and its Discontents*, New York: New Press.

Sassen, S. (1999), 'The state and the new geography of power'. In A. Calabrese and J-C. Burgelman (eds), *Communication, Citizenship and Social Policy*, New York: Rowman & Littlefield, 17–32.

Sassen, S. (2000a), *Cities and their Cross-border Networks*, Oxford: Routledge.

Sassen, S. (2000b), *Cities in a World Economy*, second edition, London: Pine Forge.

Saunders, P. and Harris, C. (1994), *Privatisation and Popular Capitalism*, Milton Keynes: Open University Press.

Savage, M. and Warde, A. (1993), *Urban Sociology: Capitalism and Modernity*, London: Macmillan.

Sawers, L. (1984), 'The political economy of urban transportation: an interpretive essay'. In W. Tabb and L. Sawers (eds), *Marxism and the Metropolis: New Perspectives on Urban Political Economy*, Oxford: Oxford University Press, 223–43.

Sawnhey, H. (1992), 'The public telephone network: stages in infrastructure development', *Telecommunications Policy*, September–October, 538–52.

Schapp, J. (1999), 'Crash (speed as engine of individuation)', *Modernism/Modernity*, 6 (1), 1–49.

Schement, J., Belinfante, A. and Povich, L. (1997), 'Trends in telephone penetration in the United States 1984–94'. In E. Noam and A. Wolfsen (eds), *Globalism and Localism in Telecommunications*, North Holland: Elsevier, 167–97.

Schiffer, S. R. (1997), 'São Paulo: the Challenge of Globalization in an Exclusionary Urban Structure'. Mimeo.

Schiller, D. (1999a), *Digital Capitalism: Networking the Global Market System*, Cambridge MA: MIT Press.

Schiller, D. (1999b), 'Deep impact: the web and the changing media economy', *Info*, 1 (1), 35–51.

Schoechle, T. (1995), 'Privacy on the information superhighway: will my house still be my castle?', *Telecommunications Policy*, 19 (6), 435–52.

Schubeler, P. (1996), *Participation and Partnership in Urban Infrastructure Management*, Urban Management Programme Policy Paper 19, Washington DC: World Bank.

Schuler, D. (1996), *New Community Networks: Wired for Change*, New York: Addison Wesley.

Schulman, R. (1999), 'Losing BID', *City Limits*, November, 8–9.

Schultz, S. and McShane, C. (1973), 'To engineer the metropolis: sewers, sanitation, and city planning in late nineteenth-century America', *AIP Journal*, January, 289–311.

Schumacher, E. F. (1973), *Small is Beautiful: a Study of Economics as if People Mattered*, London: Blond & Briggs.

Schuman, M. (1998), *Going Local: Creating Self-Reliant Communities in a Global Age*, New York: Free Press.

Schwarzer, M. (1998a), 'Ghost wards: the flight of capital from history', *Thresholds*, 16, 10–19.

Schwarzer, M. (1998b), 'Beyond the valley of silicon architecture', *Harvard Design Magazine*, winter/spring, 15–21.

Scott, A. (1997), 'The cultural economy of cities', *International Journal of Urban and Regional Research*, 21 (2), 323–39.

Seabrook, J. (1993), 'The root of all evil', *New Statesmen and Society*, 26 February, 12.

Seik, F. (2000), 'An advanced demand management instrument in urban transport: electronic road pricing in Singapore', *Cities*, 17 (1), 33–45.

Sennett, R. (1970), *The Uses of Disorder*, New York: Knopf.

Sennett, R. (1999), 'The spaces of democracy'. In R. Beauregard and S. Body-Gendrot (eds) (1999), *The Urban Moment: Cosmopolitan Essays on the late Twentieth Century City*, Thousand Oaks CA: Sage, 273–86.

Serageldin, I. (1994), *Water Supply, Sanitation, and Environmental Sustainability: Directions in Development*, Washington DC: World Bank.

Shane, D. (1995), 'Balkanization and the postmodern city'. In P. Lang (ed.), *Mortal City*, New York: Princeton Architectural Press, 55–68.

Shearer, D. (1989), 'In search of equal partnerships: prospects for progressive urban policy in the 1980s'. In G. Squires (ed.), *Unequal Partnerships: The Political Economy of Urban Redevelopment in Postwar America*, London: Rutgers University Press, 289–307.

Shields, R. (1989), 'Social spatialization and the built environment', *Environment and Planning D: Society and Space*, 7, 147–64.

Shields, R. (1995) 'A guide to urban representations and what to do about it: alternative traditions in urban theory'. In A. King (ed.), *Re-presenting the City: Ethnicity, Capital and Culture in the Twenty-first Century Metropolis*, London: Macmillan, 227–52.

Silberman, S. (1999), 'Just say Nokia', *Wired*, September, 135–47.

Silva, R. (2000), 'The connectivity of the infrastructure networks and urban space of São Paulo in the 1990s', *International Journal of Urban and Regional Research*, 24 (1), 145–64.

Simmonds, R. (1998), 'Row over road closure', *North Eastern Tribune* (Johannesburg), 5 June, 1.

Simon, D. (1996), *Transport and Development in the Third World*, London: Routledge.

Simon, J. (1993), 'The origin of US public utility regulation: elements for a history of social networks', *Flux*, 11, 33–41.

Skeates, R. (1997), 'The infinite city', *City*, 8, 6–20.

Skinner, E. (1998), 'The Caribbean data processors'. In G. Sussman and J. Lent (eds), *Global Productions: Labor in the Making of the 'Information Society'*, Cresskill NJ: Hampton Press, 57–90.

Slater, D. (1975), 'Underdevelopment and spatial inequality: approaches to the problems of regional planning in the Third World', *Progress in Planning*, 4, 97–167.

Slater, D. (1997), 'Spatial politics/social movements: questions of (b)orders and resistance in global times'. In S. Pile and M. Keith (1997), *Geographies of Resistance*, London: Routledge, 258–76.

Sleeman, J. (1953), *British Public Utilities*, London: Pitman.

Small, H. (1996), 'Learning from the phone wars', *Utility Week*, 15 March, 20–1.

Smith, C. (1999), 'The Transformative Impact of Capital and Labor Mobility on the Chinese City'. Mimeo.

Smith, D. (1996), *Third World Cities in Global Perspective: The Political Economy of Uneven Urbanisation*, Boulder CO: Westview Press.

Smith, M. (1995), 'Recourse to empire: landscapes of progress in technological America'. In M. Smith and L. Marx (eds), *Does Technology drive History? The Dilemma of Technological Determinism*, Cambridge MA:, MIT Press, 37–52.

Smith, M. and Marx, L. (eds) (1995), *Does Technology Drive History? The Dilemma of Technological Determinism*, Cambridge MA: MIT Press.

Smith, M. P. (1999), 'Transnationalism and the city'. In R. Beauregard and S. Body-Gendrot (eds) (1999), *The Urban Moment: Cosmopolitan Essays on the late Twentieth Century City*, Thousand Oaks CA: Sage, 119–40.

Smith, N. (1993), 'Homeless/global: scaling places'. In J. Bird, B. Curtis, R. Putnam and L. Tickner (eds), *Mapping the Futures: Local Cultures, Global Change*, London: Routledge, 87–119.

Smith, N. (1999), 'Which new urbanism? The revanchist '90s', *Perspecta*, 30, 98–105.

Smith, N. (2000), 'Global Seattle', *Environment and Planning D: Society and Space*, 18, 1–13.

Soja, E. (1999), 'Lessons in spatial justice', *Hunch: the Berlage Institute Report*, 1, 98–107.

Soja, E. (2000), *Postmetropolis: Critical Studies of Cities and Regions*, Oxford: Blackwell.

Solà-Morales, I. (1996), 'Presents and futures: architecture in cities'. In *Architecture in Cities: Present and Future*, Barcelona: Centre de Cultura Contemporània de Barcelona, 310–22.

Solà-Morales, I. (1998), *Les Formes de Creixement Urban*, Barcelona: Ediciones UPC.

Solnit, R. (1995), 'The garden of merging paths'. In J. Brook and I. Boals (eds), *Resisting the Virtual Life*, San Francisco: City Lights, 224–40.

Solnit, R. (2000), *Hollow City: Gentrification and the Eviction of Urban Culture*, London: Verso.

Solomon, E. (1998), *Virtual Money*, Oxford: Oxford University Press.

Solomon, L. (1996), 'Revolution on the Road', *City*, summer, 43–55.

Soo, C. (1998), 'ERP cuts congestion in Singapore', *Traffic Technology International*, June–July, 23–4.

Sorkin, M. (ed.) (1992), *Variations on a Theme Park: The New American City and the End of Public Space*, New York: Hill & Wang.

Southern, A. (2000), 'The political salience of the space of flows: information and communications technologies and the restructuring city'. In J. Wheeler, Y. Aoyama and B. Warf (eds), *Cities in the Telecommunications Age: The Fracturing of Geographies*, London and New York: Routledge, 249–66.

Southworth, M. and Ben-Joseph, E. (1997), *Streets and the Shaping of Towns and Cities*, New York: McGraw-Hill.

Spark, S. (1998), 'Rail beats road congestion', *Jane's Airport Review*, April, 19.

Sparrow, J. and Vedantham, A. (1996), 'Inner-city networking: models and opportunities', *Journal of Urban Technology*, 3 (1), 19–28.

Speak, S. and Graham, S. (1999), 'Service not included: marginalised neighbourhoods, private service disinvestment, and compound social exclusion', *Environment and Planning A*, 31, 1985–2001.

Star, S. L. (1999), 'The ethnography of infrastructure', *American Behavioral Scientist*, 43 (3), 377–91.

Stedman, L. (1999), 'Why are we waiting?', *Utility Week*, 16 April, 18–19.

Storgaard, K. and Jensen, O. (1991), 'IT and ways of life'. In P. Cronberg, P. Dueland, O. Jensen and L. Qvortrup (eds), *Danish Experiments: Social Constructions of Technology*, New Social Science Monographs, Copenhagen, 123–39.

Storper, M. (1997), *The Regional World: Territorial Development in a Global Economy*, London: Guildford.

Storper, M. and Walker, R. (1989), *The Capitalist Imperative: Territory, Technology, and Industrial Growth*, Oxford: Blackwell.

Suarez-Villa, L. and Walrod, W. (1999), 'Losses from the Northridge earthquake: disruptions to high-technology industries in the Los Angeles basin', *Disasters*, 21 (1), 19–44.

Sudjic, D. (1995), *The 100 Mile City*, London: Flamingo.

Sum, N-L. (1999), 'Rethinking globalization: re-articulating the spatial scale and temporal horizons of trans-border spaces'. In K. Olds (ed.), *Globalization in the Asia Pacific*, London: Routledge, 129–45.

Summerton, J. (ed.) (1994a), *Changing Large Technical Systems*, Boulder CO: Westview Press, 1–24.

Summerton, J. (1994b), 'Social shaping in large technical systems', *Flux*, 17, July–September, 54–6.

Summerton, J. (1995), 'Representing users or . . . on opening the black box and painting it green, red, blue, and white'. Mimeo.

Summerton, J. (1999), 'Power plays: the politics of interlinking systems'. In O. Coutard (ed.) (1999), *The Governance of Large Technical Systems*, London: Routledge, 93–113.

Sussman, G. and Lent, J. (1998), *Global Productions: Labor in the Making of the 'Information Society'*, Cresskill NJ: Hampton Press.

Swyngedouw, E. (1993), 'Communication, mobility and the struggle for power over space'. In G. Giannopoulos and A. Gillespie (1993), *Transport and Communications in the New Europe*, London: Belhaven, 305–25.

Swyngedouw, E. A. (1995a), 'The City as Hybrid: In Nature, Society and Cyborg Urbanisation'. Mimeo.

Swyngedouw, E. A. (1995b), 'Power, nature and the city: the conquest of water and the political ecology of urbanisation in Guayaquil, Ecuador, 1880–1990', *Environment and Planning A*.

Swyngedouw, E. A. (1995c), 'The contradictions of urban water provision: a study of Guayaquil, Ecuador', *Third World Planning Review*, 17 (4), 387–405.

Swyngedouw, E. A. (1998), 'The specter of the phoenix: reflections on the contemporary urban condition'. In K. Bosma and H. Hellinga (eds), *Mastering the City: North European City Planning 1900–2000*, Rotterdam: NAI Publishers, 104–21.

Tadiar, N. (1995), 'Manila's new metropolitan form'. In V. Rafael (ed.), *Discrepent Histories*. Philadelphia: Temple University Press, 285–313.

Taffe, E. J., Morrill, R. L. and Gould, P. R. (1963), 'Transport expansion in underdeveloped countries: a comparative analysis', *Geographical Review*, 53 (4), 503–29.

Tan, A. and Low, L. (1998), 'Money: Shanghai World Financial Centre'. In R. Moore (ed.), *Vertigo: The Strange New World of the Contemporary City*, London: Laurence King, 142–57.

Tanner, J. (2000), 'New life for old railroads', *New York Times*, 6 May, B1.

Tarr, J. (1984), 'The evolution of urban infrastructure in the nineteenth and twentieth centuries'. In R. Hanson (ed.), *Perspectives on Urban Infrastructure*, Washington DC: National Academy Press, 4–62.

Tarr, J. (1989), 'Infrastructure and city building in the nineteenth and twentieth centuries'. In S. Hays (ed.), *City at the Point: Essays on the Social History of Pittsburgh*, Pittsburgh PA: University of Pittsburgh Press, 213–59.

Tarr, J. and Dupuy G. (eds) (1998), *Technology and the Rise of the Networked City in Europe and North America*, Philadelphia: Temple University Press, 322–38.

Taverne, E. (1998), 'Havens in a heartless world'. In C. Davidson (ed.), *Anyhow*, Cambridge MA: MIT Press, 81–7.

Taylor, P. (1994), 'The state as container: territoriality in the modern world-system', *Progress in Human Geography*, 18 (2), 151–62.

Technology Foresight Transport Panel (1995), *Progress through Partnership* 5, *Transport*, Technology Foresight, London: Office of Science and Technology, HMSO.

Tella, R. and Foster, V. (1996), *Guide to the Economic Regulation of the Latin American Utilities*, Oxford: OXERA Press.

Thomas, R. (1996), 'Rich – and excluded', *Observer*, 20 September, 11.

Thrift, N. (1990), 'Transport and communications 1730–1914'. In R. Dogshon and R. Butlin (eds), *An Historical Geography of England and Wales*, London: Academic Press, 453–84.

Thrift, N. (1995), 'A hyperactive world'. In R. Johnston, P. Taylor and M. Watts (eds), *Geographies of Global Change*, Oxford: Blackwell, 18–35.

Thrift, N. (1996a), *Spatial Formations*, London: Sage.

Thrift, N. (1996b), 'New urban eras and old technological fears: reconfiguring the goodwill of electronic things', *Urban Studies*, 33 (8), 1463–93.

Thrift, N. (1996c), '"Not a straight line but a curve": or, cities are not mirrors of modernity'. Mimeo.

Thrift, N. (1997a), 'Cities without modernity, cities with magic', *Scottish Geographical Magazine*, 113 (3), 138–49.

Thrift, N. (1997b), 'The still point: resistance, expressive embodiment and dance'. In S. Pile and M. Keith (eds) (1997), *Geographies of Resistance*, London: Routledge, 124–51.

Thrift, N. (2000), 'Less mystery, more imagination: the future of the City of London', *Environment and Planning A*, 32, 381–90.

Thrift, N., Carlstein, T. and Parkes, D. (1978), *Timing Space and Spacing Time* (3 vols), London: Edward Arnold.

Toon, I. (2000), '"Finding a place on the street": CCTV surveillance and young people's use of urban public space'. In D. Bell and A. Haddour (eds), *City Visions*, London: Longman, 141–65.

Trench, R. and Hillman, E. (1984), *London under London: A Subterranean Guide*, London: John Murray.

Trenor, P. (1997), 'An Urban Ethic for Europe'. Mimeo.

Triche, T. (1990) *The Institutional and Regulatory Framework for Water Supply and Sewerage, Public and Private Roles*, Infrastructure Notes WS-9, Washington DC: World Bank.

Tsagarousianou, R., Tambini, D. and Bran, C. (eds), *Cyberdemocracy: Technology, Cities and Civic Networks*, London: Routledge.

Tschumi, B. (1996), 'Some urban concepts'. In *Architecture in Cities: Present and Future*, Barcelona: Centre de Cultura Contemporània de Barcelona, 40–3.

Tseng, E. (2000), 'The Geography of Cyberspace'. Mimeo.

Tyler, P. (2000), 'Theives looting Russia of even its power lines', *New York Times*, 18 April, A1–A10.

Tyne & Wear Economic Development Company (1999), 'Boom at Doxford continues', *The Investor*, March, 6.

United Nations Centre for Human Settlement (UNHCS) (1991), 'Urbanization: Water Supply and Sanitation Sector Challenges', paper presented at the Water Supply and Sanitation global forum, Oslo, 18–20 September.

United Nations Development Programme (1999), *Human Development Report*, New York: Oxford University Press.

United States Congress (1984), *Hard Choices: A Report on the Increasing Gap between America's Infrastructures and our Ability to Pay for Them*, Washington DC: Government Printing Office.

Urry, J. (1999), 'Automobility, Car Culture and Weightless Travel: A Discussion Paper'. Available at http://www.comp.lancs.ac.uk/ sociology/soc008ju.html.

Urry, J. (2000a), 'Mobile sociology', *Sociology*, 51 (1), 185–203.

Urry, J. (2000b), *Sociology Beyond Societies: Mobilities for the Twenty-first Century*, London: Routledge.

Utility Week (1995), special issue, *IT in Utilities*, 19 November.

Vale, L. (1999), 'Mediated monuments and national identity', *Journal of Architecture*, 4, 391–408.

Vallone, P. and Berman, H. (1995), *Cities Within Cities: Business Improvement Districts and the Emergence of the Micropolis*. Report to the City of New York.

Van Grunsven, L. (1998), 'The sustainability of urban development in the "SIJORI" growth triangle', *Third World Planning Review*, 20 (2), 179–201.

Van Grunsven, L. and Van Egeraat, C. (1999), 'Achievements of the industrial "high road" and clustering strategies in Singapore and their relevance to European peripheral economies', *European Planning Studies*, 7 (2), 145–73.

Van Toorn, R. (1999), 'The society of the and (an introduction)', *Hunch: the Berlage Institute Report*, 1, 98–7.

Varsanyi, M. (2000), 'Global cities from the ground up: a response to Peter Taylor', *Political Geography*, 19, 33–8.

Vedel, T. (1997), 'The ecology of games', *Trends in Communication*, 3, 34–45.

Veltz, P. (1996), *Mondialisation: Villes et Territoires*, Paris: Presses Universitaires de France.

Veltz, P. (2000), 'European cities in the world economy'. In A. Bagnasco and P. LeGalès (eds), *Cities in Contemporary Europe*, Cambridge: Cambridge University Press, 33–47.

Vergara, C. (1995), 'Bunkering the poor: our fortified ghettoes'. In P. Lang (ed.), *Mortal City*, New York: Princeton Architectural Press, 19–27.

Vergara, C. (1997), *The New American Ghetto*, New Brunswick NJ: Rutgers University Press.

Virilio, P. (1987), 'The overexposed city', *Zone*, 1 (2).

Virilio, P. (1991), *The Lost Dimension*, New York: Semiotext(e).

Virilio, P. (1994), 'The third interval: a critical transition'. In V. Andermatt-Conley (ed.), *Rethinking Technologies*, Minneapolis MN and London: University of Minnesota Press, 3–10.

Wacquant, L. (1996), 'The rise of advanced marginality: notes on its nature and implications', *Acta Sociologica*, 39, 121–39.

Waites, B., Bessel, R. and Moore, J. (1989), 'Everyday life and the dynamics of technological change'. In C. Chant (ed.), *Science, Technology and Everyday Life*, London: Routledge, 9–38.

Wajcman, J. (1991), *Feminism Confronts Technology*, Cambridge: Polity Press.

Waley, P. (2000), 'Tokyo: patterns of familiarity and partitions of difference'. In P. Marcuse and R. van Kempen (eds), *Globalizing Cities: A New Spatial Order?*, Oxford: Blackwell, 127–57.

Wall, A. (1994), 'The dispersed city'. In J. Woodroffe, D. Papa and I. MacBurnie (eds), *The Periphery*, London: Wiley/Architectural Design, 108, 8–11.

Wall, A. (1996), 'Flow and interchange: mobility as a quality of urbanism'. In *Architecture in Cities: Present and Future*, Barcelona: Centre de Cultura Contemporània de Barcelona, 158–66.

Warf, B. (1995a), 'Separated at birth? Regional science and social theory', *International Regional Science Review*, 18 (2), 185–94.

Warf, B. (1995b), 'Telecommunications and the changing geographies of knowledge transmission in the late twentieth century', *Urban Studies*, 32 (2), 361–78.

Warf, B. (1998), '"Reach out and touch someone": AT&T's global operations in the 1990s', *Professional Geographer*, 50 (2), 255–67.

Warf, B. (2000), 'Compromising positions: the body in cyberspace'. In J. Wheeler, Y. Aoyama and B. Warf (eds), *Cities in the Telecommunications Age: The Fracturing of Geographies*, London and New York: Routledge, 54–70.

Wark, M. (1998), 'On technological time: Virilio's overexposed city', *Arena*, 83, 1–21.

Warwick, D. (1999), 'The phone box war', *Utility Week*, 18 April, 42–4.

Waterhouse, A. (1996), *Boundaries of the City: The Architecture of Western Urbanism*, Toronto: University of Toronto Press.

Watson, S. (1999), 'Background Paper for Seminar on City Politics, Durham, July 8th and 9th'. Mimeo.

Webber, M. (1964), 'The urban place and the non-place urban realm'. In M. Webber, J. Dyckman, D. Foley, A. Guttenberg, W. Wheaton and C. Whurster (eds), *Explorations into Urban Structure*, Philadelphia: University of Pennsylvania Press, 79–153.

Weisman, L. (1994), *Discrimination by Design: A Feminist Critique of the Man-made Environment*, Chicago: University of Illinois Press.

Wetzler, B. (2000), 'Boomgalore', *Wired*, March, 152–69.

Wheeler, J., Aoyama, Y. and Warf, B. (eds) (2000), *Cities in the Telecommunications Age: The Fracturing of Geographies*, London and New York: Routledge.

Whitelegg, J. (1993), 'The conquest of distance by the destruction of time'. In J. Whitelegg, S. Hultén and T. Flink (eds), *High Speed Trains: Fast Tracks to the Future?*, Hawes: Leading Edge, 203–12.

Whiteman, J. (1990), 'Responding to Fritz: on disarming words and pictures in the telling of teleological stories'. In M. Angeli (ed.), *On Architecture, the City and Technology*, Washington DC: Butterworth, 26–32.

Wieners, B. (1999), 'Arcadia', *Wired*, November, 304–8.

Wigley, M. (1996), 'Lost in space'. In M. Speaks (ed.), *Critical Landscape*, Rotterdam: 010 Publishers, 30–57.

Williams, R. (1973), *The Country and the City*, London: Hogarth Press.

Willoughby, K. W. (1990), *Technology Choice: A Critique of the Appropriate Technology Movement*, Boulder CO: Westview Press.

Willson, W. (2000), letter to *Guardian Weekly*, 4–10 May, 15.

Wilson, E. (1995), 'The rhetoric of urban space', *New Left Review*, 209, 146–60.

Wilson, M. (1998), 'Information networks: the global offshore labor force'. In G. Sussman and J. Lent (eds), *Global Productions: Labor in the Making of the 'Information Society'*, Cresskill NJ: Hampton Press, 39–56.

Winner, L. (1980), 'Do artefacts have politics?', *Daedalus*, 109 (1), 121–36.

Winner, L. (1992), 'Silicon Valley mystery house'. In M. Sorkin (ed.), *Variations on a Theme Park: The New American City and the End of Public Space*, New York: Hill & Wang.

Winpenny, J. (1994), *Managing Water as an Economic Resource*, London: ODI.

Winsbury, R. (1997), 'How grand are the grand alliances?', *Intermedia*, 25 (3), 26–31.

Winter, B. (1995), 'Getting to know you', *Utility Week*, 9 June, 14–15.

Wodiczko, K. (1999), *Critical Vehicles*, Cambridge MA: MIT Press.

Wolf, G. (2000), 'The unmaterial world', *Wired*, June, 308–19.

Wolff, H. (1997), 'The future of hospitality design and development', *Urban Land*, August, 54–8.

Woo, E. (1994), 'Urban development'. In Y. Teiung and D. Chu (eds), *Guangdong: Survey of a Province Undergoing Rapid Change*, Hong Kong: Chinese University Press, 327–54.

Woodroffe, J., Papa, D. and MacBurnie, I. (1994), 'An introduction'. In J. Woodroffe, D. Papa and I. MacBurnie (eds), *The Periphery*, London: Wiley/Architectural Design.

Woods, L. (1995), 'Everyday war'. In P. Lang (ed.), *Mortal City*, New York: Princeton Architectural Press, 48–53.

Woolgar, S. (1991), 'The turn to technology in social studies of science', *Science, Technology and Human Values*, 16 (1), 20–50.

World Bank (1994), *World Development Report 1994: Infrastructure for Development*, Oxford: Oxford University Press.

World Bank (1995), *Meeting the Infrastructure Challenge in Latin America and the Caribbean*, Directions in Development, Washington DC: World Bank.

World Bank (1996), *Livable Cities for the Twenty-first Century: Directions in Development*, Washington DC: World Bank.

World Teleport Association (1999), corporate web site, available at http://www.worldteleport. org/.

Wu, C. (1998), 'Globalisation of the Chinese countryside: international capital and the transformation of the Pearl river delta'. In P. Rimmer (ed.), *Pacific Rim Development*, London: Allen & Unwin, 57–81.

Wu, F. (1996), 'Urban restructuring in China's emerging market economy: towards a framework for analysis', *International Journal of Urban and Regional Research*, 20 (4), 640–63.

Wu, V. (1998), 'The Pudong development zone and China's economic reforms', *Planning Perspectives*, 13, 133–65.

Wurtzel, A. and Turner, C. (1977), 'Latent functions of the telephone: what missing the extension means'. In I. de Sola Pool (ed.), *The Social Impact of the Telephone*, Cambridge MA: MIT Press, 246–61.

Yearly, S. (1992) 'Environmental challenges'. In S. Hall, D. Held and T. McGrew (eds), *Modernity and its Futures*, Milton Keynes: Open University Press, 117–68.

Yeoh, B. (1996), *Contesting Space: Power Relations and the Urban Built Environment in Colonial Singapore*, Kuala Lumpur: Oxford University Press.

Yepes, G. (1992), *Infrastructure Maintenance in Latin America and the Caribbean: The Costs of Neglect and Options for Improvement*, Latin American and Caribbean Technical Department Report 17, Washington DC: World Bank.

Yeung, H. (1999), 'Regulating investment abroad: the political economy of the regionalization of Singaporean firms', *Antipode*, 31 (3), 245–73.

Yeung, Y. (1997), 'Geography in an age of mega-cities', *International Social Sciences Journal*, 151, 91–105.

Young, I. M. (1990), *Justice and the Politics of Difference*, Princeton NJ: Princeton University Press.

Zaner, P. (1997), 'The multifamily housing industry embraces high-tech amenities', *Urban Land*, April, 63–6.

Zhang, Y. (1996), 'The entrepreneurial role of local bureaucracy in China: a case study of Shandong province', *Issues and Studies*, 32 (12), 89–110.

Zook, M. (2000), 'The web of production: the economic geography of commercial Internet content in the United States', *Environment and Planning A*, 32, 411–26.

Zukin, S. (1982), *Loft Living*, Baltimore MD: Johns Hopkins University Press.

Zukin, S. (1995), *The Cultures of Cities*, Oxford: Blackwell.

Plate 23 Looking north across Los Angeles to Bunker Hill.
Photograph: Stephen Graham